Weak and Electromagnetic Interactions at High Energies

Cargèse 1975

Part A

NATO ADVANCED STUDY INSTITUTES SERIES

A series of edited volumes comprising multifaceted studies of contemporary scientific issues by some of the best scientific minds in the world, assembled in cooperation with NATO Scientific Affairs Division.

Series B: Physics

RECENT VOLUMES IN THIS SERIES

Volume 7 – Low-Dimensional Cooperative Phenomena
edited by H. J. Keller

Volume 8 – Optical Properties of Ions in Solids
edited by Baldassare Di Bartolo

Volume 9 – Electronic Structure of Polymers and Molecular Crystals
edited by Jean-Marie André and János Ladik

Volume 10 – Progress in Electro-Optics
edited by Ezio Camatini

Volume 11 – Fluctuations, Instabilities, and Phase Transitions
edited by Tormod Riste

Volume 12 – Spectroscopy of the Excited State
edited by Baldassare Di Bartolo

Volume 13 – Weak and Electromagnetic Interactions at High Energies – *Cargèse 1975*
(Parts A and B)
edited by Maurice Lévy, Jean-Louis Basdevant,
David Speiser, and Raymond Gastmans

Volume 14 – Physics of Nonmetallic Thin Films
edited by C. H. S. Dupuy

Volume 15 – Nuclear and Particle Physics at Intermediate Energies
edited by J. B. Warren

Volume 16 – Structure and Reactivity of Metal Surfaces
edited by E. G. Derouane and A. A. Lucas

The series is published by an international board of publishers in conjunction with NATO Scientific Affairs Division

A	**Life Sciences**	Plenum Publishing Corporation
B	**Physics**	New York and London
C	**Mathematical and Physical Sciences**	D. Reidel Publishing Company Dordrecht and Boston
D	**Behavioral and Social Sciences**	Sijthoff International Publishing Company Leiden
E	**Applied Sciences**	Noordhoff International Publishing Leiden

Weak and Electromagnetic Interactions at High Energies

Cargèse 1975

Part A

Edited by

Maurice Lévy and Jean-Louis Basdevant

Laboratory of Theoretical Physics and High Energies
Université Pierre et Marie Curie
Paris, France

David Speiser

Institute of Theoretical Physics
Université Catholique de Louvain
Louvain-la-Neuve, Belgium

and

Raymond Gastmans

Institute of Theoretical Physics
Katholieke Universiteit Leuven
Leuven, Belgium

PLENUM PRESS • NEW YORK AND·LONDON
Published in cooperation with NATO Scientific Affairs Division

Library of Congress Cataloging in Publication Data

Cargèse Summer Institute on Weak and Electromagnetic Interactions at High Energies, 1975.
Weak and electromagnetic interactions at high energies.

(NATO advanced study institutes series: Series B, Physics; v. 13)
"Organized by the Université Pierre et Marie Curie. . . the Katholieke Universiteit te Leuven. . . and the Université Catholique de Louvain."
Includes bibliographies and index.
1. Weak interactions (Nuclear physics)–Congresses. 2. Electromagnetic interactions–Congresses. I. Lévy, Maurice, 1922- II. Université Pierre et Marie Curie. III. Louvain. Université catholique. IV. Katholieke Universiteit te Leuven. V. Title. VI. Series.
QC794.8.W4C37 1975 539.7'54 76-3672
ISBN 0-306-35795-X

First half of the Proceedings of the Summer Institute held at Cargèse, France, June 30-July 25, 1975, sponsored in part by NATO

A Division of Plenum Publishing Corporation
227 West 17th Street, New York, N.Y. 10011

United Kingdom edition published by Plenum Press, London
A Division of Plenum Publishing Company, Ltd.
Davis House (4th Floor), 8 Scrubs Lane, Harlesden, London, NW10 6SE, England

Printed in the United States of America

Preface

The Cargèse Summer Institute 1975 on Weak and Electromagnetic Interactions at High Energies was organized by the Université Pierre et Marie Curie (M. LEVY et J.L. BASDEVANT), the Katholieke Universiteit te Leuven (R. GASTMANS) and the Université Catholique de Louvain (D. SPEISER et J. WEYERS) who made in 1973 the first contacts with some lecturers, who, on the advice of NATO, joined their efforts and worked in common. It was the 16th Summer Institute held at Cargèse and the 3rd one organized by the two departments of Theoretical Physics at Leuven and Louvain-la-Neuve.

When the two groups decided (independently) on the subject of the school, they could not know how lucky their choice eventually would turn out to be : rarely has it been possible to present an audience with such a great number of new and decisive discoveries who are likely to stimulate the imagination of theoreticians and the research projects of experimentalists alike. Such were the decisive confirmation of the neutral currents, the di-muon events, the slowly decaying new particles, etc. The organizers were grateful indeed that they could find physicists from almost all great centers of high energy physics who had themselves participated in these discoveries. Although the theorists could not match during the last two years the spectacular success of their experimental colleagues, there has been enough important programs, especially in field theory : renormalization of gauge theories, the Brout-Englert-Higgs mechanism, etc... All lectures and seminar talks were related to the main topics.

It is a pleasure to thank all those who made this summer institute possible !

Thanks are due to the Scientific Committee of NATO for a generous grant and its President, and especially to its scientific secretary, Dr. T. KESTER, for his constant help and encouragement.

We also thank the C.N.R.S., the D.G.R.S.T. (France) and the N.S.F. (U.S.) for travel grants as well as the Krediet Bank (Belgie) for special contribution for fellowship for members coming from non NATO countries.

Special thanks are due to the Université of Nice for having put at our disposal the Institut at Cargèse, and the Université Pierre et Marie Curie, Paris, the K.U.L. and the U.C.L. for putting at our disposal their facilities.

We wish to thank M.F. HANSELER, R. NARTUS, N. RIBET and other helps at Paris, Leuven, Louvain-la-Neuve and especially Cargèse for their collaboration as secretaries.

We thank Mrs. B. STEYAERT for typing the manuscript and Miss M. de CROMBRUGGHE, MM. HAUT, LEROY, URIAS for their help and correcting proofs. We are grateful to Plenum Press and Mrs. L.A. COLEMAN for their collaboration.

Mostly however, we would like to thank all lecturers and participants who came from over 20 countries : the interest of the latter and their questions and the willingness of the former to answer them all, provided the stimulus who made (we hope) this institute a success.

M.LEVY
J.L.BASDEVANT
D.SPEISER
R.GASTMANS

Contents

ASPECTS OF QED

B. LAUTRUP

The Niels Bohr Institute, University of Copenhagen
DK-2100 Copenhagen Ø, Denmark

1. WHAT IS QED ?

It is now 75 years ago that Planck [1] discovered that light is emitted and absorbed in small packets and not in a continuous process. Quantum electrodynamics, if taken in its widest sense as the quantum theory of the interaction of light with matter, lies thus at the root of quantum theory. The behaviour of light was thus decisive for the development of both the theory of relativity and quantum mechanics.

A satisfactory formulation of quantum electrodynamics was not obtained before the introduction of quantized fields by Dirac [2], and Heisenberg and Pauli [3]. This happened only shortly after quantum mechanics itself had found its modern formulation in the mid-twenties. Since then quantum electrodynamics has almost always been taken in the narrow sense of the interaction between photons and electrons, with an occasional external Coulomb field and perhaps a recoil correction thrown in.

Today we may (a bit circularly) characterize quantum electrodynamics as the theory of the interaction between photons and all those particles for which the dominant interaction is electromagnetic. This leaves us primarily with electrons and muons. Other particles may be included either if one can disregard the internal structure due to the stronger interactions or if it is known sufficiently well to be included as a small correction. This is for instance the case with the energy levels of the atoms, where the electromagnetic structure of the nuclei only play a minor role, but it is not the case for electron-proton scattering at high momentum transfer. Here one rather uses the well-known electron as a probe to explore certain aspects of the complicated electromagnetic properties of the proton.

The above definition of quantum electrodynamics is perhaps wrong. It depends on whether we know what a photon is, i.e. whether we can construct apparatus that distinguishes between photons and other particles. With the advent of the modern unified gauge field theories some doubt has been cast on exactly this point. At very high energies weak and electromagnetic interactions mix up with each other and make it impossible to say whether a photon or a neutral vector boson was exchanged. If for example we want to explore the electromagnetic structure of the neutrino we might do it in the same way as we study the electromagnetic structure of the proton, i.e. scatter electrons off it. But the energies needed to see any structure are so large that the electrons'own weak structure plays a disturbing role. Even if the theory does not contain any neutral vector boson there will be competition between the photon and two charged vector bosons. And since renormalizability requires that all matter obeys the gauge invariance no probe can ever be found which only emits photons. As a consequence a neutrino does not have a calculable charge radius [4].

The early QED of the '30s was only a partially successful theory. Although it was possible to calculate the leading behaviour of many processes one could not calculate non-leading corrections, because the integrals diverged. In the late '40s this difficulty was overcome. It was recognized that the divergences occurred according to a well-defined pattern which allowed them to be collected into corrections to the mass and the charge of the electron. Other divergences cancelled among themselves, when calculations were done carefully. Only finite remainders were left when physical quantities were expressed in terms of the corrected, renormalized, mass and charge.

Without belittling the efforts and the genius of the people involved in this development one may in retrospect say that the advance was of a rather technical nature and did not really add to the fundamental understanding of physics. The QED of the thirties contained all the physics that the modern QED describes, but in an - at that moment - inaccessible form. In the last few years this "technicality" has been elevated into a principle of Physics. The requirement of renormalizability severely limits the number of acceptable theories, even though a whole new class of theories, the gauge theories, have been drawn into the game. It is by no means clear, however, that this is a valid principle. Renormalizability i.e. the property that the divergences disappear when the parameters of the theory are suitably defined is inextricably tied to perturbation theory and it is an open question whether the theories we today call non-renormalizable may make sense in some non-perturbative way.

The divergences that are removed by renormalization happen at large internal momenta. In order that they be collectible into mass and charge corrections they must be highly correlated between different orders of perturbation theory. It is then perhaps not surprising that the behaviour of the amplitudes at large external momenta is also highly correlated between different orders of perturbation theory. This was recognized very early [5)] and formalized in the renormalization group approach [6)] and the Callan-Symanzik equations [7-9)].

The comparison of QED with experiments has followed three paths. Firstly there are the high energy scattering experiments that test the general behaviour of the theory, essentially only the Born approximation. One might here expect to see gross deviations between theory and experiment and although at times such deviations have been claimed to have been seen, none have remained at the present time. Secondly there are the atomic physics tests where tiny corrections to bound state energies are both measured and calculated. Here one tests the effects of higher order virtual corrections. Again the general result is that the agreement is surprisingly good although untill recently some problems existed in heavy muonic atoms. Thirdly, there remain the anomalous magnetic moments of the charged leptons, measurable to a fantastic precision, and calculable to the same precision or even better. Again there is an almost chilling agreement between theory and experiment.

The story of QED is a story of continued success beyond the wildest imaginations. This seems even stranger because it stands alone. It is the only relativistic quantum field theory that has allowed detailed comparisons with experiment - and never let us down. Perhaps we in the next years will see the weak interactions attain a similar status.

In writing these lecture notes I have been torn between my desire to be a teacher and my desire to report faithfully on a very complicated subject. To teach a good and penetrating course on QED one needs between one and two years of weekly lectures. To report on the present status of QED one needs between one and two hours. So what does one do with 5 x 1 1/2 hours ? With an audience consisting of unhatched as well as long-grown-up physicists ?

In the following you will see that I've tried to satisfy all of you. There are simple calculations illustrating fundamental points, compressed review of the current status and general structural descriptions of QED. In the end I may only have succeeded in producing confusion (and a lot of paper). For the simple calculations are not really simple, the reviews are not really complete and the structural descriptions not really finished.

In the first lecture I shall talk about a nice effect of the zero-point oscillations of the electromagnetic field, and also talk about the basic definition of QED. In my second lecture, I shall discuss tests of the Born approximations and derive the equivalent photon approximation. In the third lecture I shall discuss renormalization and as a crowning effect derive the Callan-Symanzik equation for the photon propagator. The fourth lecture is dedicated to the structure of bound state level corrections and a discussion of the experimental news. Finally the fifth lecture contains a discussion of the anomalous magnetic moment of the leptons, including a calculation of a contribution to the eigth order correction due to 304 diagrams by means of the Callan-Symanzik equation.

2. THE CASIMIR EFFECT

As a soft start I would like to tell you about a phenomenon which has always fascinated me. It has to do with the zero-point energy of the electromagnetic field. One often hears the statement that it can be disregarded in the Hamiltonian, but we shall see that this is not always the case.

Consider classically a cavity in a perfectly conducting metal. If we let some electromagnetic radiation into this cavity, it will be reflected from the walls and interfere with itself. The result-

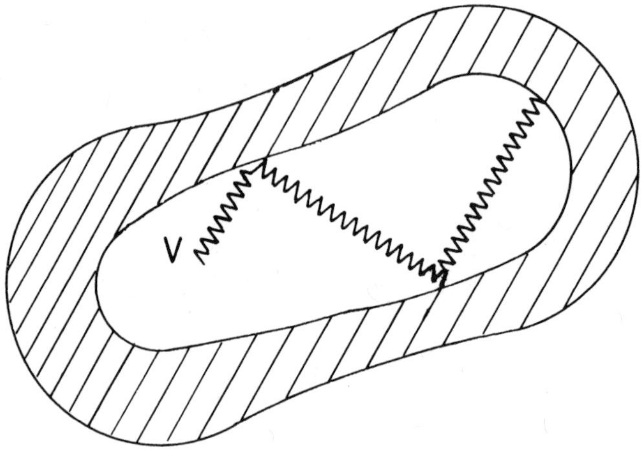

ing wave pattern is very complicated. But since the boundary conditions are independent of time, the field will be a superposition of standing waves. A standing wave retains its form at all times, while its amplitude oscillates. But only a discrete set of possible standings wave forms and corresponding resonance frequences can exist in a given cavity. These are called the modes of the cavity. They are completely determined by the form of the cavity. The true dynamical parameters (i.e. the generalized coordinates and momenta) for the field in the cavity are therefore the various amplitudes of the harmonic oscillations of all the modes. The field in the cavity is consequently interpretable as a collection of harmonic oscillators, (one for each mode) in various degrees of excitation. The quantization of the electromagnetic field in the cavity is traditionally done by quantizing these harmonic oscillators. Let their frequences be $_n$. Then the total energy of the field in the cavity is

$$E = \sum_n (N_n + \frac{1}{2})\hbar\omega_n \tag{2.1}$$

where the N_n indicates the level to which the n'th oscillator (mode) is excited. Even if none of the oscillators are excited, the energy will be non-zero. A quantum oscillator also oscillates in its ground state. This means that the cavity contains a fluctuating electromagnetic field even if no radiation has been let into it. Conversly, if we open a little hole in the cavity, no radiation comes out. The field nevertheless fluctuates in the cavity.

Does the zero-point energy

$$E_o = \sum_n \frac{1}{2}\hbar\omega_n \tag{2.2}$$

have physical significance ? If the cavity is the whole universe, the zero-point energy becomes an embarrassment to the relativistic invariance of the theory since there is no corresponding zero-point momentum (unless one really thinks there are metal walls "out there" defining a preferential reference frame). Accordingly one drops it and forgets about it. But if the cavity is an ordinary down-to-earth-cavity of finite size, Casimir [10] showed that the zero-point energy indeed has physical significance. It gives rise to an attractive force between two metal plates in vacuum. One has even attempted to measure it [11].

Intuitively one may understand it in the following way. The zero-point energy depends on the form of the cavity, particularly through the lowest-lying modes (i.e. those with the lowest frequency for which the wave length is of the order of the size of the cavity). This means that the energy changes if we deform the cavity, and from this it follows that the vacuum in the cavity must act upon the metal with certain forces. We may even understand intuitively that these forces act as a negative pressure i.e. as a pull from the inside. This is simply a consequence of the classical result that the force on a metallic conductor per unit of area is 1/2 $\bar{E}^2$ directed away from the metal where $\bar{E}$ is the electric field strength immediately outside the metal. Thus the direction of this force is always inwards towards the cavity independently of the direction of the fluctuating electric field.

In calculating the total zero-point energy (2.2) for any cavity we run into the problem that there are infinitely many modes with infinitely high frequencies. In other words the sum diverges. We shall make it convergent by means of an exponential cut-off factor $e^{-\omega_n/\Lambda}$ so that

$$E_o = \frac{1}{2}\hbar \sum_n \omega_n e^{-\omega_n/\Lambda} \tag{2.3}$$

instead of (2.2). This means that frequencies greater than Λ are essentially excluded. This is a normal procedure for regulating divergences in quantum field theory and can be "physically" realized by imparting a granular structure to space-time (for instance by assuming the vacuum to be a lattice) with characteristic fundamental length $1/\Lambda$ (we take C = 1). We must naturally require that $1/\Lambda$ is much smaller than the size of the cavity.

In a rectangular box of dimensions L_1, L_2 and L_3 the resonance frequencies are

$$\omega_{n_1n_2n_3} = \sqrt{\left(\frac{n_1\pi}{L_1}\right)^2 + \left(\frac{n_2\pi}{L_2}\right)^2 + \left(\frac{n_3\pi}{L_3}\right)^2} \tag{2.4}$$

where n_1, n_2, n_3 are integers. There are two types of modes. Doubly degenerate modes for $n_1 \geq 1$, $n_2 \geq 1$, $n_3 \geq 1$, and nondegenerate modes for $n_1 = 0$, $n_2 \geq 1$, $n_3 \geq 1$, $n_1 \geq 1$, $n_2 = 0$, $n_3 \geq 1$ and $n_1 \geq 1$, $n_2 \geq 1$, $n_3 = 0$. The doubly degenerate modes have orthogonal polarization vectors orthogonal to the vector $(\frac{n_1}{L_1}, \frac{n_2}{L_2}, \frac{n_3}{L_3})$. The nondegenerate modes have polarization vectors parallel to the axes.

We are interested in the zero-point energy between two large parallel metal plates with distance a which is small compared to the size of the plates. Hence we want to let $L_1, L_2 \to \infty$ for fixed $L_3 = a$, keeping only the leading terms. When $L_1, L_2 \gg L_3 = a$ we may replace the summation over n_1 and n_2 by an integral over $k_1 = n_1 \pi/L_1$ and $k_2 = n_2 \pi/L_2$. Then the zero-point energy from the doubly degenerate modes becomes

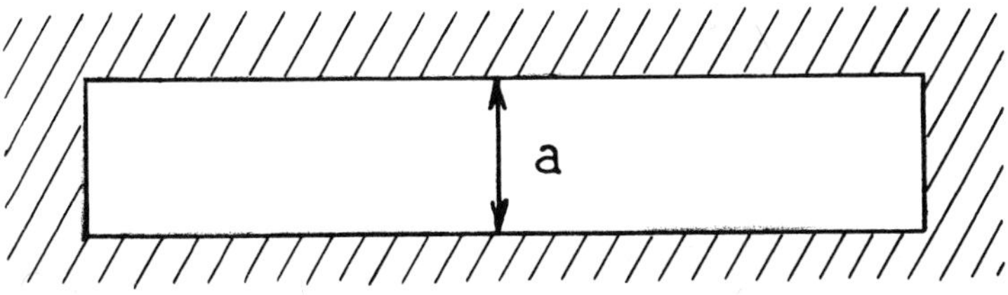

$$E_0 = \hbar \frac{L_1 L_2}{\pi^2} \sum_{n=1}^{\infty} \int_0^{\infty} dk_1 \int_0^{\infty} dk_2\, \omega_n(k_1 k_2)\, e^{-\omega_n(k_1 k_2)/\Lambda} \tag{2.5}$$

where

$$\omega_n(k_1 k_2) = \sqrt{k_1^2 + k_2^2 + (\frac{n\pi}{a})^2} \tag{2.6}$$

The singly degenerate solutions with $n_1 = 0$ or $n_2 = 0$ are not proportional to L_1L_2 and hence can be disregarded. The singly degenerate solutions with $n_3 = 0$ do not depend on a, and may also be disregarded, since they cannot give rise to a force in a's direction.

The integral is easy to carry out. First one introduces polar coordinates in the k_1k_2 plane and integrates trivially over the polar angle. Next one introduces the variable $x = \frac{a}{n\pi}\omega_n$ and find

$$E_0 = \hbar \frac{L_1 L_2}{\pi^2} \sum_{n=1}^{\infty} \frac{\pi}{2} (\frac{n\pi}{a})^3 \int_1^{\infty} dx\, x^2\, e^{-\frac{n\pi}{a\Lambda}x}$$

$$= \hbar \frac{L_1 L_2}{\pi^2} \sum_{n=1}^{\infty} \frac{\pi}{2} (-\Lambda)^3 \int_1^{\infty} dx\, x^2 \frac{\partial^3}{\partial x^3} e^{-\frac{n\pi}{a\Lambda}x}$$

$$= \hbar \frac{L_1 L_2}{\pi^2} \frac{\pi}{2} (-\Lambda)^3 \int_1^{\infty} dx\, x^2 \frac{\partial^3}{\partial x^3} \frac{1}{e^{\frac{\pi}{a\Lambda}x} - 1}$$

$$= \hbar \frac{L_1 L_2}{\pi^2} \frac{\pi}{2} \Lambda^3 (\frac{\partial^2}{\partial x^2} - 2\frac{\partial}{\partial x} + 2) \frac{1}{e^{\frac{\pi x}{a\lambda}} - 1} \Bigg|_{x=1}$$

In this formula we may let $\Lambda \to \infty$ using

$$\frac{1}{e^t - 1} = \frac{1}{t} - \frac{1}{2} + \frac{1}{12} t - \frac{1}{720} t^3 + \ldots \tag{2.7}$$

and find to $O(1/\Lambda)$

$$E_o = \hbar \frac{L_1 L_2}{\pi^2} \left[3a\Lambda^4 - \frac{1}{720} \frac{\pi^4}{a^3} \right] \tag{2.8}$$

The pressure is

$$p \equiv - \frac{1}{L_1 L_2} \frac{\partial E_o}{\partial a} = - \frac{3\hbar\Lambda^4}{\pi^2} - \frac{\hbar\pi^2}{240\ a^4} \tag{2.9}$$

As expected it is negative, i.e. the two metal walls of the cavity attract each other. The appearance of the cut-off Λ in the first term indicates that the pressure is infinite while the second term is finite. It seems as if we have arrived at a non-sensical result. Two metal plates in vacuum attract each other with an infinite force ! The reason for our error lies in the assumption that the cavity between the plates is the only one present, i.e. that the plates each fill out half the universe. We have here a case of an irresistable force meeting an immovable obstacle !

Observe, however, that the first term in the pressure is independent of the distance a. It is in fact completely universal acting on all surfaces of a metallic body, internal as well as external. Hence its effect cancels itself out completely. Consider for instance a cavity with a metallic diaphragm in the middle

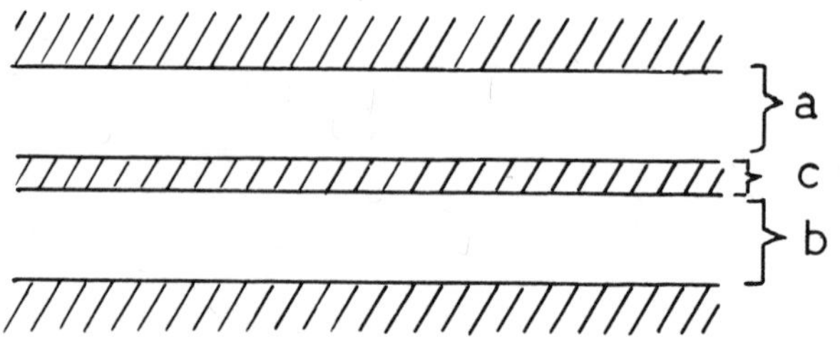

The overall pressure felt by the diaphragm is

$$p = - \frac{\hbar c \pi^2}{240} \left(\frac{1}{a^4} - \frac{1}{b^4} \right)$$

and is perfectly finite. One might ask whether the infinite universal pull does not create infinite tensions in the metal, tensions that ought to tear it apart. It would, if a metal stayed ideal for arbitrary high frequencies. But that is not the case. Natural cut-off frequencies occur long before any appreciable tension builds up.

Finally it should be mentioned that for a metal cavity with the walls at temperature T the pressure has been evaluated by Fierz [12]. The result is $(\hbar = C = k = 1)$

$$p = -\frac{3\Lambda^4}{\pi^2} - \frac{T^4}{4\pi a^3} \frac{\partial^3}{\partial T^3} \sum_{n=1}^{\infty} \frac{1}{n^3} \frac{1}{e^{4\pi naT} - 1}$$

The reason I have chosen to talk in such detail about this effect is that it illustrates in a relatively simple context various of the aspects of quantum field theory that are normally more difficult calculationally. Firstly it shows that divergences may occur in the high frequency domain (ultraviolet divergences) and that they may be regulated by means of a suitable cut-off. Secondly, it shows that in a naive calculation one finds a physical effect that depends on the cut-off, but that on closer inspection this effect is not observable. If a proprer physical situation is analyzed the effect is finite. This is entirely parallel to the unobservable infinite dielectric constant of the vacuum that arises due to vacuum polarization or to the unoberservable infinite self-energy of the electron.

3. CONVENTIONAL QED

We define conventional QED by postulating a Lagrangian for it. The Lagrangian consists of three parts

$$L = L_0 + L_1 + L_2 \tag{3.1}$$

The first part L_0 is the Lagrangian of the free non-interacting photons and electrons

$$L_0 = -\frac{1}{4} F_{\mu\nu}F^{\mu\nu} - \frac{1}{2} (\partial_\mu A^\mu)^2 + \bar{\psi} (i\not{\partial} - m)\psi \tag{3.2}$$

Here A_μ is the covariant four-potential (the Maxwell field) and $F_{\mu\nu} = \partial_\mu A_\nu - \partial_\nu A_\mu$ the covariant field strength tensor. The electron field is denoted by ψ and m is the electron mass. The first term is the usual Lagrangian of classical electrodynamics $\frac{1}{2}(\bar{E}^2 - \bar{B}^2)$ and the second is a gauge fixing term first introduced by Fermi [13]. The third term is the Dirac Lagrangian (*).

(*) We use a metric with signature (1,-1,-1,-1) and γ-matrices satisfying $\gamma_\mu^+ = \gamma_0\gamma_\mu\gamma_0$ (See Bjorken and Drell [14]).

The second part of the total Lagrangian is the interaction

$$L_1 = -eA_\mu \bar{\psi} \gamma^\mu \psi \tag{3.3}$$

where $e = -|e|$ is the electron charge. This interaction is the minimal gauge invariant coupling between electrons and photons. Any other gauge invariant coupling would give additional terms. Were it not for the Fermi term in (3.2) the Lagrangian $L_0 + L_1$ would be invariant under the local gauge transformations

$$\begin{aligned} \psi(x) &\to e^{i\theta(x)}\psi(x) \\ \bar{\psi}(x) &\to e^{-i\theta(x)}\,\bar{\psi}(x) \\ A_\mu(x) &\to A_\mu(x) - \frac{1}{e}\,\partial_\mu\theta(x) \end{aligned} \tag{3.4}$$

Finally there is the term L_2. This term is necessary in order to guarantee that the parameters m and e are the physical parameters of the electron. The interaction L_1 gives by itself rise to changes in mass and charge of the electron. These are called self-mass and self-charge. If we only had L_1, m and e would not be what we measured in the laboratory but rather the mechanical mass of the electron excluding the electromagnetic contribution and the bare charge of the electron, unscreened by vacuum polarization. The extra term L_2 to which we shall return in section 7, is adjusted such that it compensates these unmeasurable effects. At the moment we shall, however, disregard it.

The only way we know to solve the theory described by our Lagrangian is by means of perturbation theory. The free Lagrangian L_0 is considered to be the unperturbed system to which an exact solution is known. The exact solution is simply the non-interacting particles. The eigenstates of the Hamiltonian are the standard multiparticle states containing definite numbers of photons, electrons and positrons. The non-interacting particles are then brought to interact via L_1 and one may - in principle - calculate all physical effects as a power series in L_1 i.e. in e. When L_2 is also taken into account one must of course consider the double power series in L_1 and L_2.

It is not my intention here to go through all the gory details of canonical quantization of quantum electrodynamics. This may be found in a number of textbooks [14] and is probably well-known to all of you. But let me nevertheless make a few remarks.

In classical electrodynamics the gauge freedom requires a subsidiary condition to be imposed on the potentials. The standard covariant form is the Lorentz condition

$$\partial^\mu A_\mu = 0 \tag{3.5}$$

but more complicated covariant conditions may be envisaged, as for instance [15]

$$\partial^\mu A_\mu = g\, A_\mu A^\mu \tag{3.6}$$

Whatever form it takes, a convariant gauge condition will be a non-holonomic constraint on the theory. This means that it is a non-integrable relation between generalized coordinates and velocities. In such a case standard canonical methods do not work. This is also reflected by the fact that the time derivative of the scalar potential $\dot{A}_o(x)$ does not appear in the classical Lagrangian. Thus the momentum canonically conjugate to A_0, $\frac{\partial L}{\partial \dot{A}_0}$ vanishes. In classical electrodynamics this is not a great obstacle but in QED it is chatastrophic. The solution to this problem is to add the Fermi term to the Lagrangian. Then $\frac{\partial L}{\partial \dot{A}} = -\dot{A}_0$ and canonical quantization may proceed unhindered. It is however by no means evident that this does not change the theory into something different, something which differs from the non-covariant Coulomb gauge, in which the Fermi term is not used.

The answer is that the theory is indeed changed. The addition of the Fermi term gives the Maxwell field four degrees of freedom instead of two. Instead of having two directions of transverse polarization, photons may now also have longitudinal and scalar polarization. And what is worse, the probability for observing an odd number of scalar photons must be taken to be negative. Actually this is what saves the theory. Due to the conservation of the electromagnetic current the amplitudes for emission (or absorption) of scalar and longitudinal photons are equal. The total probability for observing a real photon that is not transversely polarized, i.e. the sum of the positive probability for emission (or absorption) of a longitudinal photon and the negative probability for emission (or absorption) of a scalar photon, is zero.

This feature is common to all gauge field theories. If we want to quantize in a covariant way it is necessary to augment the Hilbert space with non-physical states of which some have negative metric. Because of current conservation, or rather because of the gauge invariant nature of the classical Lagrangian, and the indefinite metric, the unphysical states cancel against each other in the total probability. Provided there is no way experimentally to distinguish the non-physical states with positive and negative metric from each other, it will always look as if the physical states are the only ones present since their probability alone sums up to one.

There is another feature of gauge field theories which is usually not connected with QED. It is the phenomenon of Feynman-Fadeev-Popov ghosts [16]. In non-abelian gauge field theories [17] such ghosts always arise even with a simple quadratic gauge fixing (Fermi) term. They express that the naive Feynman rules derived in such a theory do not give rise to a gauge invariant S-matrix. In order to obtain a gauge invariant S-matrix one must subtract a ce tain term for each closed gauge field loop. This term looks as if a scalar particle propagates around the loop, and to get the sign right one must take it to be a Fermion. In Abelian gauge field theories like QED such ghosts do not occur provided the gauge condition is linear so that the gauge fixing term is quadratic. Then the S-matrix is indeed gauge invariant between physical states. It was pointed out by 'tHooft and Veltman [15] that if one imposes a non-linear gauge condition (as f.inst. (3.6)) by adding a gauge fixing term $-1/2\ (\partial_\mu A^\mu - gA_\mu A^\mu)^2$ to the classical Lagrangian, Fadeev-Popov ghosts also arise in QED. They have the effect of precisely removing the influence of the extra couplings of three and four photons arising from the gauge fixing term. With the advent of non-abelian gauge theories one has thus obtained a better understanding of gauge invariance in QED itself. I shall not go further into the question of non-linear gauges but return to the Lagrangian with the standard Fermi term (3.2).

I shall assume that you are familiar with the way Feynman rules are derived in canonical field theory for an interaction without derivative couplings. Then we may directly read off the Feynman rules from the Lagrangian. From the unperturbed part we get the propagators as the inverse of the differential operators in the free equation of motion (times i). In momentum space the photon propagator becomes

$$= -i\,\frac{g_{\mu\nu}}{k^2} \tag{3.7}$$

and the electron propagator

$$= \frac{i}{\not p - m} \tag{3.8}$$

In each case the position of the pole is defined by adding a small positive imaginary part to the denominator. From the interaction part we derive the vertex amplitude

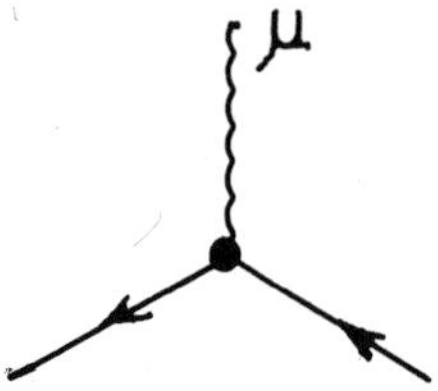

$$= -ie\gamma_\mu \tag{3.9}$$

There will be further vertices from the compensation part L_2 but we shall return to these later.

In order to calculate S-matrix elements we also must state the amplitudes for incoming and outgoing particles. They are

$= \varepsilon_\mu$ incoming photon (3.10)

$= \varepsilon^*_\mu$ outgoing photon (3.11)

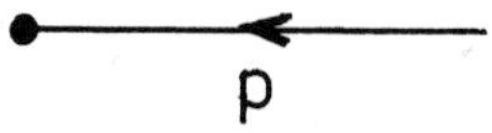

$= u$ incoming electron (3.12)

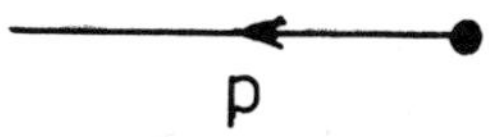

$= \bar{u}$ outgoing electron (3.13)

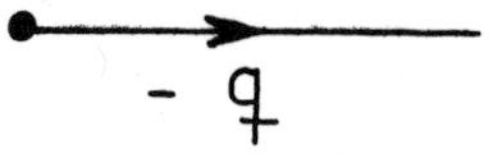

$= \bar{v}$ incoming positron (3.14)

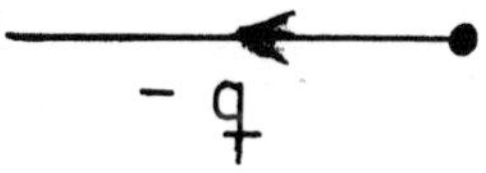

$= v$ outgoing positron (3.15)

Here ε_μ is the polarization vector of the photon, satisfying $k \cdot \varepsilon = 0$; u is the spinor of the electron satisfying $(\not{p}-m)u = 0$; and v is the spinor of the positron satisfying $(\not{q}+m)v = 0$. The momenta k, p and q are the physical momenta of the particles with positive fourth components.

The S-matrix element for a given process is now written as a sum of diagrams (Feynman graphs) built up of the compononents given above. There are a few rules more, before one can actually write down the amplitude. The sum should only include the topologically inequivalent diagrams. Two diagrams are topologically equivalent if they only differ by a permutation of the vertices. Thus

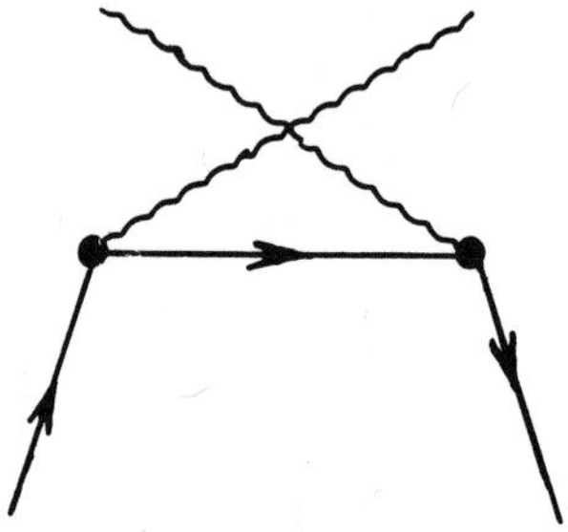

and

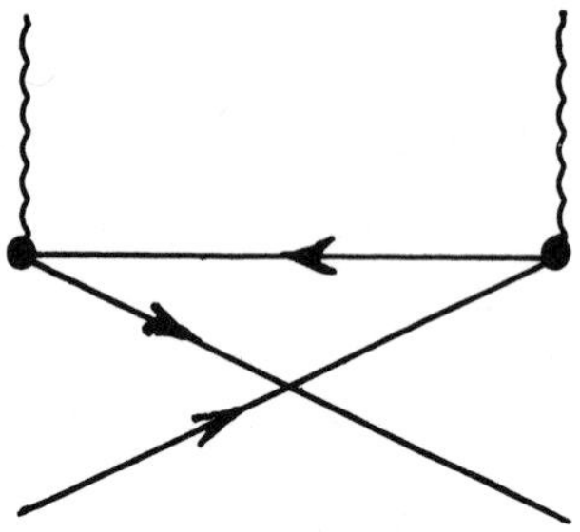

are topologically equivalent while

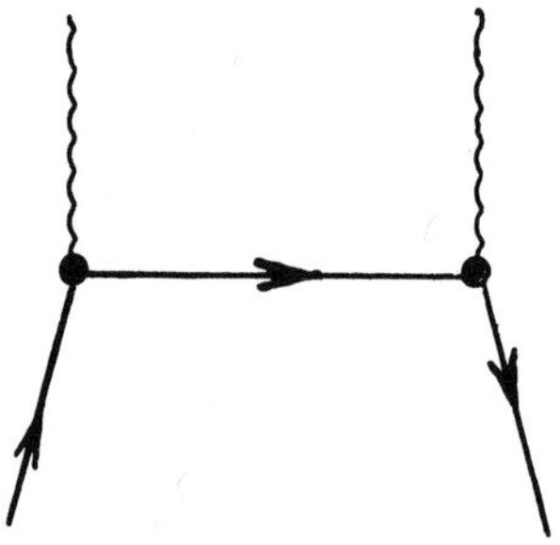

and

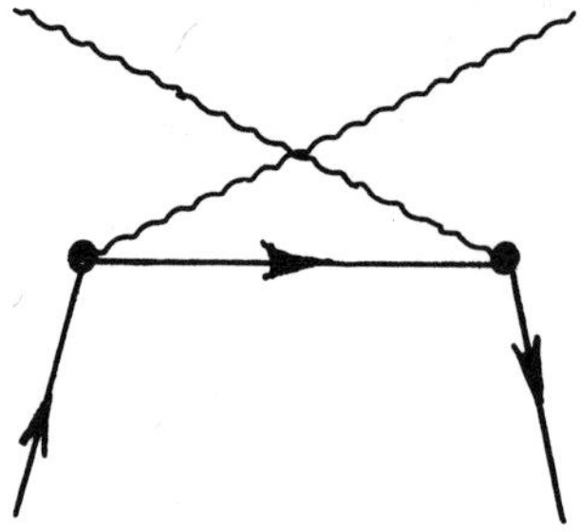

are topologically inequivalent. In other words if one diagram can be deformed into another while keeping the external linear fixed they are topological equivalent.

The amplitude for a diagram is simply the product of all the components, times a sign factor (∗). The relative sign between all diagrams may be calculated from the rules, that each closed fermion loop gives rise to a factor -1, and that two diagrams which differ from each other by exchange of a pair of identical fermion lines have a relative factor -1. Thus

(∗) and times a momentum conserving δ-function for each disconnected part.

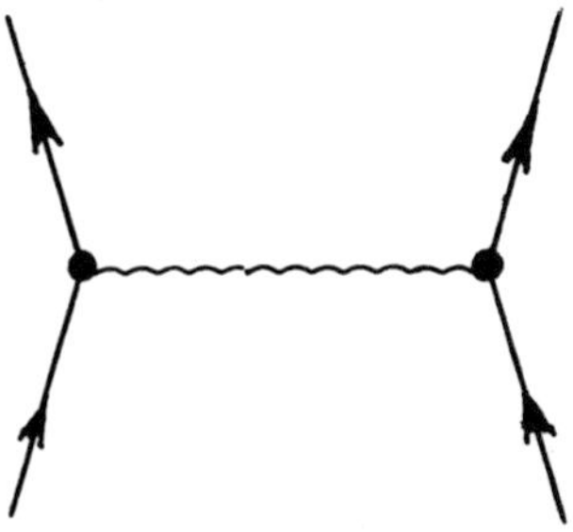

and

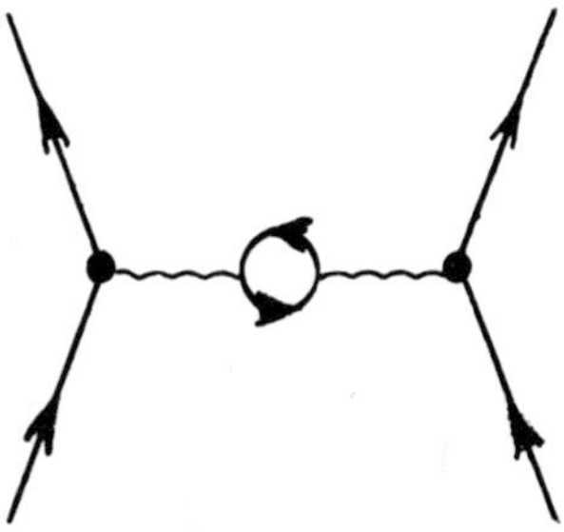

have oppositve sign. This is also the case for

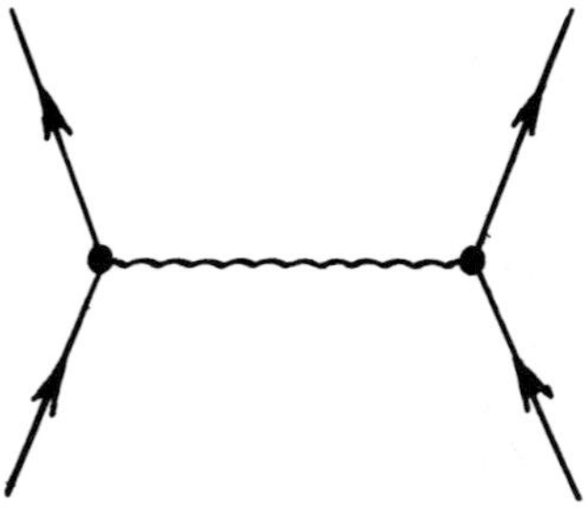

and

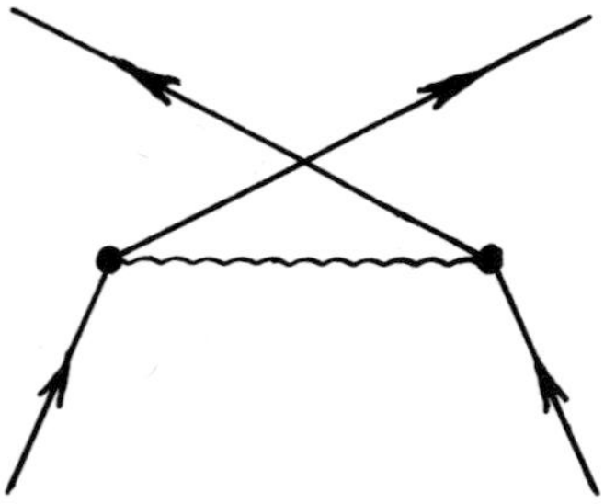

In all other cases the diagrams have the same sign. Finally there is the overaal (absolute) sign of all the diagrams. It is generally of little interest and may best be determined by carrying out the Wick pairings for the simplest diagram contributing to the process.

There are four extensions of this theory that should be mentioned. The first has to do with inclusion of muons. This is simply done by adding an extra piece to the free Lagrangian and an extra piece to the interaction of exactly the same form as the electron terms except for the mass, which so far is the only way in which muons seem to differ from electrons.

Secondly one needs to include an external field, $a_\mu(x)$ for many applications. If the electrons or positrons form bound states with the external field, as is for instance the case with a Coulomb field, one must include the interaction in the free Lagrangian, adding a term $-e\, a_\mu(x)\bar{\psi}\gamma^\mu\psi$ to L_0. It is not possible to perturb free particles to form bound states, as one would have to, if the external field term was added to the interaction L_1. This leads to the Furry picture in which an electron, while propagating, interacts an indefinite number of times with the external field. We shall get back to the discussion of bound states in Section 9.

Thirdly, there are divergence problems associated with this Lagrangian that are due to the fact that the photon mass vanishes.

Whether or not the photon mass actually vanishes, is an experimental question [18]. At any rate it is so small $\lesssim 10^{-49}$g that it should be legal to take it equal to zero in all but the most refined physical situations. If the photon mass is non-zero one must add a term $+1/2\ \mu^2 A_\mu A^\mu$ to the free Lagrangian. Then the propagator becomes $-ig\mu_\nu/k^2-\mu^2$ instead of (3.7). This, however, destroys gauge invariance in a manifest way. For a non-Abelian gauge theory a mass term added in by hand would destroy the renormalizability (or unitarity) which is tied closely to the gauge invariance in this case. For the case of QED this is actually not so, and the massive theory is unitary as well as renormalizable. The divergences that occur for $\mu \to 0$ are called infrared divergences and show up in the total Coulomb scattering cross section and in radiative corrections.

Actually if one asks the proper physical questions of the theory one finds that there is no true dependence on the photon mass. In the end of the calculation it may be taken to zero. It is presumably well-known to all of you that for the case of radiative corrections inclusion of real Bremsstrahlung makes the photon mass cancel out and replaces it effectively by the energy resolution of the detecting apparatus. What is perhaps not so well-known to you is that the photon mass also disappears for the case of Coulomb scattering. Here the divergence occurs in the forward direction, but that is exactly where we cannot distinguish the particles that have scattered a little, from those that haven't scattered at all. If we introduced a finite angle of resolution in the forward direction, the total cross section for scattering outside this angle is perfectly finite for $\mu \to 0$.

From this discussion it seems as if it should be possible to leave out the photon mass completely, i.e. having it equal to zero always. Various schemes have been tried [19], but the simplest seems still to be to keep a mass in intermediate calculations. Without a photon mass there are also serious problems with the definition of the one-electron state, because the one-electron-plus-many-photon state continuum goes all the way down to the one-electron pole. The electron becomes "fuzzy around the edges" [20].

Finally I shall again return to the question of gauges and gauge transformations. Since the propagator of the photon is the vacuum expectation value of the time-ordered product of the Maxwell Field in two different space-time points, $\langle 0 | J(A_\mu(x), A_\nu(y)) | 0 \rangle$, a gauge transformation (3.4), will in general lead to a propagator of the general form (*)

(*) There will be exceptions to this if the $\theta(x)$ does not commute with $A_\mu(x)$ or itself. (See eq.(3.4)).

 $$= -i\,\frac{g_{\mu\nu}}{k^2} + i(k_\mu\Lambda_\nu + k_\nu\Lambda_\mu) \qquad (3.16)$$

where Λ_ν is some function of k. There is one parameter family of covariant gauges that one often uses where the propagator is of the form

 $$= -i\,\frac{g_{\mu\nu}}{k^2} + (1-a)\,\frac{k_\mu k_\nu}{(k^2)^2} \qquad (3.17)$$

Here a is a real parameter. For a = 1 one gets the Fermi or Feynman gauge, for a=0 the Landau gauge, for a=3 the Yennie gauge and for a=∞ the unitary gauge. It may be obtained in a canonical fashion if the Fermi term in (3.2) is multiplied by 1/a. Including a mass term we accordingly have

$$L_o = -\frac{1}{4}F_{\mu\nu}F^{\mu\nu} - \frac{1}{2a}(\partial_\mu A^\mu)^2 + \frac{1}{2}\mu^2 A_\mu A^\mu + \bar{\psi}(i\partial\!\!\!/ - m)\psi \qquad (3.18)$$

which gives rise to the propagator

 $$= -i\,\frac{g_{\mu\nu} - \dfrac{k_\mu k_\nu}{k^2}}{k^2 - \mu^2} - ia\,\frac{k_\mu k_\nu}{k^2}\,\frac{1}{k^2 - a\mu^2}$$

$$= -i\,\frac{g_{\mu\nu} - (1-a)\dfrac{k_\mu k_\nu}{k^2 - a^2\mu^2}}{k^2 - \mu^2} \qquad (3.19)$$

Observe also that (3.17) is the most general parity conserving Lagrangian which is quadratic in the A-field and in the ψ field, since we may always normalize the fields in such a way that one of the terms for each field has unit coefficient.

4. THE BORN APPROXIMATION

In the Born approximation one only considers tree diagrams. The Feynman rules give a perfectly finite prescription for calculating such amplitudes. The trouble - apart from infrared divergences - first occurs in diagrams with loops.

In the Born approximation one may calculate the leading behaviour of many scattering processes, such as electron-electron scattering, electron-positron annihilation into various final states, Comton scattering, Bremsstrahlung, pair production etc.

In comparing such calculations with experiments one therefore tests primarily the coarse structure of QED, i.e. whether the free propagators indeed look the way they do, whether the interaction is the correct one (and the only one present) etc. True enough radiative corrections must be applied but they are generally small, though not vanishingly so.

It is not my intention to bore you with details of calculations. They have been carried out since the 1930's and are found in many textbooks.[14)] I am sure all of you have been through the calculation of some of the cross sections to be discussed in the following.

In the early years one could only test the low energy behavior of scattering processes. With the advent of high-energy accelerators this picture changed. In the beginning one had, however, always to use stationary targets and that made it difficult to test purely electromagnetic processes in particular at high momentum transfer. When a high energy, say 10 GeV, electron or photon hits a stationary electron the momentum transfer is maximally $\sqrt{2mE}$ = 100 MeV and no really interesting region is reached. Instead one had to resort to processes of the Bethe-Heitler type, i.e. wide angle Bremsstrahlung or pair production. They involve a nucleus and are consequently not "clean" tests of QED. At times deviations were claimed but they invariably disappeared at closer experimental scrutiny.

First when storage rings with colliding beams were taken into use did one get the chance to test clean QED processes at high momentum transfer. In the following I shall discuss these processes one-by-one and present what I think is the latest experimental information about them.

$e^+e^- \to \mu^+\mu^-$

This is undoubtedly the process with the simplest cross section. Only one graph contributes

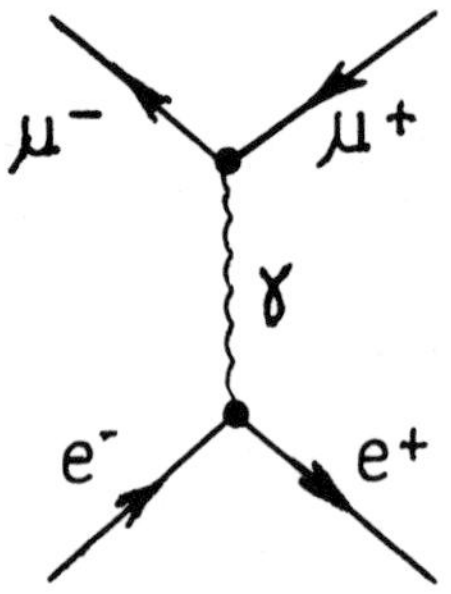

in the Born approximation. The differential CM cross section is (in the high energy limit)

$$\frac{d\sigma}{d\Omega} = \frac{\alpha^2}{16E^2}\,(1 + \cos^2\theta)$$

where θ is the scattering angle, E is the (single) beam energy and $\alpha = e^2/4\pi$ the fine structure constant. What is tested in the behaviour of the photon propagator at high virtual time-like momenta ($k^2 > 0$). Experimental results are usually in close agreement with theory and one can generally only obtain an upper limit to possible deviations. In order to have a single number which represents the agreement one deforms the photon propagator with a cut-off factor resembling the propagator for a massive particle

$$\frac{1}{k^2} \rightarrow \frac{1}{k^2}\left(1 \pm \frac{k^2}{k^2-\Lambda_\pm^2}\right) \tag{4.2}$$

The two signs are necessary in order to cover deviations above as well as below the prediction.

In a recent SPEAR experiment [21] one determined the lower limit for $\Lambda_\pm$ to be (95 % confidence level)

$$\left.\begin{array}{l}\Lambda_+ \gtrsim 35\ \text{GeV}\\ \Lambda_- \gtrsim 47\ \text{GeV}\end{array}\right\}\quad \begin{array}{c} e^+e^- \rightarrow \mu^+\mu^- \\ \text{and } e^+e^-\end{array}$$

In the comparison one made use of the $\mu^+\mu^-$ data as well as of the e^+e^- data in order to increase the sensitivity. This accounts partly for the very high cut-off limits. The experiment was normalized relative to the total number of e^+e^- events in the full angular region $|\cos\theta| \lesssim 0.6$.

The experiment was carried out at three different energies : $2E = 3.0$, 3.8 and 4.8 GeV. The ratio between the observed number of $\mu^+\mu^-$ events and the expected number was at these three energies respectively 0.95 ± 0.04, 1.05 ± 0.03 and 1.01 ± 0.03. The numbers of events were 563, 1097 and 1241.

The angular distribution is shown in Fig. 4.1 at 4.8 GeV.

Observe also that the cross section is independent of the mass of the participating particles. If for example a charged heavy lepton existed which decayed rapidly into muons one would observe a marked increase in the cross section well above threshold. Such an experiment accordingly also limits the possibilities for existence of this kind of particles, although in the experiment discussed[21] no such analysis was made.

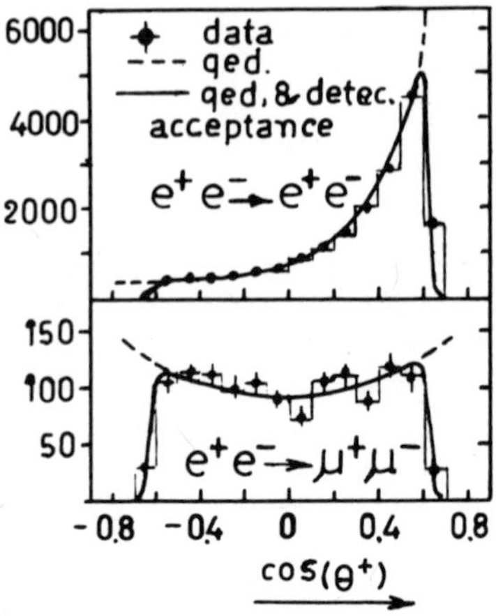

Fig. 4.1 Angular distribution of positive prongs for $e^+e^- \rightarrow e^+e^-$ and $e^+e^- \rightarrow \mu^+\mu^-$. The histograms give the observed counts alike the observed curves are QED, normalized to total e^+e^- counts within $|\cos\theta| \leq 0.6$ (taken from ref.21).

If a resonance exists with the quantum numbers of the photon and with a sufficiently strong coupling to the leptons one expects to see not only a bump in the cross section but also interference phenomena around the bump. In a recent experimental analysis [24] of the $\psi(3095)$ one found indeed such phenomena confirming that the ψ has spin-parity assignment 1^{--}.

$e^+e^- \rightarrow e^+e^-$

This is Bhabha-scattering[26] for which we have the diagrams

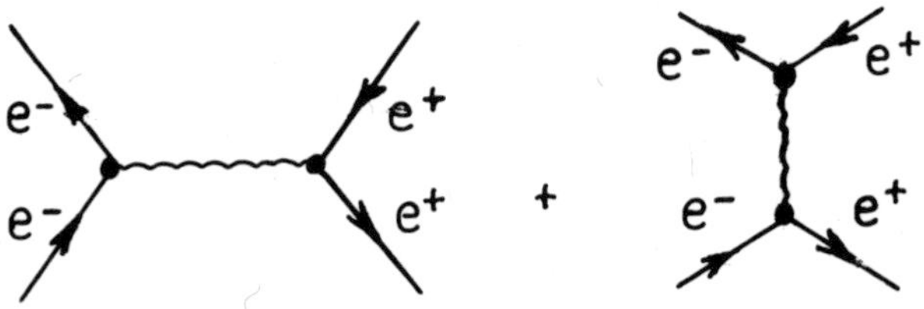

and the high energy cross section

$$\frac{d\sigma}{d\Omega} = \frac{\alpha^2}{4E^2}\left[\frac{1}{\sin^2\theta/2} - \cos^2\theta/2\right]^2 \qquad (4.3)$$

The first term which comes from the first diagram has the typical angular behaviour of Coulomb scattering with the divergence in the forward direction.

In the experiment at SPEAR [21)] one observed huge numbers of e^+e^- events. At the three energies 2E = 3.0, 3.8 and 4.8 GeV one identified respectively 7671, 13419 and 15788 pairs. No comparison of the absolute e^+e^- rate was made, only a fit to the angular distribution (Fig. 4.1) normalized to the total number of e^+e^- events. The cut-off limits for e^+e^- pairs alone were found to be

$$\left.\begin{array}{l}\Lambda_+ \gtrsim 15\ \text{GeV} \\ \Lambda_- \gtrsim 19\ \text{GeV}\end{array}\right\} \quad e^+e^- \to e^+e^- \tag{4.4}$$

Since the experiment covers angles between 50° and 130° one not only tests the dominant Coulomb diagram but also the annihilation diagram. It was even possible to obtain separate limits on space-like and time-like cut-offs. They were roughly of the same size as in (4.4).

In colliding beam processes the determination of the machine luminosity always presents a problem. The luminosity represents so to speak the number of possibilities that the particles have for interacting. Precisely we have the relation

$$\text{rate of events} = \text{cross section} \times \text{luminosity} \times \text{detection efficiency} \tag{4.5}$$

The luminosity depends on the overlap between the beams in the interaction region. If the beams do not collide the luminosity is zero. Hence the shape and extent of the interaction region are important parameters and they are not particularly well-known. Furthermore the luminosity may vary with time when the stored beams decay etc. Fortunately the cross section is independent of time such that we may write (4.5) in the form

$$R(t) = \sigma . L(t) \tag{4.6}$$

An experiment collects events over a long time so that we should really write (4.6) in the form

$$\int_{t_1}^{t_2} dt\, R(t) = \sigma \int_{t_1}^{t_2} dt\, L(t) \tag{4.7}$$

On the left hand side we have the total number of events and on the right hand side the integrated luminosity.

In order to determine the integrated luminosity one chooses a process with a known cross section and measures the total number of events during the same period in which one observes the interesting process. In the experiment discussed here[21] the monitoring process was the total number of e^+e^- counts within the observed angular region. In an earlier SPEAR experiment [22] the monitoring process was Bhabha scattering at small angles (3.7°). In this region the classical Coulomb scattering of the two charged particles will dominate and the specific QED effects can be disregarded. In a CEA experiment [27] the monitoring process was double Bremsstrahlung $e^+e^- \to e^+e^-2\gamma$ at small angles. This process is also dominated by easily calculable effects at low momentum transfer. It should be borne in mind that colliding beam measurements always are done relative to some monitoring process.

$e^+e^- \to 2\gamma$

In this annihilation process the main feature is the space-like electron propagator

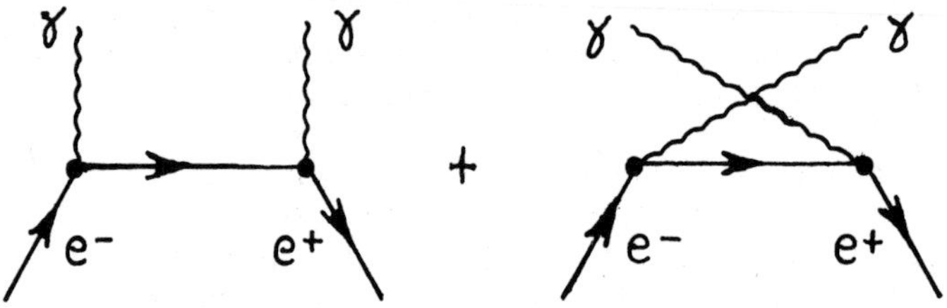

At high energy the cross section is

$$\frac{d\sigma}{d\Omega} = \frac{\alpha^2}{4E^2} \left\{ \frac{2}{\sin^2\theta} - 1 \right\} \tag{4.8}$$

The angular divergence at $\theta = 0,\pi$ is really not there. The formula is only valid for $\theta, \pi-\theta \gg m/E$. The total cross section integrated over all angles is

$$\sigma = \frac{\pi\,\alpha^2}{E^2} \left(\mathrm{Log}\,\frac{2E}{m_e} - 1/2 \right) \tag{4.9}$$

which - corresponding to the angular divergence in (4.8) - diverges for $m_e \to 0$.

In the SPEAR experiment[22] one observed 88 events of this type (at 5 GeV) against the expected number of 84. If a deformation factor

$$\frac{1}{\not{q}-m} \rightarrow \frac{1}{\not{q}-m}\,(1 \pm q^4/\,\Lambda_\pm^4) \qquad (4.10)$$

is used to parametrize the deviation one obtains

$$\left.\begin{array}{l}\Lambda_+ \gtrsim 6.2\ \text{GeV}\\ \Lambda_- \gtrsim 6.9\ \text{GeV}\end{array}\right\}\ e^+e^- \rightarrow 2\gamma \qquad (4.11)$$

It should be noticed that the deformation factor for the electron propagator takes a different form than for the photon propagator. This is motivated by the analysis of Kroll [28] who showed that a change of the electron propagator necessitates a change in the vertices if gauge invariance should be maintained. The two changes almost always cancel each other for an ordinary deformation factor.

This concludes the discussion of the experimental situation for the classical processes $e^+e^- \rightarrow \mu^+\mu^-$, e^+e^- and $\gamma\gamma$. A very thorough experimental review of colliding beam processes may be found in the article by Paoluzzi [29]. In the next section I shall discuss two-photon processes in e^+e^- collisions.

Finally let me say some words about radiative corrections. The calculation of radiative corrections is a tedious work only somewhat simplified by the use of computers for the algebra as well as for the numerics. There exist approximation schemes [30,31] based on the soft photon approximation. In realistic experiments contributions from hard photons that go undetected must also be in A complete evaluation of the radiative corrections to all QED colliding beam processes have been carried out by Berends, Gaemers and Gastmans. They have evaluated corrections to $e^+e^-\rightarrow\mu^+\mu^-$ in the c = -1 channel [32], i.e. when the charge of the muons is not detected, as well as in the c = +1 channel which is charge asymmetric [33]. They have evaluated the hard photon corrections for Bhabha scattering [34], and to $e^+e^- \rightarrow 2\gamma$ [35]. Their general conclusion is that if one wants to study QED processes with a precision at the percentage level the soft photon approximations do not give sufficiently correct results. The size of the corrections is generally of the order of a few percent although higher values may be reached in special kinematical situations, such as forward scattering.

The angular asymmetry in $e^+e^- \rightarrow \mu^+\mu^-$ has to my knowledge not been measured except in a recent experiment at SPEAR around the ψ-resonance [24]. The hard photon corrections are negligible for the pure QED asymmetry [33] and the soft photon corrections have a rather complicated history [36].

Finally one should notice that real Bremsstrahlung in $e^+e^- \to e^+e^-$ - i.e. the process $e^+e^- \to e^+e^-\gamma$ - gives rise to an acoplanarity of the final particles. The acoplanarity distribution is obtained as a byproduct to the hard photon calculation [34] and was measured recently [34]. The agreement between theory and experiment is excellent.

The general conclusion to be drawn from this discussion is that there is complete agreement between theory and experiment in the fundamental QED processes. The cut-off limits now range between 6 and 50 GeV.

5. EQUIVALENT PHOTONS

The equivalent photon approximation of Weizsäcker and Williams [37] has received new interest in recent years (*) due to the possibility of observing a process of the type

$$e^+e^- \to e^+e^- + X \tag{5.1}$$

which for small scattering angles is dominated by the graph

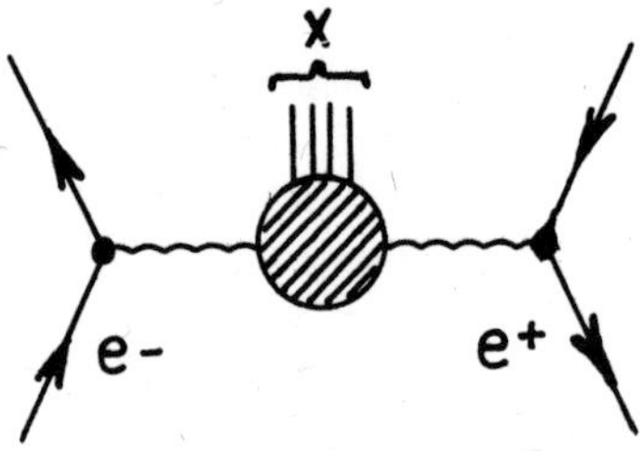

In this section, I shall give a modern derivation[39] of the approximation, a derivation which directly aims at scattering processes like the one above. Consider an arbitrary process in which a photon may be absorbed with the amplitude (matrix element of the electromagnetic current)

 $= J_\mu(k)$ (5.3)

(*) The history is rather complicated. A thorough discussion is given by P. Kessler in ref. 38).

This may for instance be photoproduction $\gamma p \to X$ or some other process. The probability for absorption of a real unpolarized photon is then

$$P_\gamma = \frac{1}{2\omega}\,\frac{1}{2}\sum_{pol} |\varepsilon^\mu J_\mu|^2 = -\frac{1}{4\omega}\, J^\times_\mu J^\mu \tag{5.4}$$

with the normalization factor $(1/\sqrt{2\omega})^2$ correctly included. The late equation was derived by means of the well-known rule that a sum over transversal polarizations may be replaced by a sum over all polarizations including the unphysical ones, provided

$$k_\mu J^\mu(k) = 0 \tag{5.5}$$

But this is simply current conservation applied to the amplitude. Alternatively it may be viewed as gauge invariance since it is the necessary and sufficient condition for the invariance of $\varepsilon_\mu J^\mu$ under the transformation $\varepsilon_\mu \to \varepsilon_\mu + \lambda k_\mu$. In fact, since (5.5) may be written

$$J_o = \frac{\bar{k}}{\omega} \cdot \bar{J} \tag{5.6}$$

and since $\omega = |\bar{k}|$ for reals photons this is just the statement that the amplitude for absorption of a scalar photon is equal to the amplitude for absorption of a longitudinal photon. As discussed in section 2 the probability for absorption of the scalar photon is negative, $-|J_o|^2$ and it will therefore cancel the longitudinal probability $|\frac{\bar{k}}{\omega} \cdot \bar{J}|^2$, in (5.4).

The amplitude for the inelastic scattering of an electron by means of the exchange of a virtual photon with $J_\mu(k)$ is

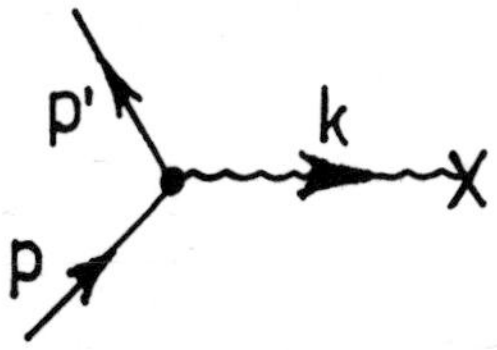

$$= e\, \frac{\bar{u}'\gamma_\mu u}{k^2}\, J^\mu(k) \tag{5.7}$$

where the incoming (outgoing) electron has momentum p(p').
The probability is accordingly

$$dP_e = \left(\frac{1}{\sqrt{2E}}\right)^2 \frac{1}{2} \sum_{spins} \left| e\, \frac{\bar{u}'\gamma_\mu u\, J^\mu}{k^2} \right|^2 \frac{d^3\bar{p}'}{(2\pi)^3 2E'} \tag{5.8}$$

where we have summed over the spins in the final state and averaged over the spins in the initial state. The last factor is the relativistically invariant phase space element.

The invariant momentum transfer or the mass of the virtual photon is

$$k^2 = (p-p')^2 = 2m^2 - 2\,EE'\,(1 - vv'\cos\theta) \tag{5.9}$$

where E(E') is the energy and v(v') the velocity of the incoming (outgoing) electron and θ the scattering angle. Due to the factor $1/k^4$ in (5.8) the probability receives its largest contribution from small k^2, i.e. for small scattering angle. Assuming $E,E' >> m$ and $\theta << 1$ we have

$$k^2 = -EE'\,(\theta^2 + (\frac{m\omega}{EE'})^2) \tag{5.10}$$

where $\omega = E - E'$ is the frequency of the virtual photon or alternatively the energy loss of the electron. This shows that the main contribution to the probability (5.8) comes from angles of the order of $m\omega/EE'$ or smaller. For θ small, E, E' large we find the component of $\bar{k}$ parallel to the velocity of the incoming electron

$$k_{//} = (\bar{p}-\bar{p}')_{//} = p-p'\cos\theta \simeq \omega + \frac{1}{2}\theta^2 E' \tag{5.11}$$

while the component perpendicular to the velocity becomes

$$|\bar{k}_\perp| = |(\bar{p}-\bar{p}')_\perp| = |\bar{p}'_\perp| = p'\sin\theta \simeq E'\theta \tag{5.12}$$

Thus we have

$$\frac{|\bar{k}_\perp|}{k_{//}} \lesssim \theta\;E'/\omega \tag{5.13}$$

The perpendicular component is much smaller than the parallel component when

$$\theta << \omega/E' \tag{5.14}$$

This in particular is the case in the region around $\theta \simeq m\omega/EE'$ i.e. where k^2 is small. In the region (5.14) we may also disregard the second term in (5.11). In the following we shall also assume $\omega << E$ i.e. that the energy loss is fairly small, and thus $E' \simeq E$.

The spin sum in (5.8) is carried out in the usual way,

$$\sum_{\text{spins}} |\bar{u}'\gamma_\mu u\, J^\mu|^2 = \mathrm{Tr}\,\left[(\not{p}'+m)\gamma_\mu(\not{p}+m)\gamma_\nu\right] J^\mu J^{\nu\times}$$
$$= 4\,(p_\mu p'_\nu + p_\nu p'_\mu + \frac{1}{2} g_{\mu\nu} k^2) J^\mu J^{\nu\times}$$
$$\simeq 8\,|p^\mu J_\mu|^2$$

where in the last line we have replaced p' by p using (5.5) and dropped the k^2 term (*). Writing out the phase space (5.8) becomes

$$d\,p_e = \frac{\alpha}{2\pi^2} \quad \frac{|p^\mu J_\mu|^2}{k^4}\, dE'\ d\cos\theta\ d\varphi \tag{5.15}$$

where φ is the asimuthal angle of $\bar{p}'$ with respect to some fixed plane containing $\bar{p}$.

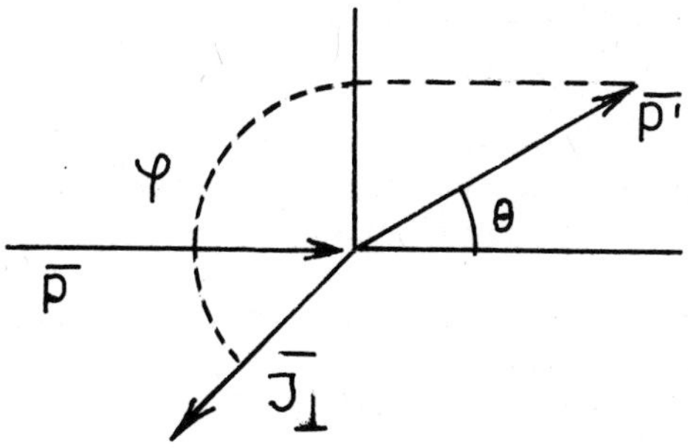

In order to evaluate $p^\mu J_\mu$ we use (5.5) in the form

$$0 = k_\mu J^\mu = \omega J_0 - k_{//} J_{//} - \bar{k}_\perp \cdot \bar{J}_\perp \tag{5.16}$$

Then since $v \simeq 1$

$$p_\mu J^\mu = E\,(j^\circ - v\,J_{//}) \simeq E(J^\circ - J_{//})$$
$$= \frac{E}{\omega}\,(\bar{k}_\perp \cdot \bar{J}_\perp + (k_{//} - \omega) J_{//})$$
$$\simeq \frac{E}{\omega}\ \bar{k}_\perp \cdot \bar{J}_\perp$$

(*) This is legal if the energy loss is not too large i.e. for $\omega \ll E$.

In the second line we have used (5.11) and (5.12) to derive that $(k_{//}-\omega)/|\bar{k}_\perp| \simeq \frac{1}{2}\theta \ll 1$ so that the second term may be dropped. This of course is true only when $|J_{//}| \simeq |\bar{J}_\perp|$ But this must in general be the case since // and $\perp$ refer to the incoming electrons direction about which the current J_μ can know nothing.

Likewise, since $|\bar{k}_\perp| \ll k_{//}$ the current $J_\mu(k)$ cannot depend on $\bar{k}_\perp$ in first approximation but only on $\bar{k}_{//}$. Thus we may take $\bar{J}_\perp$ to define our azimuthal angle so that $\bar{k}_\perp \cdot \bar{J}_\perp = |\bar{k}_\perp||\bar{J}_\perp| \cdot \cos\varphi$ Inserting this into (5.15) and carrying out the integral over φ we get

$$dp_e = \frac{\alpha}{2\pi^2} \frac{\left(\frac{E}{\omega}\right)^2 \pi \bar{k}_\perp^2 \bar{J}_\perp^2}{k^4} d\omega \, d\cos\theta \qquad (5.17)$$

Now using $J_0 \simeq J_{//}$ (see eq.(5.16)) we have $\bar{J}^2 \simeq -J_\mu J^\mu$ and then the whole expression simplifies to

$$dp_e = p_\gamma \cdot \frac{2\alpha}{\pi} \frac{\theta^3 \, d\theta}{\left[\theta^2 + \left(\frac{m\omega}{E^2}\right)^2\right]^2} \frac{d\omega}{\omega} \qquad (5.18)$$

This formula gives us the probability for scattering an electron into the angle $d\theta$ with energy loss in $d\omega$.

Since the complete angular dependence is known we may integrate over θ in the region $0 \leq \theta \leq \omega/E$, outside which our approximation breaks down. (See eq. (5.14)). This leads to the results for the leading terms

$$dp_e = p_\gamma \frac{2\alpha}{\pi} \left(\text{Log} \frac{E}{m} - \frac{1}{2}\right) \frac{d\omega}{\omega} \qquad (5.19)$$

expressing the probability that an electron interacting with the current J_μ suffers an energy loss in $\omega|\omega + d\omega$. We may interpret the probability in a different way. If instead of an electron we let an incoherent beam of photons with a frequency distribution

$$dN(\omega) = \frac{2\alpha}{\pi} \left(\log \frac{E}{m} - \frac{1}{2}\right) \frac{d\omega}{\omega}$$

interact with the current, (5.19) gives us the number of interactions for the photons with frequency in $\omega|\omega + d\omega$.
Thus an electron may be replaced by an equivalent beam of photons.

In arriving at (5.19) we have made certain approximations. We have assumed that the electron is highly relativistic (E>> m), that the scattering angle is small ($\theta \ll 1$), that it is in fact limited by $\theta \ll \omega/E'$ which is also supposed to be small. This last condition may be relaxed a bit. One then obtains a slightly better expression for the equivalent photon spectrum[40]

$$dN(\omega) = 2\,\frac{\alpha}{\pi}\;\frac{E^2+E'^2}{2E^2}\;\left(\mathrm{Log}\,\frac{E}{m} - \frac{1}{2}\right)\frac{d\omega}{\omega} \qquad (5.21)$$

In the formula relating the probabilities

$$dp_e = p_\gamma(\bar{k}_{//})\; dN(\omega) \qquad (5.22)$$

one should evaluate the photon absorption probability for real photons with momentum $\bar{k}_{//} = \omega\,\bar{v}$. The equivalent photons are approximately real, often called quasireal, because due to (5.10), (5.14)

$$-k^2 \ll \omega^2 \qquad (5.23)$$

i.e. they are only a tiny bit off-shell compared to their frequency.

We may immediately apply (5.22) to electroproduction by multiplying with the relevant phase space factors. Then we get for the

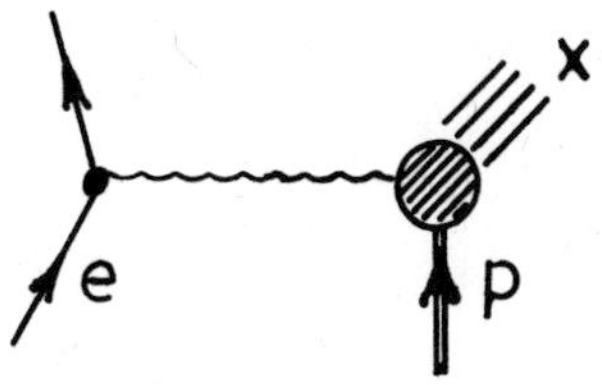

total cross section

$$\sigma_{ep\to ex}(E) = 2\,\frac{\alpha}{\pi}\;\left(\mathrm{Log}\,\frac{E}{m} - \frac{1}{2}\right)\int_{\omega_0}^{\infty}\frac{d\omega}{\omega}\;\sigma_{\gamma p\to x}(\omega) \qquad (5.24)$$

where ω_p is the photoproduction threshold. Here we have put $E' \simeq E$ since the photoproduction cross section is dominated by the region near threshold, and extended the integration to ∞ because it falls off as $1/\omega$.

In the two-photon process we may apply the equivalent photon approximation to both the colliding beams. Then the process may be interpreted as collision of the two photon beams, the inverse of an annihilation process. Using (5.19) twice we get for the total cross section

$$\sigma_{ee\to eex}(E) = [\frac{2\alpha}{\pi} (\log \frac{E}{m} - \frac{1}{2})]^2 \cdot \int \frac{d\omega_1}{\omega_1} \int \frac{d\omega_2}{\omega_2} \sigma_{\gamma\gamma\to x}(\omega_1 \bar{v}; -\omega_2 \bar{v}) \tag{5.25}$$

The $\gamma\gamma$-cross section is evaluated for the collision of a photon with momentum $\omega_1 \bar{v}$ and a photon with momentum $-\omega_2 \bar{v}$ where $\bar{v}$ is the beam direction. The total $\gamma\gamma$-cross section only depends on the invariant

$$s = (k_1 + k_2)^2 = 2\, k_1 k_2 = 4\, \omega_1 \omega_2 \tag{5.26}$$

We may accordingly integrate over the quotient

$$q = \omega_1/\omega_2 \tag{5.27}$$

which has the range $\frac{s}{4E} \leq q \leq \frac{4E}{s}$. The integral over q gives rise to a simple logarithm and the final result is

$$\sigma_{ee\to eex}(E) = [\frac{2\alpha}{\pi} (\log \frac{E}{m} - \frac{1}{2})]^2 \int_{s_0}^{4E^2} \frac{ds}{s} \log \frac{4E^2}{s} \sigma_{\gamma\gamma\to x}(s) \tag{5.28}$$

Here s_0 is the threshold and we may as before extend the integral to infinity taking in the same time the slowly varying logarithm outside the integral

$$\sigma_{ee\to eex}(E) \simeq [\frac{2\alpha}{\pi} (\log \frac{E}{m} - \frac{1}{2})]^2 \log \frac{4E^2}{s_0} \int_{s_0}^{\infty} \frac{ds}{s} \sigma_{\gamma\gamma\to x}(s) \tag{5.29}$$

This is the final compact formula which shows us that two-photon processes have a cross section which grows like $(\log E)^3$. This should be contrasted with one-photon processes for which the cross section falls like $1/E^2$. For the case of μ-pair production the two-photon process $e^+e^- \to e^+e^- \mu^+\mu^-$ overtakes the one-photon process $e^+e^- \to \mu^+\mu^-$ around 1 GeV in spite of the fact that the former has two factors of α more than the latter.

The experimental detection of two-photon processes is facilitated by the dominant forward scattering of the colliding particles. Since they suffer an energy loss they will emerge at the inside of the ring at a point which determines their momentum. The bending magnets act simultaneously as momentum analyzers. In this way one can not only tag the interesting collisions but even determine the energy losses ω_1 and ω_2 , and thereby the invariant mass $s = 4\omega_1\omega_2$ of the produced system as well as the velocity of its center of mass $(\omega_1-\omega_2)/(\omega_1+\omega_2)$.

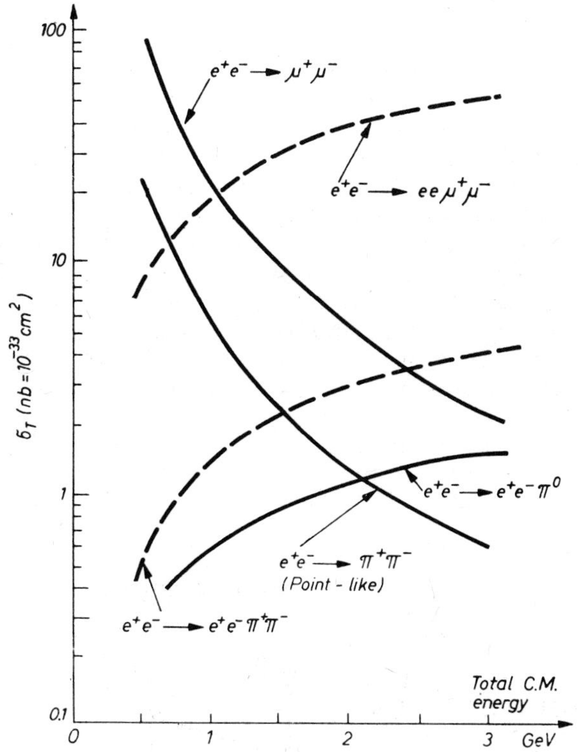

Comparison of $\gamma\gamma$-processes and annihilation processes (from ref.29).

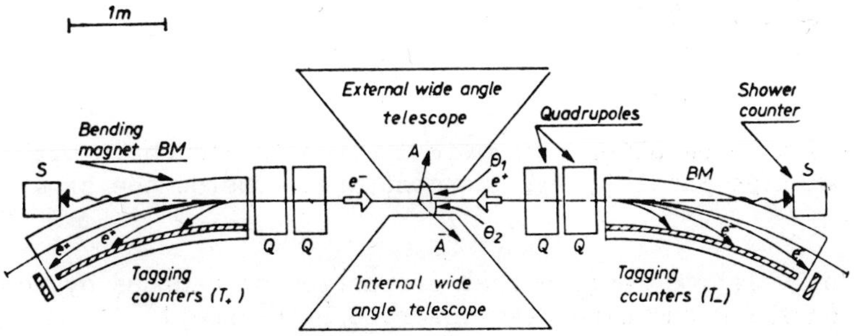

Sketch of the experimental setup for two-photon processes. (from ref.29). The shower counters S are used to veto real Bremsstrahlung processes.

If the produced system consists only of two charged particles this velocity may also be determined from their configuration. This serves as an extra check on the production mechanism[41].

Two-photon processes have been observed in Novosibirsk[42] and in Frascati[29]. The most spectacular result is presumably the identification at Frascati[41] of 34 events of the type $e^+e^- \to e^+e^-\mu^+\mu^-$. From the equivalent photon approximation one would expect 39. In the same experiment are also observed 61 events due to the process $e^+e^- \to e^+e^-e^+e^-$. In this case the equivalent photon approximation only predicts 49. Actually only 43 of the 61 candidates satisfy the kinematic requirements of production with two quasi-real photons. Of the rest 9 are interpreted to be due to a kind of mechanism proposed by Cabibbo and Parisi[43] while the rest are ambiguous. In this mechanism one of the incoming particles emits a quasi-real photon followed by conversion into a pair, one of which scatters off the other incoming particle.

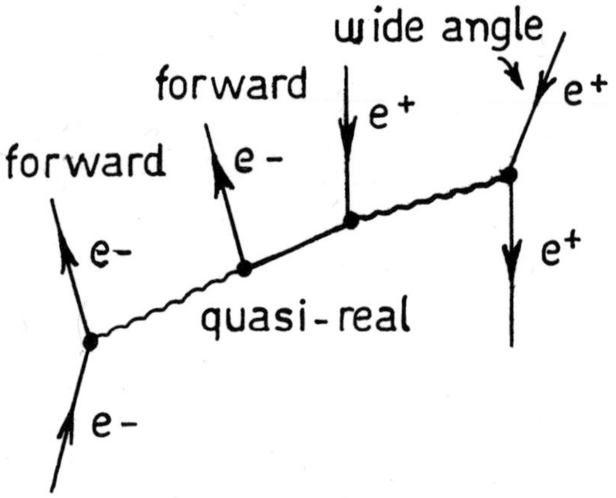

This leads to a peculiar kinematical situation in which both the forward emitted particles may be scattered towards the same tagging counter.

Finally let me mention that production of hadrons by the two-photon mechanism has also been observed at Frascati[44].

6. GREENS FUNCTIONS AND REGULARIZATION

Apart from the discussion of radiative corrections to scattering processes in section 4, we have so far avoided the problems associated with the calculation of diagrams containing loops. In the 1930's it was already recognized that divergences prevented calculation of higher order corrections to scattering processes and bound state energies. First in 1947 did it become clear through the works of Feynman, Schwinger and Tomonaga[45] that the lowest order divergences in QED only occurred in corrections to the mass and charge of the electron, and that if these corrections were included in the physical mass and charge (*), as they should be, the remainders were finite and calculable. Shortly after it was shown by Dyson[46] that this procedure could be carried through to all order. Renormalization theory was born.

The divergences happen at large values of virtual particle momenta, i.e. at short distance in configuration space. They express that something goes wrong at short distances either with our calculational techniques, i.e. perturbation expansion, or perhaps at a more fundamental level. There are some theories in which the short distance behaviour does not really influence the large distance behaviour, except in the corrections to static parameters like mass and charge. These are the renormalizable theories. Once the static parameters have been renormalized these theories become finite at large distances. Other theories, and they are by far the most numerous if "chosen at random", do not behave as nicely as that. The large distance behaviour is in general very strongly influenced by the short distance behaviour. Nobody so far knows how to make sense of non-renormalizable theories.

Maybe the short-distance behaviour of all field theories is controlled by the ultimate field theory, gravitation. At very short distances ($\sim 10^{-33}$ cm) quantum fluctuations of the gravitational field i.e. of the geometry, plays havoc with space and time. Some theorists argue [47] that gravitation is the great regularizer of all quantum field theories including itself.

For the study of divergence and convergence it is most convenient not to use the s-matrix but instead the Green's function defined as the vacuum expectation value of time-ordered field products, the most general being

---

(*) There are also other divergences but they cancel among themselves when carefully treated.

$$G_o^{E,p}{}_{\beta_1\ldots\beta_E,\alpha_1\ldots\alpha_E,\mu_1\ldots\mu_p}(y_1\ldots y_E,x_1\ldots x_E,z_1\ldots z_p)$$

$$= \langle 0|J\{\psi_{\beta_1}(y_1)\ldots\psi_{\beta_E}(y_E)\bar{\psi}_{\alpha_E}(x_E)\ldots\bar{\psi}_{\alpha_1}(x_1)A_{\mu_1}(z_1)\ldots A_{\mu_p}(z_p)\}|0\rangle$$

$$=$$

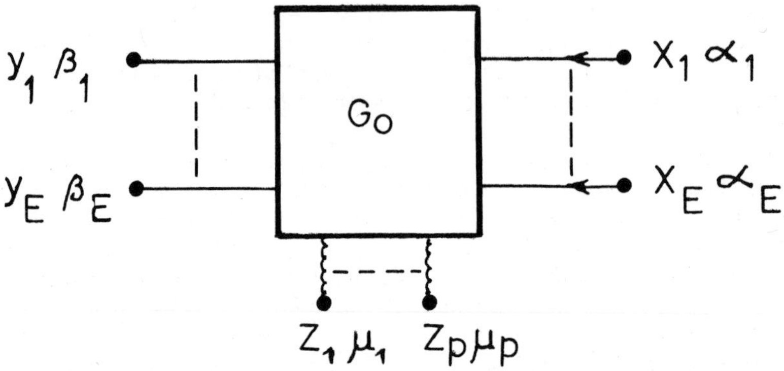

It contains E electron fields (ψ), E positron fields ($\bar{\psi}$) and P photon fields (E = 0, 1, 2,..., P = 0, 1, 2, ...). There has to be as many electron fields as positron fields, because of charge conservation. The Feynman rules are simple : A Greens' function is a sum over topologically distinct diagrams built up from propagators and vertices with the same structure as the blob diagram above. Instead of external lines one has external vertices and one must integrate over the position of all the internal vertices. The index, 0, on these Greens' functions indicates that they are calculated using the Feynman rules (3.7) - (3.9) corresponding to the Lagrangian L_o+L_1 but not including L_2. That will be discussed in section 7.

By performing a multiple Fourier transform of all the space-time arguments one obtains the momentum space Greens'functions. They are constructed by means of the elements (3.7)-(3.9) and one must integrate over the momentum of all internal lines, including a factor $(2\pi)^4 . \ \delta$ (momentum conservation) for each vertex. We denote them by

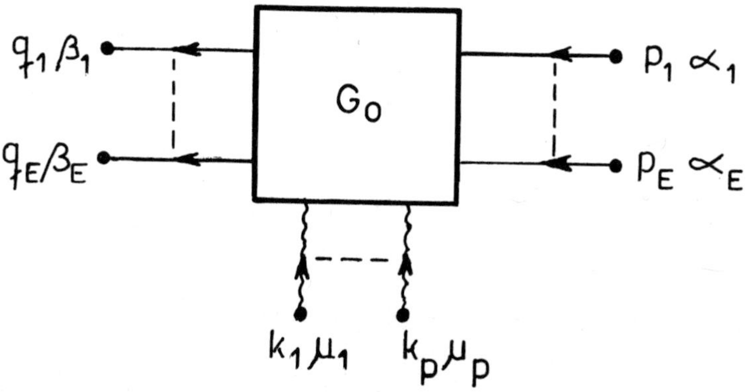

where the arrows indicate the direction of the flow of momentum. It is clear that these functions are closely related to the S-matrix elements. Roughly speaking (and precisely speaking for diagrams not containing propagators or corrections to these between external vertices) one obtains the S-matrix element by removing all external propagators, taking the external momenta on shell and multiplying with appropriate factors for the incoming and outgoing particles. If the Greens functions are calculable so are the S-matrix elements provided that there are no infrared problems (which we avoid by taking the photon mass non-zero).

All this would be very nice if the integrals converged, which they do not. To see this let us consider the lowest order correction to the electron propagator

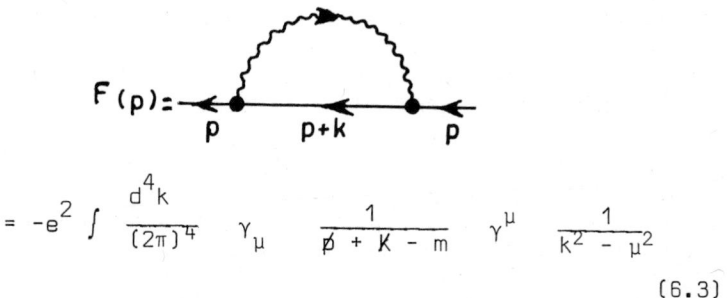

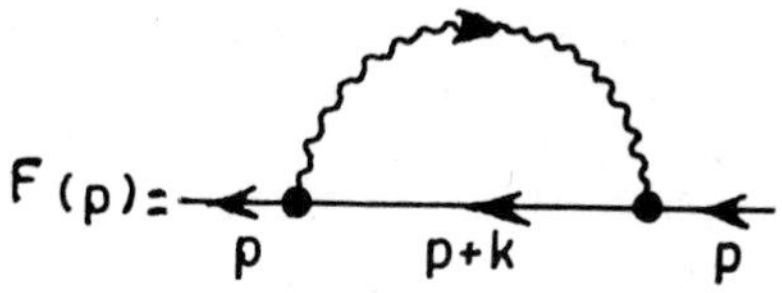

$$= -e^2 \int \frac{d^4k}{(2\pi)^4} \gamma_\mu \frac{1}{\not p + \not K - m} \gamma^\mu \frac{1}{k^2 - \mu^2} \tag{6.3}$$

where we for simplicity have taken the photon propagator in the Fermi gauge. Counting the powers of k in the numerator and denominator we see that there are 4 above and 3 below. Hence the integral is linearily divergent at large k and has no mathematical meaning.

On the other hand the integrand is a well-defined mathematical function of the external momentum p and the loop momentum k. If we formally differentiate (6.3) twice after p we get

$$\frac{\partial^2 F(p)}{\partial p^\mu \partial p^\nu} = -e^2 \int \frac{d^4k}{(2\pi)^4} \gamma_\rho \frac{1}{\not p+\not k-m} \gamma_\mu \frac{1}{\not p+\not k-m} \cdot$$

$$\cdot \gamma_\nu \frac{1}{\not p+\not k-m} \gamma^\rho \frac{1}{k^2-\mu^2} + (\mu \leftrightarrow \nu) \tag{6.4}$$

which is perfectly convergent at large k. Although (6.3) does not contain sensible information about the value of F(p) or its first derivative in any given point p it contains complete information about all the higher derivatives. The lacking information is at most a polynomium of first degree in p. If we supply a procedure by which we can calculate in a finite way a function F'(p) that has the same second derivative as F(p) then we may write

$$F(p) = A + B\not p + F'(p) \tag{6.5}$$

The two constants A and B may be chosen at will, and conversely the physical predictions of the theory must not depend on which choice we make. Usually they are fixed by requiring F(p) and its first derivative to vanish in a certain point.

The above description holds essentially true for all divergent diagrams. They consists of an indefinite polynomium in the external momenta and a finite part defined by some definite procedure. Special care has to be taken when a diagram contains subdiagrams that are also divergent. Then the finite part may itself depend on the arbitrary part of the subdiagram.

The procedure by which the finite part is defined is called the regularization procedure. Perturbation theory is not meaningful unless a regularization procedure is defined. Several such exist of which I shall only mention three.

a) Subtraction at the origin

This is the oldest, most rigorous and most beautiful technique which has reached its highest level of sophistication in Zimmermans normal product formalism[48]. If we call the integrand in (6.3) for I(p,k) then the basic idea is to replace it by

$$I(p,k) - I(0,k) - p_\mu \left[\frac{\partial}{\partial p_\mu} I(p,k) \right]_{p=0} \tag{6.6}$$

i.e. by the doubly subtracted integrand. It clearly leaves the second derivative untouched and it is easy to see that the integral over k now converges. Although very systematic it is not so easy to use it in practical calculations (*) and I shall not go further into it here.

b) Pauli-Villars regularization[49)]

Let us replace the photon propagator in (6.3) by a sum over N propagators

$$\frac{1}{k^2 - \mu^2} \rightarrow \sum_{i=1}^{N} \frac{C_i}{k^2 - \mu_i^2} \tag{6.7}$$

Then we find for large k^2

$$\sum_{i=1}^{N} \frac{C_i}{k^2-\mu_i^2} = \frac{1}{k^2} \sum_{i=1}^{N} C_i + \frac{1}{k^4} \sum_{i=1}^{N} C_i\mu_i^2 + \ldots \tag{6.8}$$

The leading behaviour may be removed by requiring

$$\sum_i C_i = 0 \tag{6.9}$$

and in that case the integral in (6.3) will become convergent. The second derivative reduces to (6.4) if we take $c_1 = 1$, $\mu_1 = \mu$ and let $\mu_i \rightarrow \infty$ (i. = 2,3,...,N). One way of solving (6.9) is by taking $c_2 = -1$, $\mu_2 = \Lambda$, so that

$$\sum_{i=1}^{N} \frac{C_i}{k^2-\mu_i^2} = \frac{1}{k^2-\mu^2} - \frac{1}{k^2-\Lambda^2} \tag{6.10}$$

This choice is sufficient to make all diagrams in QED convergent, except those that contain closed electron loops. It corresponds to having a heavy photon of mass Λ with negative metric present in the theory.

For the electron propagator one would like to do the same trick, but that gets into trouble with current conservation. Two fermion fields with different masses do not give rise to a conserved current $\bar{\psi}_1 \gamma^\mu \psi_2$ since

$$\partial^\mu(\bar{\psi}_1 \gamma_\mu \psi_2) = -i(m_2-m_1)\bar{\psi}_1\psi_2 \tag{6.11}$$

(*) Because of overlapping divergences.

It would be most convenient if the regularization procedure respected current conservation, or rather the Ward identities.

The way to regularize while respecting the Ward identities in QED is to calculate the electron loops using the same mass m_i for all the propagators and afterwards sum over the various masses with suitable coefficients. Thus an electron loop

$$\mathrm{Tr}\left[\frac{1}{\not p_1 - m} \gamma_{\mu_1} \frac{1}{\not p_2 - m} \gamma_{\mu_2} \cdots\cdots \frac{1}{\not p_k - m} \gamma_{\mu_k} \right]$$

is replaced by

$$\sum_{i=1}^{M} d_i \, \mathrm{Tr}\left[\frac{1}{\not p_1 - m_i} \gamma_{\mu_1} \cdots \frac{1}{\not p_k - m_i} \gamma_{\mu_k} \right]$$

All diagrams are then convergent provided

$$\sum_i d_i = \sum_i d_i m_i^2 = 0 \tag{6.12}$$

c) Complex dimension regularization

This is a very elegant regularization scheme invented by 'tHooft and Veltman[50]. In short it consists in continuing the Greens functions into complex space-time dimension. This is possible because (a) the Feynman rules do not explicitly refer to the dimension (b) all loop diagrams may by standard methods be reduced to the integral

$$\int \frac{dnk_1 \cdots dnk_\ell}{(2\pi)^{n\ell}} \frac{1}{\left[L - \sum_{r,s} U_{rs} k_r . k_s \right]^m} = \frac{i^\ell}{(4\pi)^{n\ell/2}} \frac{\Gamma(m - n\ell/2)}{\Gamma(m)\, U^{n/2}} \frac{1}{L^{m-n\ell/2}} \tag{6.13}$$

where $U = \det\{U_{rs}\}$, n the dimension of space-time, ℓ the number of loops, and L, U_{rs} functions of the Feynman parameters, (c) the numerator algebra at most gives rise to a polynomial in n due to

$$g_{\mu\nu} g^{\mu\nu} = n \tag{6.14}$$

Since the right hand side of (6.13) has an obvious analytic extension to complex n, this is then also possible for an arbitrary Feynman graph whether it is convergent or not. For divergent graphs (6.13) will give rise to a pole for n=4 because $\ell n \geq 2m$ in that case.

Kinoshita and Cvitanovic[51)] have given a precise definition of the mamplitude in complex dimension which answers all questions about uniqueness, independence of the choice of loop momenta etcetera by formulating the Feynman rules entirely in the space of Feynman parameters.

The method of complex dimension has the practical advantage that the integrand is left unchanged and that all "natural" manipulations are permitted. This is not the case for any other method, and has the consequence that naive Ward identities are valid and derivable by purely formal manipulations.

7. RENORMALIZATION

We have now established a method for calculating the amplitude for any Feynman diagram in QED using one of the above mentioned regularization methods. But this is not sufficient. We started out with certain free physical particles with masses and charges known from experiment. The masses and the gauge parameter were punt into the theory via the free Lagrangian (eq.(3.17))

$$L_o = -\frac{1}{4} F_{\mu\nu}F^{\mu\nu} - \frac{1}{2a} (\partial_\mu A^\mu)^2 + \frac{1}{2} \mu^2 A_\mu A^\mu + \bar{\psi}(i\not{\partial} - m)\psi \quad (7.1)$$

and the charge was introduced as the coupling constant in the interaction (eq.(3.3))

$$L_1 = -e A^\mu \bar{\psi} \gamma_\mu \psi \quad (7.2)$$

As we discussed in Section 4 this interaction has been verified to high precision in scattering experiments. But that is not its only effect. Due to the particles'interaction with their own radiation field there will be changes to their static properties such as mass, charge and magnetic moment.

There are now two ways of taking this into account. They are perfectly equivalent but while the first is most convenient from a "philosophical" point of view, the second is most convenient from a calculational point of view.

In the first way one accepts that the parameters with which the theory was first formulated, m and e, are not really the observable mass and charge. The observable mass and charge are however calculable functions of the input parameters given as a perturbation series. Inverting these functions one may solve for the input parameters in terms of the physical parameters, and finally reexpress all physical quantities such as cross sections, energy levels etc. in terms of the physical parameters. The unpleasant feature of this way of renormalizing is that the incoming and outgoing free particles have the wrong mass and charge. The convulsion whereby the theory makes this come out right in the end, as it will do, is rather unpleasant mathematically.

In the second way which I shall describe for you in some detail one insists on the interpretation of the parameters in the Lagrangian. Thus m is really the mass of the electron, e is really its charge, μ is "really" the mass of the photon and a is "really" its gauge parameter. Since the interaction L_1 by itself would give rise to changes of all these parameters, it is necessary to introduce an extra piece, L_2, in the Lagrangian which has no influence on the Born approximation but which cancels out all changes to the input parameters. It is almost obvious and we shall justify it below that this piece must have the form

$$L_2 = C_1\, e\, A_\mu\, \bar{\psi}\, \gamma^\mu \psi \;-\; C_2\, \bar{\psi}\, i\, \not{\partial}\psi \;+\; C_3\, \frac{1}{4} F_{\mu\nu}F^{\mu\nu} \;+\; C_4\, m\, \bar{\psi}\, \psi$$

$$-\; C_5\, \frac{1}{2}\mu^2\, A_\mu A^\mu \;+\; C_6\, \frac{1}{2a}\, (\partial_\mu A^\mu)^2 \qquad (7.3)$$

where the C_i are dimensionless parameters, which are to be adjusted such as to cancel the unwanted effects. We have included C_2 and C_3 since the 1's in front of $-\frac{1}{4} F_{\mu\nu}F^{\mu\nu}$ and $\bar{\psi}\, i\not{\partial}\psi$ in (7.1) are in fact also input parameters to the theory fixing the normalization of the fields. The numbering of the C_i is traditional for the case of C_1, C_2 and C_3.

Now it is clear that this extra piece, considered as a perturbation, gives rise to some new vertices. They are

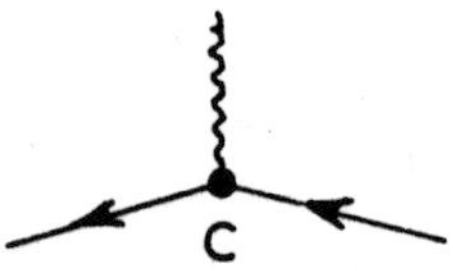

$$= i\, e\, C_1\, \gamma_\mu \qquad (7.4)$$

$$= i(-\not{p}\, C_2 + m\, C_4) \tag{7.5}$$

$$= i[\, C_3 (g_{\mu\nu} k^2 - k_\mu k_\nu) - C_5\, \mu^2 g_{\mu\nu} + C_6\, \frac{1}{a}\, k_\mu k_\nu] \tag{7.6}$$

The new Green's functions are calculable as a double perturbation expansion in the C's and in e by means of these Feynman rules. Suppressing all spin indices and calling all the momenta p we denote them by

$$G_C^{EP}(p,g) = \tag{7.7}$$

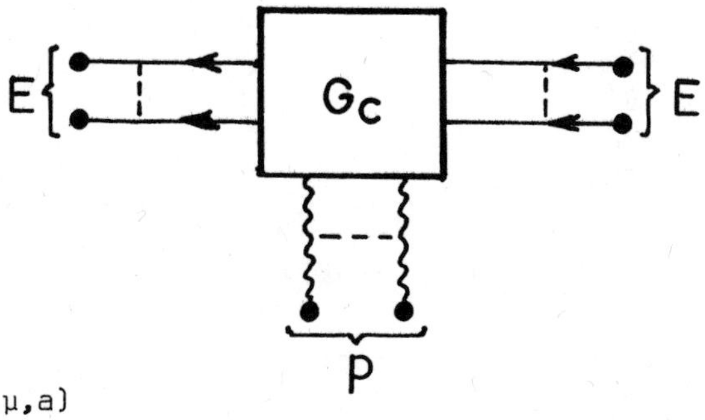

where

$$g = (e, m, \mu, a) \tag{7.8}$$

is a collective way of writing all the input parameters. If we take $C_i = 0$ in (7.7) we obtain the uncompensated Greens'functions defined in (6.2).

There is a simple functional relationship between the compensated and uncompensated Green's functions. The total lagrangian may be written

$$L = L_0 + L_1 + L_2$$

$$= -\frac{Z_3}{4} F_{\mu\nu}F^{\mu\nu} - \frac{Z_6}{2a} (\partial_\mu A^\mu)^2 + \frac{Z_5}{2} \mu^2 A_\mu A^\mu$$

$$+ Z_2\, \bar{\psi}\, i\, \not{\partial}\, \psi - Z_4\, m\, \bar{\psi}\psi - e\, Z_1\, A_\mu\, \bar{\psi}\, \gamma^\mu\, \psi \tag{7.9}$$

where

$$Z_i = 1 - C_i \tag{7.10}$$

By a redefinition of the parameters and the fields this Lagrangian may be reduced to the old $L_0 + L_1$ without L_2. The redefinition is

$$A_o = \sqrt{Z_3}\; A_o^{\mu} \tag{7.11}$$

$$\psi_o = \sqrt{Z_2}\; \psi \tag{7.12}$$

$$e_o = e \frac{Z_1}{\sqrt{Z_3} Z_2} \tag{7.13}$$

$$m_o = m \frac{Z_4}{Z_2} \tag{7.14}$$

$$\mu_o = \mu \sqrt{\frac{Z_5}{Z_3}} \tag{7.15}$$

$$a_o = a \frac{Z_3}{Z_6} \tag{7.16}$$

Hence it follows that

$$G_C^{EP}\,(p,g) = Z_3^{-P/2}\; Z_2^{E}\;\; G_0^{EP}\,(p,g_o) \tag{7.17}$$

This should of course be understood order by order in perturbation theory in e and the C's. The Z's in front of G_o comes from the rescaling of the fields in the time-ordered product. One may rather easily demonstrate (7.17) by combinatorics of Feynman graphs.

Eq.(7.17) together with (7.13)-(7.16) are the fundamental equations that tell us the relationship between renormalized and unrenormalized QED. So far the C's, and thereby the Z's are arbitrary quantities of the same nature as the parameters e, m, a, and μ but we shall now see that it is possible to choose the C's as functions of the parameters in such a way that the parameters themselves retain their original meaning.

Consider the electron propagator and its corrections

$S_c(\not{p}) \equiv$

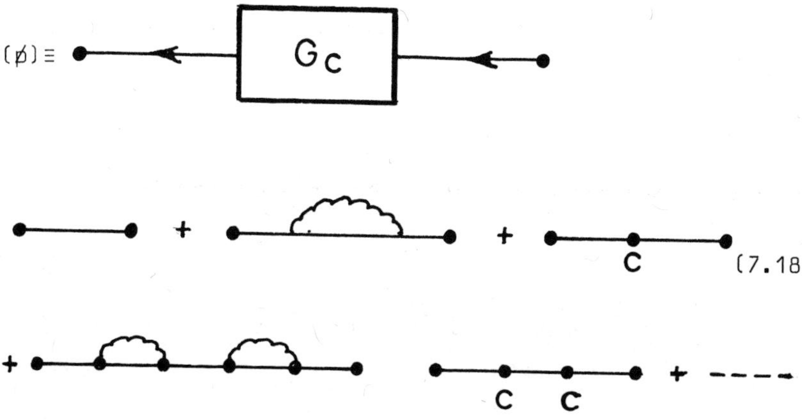

(7.18)

Some of the corrections may be divided into two parts connected only by a single propagator, such as the last two of the diagrams displayed. They are called one-particle-reducible. Let

$-i\Sigma_c(\not{p}) \equiv$

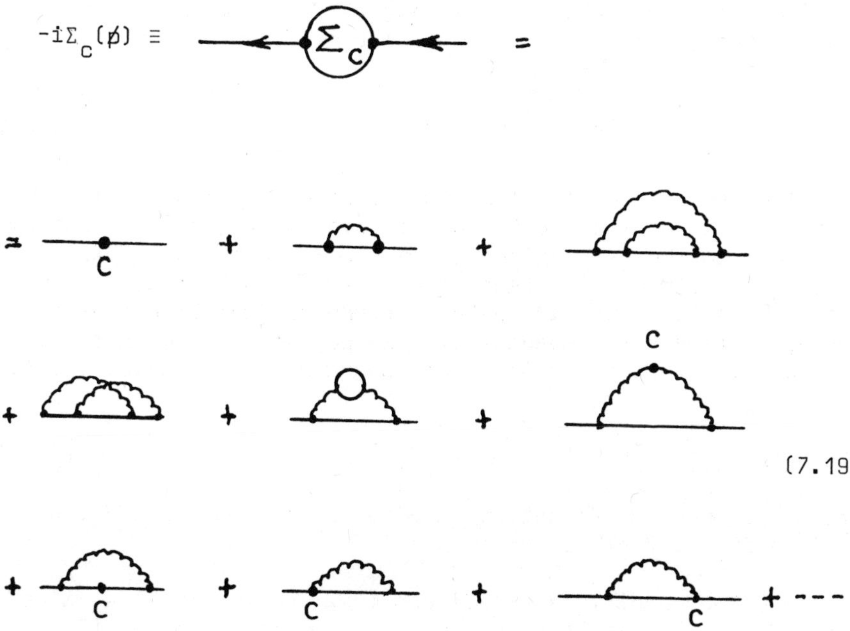

(7.19)

by the sum of all one-particle-irreducible diagrams with counter-terms included. Then clearly

$S_c(\not p) =$

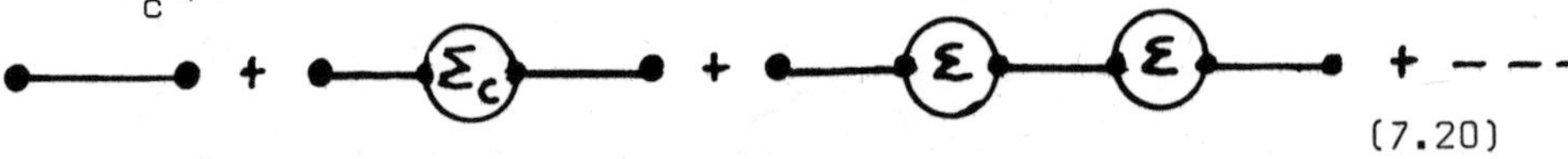

(7.20)

$$= \frac{i}{\not p - m} \sum_{n=0} \left(\frac{\Sigma_c(\not p)}{\not p - m} \right)^n$$

i.e.

$$S_c(\not p) = \frac{i}{\not p - m - \Sigma_c(\not p)} \tag{7.21}$$

This is the full electron propagator including all the radiative corrections. It behaves no more as a free propagator. A dressed electron, i.e. an electron with its associated electromagnetic field, behaves differently off-shell than a free electron. We must, however, insist that the propagator has a pole at $\not p = m$, defining the mass of the dressed electron. Thus we get the condition

$$\Sigma_c(\not p)\big|_{\not p = m} = 0 \tag{7.22}$$

The residue at the pole in the free electron propagator is i. This residue is defined by the coefficient of the kinetic energy term $\bar\psi i \not\partial \psi$ in the Lagrangian. This coefficient, which has been set equal to 1 from the outset, could have been left arbitrary in which case it would have been a parameter just like e or m. In order to retain the meaning of this parameter we must require that the residue at the pole is unchanged by the interaction, i.e. that

$$\dot\Sigma_c(\not p)\big|_{\not p = m} = 0 \tag{7.23}$$

where a dot means differentiation after $\not p$ (remembering of course that $p^2 = \not p^2$ so that $\frac{\partial}{\partial \not p} p^2 = 2\not p$).

These two conditions (7.22) and (7.23) determine the values of C_2 and C_4. To see this we separate the first term in (7.19) writing

$$\Sigma_c(\not p) = \not p\, C_2 - mC_4 + \Sigma'_c(\not p) \tag{7.24}$$

where $(-i)\Sigma'_c(\not p)$ is the sum of all graphs in (7.19) except the first for which we used (7.5). Hence the conditions become

$$m(C_4 - C_2) = \Sigma'_c(\not p)\big|_{\not p=m} \tag{7.25}$$

$$-C_2 = \dot{\Sigma}'_c(\not p)\big|_{\not p=m} \tag{7.26}$$

Since all the terms except the first in (7.19) are proportional to e^2 this allows us to determine the C's as a perturbation expansion in e^2

$$C_i = \sum_{n=1}^{\infty} C_i^{(2n)}\, e^{2n} \tag{7.27}$$

beginning with the e^2 term.

For the photon propagator a similar procedure may be carried through determining C_3, C_5 and C_6. It then turns out that C_5 and C_6 are proportional to μ^2 such that in the limit of $\mu \to 0$

$$C_5 = C_6 = 0 \qquad (\mu = 0) \tag{7.28}$$

They are accordingly not necessary for the renormalization of QED in the massless limit. If we define the one-particle irreducible photon graphs

$-i\Pi_c^{\mu\nu}(k) =$

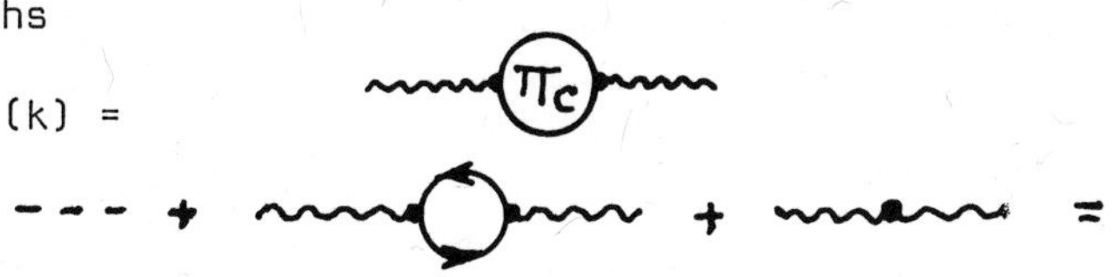

(7.29)

and use gauge invariance $k_\mu \Pi_c^{\mu\nu} = 0$ to obtain

$$\Pi_c^{\mu\nu}(k) = (g^{\mu\nu}k^2 - k^\mu k^\nu)\, \Pi_c(k^2) \tag{7.30}$$

we find the dressed photon propagator (for $\mu = 0$) in the same way as before

$$D_c^{\mu\nu}(k) = -i\left(g^{\mu\nu} - \frac{k^\mu k^\nu}{k^2}\right)\frac{1}{k^2(1+\Pi_c(k^2))} - i\,\frac{k_\mu k_\nu}{k^2} \cdot \frac{a}{k^2} \tag{7.31}$$

We see that the pole has remained in the same place $k^2 = 0$ (for $\mu=0$) due to gauge invariance (7.30). This point is actually a bit more subtle than one might think at first sight. One has to prove that no pole type singularity remains in $\Pi_c(k^2)$ at $k^2 = 0$. That would give the photon a mass. Such a mechanism actually plays a part in the Schwinger model which is QED in two space-time dimensions[52]. It has been demonstrated by Brandt and Ng[53] that the absence of a pole in $\Pi_c(k^2)$ for $k^2=0$ depends on the existence of gauge transformations of the type (3.4) with $\theta(X) = r_\mu X^\mu$ for arbitrary r_μ. This is the case in QED, but not in the Schwinger model.

The renormalization condition is then

$$\Pi_c(k^2)\Big|_{k^2=0} = 0 \tag{7.32}$$

Since by (7.29) and (7.6)

$$\Pi_c(k^2) = -C_3 + \Pi'_c(k^2) \tag{7.33}$$

we have

$$C_3 = \Pi'_c(k^2)\big|_{k^2=0} \tag{7.34}$$

Also this renormalization constant is given by a series like (7.27) and calculable order by order.

Finally there is C_1. Let the one-particle-irreducible vertex function be

$$-ie\Gamma^\mu_c(\not{p}',\not{p}) \equiv -ie(\gamma^\mu + \Lambda^\mu_c(\not{p}',\not{p})) \tag{7.35}$$

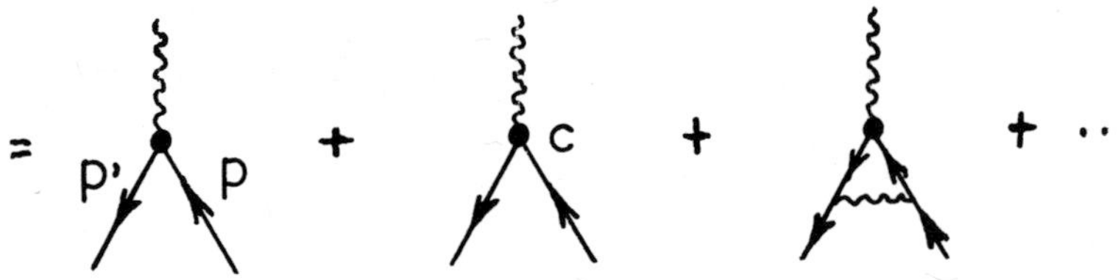

The renormalization condition is in this case

$$\Lambda^\mu_c(\not{p}',\not{p})\big|_{\not{p}'=\not{p}=m} = 0 \tag{7.36}$$

expressing that for an on-shell electron at large distances ($p'-p=0$) we see the Coulomb field from the charge e. Since

$$\Lambda^\mu_c = -C_1\gamma^\mu + \Lambda'^\mu_c \tag{7.37}$$

it determines C_1 by

$$\gamma^\mu C_1 = \Lambda'^\mu_c(\not{p}',\not{p})\big|_{\not{p}'=\not{p}=m} \tag{7.38}$$

Actually it is not necessary to calculate C_1 and C_2 separately since the Ward identity[54)] tell us that

$$C_1 = C_2 \tag{7.39}$$

So there are only three constants it is necessary to determine : C_2, C_3 and C_4. They are all given by a perturbation series (7.27) as functions of the parameters g

$$C_i = C_i(g) \tag{7.40}$$

determined by the renormalization conditions (7.22), (7.23) and (7.36).

The renormalized Greens functions are now defined by means of the functions $C_i(g)$ via the equation

$$G^{EP}(p,g) = G_c^{EP}(p,g)\big|_{C\,=\,C(g)} \tag{7.41}$$

Remember that G_c were defined for all values of the counterterms C_i. They only become functions of the momenta and the parameters g when we substitute the C's determined by the renormalization conditions.

Although it has not been explicitly shown in (7.41) the renormalized functions also depend on the parameter (s) that cut off the divergences. The remarkable fact about QED and other renormalizable field theories is that when the renormalization has been performed, one may let the cut-off parameters go towards their limiting values. Thus (7.41) is finite if we let the cut-off masses go to infinity or the complex dimension towards 4. In other words, when the unwanted effects have been compensated by means of suitable counterterms the divergences have disappeared from the theory.

I shall only sketch the proof of this theorem. First one makes the observation that the renormalized Greens functions are functionals of the renormalized propagators and vertex

$$G^{EP} = G^{EP}[S,D,\Gamma] \tag{7.42}$$

Then each Greens'function is a sum over skeletons in which the propagators and vertices are dressed. For example in Compton scattering we have

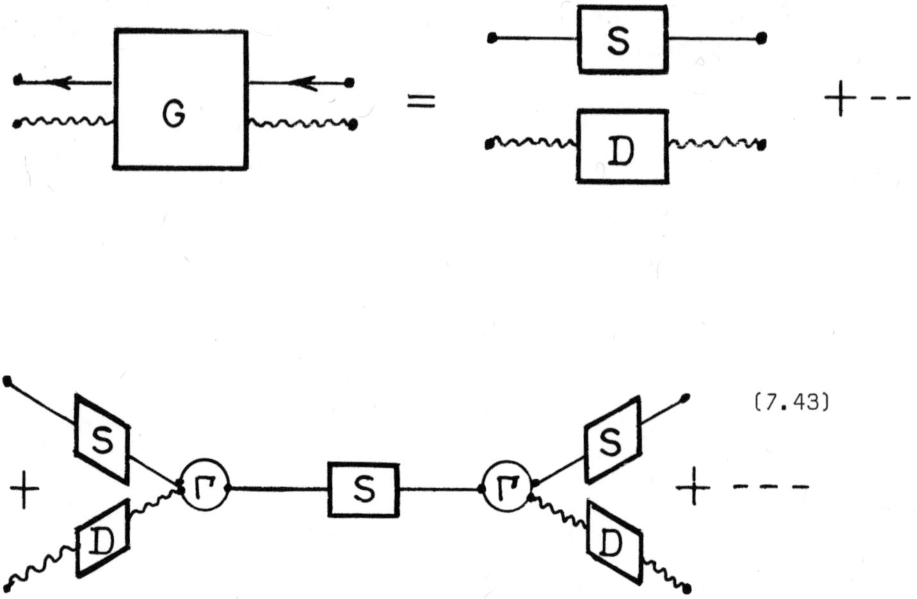

(7.43)

This is rather intuitively clear and needs no further justification. For the propagators and the vertex, (7.42) reduces to trivial identities. Next one proves (by power counting) that provided the dressed propagators and vertex are finite and behave for large momenta in the same way as the free propagators and the free vertex (apart from logarithms) then the skeleton loops contain no divergences. Hence all divergences arise from the propagators and the vertex (*).

Finally one proves by induction that the renormalized propagators and the vertex order by order indeed are finite and behave correctly for large momenta. The difficult part of the proof is to show that the counterterms remove the overlapping divergences. An overlapping divergence is seen in the fourth diagram of eq. (7.19). It is impossible to decide whether it should be read as a correction to the first vertex or to the second vertex inserted into the second diagram. The two last diagrams show that there is a counterterm for each interpretation and one can show that this is precisely enough to remove the divergences.

(*) Photon-photon scattering subgraphs contain logarithmic divergences that cancel among themselves.

I have spent quite some time on the question of renormalization well knowing that these or similar arguments may be found in the standard textbooks on field theory[55]. The reason is that I hope to clarify the different stages of the proof of renormalizability and also demonstrate how to do it in practice. The stages were

1) Make the perturbation expansion mathematically well-defined by means of a regularization procedure.
2) Introduce counterterms such that the parameters retain their initial "physical" meaning.
3) Show that the resulting theory is finite.

Point 2) may be relaxed somewhat. It is sufficient for making the theory finite to specify finite values of Σ_c, $\dot{\Sigma}_c$, π_c and Λ_c at some finite fixed momentum. Then there will be finite renormalizations left to perform afterwards. This is called intermediate renormalization.

There is also a question about which counter terms to introduce. We saw that we had introduced C_5 and C_6 and that they disappeared when the photon mass vanished. A term in a Lagrangian is called renormalizable if its coupling constant has the dimension of a mass to a non-negative power. A renormalizable Lagrangian only contains renormalizable terms. The dimension of a field is determined from the kinetic energy term ($-1/4\ F_{\mu\nu}F^{\mu\nu}$ and $\bar{\psi}i\not{\partial}\psi$) by giving it the coefficient 1. Since the Lagrangian has dimension 4 we find $\dim A_\mu = 1$ and $\dim\psi = 3/2$. Knowing the dimension of all fields we can calculate the dimension of any coupling constant. If the original Lagrangian possesses a symmetry one should introduce all renormalizable counter terms consistent with this symmetry, and conversely only those. This sometimes leads to the necessary introduction of counter terms that do not correspond to terms in the original interaction. In scalar QED, for example, it is necessary to introduce a counter term quartic in the scalar field although the classical Lagrangian of scalar QED does not contain such an interaction. When that happens renormalization may not be nicely multiplicative as in eqs.(7.11)-(7.17).

In practice renormalization proceeds along the lines that were indicated above. For scattering processes eq. (7.42) tells us that we should calculate the skeletons with the renormalized propagators and vertices, of course expanded to the desired order. In calculating the next order in the propagators and the vertex care must be taken of including all the relevant counter terms to the desired order. Let us again use the corrections to the electron propagator as an example. In second order we have

determining the second order approximation to C_2 and C_4. If the divergence is cut-off by means of Pauli-Villars or dimensional regularizations one may integrate the loop first and determine $C^{(2)}_{2,4}$ afterwards. In fourth order we have

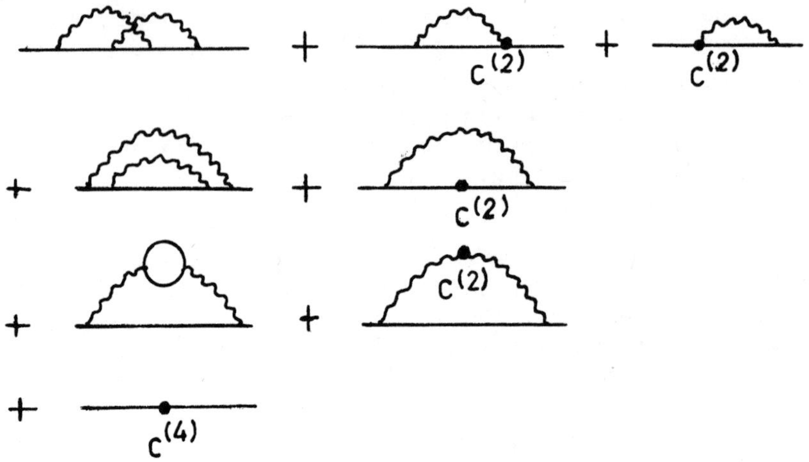

The first diagram contains an overlapping divergence and should be completely evaluated by use of, say, dimensional regularization. The second and third diagrams are known from the second order calculation. In second line the diagrams collaborate to make a renormalized second order propagator correction, and the same is true in the third line. The last diagram cancels the remaining divergences and determines $C^{(4)}_{2,4}$ via the renormalization conditions.

Finally I shall mention that practical calculation of higher order corrections to s-matrix elements are simplified by the on-shell renormalization conditions (7.22), (7.23) and (7.32), because renormalized corrections to external legs vanish for on-shell momenta. One may simply disregard all corrections to the external legs in the renormalized theory.

8. CALLAN-SYMANZIK EQUATION FOR THE PHOTON PROPAGATOR

In the preceding section we saw how the Lagrangian including counterterms could be expressed as a Lagrangian without counterterms but different (unrenormalized) parameters. In particular, the unrenormalized mass and the charge of the electron were given by eqs. (7.13) and (7.14). When inserting the correct counterterms we obtain functions of the form

$$e_o = e_o(e,m) \tag{8.1}$$

$$m_o = m_o(e,m) \tag{8.2}$$

since one can show that it is alright to let $\mu \to 0$ and that these quantities do not depend on the gauge parameter a (as they oughtn't).

The same is true for the photon propagator which may be written in the form

$$D_{\mu\nu}(k) = -i(g_{\mu\nu} - \frac{k_\mu k_\nu}{k^2}) \frac{d(k^2,e,m)}{k^2} - ia \frac{k_\mu k_\nu}{k^4} \tag{8.3}$$

where d is a gauge independent, infrared convergent function. According to eq.(7.17) it is related to its unrenormalized counterpart by

$$d(k^2,e,m) = Z_3^{-1} d_o (k^2,e_o,m_o) \tag{8.4}$$

Using eq.(7.13) and $Z_1 = Z_2$ (eq.(7.39)) this may be written

$$e^2 d(k^2,e,m) = e_o^2 d_o(k^2,e_o,m_o) \tag{8.5}$$

where on the right hand side we should insert (8.1) and (8.2). Implicit in this equation there is of course the dependence on the cut-off parameter.

Following Adler[9] we ask how differentiation after m with e_o held constant works on the two sides of (8.5). We denote this differential operator with $(\frac{\partial}{\partial m})_{e_o}$ as in thermodynamics. By the rules of differential calculus we have

$$(\frac{\partial}{\partial m})_{e_o} F(m_o,e_o) = (\frac{\partial m_o}{\partial m})_{e_o} \frac{\partial}{\partial m_o} F(m_o,e_o) \tag{8.6}$$

$$(\frac{\partial}{\partial m})_{e_o} F(m,e) = (\frac{\partial}{\partial m} + (\frac{\partial e}{\partial m})_{e_o} \frac{\partial}{\partial e}) F(m,e) \tag{8.7}$$

Applying this to (8.5) we have the identity (in a form determined by convention)

$$(m \frac{\partial}{\partial m} + \beta e^2 \frac{\partial}{\partial e^2}) e^2 d(k^2,e,m) = e^2 \Delta(k^2,e,m) \tag{8.8}$$

where

$$e^2 \Delta(k^2,e,m) = m (\frac{\partial m_o}{\partial m})_{e_o} \frac{\partial}{\partial m_o} e_o^2 d_o(k^2,e_o,m_o) \tag{8.9}$$

and

$$\beta(e,m) = (\frac{m}{e^2} \frac{\partial e^2}{\partial m})_{e_o} \tag{8.10}$$

Both of these functions should be expressed in terms of e and m as indicated.

Eq.(8.8) is the Callan-Symanzik equation for the photon propagator, and as it stands absolutely useless. The remarkable fact is however that one can prove that the right hand side is finite. There are various ways of doing this[7-9] and I shall just give a very short indication of one way to do it. Since for fixed f_o

$$f_o \frac{\partial}{\partial m_o} \frac{i}{\not p - m_o} = \frac{i}{\not p - m_o}(-if_o)\frac{i}{\not p - m_o} \qquad (8.11)$$

looks like the unrenormalized coupling of a scalar particle, θ, to the electron and since such a coupling $(-f_o\psi_o\bar\psi_o\psi_o)$ is renormalizable, it follows that the properly renormalized $\theta\gamma\gamma$ vertex must be finite. It then turns out that the righ hand side of (8.8), i.e. (8.9) is just proportional to this vertex at zero momentum transfer.

Since $e^2\Delta$ and e^2d are both finite β must be finite, and since it is dimensionless it can only depend on $\alpha = e^2/4\pi$ when the cut-offs are taken to their limits.

Although (8.8) is an equation among finite quantities it is still not very useful. But because of (8.11) there will be a propagator more in every diagram that contributes to $e^2\Delta$ as compared to e^2d. This has the effect that for $k^2 \to -\infty$, $\frac{e^2\Delta}{e^2d} \to 0$ least as fast as $1/\sqrt{-k^2}$. Such a statement depends on a careful study of the asymptotic behaviour of Feynman graphs[56].

For $k^2 \to -\infty$ the right hand side may accordingly be disregarded compared with e^2d. Hence the asymptotic photon propagator satisfies the equation

$$\left(m\frac{\partial}{\partial m} + \beta(\alpha)\alpha\frac{\partial}{\partial\alpha}\right)\alpha d_\infty(k^2/m^2,\alpha) = 0 \qquad (8.12)$$

We shall put this equation to explicit use in section 11.

The Callan-Symanzik equations which can be proved for all the Greens functions have played an immense role for recent developments of field theory. They were antedated in QED by the Peterman, Stueckelberg[5], Gell-Mann-Low[6] renormalization group approach. A review of their use in QED may be found in ref.9.

The contents of eq.(8.12) is that the function $\alpha d_\infty(k^2/m^2,\alpha)$ which <u>a priori</u> could depend arbitrarily on two variables $\alpha, k^2/m^2$, must satisfy a partial differential equation. The solution can only depend on $\beta(\alpha)$ and an arbitrary function of one variable. It is not difficult to see that the solution is

$$\alpha d_\infty\left(\frac{k^2}{m^2},\alpha\right) = \left(-\frac{k^2}{m^2}\right)^{\frac{1}{2}\beta(\alpha)\alpha\frac{\partial}{\partial\alpha}} F(\alpha) =$$

$$\sum_{n=0}^{\infty} \frac{\mathrm{Log}^n\left(\frac{-k^2}{m^2}\right)}{n!} \left(\frac{1}{2}\beta(\alpha)\alpha\frac{\partial}{\partial\alpha}\right)^n F(\alpha) \qquad (8.13)$$

where $F(\alpha)$ is arbitrary.

By inserting $e^2 = e_o^2 Z_3$ in (8.10) we obtain by making use of (8.7)

$$\left(m\frac{\partial}{\partial m} + \beta\left(e^2\frac{\partial}{\partial e^2} - 1\right)\right) Z_3 = 0 \qquad (8.13')$$

i.e. a Callan-Symanzik equation for Z_3. Although β is finite when the cut-off goes to its limit we cannot take this limit in this equation, because Z_3 will not be finite. If we use complex dimension regularizations the dimension of the fields and thereby of e depends on n. In fact dim $A_\mu = (n-2)/2$, dim $\psi = (n-1)/2$ and thus dim $e = n$ - dim A - 2 dim $\psi = 2 - n/2$. Hence the only dimensionless variable we can make of e and m and n is

$$\alpha = \frac{e^2\, m^{n-4}}{(2\sqrt{\pi})^{n-2}} \qquad (8.14)$$

This is the generalization of the fine-structure constant α to complex dimension. A dimensionless function of e and m can only depend on α. Thus (8.13') becomes an ordinary differential equation in α

$$\left[(n-4)\alpha\frac{\partial}{\partial\alpha} + \beta(\alpha,n)\left(\alpha\frac{\partial}{\partial\alpha} - 1\right)\right] Z_3(\alpha,n) = 0 \qquad (8.14)$$

which has the solution

$$Z_3(\alpha,n) = e^{\int_0^\alpha \frac{dx}{x}\frac{\beta(x,n)}{n-4+\beta(x,n)}} \qquad (8.15)$$

when we use the boundary condition $Z_3\ (\alpha=0) = 1$. Although very beautiful, this equation is not particularly useful. Similar equations may be obtained for the other renormalization constants[57].

For n=4 the β-function has been determined up to sixth order[78)]

$$\beta(\alpha) = \frac{2}{3}\,\frac{\alpha}{\pi} + \frac{1}{2}\left(\frac{\alpha}{\pi}\right)^2 - \frac{121}{144}\left(\frac{\alpha}{\pi}\right)^3 + \dots \qquad (8.16)$$

It is not expected that the coefficients are going to continue being rational in higher orders [82)].

9. THE BOUND STATE PROBLEM

In this section I shall discuss for you the fundamental steps in bound state calculations. It is not my intention to do detailed calculations, only outline the main arguments.

Bound systems divide themselves into two groups. One in which there is a heavy central particle with a light particle orbiting it. This is the case for Hydrogen, muonium and Z-1 times ionized elements He^+, Li^{++}, C^{5+} etc. Because a muon bound to a nucleus has an orbit which lies much closer to the nucleus than the surrounding electron cloud, one may also count muonic atoms into this category. The other group consists of those two-particle systems in which the masses are nearly equal. Of these we only know positronium and in the future perhaps the $\pi^+\mu^-$ state. In the first group the main calculational simplification is due to the heaviness of the central particle which permits us to use the external potential approximation. In the second group, one has to begin with the fully relativistic Bethe-Salpeter equation. Actually, there is also a third group of bound states, namely the many-electron atoms. Although very important (we are all made of them) they are of less fundamental interest (except perhaps Helium) because of calculational difficulties.

a) Hydrogen-like systems

As mentioned in section 3 we may <u>formally</u> include an external potential $a_\mu(x)$ in the theory by adding the term $-e\, a_\mu(x)\, \bar{\psi}\, \gamma^\mu \psi(x)$ to the interaction L_1. This corresponds to the vertex (in coordinate space)

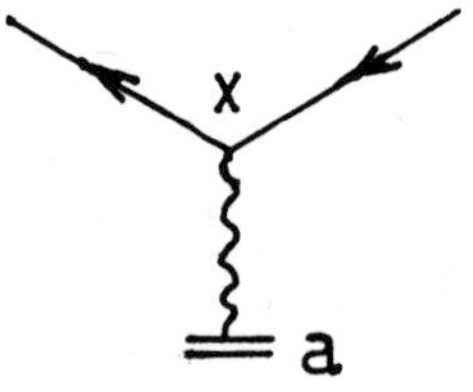

$$= -i\, e\, \not{a}(x) \qquad (9.1)$$

In momentum space the potential is replaced by its Fourier transform.

We must also expect that there will be counterterms involving the potential. Most of the divergences are cancelled by the ordinary counterterms but there will be some that aren't. They occur as corrections to the vertex (9.1), for example

In order to cancel the divergences it is necessary to introduce counterterms of the form

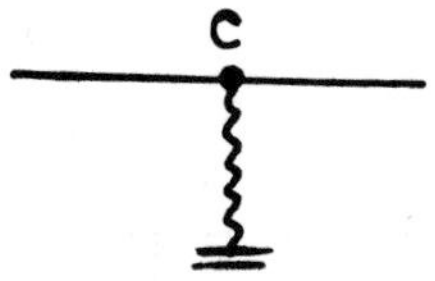

$$= i\, C_1\, e\, \not{a} \qquad (9.2)$$

$$= i\, C_3\, (g_{\mu\nu}\Box + \partial_\mu \partial_\nu) a^\nu \qquad (9.3)$$

This corresponds to adding the terms $C_1 e\, a^\mu\, \bar{\psi}\, \gamma_\mu \psi$ and $C_3 A^\mu\, (\Box g_{\mu\nu} + \partial_\mu \partial_\nu) a^\nu$ to L_2. The latter is an example of a counterterm for which there is no corresponding term in the original Lagrangian. We also see that these two terms are the only renormalizable ones one can add to L_2. The reason that C_1 and C_3 are the same as in the theory without external potentials follows from the requirement that the Coulomb scattering amplitude is not changed at large distances.

The electron propagator must also in the presence of an external field be decomposable into one-particle irreducible graphs, and satisfy an equation like (7.20).

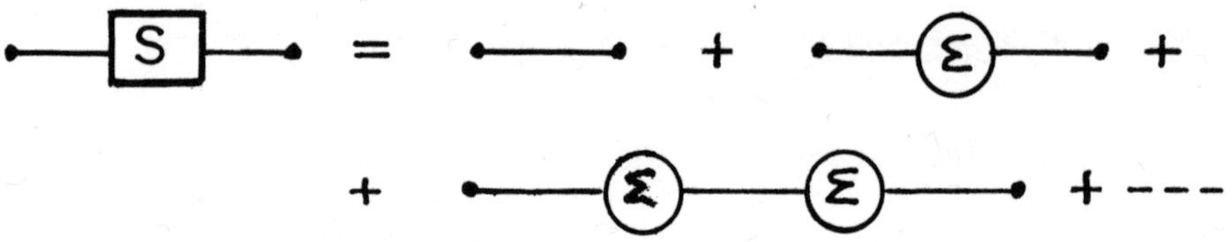

This can be written in the form

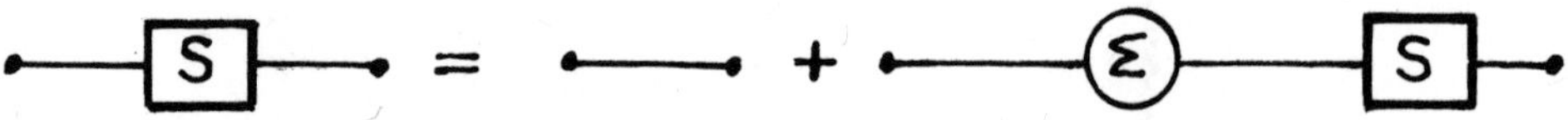

The irreducible graphs have the expansion

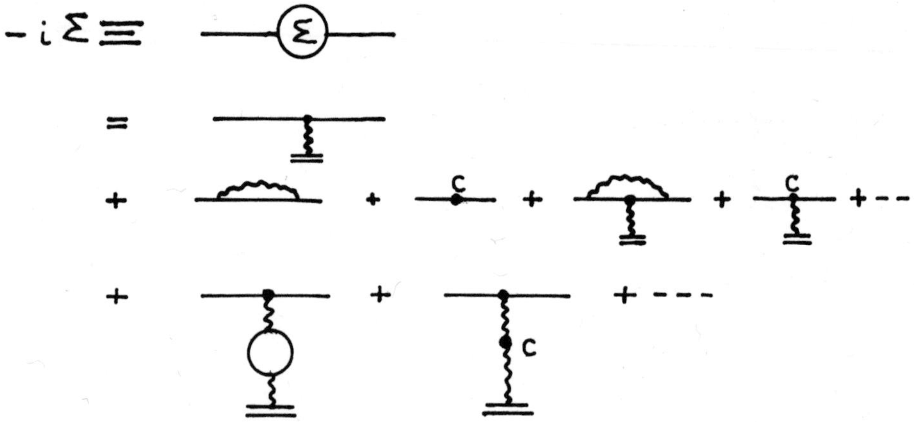

If we write (9.5) in coordinate space and multiply from the left with the inverse of the free propagator $(i\not\partial-m)$, we obtain the equation

$$(i\not\partial_x - m)\, S(xy) - \Sigma(xz)S(zy) = i\delta(xy) \tag{9.7}$$

where we integrate over repeated variables.

Assume now that the potential does not depend on time. Then $S(xy)$ and $\Sigma(xy)$ can only depend on x_0-y_0 and it will be convenient to introduce the Fourier transform

$$S_E(\bar{x}\ \bar{y}) = \int_{-\infty}^{\infty} dx_o\ e^{iE(x_o-y_o)}\ S(xy) \tag{9.8}$$

and similarly for Σ. Using the notation from ordinary one-particle quantum mechanics we can write (9.7) in operator form

$$(\not{p} - m - \Sigma_E)S_E = i \tag{9.9}$$

where $p_o = E$ and $\bar{p} = -i\nabla_x$.

Let us furthermore assume that a non-degenerate bound state $|B\rangle$ exists with energy $E = B$, and charge e. Then we define its wave function by

$$\langle 0|\psi(x)|\ B\rangle = N\ e^{-iBt}\ \psi_B(\bar{x}) \tag{9.10}$$

$$\psi_B^+\ \psi_B \equiv \int d^3\bar{x}\ \psi_B^+\ \psi(\bar{x})\ \psi_B(\bar{x}) = 1 \tag{9.11}$$

where N is a normalization constant and where we have used that $|B\rangle$ is an eigenstate of energy. This bound state will give the contribution

$$\theta(x_o-y_o)\ \langle 0|\psi(x)|B\rangle\langle B|\bar{\psi}\ (x)|0\rangle$$

to the propagator $S(xy) = \langle 0|\ T(\psi(x)\bar{\psi}(y))\ |0\rangle$

Inserting this into (9.8) we get for $E \simeq B$

$$S_E(\bar{x}\ \bar{y}) = \frac{i\ N^2\ \psi_B(\bar{x})\ \bar{\psi}_B(\bar{y})}{E - B + i\ 0} + \ldots \tag{9.12}$$

where the terms that are left out are regular for $E = B$. Inserting this into (9.9) and taking $E = B$ we get

$$(\not{p}_B - m - \Sigma_B)\psi_B = 0 \tag{9.13}$$

where $\not{p}_B = p_o\gamma_o - \bar{\gamma}\cdot\bar{p}$. Similarly, by summing (9.4) → (9.5) with S on the other side of Σ we get

$$\bar{\psi}_B(\not{p}_B - m - \Sigma_B) = 0 \tag{9.14}$$

The normalization constant may be determined by expanding (9.9) around the pole. Then we get using (9.14)

$$\frac{1}{N^2} = \bar{\psi}_B\,(\beta - \dot{\Sigma}_B)\,\psi_B = 1 - \bar{\psi}_B\,\dot{\Sigma}_B\,\psi_B \tag{9.15}$$

where $\dot{\Sigma}_E = \dfrac{\partial \Sigma_E}{\partial E}$.

If we split off the first term in (9.6) by writing

$$\Sigma_E = e\not{A} + M_E \tag{9.16}$$

we find the wave equation

$$(\not{p}_B - e\not{A} - m - M_B)\psi_B = 0 \tag{9.17}$$

which is the field theoretical replacement for the Dirac eigenvalue equation. The operator M_E is called the mass operator[61].

Eq.(9.17) may be attacked by ordinary perturbation theory. For $M_B = 0$ we have the ordinary Dirac equation and to first order in M_B the correction to the Dirac energy levels are

$$\Delta E = \bar{\psi}_B^{(0)}\, M_B\, \psi_B^{(0)} \tag{9.18}$$

where $\psi_B^{(0)}$ are the Dirac wave functions. One should of course choose a potential for which it can be solved exactly as for instance the Coulomb potential $e\not{A} = -Z\alpha/r$.

The expansion of M_B follows from (9.6). If we introduce the Dirac propagator in the presence of the Coulomb field

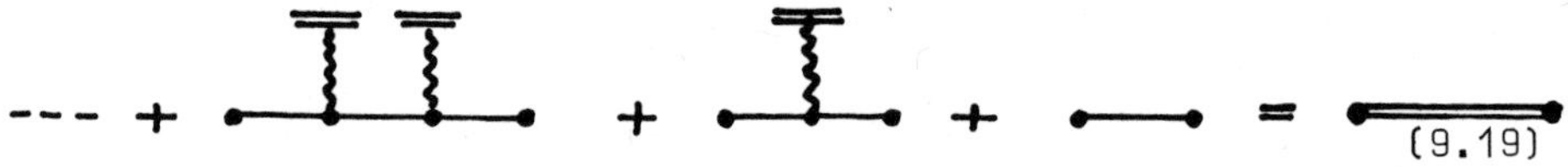

(9.19)

we may write it to all orders in $Z\alpha$ and to first in α

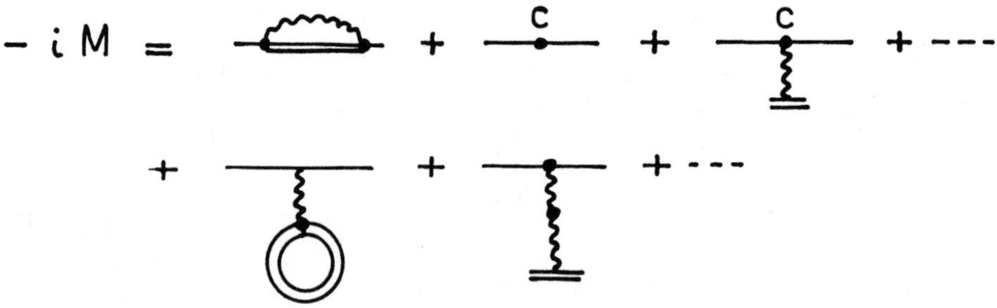

This is actually the Furry picture mentioned in section 3.

We are now in a position to make a short discussion of the contributions to the energy levels in Hydrogen-like atoms[62]

1) Schrödinger levels
The well-known result is

$$E_n = -\frac{1/2\, m(Z\alpha)^2}{n^2} \tag{9.21}$$

with n^2-fold degeneracy in each level :
1S, 2S = 2P, 3S = 3P = 3D, etc.

2) Dirac levels
Here some of the degeneracy is lifted. The levels depend now also on the total angular momentum, j, of the electron $(1/2 \leq j \leq n-1/2)$

$$E_{nj} = \frac{m}{\sqrt{1 + \frac{(Z\alpha)^2}{\left(\sqrt{(j+1/2)^2-(Z\alpha)^2} + n-(j+1/2)\right)^2}}} \tag{9.22}$$

If we expand in $(Z\alpha)$ we get (apart from the first m) the Schrödinger levels (9.21) plus a correction of the form

$$\Delta E_0 = m\,[\,C_{04}(Z\alpha)^4 + C_{06}(Z\alpha)^6 + \ldots\,] \tag{9.23}$$

with easily calculable coefficients. The spectrum has acquired fine structure. The degeneracy is partly lifted. The levels occur in doublets (except for j = n-1/2) :
$1\,S_{1/2}$, $2\,S_{1/2} = 2\,P_{1/2}$, $2\,P_{3/2}$, $3\,S_{1/2} = 3\,P_{1/2}$; $3\,P_{3/2} = 3D_{3/2}$ etc.

3) Radiative corrections (second order)
These are to first order in α and all order in Zα given by (9.18) and the diagrams in (9.20). In the first line we see the self-energy of an electron in an external field plus counter-terms and in the second the influence of the general vacuum polarization correction to the external field. Both of these diagrams have been subject to intense study during the last 25 years. The classic treatment of the self-energy diagram has been given by Ericson and Yennie[58] and of the vacuum polarization diagram by Wichmann and Kroll[63].

The structure of the self-energy diagram is a bit more involved than the expansion

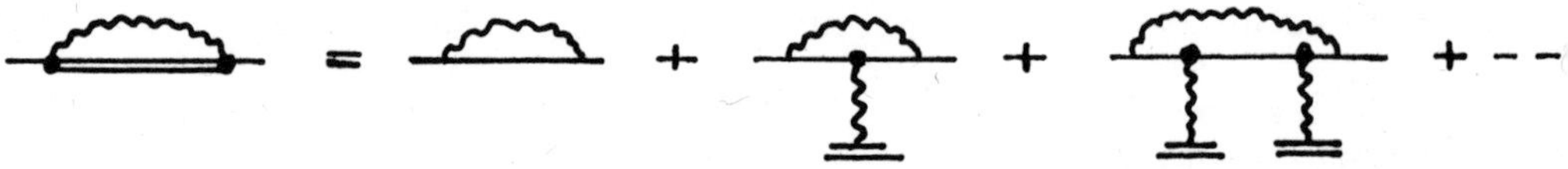

(9.25)

might lead us to think. Let us first be naive and try to estimate the powers of $Z\alpha$ that are involved. In the first diagram the static effects are cancelled off by the counterterm and it behaves like $(\not p - m)^2$ close to the mass shell. Now from the Dirac equation $\not p - m \simeq e\not A \simeq \frac{Z\alpha}{r} \simeq m(Z\alpha)^2$ because a typical bound state in Hydrogen has radius $r \simeq \frac{1}{mZ\alpha}$. Thus the first diagram is of order $m\alpha(Z\alpha)^4$. In the second diagram the static effect is also cancelled off by the counterterms. Since  is a correction to the form factor of the electron, it will go like $\alpha q^2/m^2$ for small momentum transfer. Typical momenta are $q \simeq mZ\alpha$ in Hydrogen, and as the potential is $e\not A \simeq m(Z\alpha)^2$ we get again $m\alpha(Z\alpha)^4$.

We have learnt two things by these considerations. Firstly that the energy levels are of the forms

$$\Delta E_1^{SE} = m\alpha\,[\,C_{14}^{SE}(Z\alpha)^4 + C_{15}^{SE}(Z\alpha)^5 + C_{16}^{SE}(Z\alpha)^6 + \ldots\,] \tag{9.26}$$

beginning with $\alpha(Z\alpha)^4$ and secondly that we probably can not expand in $Z\alpha$ since the two first diagrams give contributions of the same order of magnitude. This is also reflected by the fact that these diagrams are infrared divergent (*). If we calculate naively,

---

(*) The infrared divergence of the second diagram is related to the infrared divergence of the charge radius of the electron. The anomalous magnetic moment of the electron gives an infrared finite contribution truly of order $m\alpha(Z\alpha)^4$.

The Lamb shift due to the charge radius is positive since a spreading of the charge lessens the binding. Conversely, vacuum polarization (see below) tightens the binding and gives a negative contribution. For most of the effects discussed the influence on the S-levels is by far the largest because S-wave electrons penetrate to small distances of the order of one electron Compton wave length where radiative effects become significant. This consideration permits another estimate of Lamb shift

$\Delta E \simeq \alpha \frac{Z\alpha}{\lambda_c} (\lambda_c/r_B)^3$ which is a radiative correction (α) to the potential ($Z\alpha/\lambda_c$) seen by the electron times ../..

the coefficients above will depend on the photon mass. The infrared divergence is spurious and does not appear if we do not expand. This is also intuitively understandable since a bound electron can only emit real photons with definite frequencies, not arbitrarily small. They are in fact of order $m(Z\alpha)^2$, the energy separation between the levels. This explains the form of the "constants".

$$C_{14}^{SE} = C_{141}^{SE} \log \frac{1}{(Z\alpha)^2} + C_{140}^{SE} \tag{9.27}$$

$$C_{16}^{SE} = C_{162}^{SE} \log^2 \frac{1}{(Z\alpha)^2} + C_{161}^{SE} \log \frac{1}{(Z\alpha)^2} + C_{160}^{SE} \tag{9.28}$$

while C_{15} has no logarithms. The name of the game in the higher order calculations is to avoid false expansions that lead to spurious divergences[58]. The coefficients C_{141}, C_{162} and C_{161} are known analytically[58] while C_{140} which involves the famous Bethe-logarithm is known to a high precision numerically[58,59,74]. In recent years the main effort has been spent in trying to calculate C_{160} and higher coefficients with reasonable precision[60,64]. The method used[64] is not to expand at all, but to calculate all the remaining terms as a function of $Z\alpha$ for arbitrary $Z\alpha$.

The vacuum polarization diagram has the expansion

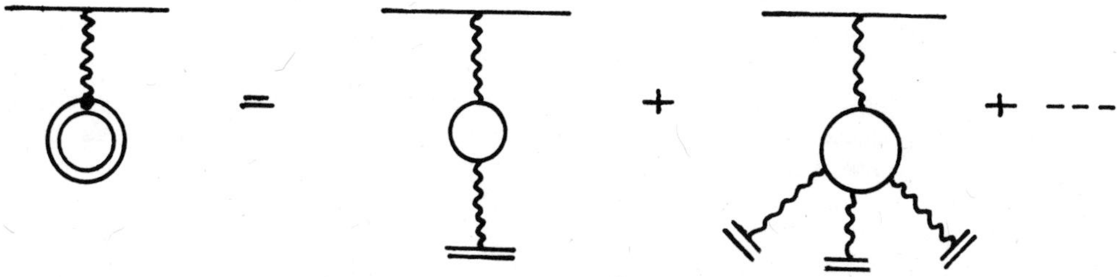

because of Furry's theorem which excludes even powers of $Z\alpha$. The first diagram is the famous Uehling potential[65] and contains no infrared divergence in its lowest order terms. The form of the correction is[63,64] for electronic atoms

$$\Delta E_1^{VP} = m_e\alpha\, [C_{14}^{VP}(Z\alpha)^4 + C_{15}^{VP}(Z\alpha)^5 + C_{16}^{VP}(Z\alpha)^6] \tag{9.27}$$

(*) .....
the probability $(\lambda_c/r_B)^3$ that the electron is within one Compton wave length of the nucleus. (r_B is the Bohr radius).

$$C_{16}^{VP} = C_{161}^{VP} \log \frac{1}{(Z\alpha)^2} + C_{160}^{VP} \tag{9.28}$$

where C_{14}, C_{15} and C_{161} are known analytically.

The second diagram has very little influence for low $Z\alpha$. It contributes only to the term C_{160} and higher. For higher $Z\alpha$ its influence, however, increases.

For electronic atoms the vacuum polarization contribution is generally small compared to the self-energy contribution mainly because of the lack of the log. For muonic atoms this contribution becomes much more important. The reason is that a muon is tied to an atomic nucleus at at typical distance $1/m_\mu Z\alpha$ while the space charge density due to vacuum polarization extends about one electron Compton wave length $(1/m_e)$ away from the nucleus. Thus, since $m_e/m_\mu\alpha \simeq O(1)$ the muon sees this cloud clearly. We can estimate the leading effects in the same way as before. The Uehling potential will at short distances increase the strength of the Coulomb potential logarithmically

$$-\frac{Z\alpha}{r} \to -\frac{Z\alpha}{r}\left[1 + \frac{2}{3}\frac{\alpha}{\pi}\log\frac{1}{m_e r} + \dots\right] \tag{9.29}$$

because at close range one sees more and more of the bare unscreened charge of the nucleus. We therefore estimate that in a muonic atoms $r \simeq 1/m_\mu Z\alpha$, and the energy levels will have the form

$$\Delta E_1^{VP,\mu} = m_\mu \alpha \left[C_{12}^{VP} (Z\alpha)^2 + \dots\right] \tag{9.30}$$

where C_{12} depends on the details of the state. Unfortunately the parameter $m_\mu Z\alpha/m_e$ is neither large nor small so that one has to calculate numerically[67]. But it is clear that the nominal order is such that vacuum polarization must dominate the radiative level shifts even in light muonic atoms. In heavy atoms the higher order vacuum polarization diagrams such as the second one in (9.29) play a significant role. Their dominant contribution to the potential is just a constant times the Coulomb potential[68] such that the nominal order of the contribution to the energy levels is $m_\mu\alpha\,(Z\alpha)^4$.

4) Radiative corrections (4th order)

At the next level of approximation one includes the fourth order radiative corrections in (9.10). They are for the case of the self-energy

plus counter-terms, and for the case of vacuum polarization

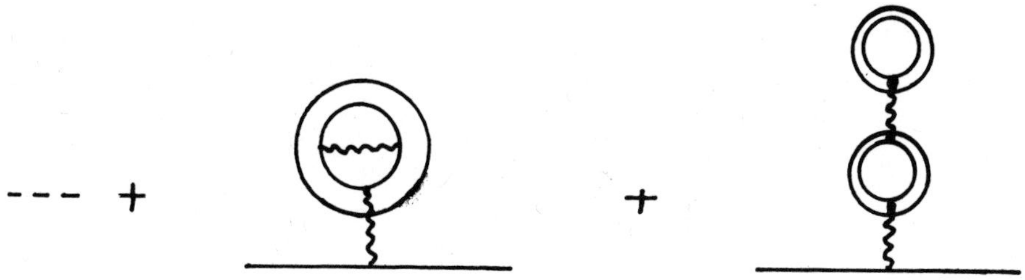

The leading contributions are of nominal order

$$\Delta E_2 = m\,\alpha^2[\,C_{24}(Z\alpha)^4 + \ldots\,] \qquad (9.31)$$

and contain no logarithms. C_{24} is known analytically[58,23,69].

5) Hyperfine structure

In hydrogenic atoms the interaction of the magnetic moment of the orbital electron with the magnetic moment of the nucleus leads to a hyperfine splitting of all the levels.

In order to calculate it one uses the perturbation-theory developed above including the potential due to the magnetic moment which is $\delta\bar{a} = \bar{\mu} \times \bar{r}/r^3$. Thus we may use (9.18) with δM instead of M where

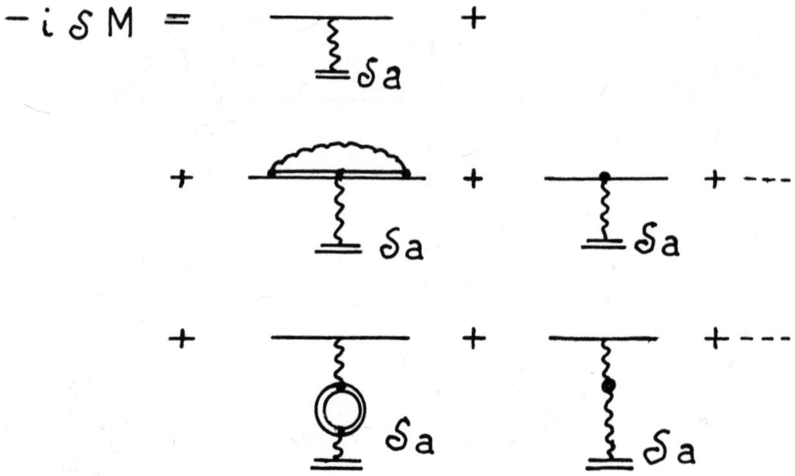

We may estimate the nominal order of the leading term. Since the interaction energy is $\bar{\mu}_e \cdot \bar{B}$ where $\bar{B} \sim \frac{\mu_N}{r^3}$ we estimate it to be of order

$$\mu_e \mu_N \,(m_e Z\alpha)^3 = \frac{\mu_N}{\mu_e}\, m_e \,\alpha (Z\alpha)^3 \qquad (9.33)$$

where we have used $\mu_e^o = \frac{e}{2m_e} = \mu_e$, $r \simeq \frac{1}{m_e Z\alpha}$. Thus the Hyper-fine corrections must be of the form

$$\begin{aligned} \Delta E^{Hfs} &= \mu_e^o \,\mu_N (m_e Z\alpha)^3 \{ C_{00} + C_{02}(Z\alpha)^2 + \ldots \\ &+ \alpha \,[C_{10} + C_{11} \,Z\alpha + C_{12}(Z\alpha)^2 + \ldots] \\ &+ \alpha^2 [C_{20} + C_{21} \,Z\alpha + \ldots] \\ &+ \quad \ldots \} \end{aligned} \qquad (9.34)$$

Here C_{00} is the Fermi term and C_{10} and C_{20} arise from the corrections to the electrons'magnetic moment. C_{02} is the relativistic correction due to the Dirac wave functions in the first term in (9.32). The other terms are various binding corrections. The coefficient C_{12} contains logarithms

$$C_{12} = C_{122} \log^2 \frac{1}{(Z\alpha)^2} + C_{121} \log \frac{1}{(Z\alpha)^2} + C_{120} \qquad (9.35)$$

The constant term C_{120} is only known numerically[70]. Hyperfine structure is reviewed in Brodsky and Ericson[70] and some newer results in ref.23.

6) Reduced mass corrections

The main effect of the motion of the nucleus is of course the replacement of m by $\mu = \frac{mM}{m+M}$ in the Schrödinger levels. The same may be done but without much justification in the Dirac levels. There will be additional relativistic corrections from the Bethe-Salpeter equation. In the second order radiative corrections one should multiply with a factor $1/(1+m/M)^3$. This follows from the intuitive argument by which we estimated these corrections, since they were of the order of $\alpha\, q^2/m_e^2 \sim Z\alpha/r$ where q and r were typical Schrödinger momenta and distances ($q \simeq \mu Z\alpha$, $r \simeq 1/q$). For the Hyperfine structure we also find a reduced mass effect to be a factor $(1 + m/M)^{-3}$.

7) Recoil corrections

These are the corrections due to the nuclear motion not included in the reduced mass corrections. The vehicle for the calculation of these effects is the fully relativistic Bethe-Salpeter equation. We expect the leading term to be of the form [71,73]

$$\Delta E^{rec} = \frac{m}{M}\, m\, [C_{05}(Z\alpha)^5 + C_{14}\, \alpha(Z\alpha)^4 + \dots] \qquad (9.36)$$

where C_{05} contains a $\log 1/(Z\alpha)^2$. In the case of Hyperfine structure recoil corrections have been evaluated to even higher approximations[72].

8) Nuclear structure effects

The main effect is due to the spatial extension of the static charge distribution of the nucleus. If R_N is the charge radius we estimate the effect to be of the size $Z\alpha/R_N\ (R_N/R_B)^3$ where $r_B = 1/m_e Z\alpha$ is the Bohr radius, because this is the potential within the nucleus times the probability that the electron is there. Thus we have [71]

$$\Delta E^{NS} = m_e(Z\alpha)^4 (m_e R_N)^2 C_{04} \qquad (9.37)$$

9) Nuclear polarizability

The electric field of the atomic electron may excite the nucleus to other states and this gives rise to a dynamically induced dipole moment. This leads to tiny corrections that are almost only of importance for the case of the Hyperfine structure of Hydrogen (see ref.23, and also ref.67 for the case of muonic atoms).

10) Electronic screening

In muonic atoms the atomic electrons, particularly from the 1S level, will screen a fraction of the nuclear charge. It is a hard contribution to evaluate. A review of the current situation may be found in ref.67.

b) Positronium-like systems

In systems like positronium (e^+e^-) and pionic muonium ($\mu^-\pi^+$) the problem is that the masses are of the same size. This means that the relativistic Bethe-Salpeter equation must be used from the outset, and not just to calculate some tiny recoil corrections. Let me indicate a few relevant features.

The Bethe-Salpeter equation for the $\pi\mu$ system (which in some respects is easier than positronium because of the lack of annihilation terms) is a recursive equation for the 4-point function

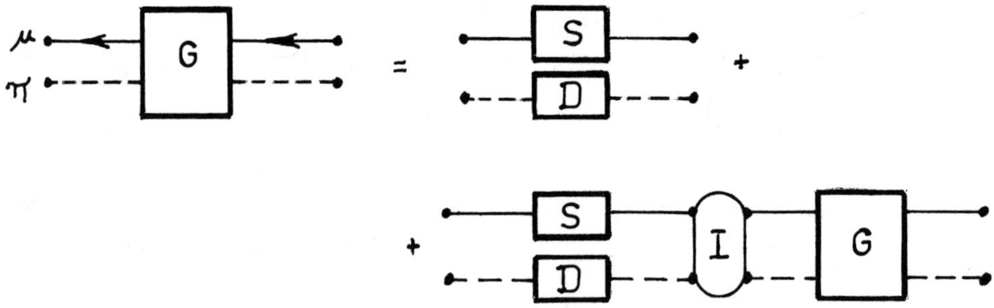

where I are the two-particle irreducible diagrams

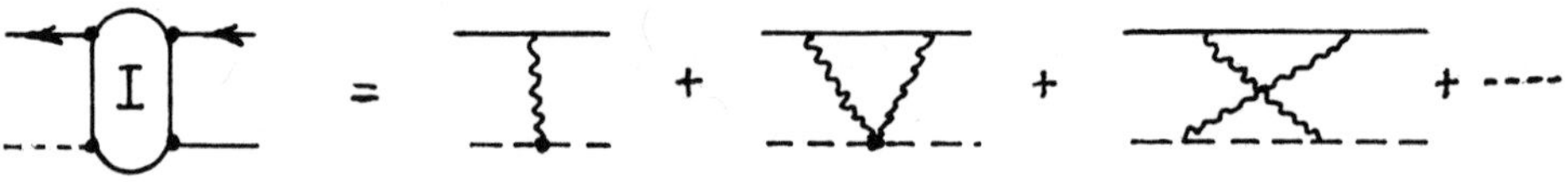

Eq.(9.38) is entirely analogous to eq. (9.5) and the reduction proceeds in the roughly same way. One projects out the total momentum and defines a wave function which satisfies an energy eigenvalue equation in the center of mass system. One would like to solve it in the ladder approximation by using the first diagram in (9.39). This would correspond to the separation of the external potential in eq.(9.16). This is however where the trouble starts. For the ladder approximation does not reduce to the Dirac equation when the mass of the pion goes to ∞ , or to the Klein-Gordon equation when the mass of the muon goes to ∞ . It is necessary to include all crossed graphs in order to get the correct limits. Furthermore, and related to this, the ladder approximation is not by itself gauge invariant. It is understandable that the calculation of higher order corrections using the Bethe-Salpeter equation requires great technical skill[75]. The radiative corrections to the $\pi\mu$ atom have recently been calculated by Cho[77] . It is not impossible that it might be observed in $K^{\circ} \to \mu\pi\nu$.

10. SOME BOUND STATE EXPERIMENTS

This section contains a discussion of some of the most recent developments in the comparison of bound state theory and experiment for fundamental systems. It is by no means complete nor exhaustive.

a) Lamb shift in H (n=2)

The $2S_{1/2} - 2P_{1/2}$ energy difference in Hydrogen is entirely due to radiative effects and consequently well suited for tests of our understanding of these matters. Experiment are made possible by the metastability of the $2S_{1/2}$ state as compared with the instability of the $2P_{1/2}$ state against one-photon decay to the ground state $1S_{1/2}$. The $2S_{1/2}$ component of a Hydrogen beam may be quenched by means of an applied rf electric field (possibly in the presence of a magnetic field), which induces resonant transitions to the $2P_{1/2}$ state, which promptly decays. The completion of the $2S_{1/2}$ states may be detected by the consequent decrease in Lyman-α radiation in another quenching region. The center of the resonance curve obtained by sweeping the frequency over it determines the Lamb shift. In a recent experiment[79] a sophisticated technique has been used to reduce the width of the resonance, which is primarily due to the natural width of the $2P_{1/2}$ state. The metastable beam is passed through two separate coherent rf electric fields that are either in the same phase or precisely in opposite phase. The difference between the resonance shapes in the two cases shows a more sharply defined peak structure which enables a better determination of the center. The result[79] which is the best ever obtained is

$$S_{exp} = (E(2S_{1/2}) - E(2P_{1/2}))_{exp} = 1057.893(20)\text{MHZ} \qquad (10.1)$$

The corresponding latest theoretical result is[64)]

$$S_{th} = 1057.864(14)\text{MHZ} \qquad (10.2)$$

in reasonable agreement. The various theoretical contributions are shown in Table 10.1. It includes the recently calculated higher order binding corrections by Mohr[64)] which disagree with the ones previously calculated by Erickson[60)]. The uncertainty may be reduced experimentally by a factor of 3 and that would demand the reconciliation of the theoretical values.

Lamb shift i.e. $n(S_{1/2}\rightarrow P_{1/2})$ energy differences have been measured in many atoms[80)] (see also ref.23). Recently[81)] it was measured in $^6Li^{2+}$ with the result $S_{exp} = 62765 \pm 21$ MHZ as compared with the theoretical result $S_{th} = 62762 \pm 9$ MHZ.

b) Hyperfine structure of positronium

Positronium (e^+e^-) is one of the cleanest QED systems. In the non-relativistic Schrödinger approximations its level structure is analogous to that of hydrogen except for the fact that the reduced mass is 1/2 m_e which halves all binding energies. The ground state is split into a triplet orthopositronium state ($^3S_1$) and a singlet parapositronium state ($^1S_0$) by the spin-spin interaction and by the annihilation term which is peculiar to this system. The orthopositronium state lies highest. The splitting is of order $m\alpha^4$ (see eq.(9.34)) and is of the same order of magnitude as the relativistic corrections. In fact positronium presents a serious theoretical challenge because of its lack of expansion parameters. All effects mix with each other at every level. The present theoretical value is of the form

$$\Delta\nu_{th} \equiv E(1^3\,S_1 - 1^1\,S_0)$$

$$= \frac{1}{2} m\alpha^4 \left[\frac{7}{6} - (\frac{16}{4} + \log 2) \frac{\alpha}{\pi} + \frac{1}{2} \alpha^2 (\log \frac{1}{\alpha} + C)\right]$$

$$= (203404.0(6) + 4.66C)\ \text{MHZ} \qquad (10.3)$$

The constant C is only partly known. It contains fourth order binding, self-energy, vacuum polarization recoil, and annihilation corrections all terribly mixed up with each other. The term involving $\log 1/\alpha$ was formerly thought to have the coefficient 3/4 arising from recoil terms only but it was shown[84,85,86)] that

Table 10.1 Contributions to Lamb shift in H (n=2) (compiled from ref.23 and 64)

Nominal order	Effect	Volume (MHZ)
$m\alpha(Z\alpha)^4$	2nd order self-energy	1009.920
$m\alpha(Z\alpha)^4$	2nd order magnetic moment	67.720
$m\alpha(Z\alpha)^4$	2nd order vacuum polarization	-27.084
$m\alpha(Z\alpha)^5$	Binding corrections	7.140
$m\alpha(Z\alpha)^{\geq 6}$	Higher order binding corrections	-0.419
$m\alpha^2(Z\alpha)^4$	4th order self-energy	0.444
$m\alpha^2(Z\alpha)^4$	4th order magn.	-0.102
$m\alpha^2(Z\alpha)^4$	4th order vacuum polarization	-0.239
$m\,\frac{m}{M}\,(Z\alpha)^5$	Recoil corrections	0.359
$m(Z\alpha)^4(mR_p)^2$	Proton size	0.125
Total		1057.864(14)

The uncertainty is composed of many contributions of almost equal size. It will be difficult to reduce it much further. Reduced mass effects have been included in all contributions.

the threshold singularity in fourth order vacuum polarization annihilation terms also gave rise to a logarithm. The contribution to C was also evaluated for these diagrams[85,86].

The most recent experiment[87] gives the result

$$\Delta\nu_{exp} = 203387.0(1.0)\text{MHZ} \tag{10.4}$$

in rather violent disagreement with theory. It corresponds to C = -3.6(4) which is however not unreasonable.

It is interesting to note that an improvement in experiment by a factor of 3 would allow us to deduce α to a high precision (1/2 ppm) provided all the α^2 terms were known.

Let me also remark that the radiative corrections to the 3γ decay of orthopositronium have also recently been evaluated[88]. The result is

$$\Gamma^{th}_{ortho} = m\alpha^6 \; \frac{2(\pi^2-9)}{9\pi} \left[1 + \frac{\alpha}{\pi}(1.8 \pm 0.6)\right]$$

$$= 0.7241\ (10) \times 10^7 \text{ sec}^{-1} \tag{10.5}$$

while the experimental value is[89]

$$\Gamma^{exp}_{ortho} = 0.7275(15) \times 10^7 \text{ sec}^{-1} \tag{10.6}$$

There is a 2.5 s.d. discrepancy which ought to be cleared up by further theoretical and experimental work.

Finally there has been a very recent measurement of the fine structure of the n = 2 level in Positronium[115]. The theoretical splitting between the 2^3S_1 and 2^3P_2 levels is [116]

$$\Delta E_{th} \equiv E(2^3S_1) - E(2^3P_2)$$

$$= \frac{23}{960}\, m\alpha^4 \left[1 + 3.766\ \alpha\right] \tag{10.7}$$

$$= 8625.14 \text{ MHZ}$$

while the experiment[115] gave

$$\Delta E_{exp} = 8628.4(2.8)\text{MHz} \tag{10.8}$$

The discrepancy is less than 2s.d. Again it would be interesting to know the α^2 corrections.

c) The muonic Helium ion $(\mu-{}^4He)^+$

This atom is Hydrogen-like but because the muon orbit lies so close to the nucleus, the vacuum polarization effects due to electrons dominate the level corrections. Thus the $2S_{1/2}$ level is pulled down below the $2P_{1/2}$ level. The $2P_{3/2}$ level lies above the $2S_{1/2}$ level due to relativistic corrections. The theoretical value for the fine structure interval is [117,118]

$$\Delta E_{th} \equiv E(2P_{3/2}) - E(2S_{1/2}) = 1.5251(87)\,\text{eV} \qquad (10.9)$$

The various contributions are displayed in Table 10.2, where the nuclear radius $\langle r^2\rangle$ has been kept arbitrary.

Table 10.2 Contributions to ΔE (from ref.118)

			ΔE(eV)
Relativistic fine structure			0.1457
Muonic radiative corrections			-0.0143
2nd order electronic vac.pol.			1.6659
Fourth order electronic vac.pol.			0.0115
Nuclear size (r in fm)		$-\langle r^2\rangle$	0.1053
Nuclear Polarizability			0.0031
Total theory	1.8119	$-\langle r^2\rangle$	0.1053
for $\sqrt{\langle r^2\rangle}$ = 1.650(25)fm			1.5251(87)
Experiment[118] (CERN)			1.5274(9)

The main theoretical uncertainty is due to the nuclear radius which is only known to about 2 % . If one considers this experiment to be a determination of the nuclear radius the result [118] is $\sqrt{\langle r^2\rangle} = 1.644(5)$ fm i.e. an improvement of a factor of 5 in the uncertainty. But it is perhaps more correct to view the experiment as a direct measurement of the electronic vacuum polarization which as the table shows by far dominates the fine structure.

d) Heavy muonic atoms

For about four years there has existed some discrepancy between theory and experiment in heavy muonic atoms for transitions between high-lying circular orbits[119]. The discrepancy triggered a wealth of theoretical work on the subject (see ref.67 for a review). The theorists were however not really able to explain away the deviations, although several erros were discovered in previous calculations[68].

The largest discrepancies were reported[119] for the $4f_{7/2} \rightarrow 3d_{5/2}$ and $4f_{5/2} \rightarrow 3d_{3/2}$ transitions in Ba^{56} and for the $5g_{9/2} \rightarrow 4f_{7/2}$ and $5g_{7/2} \rightarrow 4f_{5/2}$ transitions in Pb^{82}. They amounted to 2-4 s.d. Smaller deviations were found in other atoms.

In a very recent CERN experiment these transitions have been remeasured[120] and the discrepancies have disappeared ! In table 10.3 the comparison with experimental is displayed.

Table 10.3 Comparison with experiment (from ref. 120)

Transition	E^{exp} (keV)	E^{calc} (keV)	$E^{calc} - E^{exp}$ (eV)
Ba $4f_{5/2}-3d_{3/2}$	441.366 ± 13	441.368	+ 2 ± 13
Ba $4f_{7/2}-3d_{5/2}$	433.916 ± 12	433.912	- 4 ± 12
Pb $5g_{7/2}-4f_{5/2}$	437.744 ± 16	437.748	+ 4 ± 16
Pb $5g_{9/2}-4f_{7/2}$	431.353 ± 14	431.334	-19 ± 14

The main reason for the improvement of the experiment is the relative and absolute calibration which is done primarily with a ^{137}Cs line at 455 keV (which arises from muon capture in ^{138}Ba) supported by the 411 keV line from radioactive ^{198}Au. The interesting lines all lie between these two calibration points. Several other calibration lines of less importance were also used.

Theoretically the largest correction to these lines is due to second order vacuum polarization (the Uehling term) as discussed in section 9. It amounts to about 2 keV. Fourth order vacuum polarization, light-by-light and higher amount to 10 - 100 eV. Likewise electron screening gives contributions of the order of 100 eV. Finite size, nuclear polarization, relativistic recoil etc. amount to less than 10 eV. The detailed breakdown of the various terms may be found in ref.67. We see thus that these measurements verify a number of effects but in particular the vacuum polarization contributions.

There has recently been theoretical disagreements about the contribution from graphs of the type

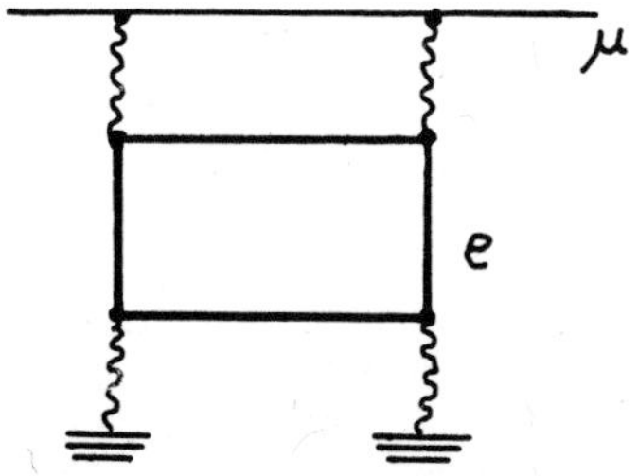

which are of nominal order $\alpha^2(Z\alpha)^2$. In one paper[121)] these graphs were claimed to contribute up to -35 eV to the transitions, but the concensus is now, supported by the experiment, that the other calculations[122,123)] which predict less than 1 eV, are correct.

As far as I know these are all the news that are in this field at this moment. Other fundamental systems, such as hyperfine structure in Muonium and Hydrogen, fine structure in Hydrogen and Helium have not been touched upon here since they are adequately described in previous reviews[23,92,124)].

The over-all impression is that there are no more discrepancies floating around, and that the comparison of theory and experiment has been moved one notch further along into the ppm region[126)].

11. THE ANOMALOUS MAGNETIC MOMENTS OF THE LEPTONS

The anomalous magnetic moments of the electron and muon stand out as the foremost tests of our understanding of QED. Experiments have been carried to the ppm level and theory has followed suit.

The intrinsic magnetic moment of a spin 1/2 lepton is of the form

$$\bar{\mu} = g\,\frac{e}{2m}\,(\frac{1}{2}\,\bar{\sigma}) \qquad (11.1)$$

where the g factor according to the Dirac equation should be equal to 2. Radiative effects give rise to an anomaly, a, defined by

$$g = 2(1+a) \qquad (11.2)$$

The anomaly as about 10^{-3}.

a) Experiment

Experiments are made possible by an accident of Nature resembling the one that makes Lamb shift a purely radiative correction. In a magnetic field the Larmor precession of the magnetic moment and the relativistic Thomas precession collaborate to make the spin of a Dirac electron exactly follow its momentum. If they are parallel to begin with they stay parallel. Furthermore, if the electron has an anomalous magnetic moment, the precession rate of the spin relative to the momentum is independent of the velocity of the particle, even relativistically. This is only true for motion in a magnetic field. If an electric field is also present there will be velocity dependent effects. The complete formula is for the precession rate[90]

$$\bar{\omega}_a = \frac{e}{m} \left[-a\bar{B} + \left(a - \frac{1}{\gamma^2 - 1} \right) \bar{V} \times \bar{E} \right] \tag{11.3}$$

provided $\bar{V}.\bar{E} = \bar{V}.\bar{B} = 0$ where $\bar{V}$ is the velocity and $\gamma = (1 - V^2)^{-1/2}$.

The best experimental values for the electron anomaly[91] and the muon anomaly[90] are

$$a_e^{exp} = 0.001159\,6567(35) \tag{11.4}$$

$$a_\mu^{exp} = 0.001165\,895\,(27) \tag{11.5}$$

The electron experiment is fairly old and is amply commented in the various reviews[23,92,93,94]. The muon experiment is completely new.

The drastic reduction of the uncertainty by a factor of 10 since the last experiment is mainly due to another accident of Nature, namely the factor in front of the $\bar{V} \times \bar{E}$ term in eq.(11.3). In order to avoid uncertainty due to inhomogeneities in $\bar{B}$ it is important that $\bar{B}$ is kept as homogeneous as technically possible. But then vertical focusing in the storage ring becomes a problem. The particles would like to drift up or down along the magnetic flux lines. To keep them in place electric quadrupoles are used. The effect of the electric field may however be eliminated by choosing the coefficient in (11.3) to be zero. This occurs for $\gamma = \sqrt{1/a+1} \simeq 30$ which corresponds to $p_\mu = 3.094$ GeV/c, a very handy momentum, indeed. This is the third lucky accident. Several more factors contribute to the improvement which eventually may reach a factor of 30. The time dilation of the relativistic muons allows the study of them for many more g - 2 cycles than if they moved more slowly. The pion beam is momentum selected before it is injected into the storage ring. This improves the initial polarization of the muons.

On the inside of the ring one detects the decay electrons. Those of higher energy comes from forward μ-decay which is sensitive to the angle between spin and polarization. Hence the rate of the highest energy decay electrons will be modulated by the g - 2 frequency (see Fig. 11.1). Having measured ω_a the anomaly may be deduced if B is known. The magnetic field is monitored by nuclear magnetic resonance probes. The resonance frequency is

$$\omega_p = \mu_p B \tag{11.6}$$

Hence

$$\frac{\omega_a}{\omega_p} = \frac{a \quad e/m_\mu}{\mu_p} = \frac{a}{1+a}\lambda \tag{11.7}$$

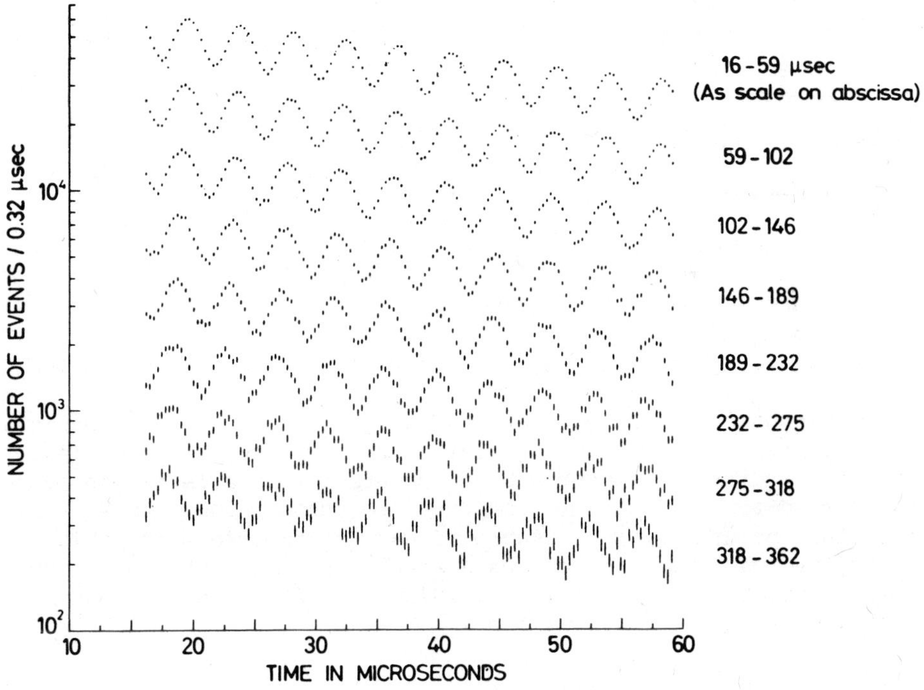

Fig. 11.1 g-2 oscillations (from ref.90)

where $\lambda = \frac{\mu_\mu}{\mu_p} = \frac{\omega_\mu}{\omega_p}$ is the ratio of the Larmor precession frequencies for muons and protons at rest. This quantity is measured independently[95] and known with a precision of 3 ppm. Measuring ω_a and ω_p the anomaly may be derived from this equation.

b) Theory

The theoretical evaluation is rather easy because it can be related directly to Feynman diagrams and because it has a strict expansion in powers of α

$$a = a_1 \frac{\alpha}{\pi} + a_2 \left(\frac{\alpha}{\pi}\right)^2 + a_3 \left(\frac{\alpha}{\pi}\right)^3 + a_4 \left(\frac{\alpha}{\pi}\right)^4 + \ldots \qquad (11.8)$$

Unlike the bound state problem there are no log α's appearing in the constants a_n.

The diagrams that are involved are of the vertex type with on-shell electrons and an off-shell photon. Such diagrams may be expressed in terms of form factors $F_{1,2}(k^2)$

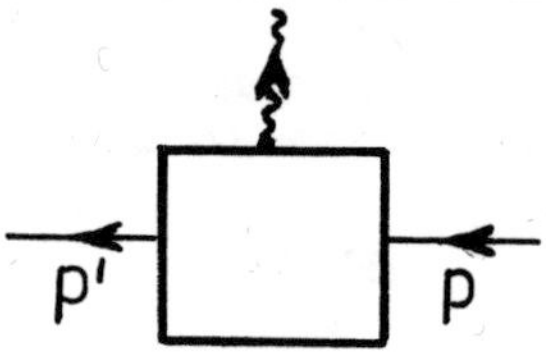

$$= -ie\bar{u}' \left\{\gamma_\mu (F_1+F_2) - \frac{(p+p')_\mu}{2m} F_2\right\} u \qquad (11.9)$$

where $k = p - p'$. The anomaly is then given by

$$a = F_2(k^2=0) \qquad (11.10)$$

I shall assume that you are familiar with the way this is derived.

In practice there are two ways of getting hold of the anomaly. In calculations where the algebra is done by hand, it is easiest to evaluate a diagram using the Dirac equation to the left and to the right as often as one can, dropping terms proportional to γ_μ whenever they occur. If the algebra is done by computer, and this is often the case, it is sometimes easier to project out the second form factor F_2 by multiplying (11.9) with $\bar{u}\,\Lambda^\mu_z u'$ and summing over spins, where ($k = p-p'$, $p = p+p'$)

$$\Lambda^\mu_z = \frac{m^2}{k^2 p^2} \gamma^\mu - \frac{m(m^2 + \frac{1}{2} k^2)}{k^2 p^2} \frac{1}{2} p^\mu \qquad (11.11)$$

In the following I shall review the various contributions to the electron and muon anomalies. They differ from each other for the following intuitive reasons. When an electron interacts with its own radiation field it happens typically over distances of one electron Compton wave length ($1/m_e$). At such distances the virtual muons are almost not excited and have no effect on the anomaly[96]. On the other hand when a muon interacts with its own radiation field it happens at distances of the order of one muon Compton wave length ($1/m_\mu$) and this is far inside the distances ($1/m_e$) where the virtual electrons are excited. Consequently there is a very significant effect on the muon anomaly.

Actually these considerations may be extended to other effects as well. Assume that a charged lepton of mass m interacts with itself via an interaction that requires excitation of states with mass scale M (exchange of a particle of mass M, forces of range 1/M, etc.). Then the contributions to the anomaly usually vanishes as $1/M^2$ [98]. One may understand this intuitively by noticing that a self-interaction always occurs in third order perturbation theory, involving two energy denominators, each of size M. Thus the extra contribution to the anomaly is of the form

$$\delta a = C \frac{m^2}{M^2} \qquad (11.12)$$

and this leads to the important scaling law

$$\delta a_e = \frac{m_e^2}{m_\mu^2} \frac{C_e}{C_\mu} \delta a_\mu \qquad (11.13)$$

which is independent of M. If the interaction is universal for muons and electrons we have $C_e = C_\mu$ and the scaling law becomes universal. Since $m_e^2/m_\mu^2 \simeq 2 \times 10^{-5}$ it means that the electron anomaly is untouched by effects that would be significant for the muons. The electron g - 2 is a clean QED quantity. Conversely agreement between theory and experiment for the muon places much stronger constraints on exotic interactions than is the case for the electron.

<u>The electron anomaly</u>. The first two coefficients have been known for many years [99]. They are

$$a_1^e = 1/2 \qquad (11.14)$$

$$a_2^e = \frac{197}{144} + \frac{\pi^2}{12} - \frac{1}{2}\pi^2 \log 2 + \frac{3}{4}\varphi(3) \qquad (11.15)$$

$$= -0.32848\ ..$$

They are obtained from the single second order diagram

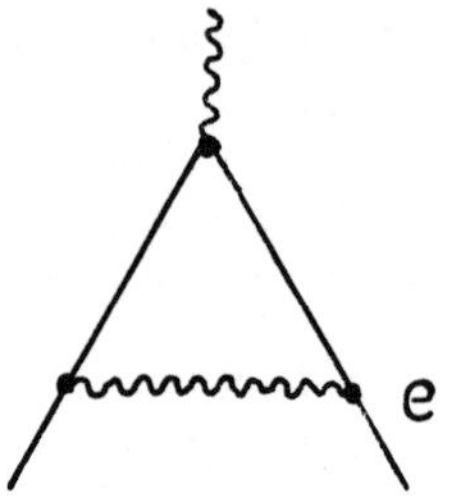

and the 7 fourth order diagrams

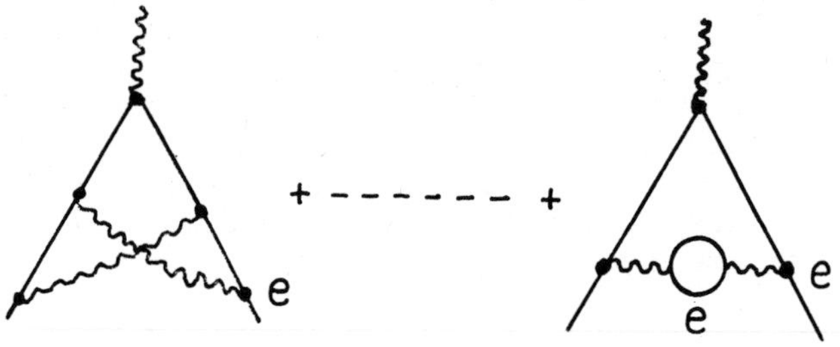

The evaluation of the sixth order coefficient has been the result of the heroic efforts of a number of people. The values quoted by the various groups are

		1.195(26)	Cvitanovic and Kinoshita[100]
a_3^e	=	1.21 (7)	Levine and Wright[101]
		1.01 (6)	Carroll and Yao[102]

There seems to be no simple intuitive way of understanding the values of any of the coefficients a_i^e , not even the first. They are therefore of little pedagogical value. Let me anyhow make a few remarks about the technical details. There are altogether 72 diagrams that give non-zero contributions to the anomaly :

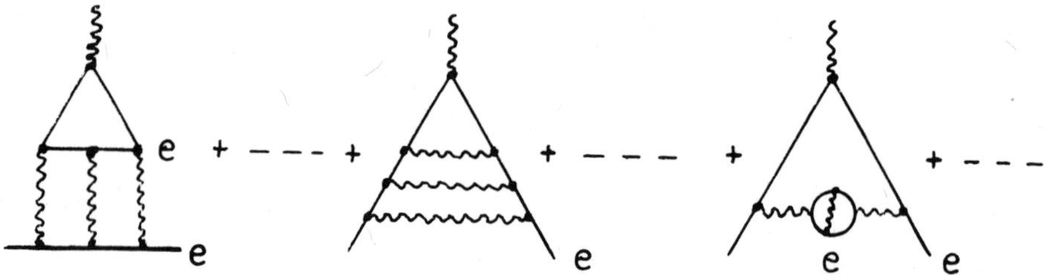

They involve 9 propagators and therefore 8 Feynman parameters of which 1 is trivial. Most of the diagrams consequently lead to 7 dimensional integrals. Neither the numerator algebra nor the numerical integration over the parameter space are trivial. The algebraic work is usually aided by large scale algebraic manipulation programs[99] (Schoonschip, Reduce, Ashmedai) with special built-in facilities for handling γ-matrices. General highly efficient methods for parametrizations[51], subtraction of ultraviolet divergences and cancellation of infrared divergences[103] have been developed. The numerical integration is mostly done by Monte Carlo type methods with selective sampling using adaptive algorithms (Riwiad[104]). The world total cost in computer time for the theoretical evaluation of a_3^e must today rival the cost of the new CERN experiments (megadollars)!

The various contributions to the anomaly are displayed in Table 11.1. We see that the largest uncertainty comes from α in the 2nd order term. If the experiment is improved by a factor of 3 or better this will lead to a competitive value of α. The largest theoretical uncertainty is the one of the sixth order coefficient. If that could be eliminated by analytic calculations[105] the theoretical value of the electron anomaly is ready for a 100 fold increase in the experimental accuracy.

The muon anomaly. Every charged particle polarizes the vacuum around itself. A positively charged particle surrounds itself with a cloud of negative virtual electrons (the positrons are pushed off to infinity). Therefore the effective charge seen at a finite distance is always larger than the physical value of the charge which is seen at very large distances. This explains that vacuum polarization strengthens the Coulomb potential as shown in eq. (9.29). From that equation we deduce that the effective fine structure constant at distance $r \ll 1/m_e$ is

$$\bar{\alpha} = \alpha\left(1 + \frac{2}{3}\,\frac{\alpha}{\pi}\log\frac{1}{m_e r} + \ldots\right) \tag{11.17}$$

It diverges at $r \to 0$, i.e. the bare charge is infinite. The size of the cloud is $1/m_e$.

As I mentioned before a muon interacts with its own radiation field at typical distances $r = 1/m_\mu$ i.e. deeply inside the vacuum polarization cloud. We therefore expect that its anomaly should be calculated using the effective charge corresponding to this distance, i.e.

$$a^\mu \simeq \frac{\bar{\alpha}}{2\pi} = \frac{\alpha}{2\pi} + \frac{1}{3}\left(\frac{\alpha}{\pi}\right)^2 \log\frac{m_\mu}{m_e} + \ldots \tag{11.18}$$

Our intuition does not fail us here. This is indeed the correct result.

The two first coefficients are[99]

$$a_1^{\mu} = a_1^{e} = 1/2 \tag{11.19}$$

$$a_2^{\mu} = a_2^{e} + \frac{1}{3}\log\frac{m_{\mu}}{m_e} - \frac{25}{36} + O\left(\frac{m_e}{m_{\mu}}\right)$$

$$= 1.09426 + a_2^{e} \tag{11.20}$$

The second order coefficient stems from the same diagram as a_1^e. It is a dimensionless number and can consequently not depend on the value of the lepton mass. This is also the reason that a_2^e appears in the 4th order coefficient. The other terms come from the diagram

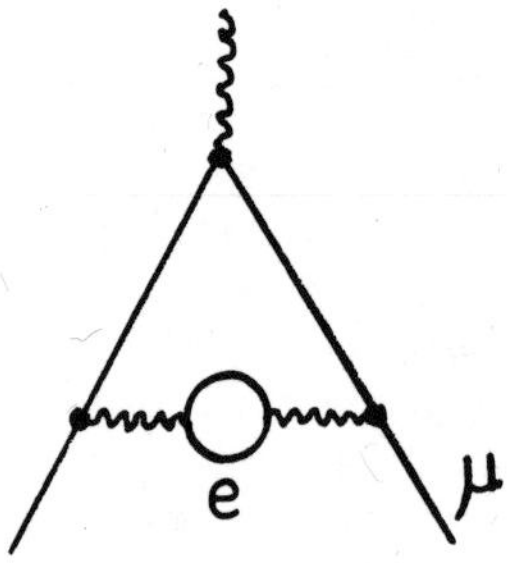

which mixes muons and electrons.

It is possible to formalize the relation (11.18) such that it may be applied in all orders. Consider the diagrams that do not mix muons and electrons and let us replace the free photon propagators by the full photon propagator including all corrections due to electron vacuum polarization (but no corrections due to muons). For example

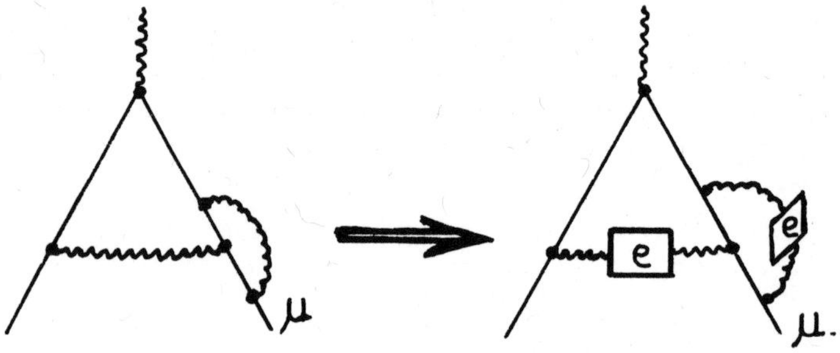

etc. Then the anomaly due to the class of diagrams generated in this way becomes a functional of the function $d(k^2,e,m_e)$ defined in eq.(8.3)

$$a^{VP}(m_\mu/m_e,\alpha) = \varphi\,[m_\mu,\alpha d(k^2,\alpha,\ m_e)] \tag{11.27}$$

This functional depends on m_μ through the muon propagators but only on α through the combination αd because each propagator attaches to two vertices.

We now let $m_e \to 0$ and keep all non-vanishing terms. This defines the asymptotic anomaly $a_\infty^{VP}(m_\mu/m_e,\alpha)$. It is intuitively clear that this is obtained by replacing αd by αd_∞ which satisfies eq.(8.12). The actual proof is rather involved[106]. But then by applying the differential operator $m_e \frac{\partial}{\partial m_e} + \beta(\alpha)\frac{\partial}{\partial\alpha}$ to (11.27) we get

$$\left(m_e \frac{\partial}{\partial m_e} + \beta(\alpha)\alpha\frac{\partial}{\partial\alpha}\right) a_\infty^{VP}\left(\frac{m_\mu}{m_e},\alpha\right) = 0 \tag{11.28}$$

i.e. a Callan-Symanzik equation for the anomaly.

This equation may be solved as in eq. (8.13)

$$a_\infty^{VP}\left(\frac{m_\mu}{m_e},\alpha\right) = \left(\frac{m_\mu}{m_e}\right)^{\beta(\alpha)\alpha\frac{\partial}{\partial\alpha}} b(\alpha)$$

$$= \sum_{n=0}^{\infty} \frac{1}{n!}\log^n\left(\frac{m_\mu}{m_e}\right)\left(\beta(\alpha)\alpha\frac{\partial}{\partial\alpha}\right)^n b(\alpha) \tag{11.29}$$

where $b(\alpha) = a_\infty(1,\alpha)$. From the coefficients of b

$$b(\alpha) = b_1\frac{\alpha}{\pi} + b_2\left(\frac{\alpha}{\pi}\right)^2 + b_3\left(\frac{\alpha}{\pi}\right)^3 + \ldots \tag{11.30}$$

and the coefficients of $\beta(\alpha)$ given in (8.16) we may now calculate the vacuum polarization contribution to the muon anomaly.

$$a_1^{VP} = b_1 \tag{11.31}$$

$$a_2^{VP} = b_2 + C_2\log\frac{m_\mu}{m_e} \tag{11.32}$$

$$a_3^{VP} = b_3 + C_3\log\frac{m_\mu}{m_e} + d_3\log^2\frac{m_\mu}{m_e} \tag{11.33}$$

$$a_4^{VP} = b_4 + C_4 \log \frac{m_\mu}{m_e} + d_4 \log^2 \frac{m_\mu}{m_e} + e_4 \log^3 \frac{m_\mu}{m_e} \tag{11.34}$$

where

$$C_2 = \beta_1 b_1 \tag{11.35}$$

$$C_3 = \beta_2 b_1 + \beta_1 2b_2 \tag{11.36}$$

$$C_4 = \beta_3 b_1 + 2\beta_2 b_2 + 3\beta_1 b_3 \tag{11.37}$$

$$d_3 = \beta_1^2 b_1 \tag{11.38}$$

$$d_4 = \frac{5}{2} \beta_1 \beta_2 b_1 + 3 \beta_1^2 b_2 \tag{11.39}$$

$$e_4 = \beta_1^3 b_1 \tag{11.40}$$

From (11.19) and (11.20) we see that

$$b_1 = a_1^e \tag{11.41}$$

$$b_2 = a_2^e - 25/36 \tag{11.42}$$

Knowing these we may evaluate all logarithmic coefficients except C_4, which depends on b_3.

This coefficient is given by[107)]

$$\begin{aligned} b_3 = a_3^e &+ \frac{1075}{216} - \frac{25}{3} \zeta(2) + 10 \zeta(2) \mathrm{Log}\, 2 \\ &-3\zeta(3) + \frac{11}{6} \zeta^2(2) - \frac{4}{3} \zeta(2) \mathrm{Log}^2 2 \\ &- \frac{1}{9} \log^4 2 - \frac{8}{3} \mathrm{Li}_p (1/2) \end{aligned} \tag{11.43}$$

$$= a_3^e + 1.56602 \ldots = 2.761(26)$$

where $\zeta(p) = \sum_{n=1}^{\infty} \frac{1}{n^p}$ and $\mathrm{Li}_p(x) = \sum_{n=1}^{\infty} \frac{X^n}{n^p}$

The uncertainty is entirely due to the uncertainty in a_3^e .

Inserting these in (11.31) - (11.40) we obtain for the sixth and eighth order cases

$$a_3^{\mu,VP} = a_3^e + 1.94404 = 3.139(26) \tag{11.44}$$

$$a_4^{\mu,VP} = b_4 + 17.27(3) \tag{11.45}$$

The constant b_4 is unknown and will probably remain so. Presumably it is not too large. The number of diagrams contributing to $a_4^{\mu,VP}$ is 304.

But this is not all. In sixth order there are diagrams containing light-by-light scattering insertions of the type

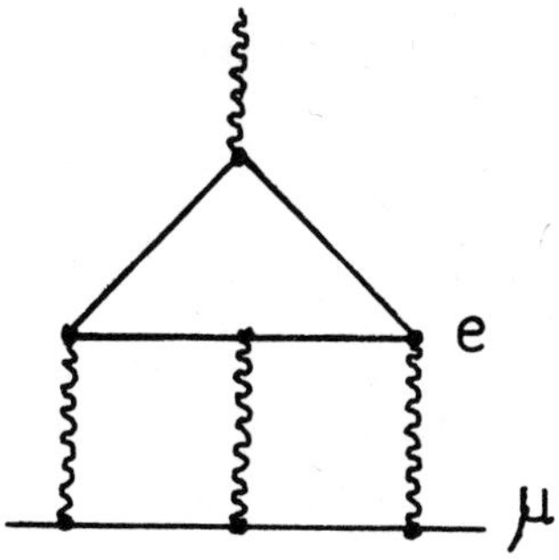

They were surprisingly discovered[108)] to give quite a large contributions to the anomaly. The latest evaluation yields[109)]

$$a_3^{\mu,LL} = 19.79(16) \tag{11.46}$$

an thus adding this to (11.44)

$$a_3^{\mu} = 22.93(16) \tag{11.47}$$

In eight order[110,106,111,107)] the main correction probably comes from diagrams of the type

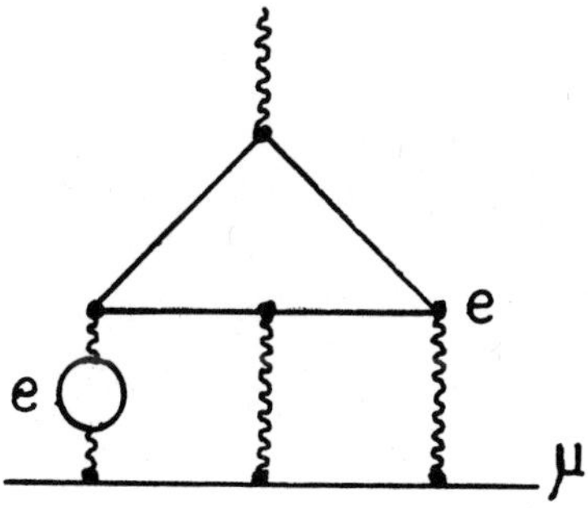

We may try to estimate them[110] by the same technique as was used in (11.18), i.e. replacing α by $_3\alpha$ in the sixth order light-by-light contributions $a_3^{\mu,LL}\left(\frac{\alpha}{\pi}\right) \rightarrow a_3^{\mu LL}\left(\frac{\alpha}{\pi}\right)$

This leads immediately to the estimate

$$a_4^{\mu,LL,VP} \simeq 3 \cdot \frac{2}{3} \log \frac{m_\mu}{m_e} \cdot a_3^{\mu,LL} \simeq 200 \qquad (11.48)$$

A direct numerical calculation yields a result only half as big[112]

$$a_4^{\mu,LL,VP} = 111.1(8.1) \qquad (11.49)$$

but still big enough for the eigth order contributions to compete with the weak-interactions (see table 11.1). For a detailed discussion of eighth order see ref. 106. The total number of diagrams contributing to the difference between muon and electron is 459.

In Table 11.1 strong[90,113] and weak[114] (Weinberg model) interaction contributions have also been included. They both satisfy the scaling law (11.13) with $C_e = C_\mu$. The same is true for the recently calculated gravitational contributions[125] which is of the order of 10^{-45} for the electron and 10^{-40} for the muon.

Table 11.1 Calculation of lepton anomalies

Contribution	electron × 10^9	muon × 10^9
2nd order QED	1 161 408.7 (1.3)	1 161 408.7(1.3)
4th order QED	-1 772.3	4 131.8
6th order QED	15.0 (3)	287.4 (2.0)
8th order QED	∿ -0.03	∿ 3.2
Strong interactions	0.0015	73 (10)
Weak interactions	0.00005	2.0
Total theory	1 159 651.4 (1.3)	1 165 906 (10)
Experiment	1 159 656.7 (3.5)	1 165 895 (29)
Theory-Experiment	-5.3 (3.7)	11 (30)

The value of α^{-1} used is[83)] 137.03612(15) and the value for a_3^e is the one from ref. 100. The eigth order contribution of the electron is estimated to be $\sim -(\alpha/\pi)^4$ while for the muon it is taken from ref.112. The strong interaction contribution is taken from ref.90 and the weak interaction contribution from ref.114 with sin θ_W = 0.4

REFERENCES

1. Max Planck, "On the theory of the energy distribution law in the normal spectrum", German Physical Society December 14, 1900.
2. P.A.M. Dirac, Proc.Roy.Soc. A114, 243 (1928)
3. W. Heisenberg and W. Pauli, Zeits.f.Physik 56, 1(1929)
4. W.A. Bardeen, R. Gastmans and B. Lautrup, Nuclear Physics B46, 319(1972; C. Bouchiat, J. Iliopoulos and Ph.Meyer, Phys.Lett. 42B, 91(1972)
5. A. Peterman and E. Stuckelberg, Helv.Phys.Acta 26 (1953) 499
6. M. Gell-Mann and F.E. Low, Phys.Rev. 95 (1954) 1300
7. C. Callan, Phys.Rev. D2 (1970) 1541
8. K. Symanzik, Comm.Math.Phys. 18 (1970) 227, 23(1971) 49
9. S. Adler, Phys.Rev. D5(1972) 3021
10. H.B.G. Casimir, Proc.Kon.Ned.Akad.Wetenskap 51 (1948) 793
11. M.J. Sparnay, Physica 24 (1950) 751
12. M. Fierz, Helv.Physica Acta 33 (1960) 855
13. E. Fermi, Rendiconti d.R. Acc.dei Lincei (6) 9 (1929) 881
14. J. Bjorken and S. Drell, Relativistic Quantum Mechanics and Relativistic Quantum Fields, McGraw-Hill (1964).
    See also
    G. Källén, Quanten elektrodynamik, Handbuch der Physik Vol. V-1 (1958).
    J.M. Jauch and F. Rohrlich, The theory of photons and electrons, Addison-Wesley (1955)
    E.M. Lifshitz and L.P. Pitaevskii, Relativistic quantum theory, Pergamon Press (1974)
    A.I. Akhiezer and V.B. Berestetskii, Elements of Quantum electrodynamics, Oldbourne Book Co. (1962)
15. G. 't Hooft and M. Veltman, Nucl.Phys. B50, 318 (1972)
16. R.P. Feynman, Acta Physica Polonica, 24 (1963) 697
    L.D. Fadeev and V.N. Popov, Phys.Lett. 25B, 29 (1967)
17. See f. inst. the lecture of R. Gastmans at this conference
18. J.V. Hollweg, Phys.Rev.Lett. 32 (1974) 961
19. For the more recent attempts see
    R. Gastmans and R. Meuldermans, Nucl.Phys. B63, 277 (1973)
20. D. Zwanziger, NYU-TR1/74
21. J.E. Augustin et al., Phys.Rev.Lett. 34 (1975)233
22. B.L. Beron et al., Phys.Rev.Lett. 33(1974)663
    References to previous experiments may also be found in ref.21 and in ref. 23.
23. B. Lautrup, A. Peterman and E. de Rafael, Phys.Rep. 3C(1972)193
24. A.M. Boyarski et al., Phys.Rev.Lett. 34(1975)1357
    See also ref. 25
25. R.L. Ford, Stanford HEPL-748(Jan.1975)
26. H.J. Bhabba, Proc.Roy.Soc. A154 (1935)195
27. H. Newman et al., Phys.Rev.Lett. 32(1974)483
28. N.M. Kroll, Nuovo Cimento 45A(1966) 65
29. L. Paoluzzi, Acta Phys.Polonica B5(1974)839
    See also H.L. Lynch, SLAC PUB-1536

30. N. Meister and D.R. Yennie, Phys.Rev. 130 (1963) 1210
31. E. Etim, G. Pancheri and B. Touschek, Nuovo Cim. 51B (1967)276
M. Greco and G. Rossi, Nuovo Cimento 50 (1967) 168
G. Pancheri, Nuovo Cimento 60 (1969) 321
32. F. Berends, K. Gaemers and R. Gastmans, Nucl.Phys. B57(1973)381
33. F. Berends, K. Gaemers and R. Gastmans, Nucl.Phys. B63(1973)381
34. F. Berends, and R. Gastmans, Nucl.Phys. B61 (1973) 414
35. F. Berends, K. Gaemers and R. Gastmans, Nucl.Phys. B68(1974)541
36. The history is summarized in A. Kraemmer and B. Lautrup NBI-HE-75-4.
37. C.Q. Weizsäcker and E.J. Williams, Zeits.Phys. 88 (1934) 612
38. Proceeding of the conference on two-photon processes, College de France, 1973, in Journal de Physique, supplément, Tome 35, Fasc. C2, 1974
39. See for instance S. Brodsky, T. Kinoshita and H. Terazawa, Phys.Rev. D4 (1971) 1532
40. See P. Kessler in ref. 38 p.98 for the history of this formula
41. G. Barbiellini et al., Phys.Rev.Lett. 32 (1974) 385
42. V.E. Balakin et al., Phys.Lett. 3B (1971) 328, 663
43. N. Cabibbo and G. Parisi, unpublished
44. S. Orito et al., Phys.Lett. B48 (1974) 380
45. References may be found in ref. 46
46. F.J. Dyson, Phys.Rev. 75 (1949) 1736
47. A. Salam, ICTP/74/55
48. W. Zimmermann, Brandeis Lectures, Vol.1, (1970) 397 (MIT Press)
49. W. Pauli and F. Villars, Rev.Mod.Phys. 21 (1949) 434
50. G. 't Hooft and M. Veltman, Nucl.Phys. B44 (1972) 189
51. T. Kinoshita and P. Cvitanovic, Phys.Rev. D10 (1974) 3991
52. J. Schwinger, Phys.Rev. 128 (1962) 2425
53. R.A. Brandt and Ng W-C, NYU-preprint (July 1974)
54. J.C. Ward, Phys.Rev. 78 (1950) 182
55. S.S. Schweber, An introduction to relativistic quantum fields, Harper (1964)
N.N. Bogoliubov and D.V.Shirkov, Introduction to the theory of quantized fields, Interscience (1959)
See also ref. 14.
56. S. Weinberg, Phys.Rev. 118 (1960) 838
57. B. Lautrup (to be published). For similar results see
G. 't Hooft, Nucl.Phys. B61 (1973) 455
M.J. Holwerda, W.L. van Neerven and R.P. van Royen, Nucl.Phys. B75 (1974) 302
58. G.W. Erickson and D.R. Yennie, Annals of Physics 35 (1965) 271, 477 and earlier references herein
59. C.L. Schwartz and J.J. Tiemann, Ann.Phys. (N.Y.) 6 (1959) 178
60. G.W. Ericson, Phys.Rev.Lett. 27 (1971) 780 and unpublished
61. J. Schwinger, Phys.Rev. 82 (1951) 664
62. H.A. Bethe and E.E. Salpeter, Quantum Mechanics of one- and two-electron atoms, Springer (1957)
63. E.H. Wichmann and N.M. Kroll, Phys.Rev. 101 (1956) 843

64. P.J. Mohr, Phys.Rev.Lett. 34 (1975) 1050
65. N. Uehling, Phys.Rev. 48 (1935) 49
66. There is an extensive literature on the subject. A critical review has recently been written by Sundaresan and Watson (ref. 67)
67. P.J.S. Watson and M.K. Sundaresan, Can.Journ.of Phys. 52(1974) 2037
68. J. Blomquist, Nucl.Phys. B48 (1972) 95
69. R. Barbieri, J.A. Mignaco and E. Remiddi, Nuovo Cim.Lett. 3 (1970) 588
70. S.J. Brodsky and G.W. Erickson, Phys.Rev. 148 (1966) 26
71. H. Grotch and D.R. Yennie, Rev.Mod.Phys. 41 (1969) 350
72. Th. Fulton, D. Owen and W. Repko, Phys.Rev.Lett. 26 (1971) 61
73. R.C. Barrett, D.A. Owen, J. Calmet and H. Grotch, Phys.Lett. B47 (1973) 297
    J. Friar and L. Negele, Phys.Lett. B46 (1973) 5
74. S. Klarsfeld and A. Maquet, Phys.Lett. B43 (1973) 201
75. See for instance ref. 62, 72 and also ref. 76
76. Th. Fulton, D.A. Owen and W. Repko, Phys.Rev.Lett. 24 (1970) 1035; 25 (1970) 782; Phys.Rev. A4 (1971) 1802
77. C.F. Cho, Nuovo Cimento 23A (1974) 557
78. E. de Rafael and J. Rosner, Annals of Phys. 82 (1974) 369
79. S.R. Lundeen and F.M. Pipkin, Phys.Rev.Lett. 34 (1975) 1368
80. M. Leventhal, Nucl.Instrum.Methods 110 (1973) 343
81. M. Leventhal and P.E. Havey, Phys.Rev.Lett. 32 (1974) 808
82. J. Calmet and E. de Rafael, Phys.Lett. 56B (1975) 181
83. E.R. Cohen and B.N. Taylor, J.Phys.Chem.Ref.Data 2 (1973) 663
84. D.A. Owen, Phys.Rev.Lett. 30 (1973) 887
85. R. Barbieri, P. Christillin and E. Remiddi, Phys.Lett. B43 (1973) 411
86. M.A. Samuel, Phys.Rev. A10 (1975) 1450. See also Th. Fulton, Phys.Rev. A7 (1973) 377
87. A.P. Mills, Jr. and C.H. Bearman, Phys.Rev.Lett. 34 (1975) 246
88. M.A. Stroscio and J.M. Holt, Phys.Rev. A10 (1974) 749
89. R.H. Beers and V.W. Hughes, Bull.Am.Phys.Soc. 13 (1968) 633
90. J. Bailey et al., Phys.Lett. 55B (1975) 420
91. J.C. Wesley and A. Rich, Phys.Rev. A4 (1971) 1341
92. V.W. Hughes, International Conference on Atomic Physics 3 (1973) 1.
93. A. Rich and J.C. Wesley, Rev.Mod.Phys. 44 (1972) 250
94. F. Combley and E. Picasso, Phys.Rep. 14 (1974) 1
95. References may be found in ref. 90
96. The effect is calculable (ref. 97) and of the order of 10^{-12}
97. B. Lautrup and E. de Rafael, Phys.Rev. 174 (1968) 1835
98. There are exceptions to this rule. See ref. 23 for a discussion.
99. See ref. 23 for references
100. P. Cvitanovic and T. Kinoshita, Phys.Rev. D10 (1974) 3007
101. M.J. Levine and J. Wright, Phys.Rev. D8 (1973) 3171

102. R. Carroll and Y.-P. Yao, Phys.Lett. 48B (1974) 125
103. P. Cvitanovic and T. Kinoshita, Phys.Rev. D10 (1974) 3991
104. See B. Lautrup, Proceedings of the Second Colloquium on Advanced Computer Methods in Theoretical Physics (ed. A. Visconti, Marseille, 1971)
105. Some have already been done. See for instance ref. 100 for references.
106. B. Lautrup and E. de Rafael, Nucl.Phys. B70,(1974) 317
107. See for instance R. Barbieri and E. Remiddi, CERN-TH-1895 (August 1974) and references herein.
108. J. Aldins et al., Phys.Rev. D1 (1970) 2378
109. J. Calmet and A. Peterman, CERN-TH-1978
110. B. Lautrup, Phys.Lett. 38B (1972) 408
111. M. Samuel, Phys.Rev. D9 (1974) 2913
112. J. Calmet and A. Peterman, CERN-TH-1998
113. A. Bramon, E. Etim and M. Greco, Phys.Lett. 39B (1972) 514
114. K. Fujikawa, B.W. Lee and A.I. Sanda, Phys.Rev. D6 (1972) 292
115. A.P. Mills, Jr., S. Berko and K.F. Canter, Phys.Rev.Lett. 34(1975) 1541
116. T. Fulton and P.C. Martin, Phys.Rev. 95 (1954) 811
117. E. Compani, Nuovo Cimento Lett. 4 (1970) 982
J. Bernabeu et al., Nucl.Phys. B75 (1974) 59
118. A. Bertin et al., Phys.Lett. 55B (1975) 411
119. M.S. Dixit et al., Phys.Rev.Lett. 27 (1971) 878
G. Backenstoss et al., Phys.Lett. 43B (1973) 539
120. L. Tauscher et al., CERN-preprint (June 1975)
121. M.Y. Chen, Phys.Rev.Lett. 34 (1975) 341
122. L. Wilets and R.A. Rinker, Phys.Rev.Lett. 34 (1975) 339
123. E. Borie, SIN preprint (May 1975)
124. N.M. Kroll, International Conference on Atomic Physics 3 (1973) 33
125. F.A. Berends and R. Gastmans, Phys.Lett. 55B (1975) 311
126. For a proper comparison with experiment it is still necessary to evaluate some higher order corrections. They are very difficult to calculate and will presumably require extensive use of computers for the algebra and for the numerics.

Comments

p.6 The use of an exponential cut-off in eq.(2.3) is not very well justified. In principle one ought to use the same type of cut-off in the whole theory. It is, however, not so easy to see how f.inst. dimensional regularization should be applied. The resulting pressure is presumably independent of the cut-off used. This is the case if one uses a sharp frequency cut-off. For a real cavity the cut-off naturally occurs when the metal ceases to be ideal.

STATUS OF THE ANALYTIC CALCULATIONS OF QED SIXTH ORDER RADIATIVE CORRECTIONS TO THE ELECTRON ANOMALY

E. REMIDDI [+]

Laboratoire de Physique Théorique,

Ecole Normale Supérieure, Paris [++]

Let us recall, from Lautrup's lectures[1], that in QED the anomalous magnetic moment a (anomaly) of the electron is obtained as a (formal) power series in the fine structure constant α:

1) $a = a_1 \left(\frac{\alpha}{\pi}\right) + a_2 \left(\frac{\alpha}{\pi}\right)^2 + a_3 \left(\frac{\alpha}{\pi}\right)^3 + \ldots$

The first two coefficients are

$a_1 = \frac{1}{2}$ ref.2)

2) $a_2 = \frac{197}{144} + \frac{1}{2}\zeta(2) - 3\zeta(2)\,\ell g2 + \frac{3}{4}\zeta(3) = -0.32838\ldots$ ref. 3)

where $\zeta(p) = \sum_{n=1}^{\infty} n^{-p}$ and $\zeta(2) = \pi^2/6$. The third coefficient in eq.1) is known only numerically. Its most detailed calculation was done by Levine and Wright [4], the most accurate by Cvitanovic and Kinoshita[5], giving

3) $a_3 = 1.195 \pm 0.026$ (2 % error).

Beside the numerical error in a_3, in eq. 1) one has two other sources of indeterminacy : the value of α and the remainder of the series. The latter can be estimated[1] of the order of $-(\alpha/\pi) \simeq -0.03 \times 10^{-9}$. The precision in the experimental value of α has been recently improved by an accurate measurement[6] of the proton gyromagnetic ratio. By suitably combining this quantity with other constants, such as the Rydberg constant

---

+ Permanent and from 1.11.1975 address : Istituto di Fisica, v. Irnerio 46, I 40126 Bologna.

++ Postal address : LPTENS, r. Lhomond 24, F.75231, Paris Cedex 05.

and the proton/electron mass ratio, all known to very high precision, one obtains

4) $\alpha^{-1} = 137.035987(28)$

with an error of 0.21 ppm ! Previous measurements based on the Josephon effect gave a 2 ppm error mainly due to the uncertainty in the definition of the Volt. (The Volt is presently defined[7] as the difference of potential corresponding to a Josephon frequency of 483.594000 THz).

With the above values one has

$$
\begin{aligned}
a(th) &= 1\,159652.51 \times 10^{-9} \\
&\pm \quad 0.24 \times 10^{-9} \quad \text{from } \alpha \\
&\pm \quad 0.33 \times 10^{-9} \quad \text{from } a_3 \\
&\pm \quad 0.03 \times 10^{-9} \quad \text{from the remainder in eq.1).}
\end{aligned}
\tag{5)}
$$

The present experimental value is [8]

$$a(exp) = 1159656.7\ (3.5) \times 10^{-9}\ .$$

The experimental error is expected to decrease by a factor of 10 within a couple of years[9] and hopefully even more when magnetic resonance techniques will be developped for the electron too. According to eq. 5) a measurement of the electron anomaly up to a precision of 0.1×10^{-9}, i.e. 0.1 ppm, can be regarded as an independent measure of α , provided the error in a_3 is correspondingly reduced. This seems an almost impossible task by numerical methods, so that a_3 must be evaluated analytically. There is no doubt that the analytic evaluation of a_3 is one of the biggest computational challenges as well as an excellent occasion to look in the mysteriou patterns of QED radiative corrections. The challenge and the insigh in the analytic structure of QED are, for many of the people interested in the calculation, as important as its possible practical implications.

In the usual perturbative expansion of QED, a_3 receives contributions from 72 Feynman graphs with 3 loops. Everybody knows how to evaluate one-loop Feynman graphs by means of, say, Feynman parameters techniques. Textbooks guarantee that the Feynman parameters machinery can be straightforwardly carried out for any number of internal loops : when this is done, one is left with an expression of the form

$$a = \int dx_1 \ldots dx_n\ N(x_i)/D(x_i)$$

where the x_i, usually ranging from 0 to 1, are the Feynman parameters, n is the number of propagators in the considered graph (9 for a_3), $N(x_i)$, $D(x_i)$ polynomials. Their complexity increases quickly with the number of the loops, the analytic integration becomes harder and harder and as matter of fact Feynman parameters are not being used in present attempts to evaluate analytically a_3.

They are better suited, and have been actually used for the numerical integration, where a lot of work is however to be done to get rid of ultraviolet and infrared divergences and to prepare an as smooth as possible integrand.

The main purpose of this lecture is to give a sketch of the method which has been developped in working out the analytic evaluation of some sixth order contributions[10,11] to the electron anomaly. As will be shown in the following, the essential features of the method are :

i) dispersion relations in the electron energy for suitable vertex form factors at vanishing momentum transfer;
ii) a number of recursion formulas for integrals with square roots;
iii) the theory of Nielsen polylogarithms (of which the Euler dilogarithm or Spence function is a particular case);
iv) last but not least the program SCHOONSCHIP[12], by M. Veltman, for the bookkeeping of the intermediate results, working out of algebra (e.g. traces of Dirac metrices), integrations by parts, changes of variables, etc.

To illustrate our method, let us consider the off-shell vertex amplitude $v_\mu(p,\Delta)$ for the QED vertex of fig. 1.

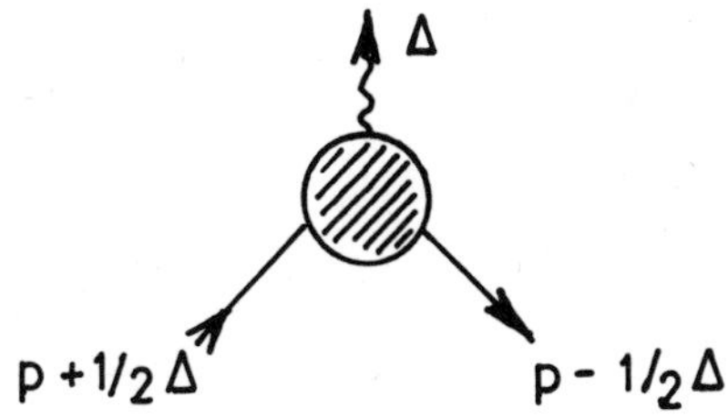

Fig. 1. The vertex amplitude.

The anomaly is a static quantity, so that we can look at it retaining only the zeroth and first order terms in Δ_ν .

6) $v_\mu(p,\Delta) = V_\mu(p) + \Delta_\nu T_{\mu\nu}(p) + \ldots$

On invariance grounds one finds the following decomposition in invariant amplitudes (electron mass = 1) :

$$V_\mu(p) = V_1\gamma_\mu + V_2(\gamma_\mu + ip_\mu) + V_3 ip_\mu(i\not{p}+1) + V_4 i[\gamma_\mu , \not{p}]$$

7)

$$T_{\mu\nu}(p) = T_1 \frac{i}{2}[\gamma_\mu,\gamma_\nu] + T_2 \frac{i}{4}\{[\gamma_\mu,\gamma_\nu],\not p\} + T_3 i(p_\mu\gamma_\nu - p_\nu\gamma_\mu)\not p$$

$$+ T_4 \frac{i}{2}[p_\mu\gamma_\nu + p_\nu\gamma_\mu, \not p] + T_5 i\delta_{\mu\nu} + T_6\delta_{\mu\nu}\not p$$

$$+ T_7\, i p_\mu p_\nu + T_8 p_\mu p_\nu \not p + T_9(p_\mu\gamma_\nu - p_\nu\gamma_\mu) + T_{10}(p_\mu\gamma_\nu + p_\nu\gamma_\mu),$$

with $[A,B] = AB-BA$.

All the above form factors depend only on the variable $-p^2=u$: $V_i \equiv V_i(u)$, $T_i \equiv T_i(u)$ (V_4 and T_j for $j \geqslant 5$ actually vanish, having wrong charge conjugation properties). They can be extracted by means of projectors which are easily worked out : for instance :

$$V_1(u) = -\frac{1}{12u}\,\mathrm{Tr}\,([\,(-u+1)\gamma_\mu + 3ip_\mu + \frac{4-u}{u}\,p_\mu\not p\,]\,V_\mu(p))$$

$$T_1(u) = \frac{i}{24u}\,\mathrm{Tr}\,([\,\frac{1}{2}\,u(\gamma_\mu\gamma_\nu-\gamma_\nu\gamma_\mu) + \not p(p_\mu\gamma_\nu-\gamma_\nu p_\mu)]\;T_{\mu\nu}(p)) \quad \text{etc.}$$

By sandwitching eqs. 6,7) between free electron spinors, the anomaly is found to be

8) $a = V_2(m^2) - 2\,T_1(m^2) - 2T_2(m^2)$.

All the form factors satisfy dispersion in u of the usual type

9) $$V_i(-p^2) = \frac{1}{\pi}\int \frac{du}{p^2+u}\ \mathrm{Im}\,V_i(u) \qquad \text{etc.}$$

($V_1(u)$ actually requires the vertex renormalization subtraction). Their discontinuities can be obtained by means of the Cutkosky rule, and they have in general the same analytic structure as the related self-mass amplitude obtained by suppressing the external photon line. To be specific, consider the fourth order graphs

Fig.2. A set of fourth order graphs.

and draw them in the $\Delta \to 0$ limit of eq. 6) as

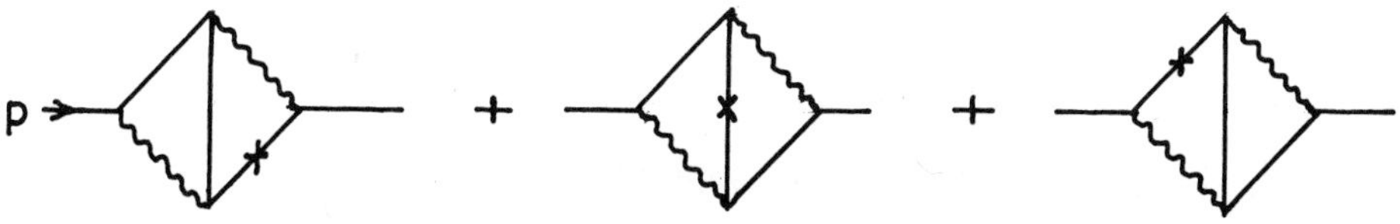

Fig.3. The graphs of fig. 2) in the $\Delta \to 0$ limit.

The related self-mass graph is

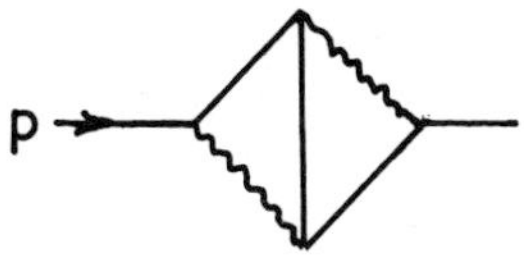

Fig.4. The self-mass corresponding to the vertex graphs of fig.3.

Let us write its amplitude as

$$10)\quad i\Sigma\,(p) = i(i\not{p}+1)\,[S_1(u) + (i\not{p}+1)S_2(u)]\,,$$

where we have performed mass renormalization, not yet the wave function renormalization. The form factors $S_i(u)$ and $V_i(u)$, $T_i(u)$ have the same analytic properties, in particular the same thresholds, and can be expressed in terms of the same mathematical functions. At variance with the self-mass, in a vertex for $\Delta \to 0$ square denominators of propagators appear; at zeroth order in Δ, the propagators just before and after the interaction with the external photon give

$$\cdot\!\!-\!\!\times\!\!-\!\!\cdot \;\simeq\; \frac{1}{(q-\frac{1}{2}\Delta)^2+1}\,\gamma_\mu\,\frac{1}{(q+\frac{1}{2}\Delta)^2+1} \;\to\; \frac{\gamma_\mu}{(q^2+1)^2} + O(\Delta^2)$$

while in first order, for any other electron propagator one has

$$\frac{1}{(q\pm\frac{1}{2}\Delta)^2+1} \;\to\; \frac{1}{q^2+1} \mp \frac{q\Delta}{(q^2+1)^2} + O(\Delta^2\,).$$

To obtain usual dispersion relations and cutting rules for squared denominators it is sufficient to put

$$11)\quad \int d^4q \ldots \frac{1}{(q^2+1)^2} = -\frac{\partial}{\partial\mu^2}\int d^4q \ldots \left(\frac{1}{q^2+\mu^2} - \frac{1}{q^2+M^2}\right)_{\mu^2=1},$$

where $-1/(q^2+M^2)$ guarantees convergence of r.h.s. before differentiating, but does not play any other role, then to write down dispersion relations for the r.h.s. of eq.11) and finally to differentiate with respect to μ^2 at $\mu^2 = 1$.

The Ward identity between vertex and self mass is

$$12)\quad \Delta_\mu V_\mu(p,\Delta) = i\sum(p + \tfrac{1}{2}\Delta) - i\sum(p - \tfrac{1}{2}\Delta)$$

It is easy to verify that it holds also if we take for the vertex and self-mass amplitudes only the graphs of figs. 3 and 4. In terms of the expansions eqs. 7,10) to zeroth order in Δ eq.12) gives

$$13)\quad \begin{aligned} V_1(u) &= -S_1(u) + 2(u-1)S_2'(u) \\ V_2(u) &= -2S_2(u) - 2(u-1)S_2'(u) \\ V_3(u) &= -2S_1'(u) - 4S_2'(u) \\ V_4(u) &= 0 \end{aligned}$$

Similar relations can be worked out at first order in Δ. Note however that the amplitudes $T_1(u)$, $T_2(u)$, necessary for determining the anomaly eq.8) do not appear in the l.h.s. of eq.12), as the corresponding tensors in the expansion eq.7) are antisymmetric in μ,ν. Eqs.13) can be of some use in an explicit calculation as they provide with checks between vertex and self-mass form factors, but they are not sufficient, of course, to express the anomaly in terms of self-mass quantities only.

Consider now any form factor, say $V(u)$, of the vertex or self-mass graphs of figs. 3,4). Its discontinuity is obtained by cutting the graph in all possible ways, i.e.

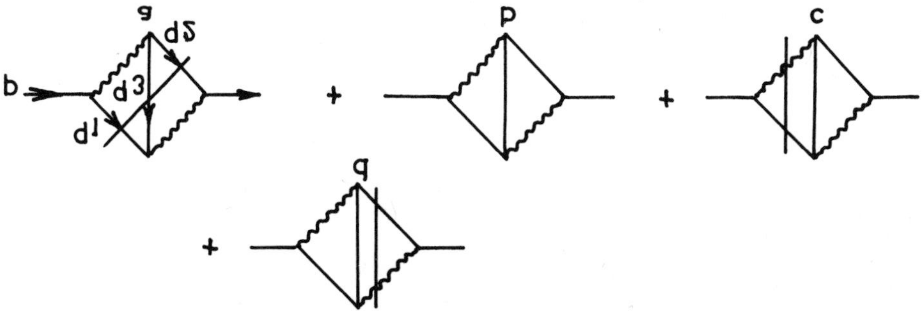

Fig.5. The way of cutting the graphs of figs. 3,4.

We will perform all the integrations in terms of the original loop momenta, never introducing Feynman parameters.
We recall that in a cut graph a cut line stands for a mass shell δ-function, so that one has to deal everywhere with phase space integrations. The basic two-body phase space in the p-rest frame for $-p^2 = u \geq (m+\mu)^2$ is

14)
$$\int d^4q\ \theta(q_o)\delta(q^2+m^2)\theta(p-q_o)\delta((p-q)^2+\mu^2) =$$
$$= \int dq_o|\vec{q}|^2\ d|\vec{q}|\ d\Omega\delta^+(|\vec{q}|^2-q_o^2+m^2)\delta^+(-u+2\sqrt{u}\ q_o-m^2+\mu^2)$$
$$= \frac{1}{8}\ \frac{R(u,m^2,\mu^2)}{u} \int d\Omega$$

where

15) $$R(u,m^2,\mu^2) = [\,u^2 + m^4 + \mu^4 - 2um^2 - 2u\mu^2 - 2m^2\mu^2\,]^{1/2}$$

Note the simple particular cases

16) $$R(u,m^2,0) = u-m^2, \qquad R(b,1,1) = \sqrt{b(b-4)}\ .$$

In the first 3-electron cut of fig.5a) it is convenient to introduce the variable $b \equiv -(q_2+q_3)^2$ and to perform the q_2 and q_1 integrations in the (q_2+q_3) and p rest-frames respectively. Angular integrations give a logarithm or are trivial and the contribution Im $V^{(3e)}(u)$ to the discontinuity of V(u) from this cut has the form

$$\text{17)}\quad \mathrm{Im}V^{(3e)}(u) = \int_4^{(\sqrt{u}-1)^2} db\; R(u,b,1)\;\sqrt{b(b-4)}\; f(u,b)$$

where f(u,b) is a rational function of u,b or a rational function times the quantity

$$\frac{1}{R(u,b,1)\sqrt{b(b-4)}}\; \lg \frac{b(u-b+3)+R(u,b,1)\;\sqrt{b(b-4)}}{b(u-b+3)-R(u,b,1)\;\sqrt{b(b-4)}}\;.$$

As eq. 17) contains the square roots of two second order polynomials in b, the b integration is ellyptic and practically impossible. Let us observe however that, if we are not interested in $\mathrm{Im}\,V^{(3e)}(u)$ itself, but rather on an integral of the form $\int du\, \varphi(u)\mathrm{Im}\, V^{(3e)}(u)$ (with $\varphi(u) = 1/(u-1)$, for instance, one gets the contribution to the form factor at u=1 from the 3-electrons cut (see also eq. 27)),one can exchange the order of integration, obtaining

$$\int du\varphi(u)\;\mathrm{Im}\; V^{(3e)}(u) = \int_9^{\infty} du\; \varphi(u) \int_1^{(\sqrt{u}-1)^2} db\; R(u,b,1)\sqrt{b(b-4)}f(u,$$

18)

$$= \int_4^{\infty} db \int_{(\sqrt{b}+1)^2}^{\infty} du\; \varphi(u)\; R(u,b,1)\; \sqrt{b(b-4)}f(u$$

The u-integration is now possible, as the integrand involves only the square root of a second order polynomial in u and one can hope that the result allows to carry out in turn also the b-integration. As a simple explicit example, consider

$$\int_9^{\infty} du \int_4^{(\sqrt{u}-1)^2} \frac{db}{R(u,b,1)\;\sqrt{b(b-4)}}\; \frac{b-1}{ub} =$$

$$\int_4^{\infty} \frac{db}{\sqrt{b(b-4)}}\; \frac{b-1}{b} \int_{(\sqrt{b}+1)^2}^{\infty} \frac{du}{R(u,b,1)}\; \frac{1}{u} = \frac{1}{2} \int_4^{\infty} \frac{db}{\sqrt{b(b-4)}}\; \frac{\lg b}{b} =$$

Working out of integrals involving the square root of a second order polynomial is greatly speeded up by the "integration by parts" formula

$$\int \frac{dv}{R(v,b,1)} \frac{1}{(v-a)^n} f(v) =$$

$$- \frac{1}{n-1} \frac{1}{R^2(a,b,1)} \left\{ \frac{R(v,b,1)}{(v-a)^{n-1}} + \int \frac{dv}{R(v,b,1)} \right. \tag{20}$$

$$\left. \left[\frac{n-2}{(v-a)^{n-2}} + (3-2n) \frac{b-a+1}{(v-a)^{n-1}} \right] - \int dv \frac{R(v,b,1)}{(v-a)^{n-1}} \frac{\partial}{\partial v} \right\} f(v).$$

It can be easily derived starting from

$$\int dv \frac{\partial}{\partial v} \left[\frac{R(v,b,1)}{(v-a)^{n-2}} f(v) \right] = \frac{R(v,b,1)}{(v-a)^{n-1}} f(v) ,$$

carrying out explicitly the derivative in the l.h.s. and then rewriting as in eq. 20). It is valid for indefinite integrals, for any value of a,n and for any function f(v) for which the formula keeps a meaning. Eq. 20) expresses the integral of the l.h.s. in terms of finite parts and "simpler" integrals which involve smaller powers of (v-a) in the denominator and/or the derivative of f(v) - if f(v) is, say, a constant or a logarithm, the derivative is zero or a fraction. Particular cases of eq. 20) for a = 0,1 are found in section 5 of ref. 10), as well as analogous formulas for ellyptic integrals. It is convenient to use eq. 20) as much as possible, until n=0,1 is obtained. To proceed further, it is necessary to remove the square root with a change of variable. For integrating in u an expression containing R(u,b,1) for $u \geq (\sqrt{b}+1)^2$ it is convenient to use the "natural" integration variable

$$x = x(u,b) = \frac{u-b-1-R(u,b,1)}{2\sqrt{b}} \tag{21}$$

giving

$$u = \frac{(\sqrt{b}+x)(1+\sqrt{b}x)}{x} , \quad R(u,b,1) = \sqrt{b} \, \frac{1-x^2}{2x} , \quad \int_{(\sqrt{b}+1)}^{U} \frac{du}{R(u,b,1)} =$$

$$\int_{x(u,b)}^{1} \frac{dx}{x} \quad \text{etc.}$$

By introducing in turn the "natural variable" for b

$$y \equiv x(b,1) = \frac{b-2-\sqrt{b(b-4)}}{2}$$

one finds for instance the remarkable simplification

$$\lg \frac{b(u-b-3)+R(u,b,1)\sqrt{b(b-4)}}{b(u-b+3)-R(u,b,1)\sqrt{b(b-4)}} = 2\lg \frac{1+x\sqrt{y}}{x+\sqrt{y}} \quad .$$

Once the square roots have disappeared, one must finally try to perform the analytic integrations. In all the cases in which we succeeded till now, the result was obtained in terms of Nielsen's generalized polylogarithms[13]. They are defined as

$$22)\quad S_{n,p}(z) \equiv \frac{(-1)^{n+p-1}}{p!(n-1)!} \int_0^1 \frac{dt}{t} \lg^{n-1}t \, \lg^p(1-tz)$$

or by the generating function

$$_2F_1(\alpha,\beta;1+\beta;z) = 1 - \sum_{p=1}^{\infty} \sum_{n=1}^{\infty} (-\beta)^n \, ^pS_{n,p}(z).$$

A particular case is the Euler dilogarithm (or Spence function)

$$23)\quad Li_2(z) \equiv S_{1,1}(z) = -\int_0^1 \frac{dt}{t} \lg(1-tz) = -\int_0^z \frac{dt}{t} \lg(1-t)$$

The basic property of these function is the differentiation rule

$$24)\quad \frac{d}{dz} S_{n,p}(z) = S_{n-1,p}(z) \quad n>1 \; , \; \frac{d}{dz} S_{1,p}(z) = \frac{(-1)^p}{p!} \lg^p(1-z)$$

which is easily obtained by differentiating and integrating by parts the definition eq. 22). All the known identities between di-logarithms and their generalizations can be proven checking them for a particular value of the argument, then differentiating and checking if the identity holds for the derivatives (which are polylogarithms with lower indices or simply logarithms) - iterating the procedure when needed. Similarly, many definite integrals involving a parameter can be most conveniently worked out by differentiating first with respect to the parameters, then integrating the (hopefully simpler) expression obtained and finally guessing the primitive with respect to the parameter. As an example, consider the "unknown" integral

$$25) \qquad X(z) \equiv \int_0^1 \frac{dt}{t+\frac{1}{z}} \lg t$$

Differentiation gives

$$X'(z) = \frac{1}{z^2} \int_0^1 \frac{dt}{(t+\frac{1}{z})^2} \lg t = (\text{by parts}) = \frac{1}{z} \lg(1+z)$$

so that, by comparison with eqs. 23,24)

$$X(z) = -Li_2(-z),$$

valid also at z=0 and so valid for any z. (In this particularly simple example, the result is immediately obtained by integrating eq.25) by parts). The (g-2) contributions which are at present known analytically all consist of Nielsen's functions of argument ± 1, 1/2; in turn, these particular values of the polylogarithms can very often be expressed in terms of Rieman ζ-functions of integer argument. More details are given in ref. 14).

The remaining cuts of fig. 5) have infrared (IR) divergences. This means that if one attempts to evaluate them with massless photon propagators, infinite meaningless results are obtained. The standard way of dealing with IR divergences is to parametrize them giving the photon a "fictitious and small" mass λ, so that everything is well defined, to carry out the calculations and then to take the $\lambda \to 0$ limit. IR divergences show up as powers of $\lg\lambda$. When collecting results from all cuts and graphs contributing to a physical (finite) quantity, the $\lg\lambda$'s compensate and the physical result is finite - as expected of course if the theory makes sense. The $\lambda \to 0$ limit is by no means trivial, especially in the dispersive approach we use. As an example, it is left to the reader to show that

26)

$$\int_{(1+2\lambda)^2}^{\infty} \frac{dv}{(v-1)^2} \int_{(1+\lambda)^2}^{(\sqrt{v}-\lambda)^2} \frac{db}{b-1}$$

$$\lg \frac{2b\lambda^2+(b+1-\lambda^2)(v-b-\lambda^2)+R(v,b,\lambda^2)R(b,1,\lambda^2)}{2b\lambda^2+(b+1-\lambda^2)(v-b-\lambda^2)-R(v,b,\lambda^2)R(b,1,\lambda^2)} =$$

$$-\lg\lambda - \frac{3}{4}\zeta(2) + \frac{1}{2}, \quad \lambda << 1 \ .$$

A more recent and very promising approach to the IR problem is based on the n-dimensional regularization scheme, which consists in suitably generalizing Feynman graphs amplitudes from 4 to n (continuous) space-time dimensions.
The method has been proposed and used by 't Hooft and Veltman[15] to work out the renormalization of gauge theories. For a sufficiently small $n\leqslant 4$, Feynman graphs integrals become Ultra-Violet convergent. The original U.V. divergences of the unrenormalized amplitudes show up as poles at integer values of n. The poles at n=4 are then removed by the usual renormalization subtractions and in the more finite $n\to 4$ limit the renormalized theory is obtained. The great advantage of the n-dimensional regularization for gauge theories consists in the fact that Ward identities hold for any value of n, and therefore also in the renormalized theory.

For n>4, conversely, the method can be used to parametrize the infrared divergences[16]. Note that, at least in QED, it is not possible to use the n-dimensional regularization for parametrizing both UV and IR divergences, as in the first case one must take n<4, in the second n>4 and the two sets of values of n do not overlap. In the dispersive approach we use to the explicit analytic evaluation of Feynman graphs the great advantage of the n-dimensional regularization consists in the use of a strictly massless photon propagator, so that in the phase space factors referring to photons square roots disappear - see eqs. 15,16). The n-dimensional regularization can be used profittably for evaluating the discontinuity from the cuts of fig. 5b,c,d) of any of the form factors of figs. 3,4), say V(u). For $u = -p^2 \neq m^2$ it is IR finite[17] - the internal fermion propagators $1/[(p+k)^2+m^2]$ develop the I.R. behaviour $1/(2pk)$ only if $p^2+m^2=0$. As the whole form factor is I.R. finite, its imaginary part ImV(u), $u>m^2$ is also finite. Its separate cuts fig. 5 b,c,d), on the contrary, have I.R. divergences, which of course cancel in the sum. We found it very convenient to work out the separate cuts with the n-dimensional regularization for $n=4+\eta$, $\eta>0$ (UV regularization and renormalization, as said above, are to be performed by some other method). In $4+\eta$ dimensions, write

$$d^{4+\eta}q = dq_0 \; |\vec{q}^{\,2}|^{2+\eta} \; d|\vec{q}| \; (\sqrt{1-\cos^2\theta})^\eta \; d(\cos\theta)\; d\varphi(\eta)$$

The basic phase space integral(eq.14) for a massless particle then becomes

$$\int d^{4+\eta}q\theta(q_0)\delta(q^2)\theta(p_0-q_0)\delta((p-q)^2+\mu^2) =$$

$$\frac{1}{8} \; \frac{(u-\mu^2)^{1+\eta}}{u^{1+\frac{1}{2}\eta}} \int(\sqrt{1-\cos^2\theta})^\eta \; d\cos\theta \; d\varphi(\eta)$$

(as a matter of fact, in our case the azimuthal angles can be always factorized). To give an idea of what is going on, note that the logarithm in the l.h.s. of eq. 26), present in the cut of fig. 5,b) in the $\lambda\to 0$ approach, becomes simply lgb. A typical integral which is now encountered is :

$$\int_1^u \frac{db}{(b-1)} \; \frac{(b-1)^\eta}{b^{\eta/2}} \; \frac{(u-b)^\eta}{u^{\eta/2}} \; \frac{\lg b}{b-1} = \frac{1}{\eta} + 2\lg(u-1) - \frac{3u-1}{2(u-1)}\lg u + 1 \; .$$

It can be simply performed by integrating by parts the "infrared factor" $(b-1)^{-1+\eta}$, so obtaining an overall factor $1/\eta$, then expanding all the rest up to first order in η. As matter of fact, in the separate cuts figs. 5b,c,d) simple poles in η are found (which compensate in the sum, as already stated), and the whole explicit expression of the discontinuities of the form factors involve at most $Li_2(1/u)$, $Li_2(-1/u)$.

The techniques discussed till here have been used to work out the sixth order graphs

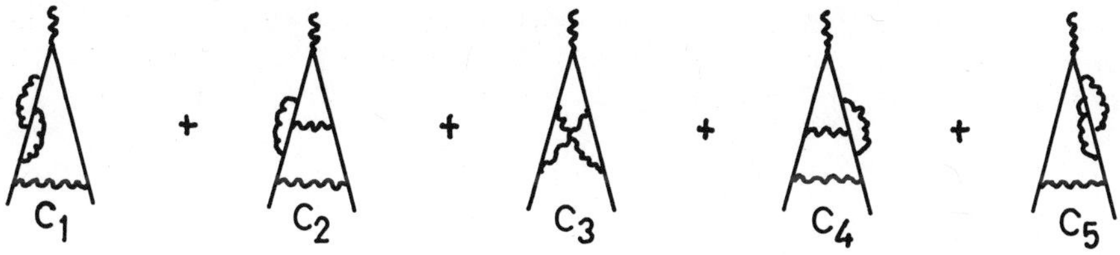

Fig. 6. A class of graphs recently worked out.

(of course $c_1=c_5$ and $c_2=c_1$). They are all of the form

Fig. 7 .

The most general graphs whose anomaly is given by eq. 27).
For graphs c_2, c_3 c_4 the inserted vertex bubble corresponds exactly to the graphs of fig. 3), described by the expansions eqs. 6,7).
The same expansion holds also for the improper vertices inserted in graphs c_1, c_5. The form factors of those improper vertex graphs are easily expressed in terms of the self-mass form factors eq. 10) as

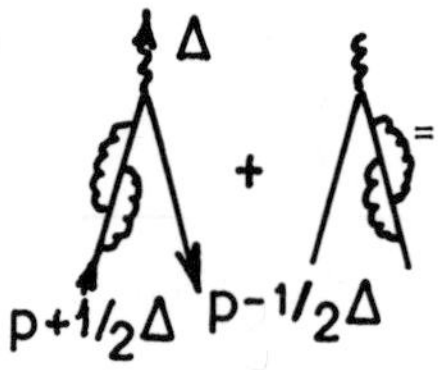

$$= 2S_1(u)\gamma_\mu + 2S_2(u)(\gamma_\mu + ip_\mu) + \Delta_\nu\{S_2(u)\frac{i}{2}[\gamma_\mu,\gamma_\nu] - S_2'(u)i(p_\mu\gamma_\nu - p_\nu\gamma_\mu)\not{p} - S_2'(u)\frac{i}{2}[(p_\mu\gamma_\nu + p_\nu\gamma_\mu),\not{p}]\} + O(\Delta^2)$$

(note that according to eq.8 these improper vertex graphs do not contribute to the anomaly - as expected, and that vertex renormalization of the improper vertex corresponds to wave function renormalization of the self-mass amplitude).
For any inserted (proper or improper) vertex amplitude, use the dispersive representation eq. 9) and insert it in the graph of fig. 7) as

$$\frac{1}{\pi}\int \frac{du}{(p-k)^2+u}\{\mathrm{Im}V_1(u)\gamma_\mu + \mathrm{Im}V_2(u)\,\frac{i}{2}[\gamma_\mu,\gamma_\nu] + \ldots \text{ etc. }\} \quad .$$

The one-loop k-integration is then easily done by any method. The result for the anomaly, using the $\lambda\to 0$ prescription for the IR divergence for easy of comparison with the existing litterature can be written as

27)

$$a = \frac{1}{\pi}\int du \; \{ \frac{lg\lambda}{u-1}[\,ImV_2(u) - 2ImT_1(u) - 2ImT_2(u)] \; + \frac{1}{2}\, ImV_1(u)\,(\frac{1}{u} - \frac{1}{u-1}) \; +$$

$$+ \; ImV_1(u)\,[-\frac{1}{3} - \frac{1}{2u} + \frac{1}{6(u-1)} + (\frac{1}{6} + \frac{u}{3} - \frac{1}{6(u-1)^2}\; lgu - \frac{2u+1}{6}\; lg(u-1)]$$

$$+ \; ImV_2(u)\,[-\frac{1}{2} + \frac{1}{2(u-1)} + (1 + \frac{u}{2} + \frac{1}{2(u-1)^2} + \frac{3}{2(u-1)} lgu$$

$$+ \; (-1 - \frac{u}{2} - \frac{1}{u-1})\; lg(u-1)]$$

$$+ \; ImV_3(u)\;\; [\,\frac{1}{4} - \frac{u}{6} + (-\frac{1}{3} - \frac{u}{3} + \frac{u^2}{6} - \frac{1}{3(u-1)})\, lgu \; + (\frac{1}{3} + \frac{u}{3} - \frac{u^2}{6})\, lg(u-1)]$$

$$+ \; ImT_1(u)\;\; [-\frac{1}{u-1} + (-1 - \frac{1}{(u-1)^2} - \frac{2}{u-1})\, lgu \; + \frac{u+1}{u-1}\; lg(u-1)]$$

$$+ \; ImT_2(u)\;\; [\,\frac{1}{2} - \frac{1}{u-1} + (-\frac{3}{2} - \frac{u}{2} - \frac{1}{(u-1)^2} - \frac{5}{2(u-1)}\;)\, lg\; u \; +$$

$$+ \; (\frac{3}{2} + \frac{u}{2} + \frac{2}{u-1})\;\; lg(u-1)] \;\; \}$$

The first term is the IR part of the graph of fig.7), equal to $lg\lambda$ times the anomaly of the inserted vertex, according to the general result of ref. 18). The next term accounts for the renormalization of the inserted fourth order vertex. It is IR divergent and its calculation for the various graphs separately is horribly involved - order of magnitude harder than the anomaly of the same graphs. On account of the Ward identities eqs. 13), wave function and vertex renormalization counterterms of the inserted self-mass and vertices of the graphs of figs. 4,3) compensate, so that in the sum of graphs c_1 to c_4 of fig. 6) those renormalization can be ignored[11)]. In the remaining part of eq. 27), the bulk of the calculation, straightforward use can be done of the (I.R. finite) form factors discontinuities obtained by the methods outlined above. Omitting for the sake of brevity the terms in $lg\lambda$, which can be found in ref. 4), the results are[19)]

$$a(c_1+c_5) = -\frac{5}{2} + \frac{25}{12}\zeta(2) + \frac{71}{24}\zeta(3) - 3\zeta(2)\lg 2 - \frac{5}{4}\zeta^2(2) - \frac{5}{3}\zeta(2)\lg^2 2$$

28) $$+ \frac{20}{3}\left(a_4 + \frac{1}{24}\lg^4 2\right) = -0.12298130$$

$$a(c_2+c_4) = \frac{235}{288} - \frac{1429}{144}\zeta(2) + \frac{11}{2}\zeta(3) + 8\zeta(2)\lg 2 - \frac{5}{6}\zeta^2(2) - \frac{19}{3}\zeta(2)\lg^2 2$$

$$+ \frac{40}{3}\left(a_4 + \frac{1}{24}\lg^4 2\right) = -0.00714606$$

$$a(c_3) = -\frac{173}{288} + \frac{287}{16}\zeta(2) - \frac{17}{3}\zeta(3) - 18\,\zeta(2)\lg 2 - \frac{37}{24}\zeta^2(2)$$

29) $$+ 11\zeta(2)\lg^2 2 - 14\left(a_4 + \frac{1}{24}\lg^4 2\right) = -1.28697677$$

summing up to

$$a(c) = -\frac{329}{144} + \frac{727}{72}\zeta(2) + \frac{67}{24}\zeta(3) - 13\,\zeta(2)\lg 2 - \frac{29}{8}\zeta^2(2) + 3\zeta(2)\lg^2 2$$

$$+ 6\left(a_4 + \frac{1}{24}\lg^4 2\right) = -1.41710413.$$

All compare very well with known numerical values[4,5]. In above formulas,

$a_4 \equiv Li_4(\frac{1}{2}) = \sum_{n=1}^{\infty} 1/(2^n n^4) = 0.517479061\ldots$. Note the appearance of the combination $a_4 + \frac{1}{24}\lg^4 2$ and the absence of $\zeta(3)\lg 2$, which is however present in intermediate results not shown here.

To conclude, let us review what has been done till now in the analytic evaluation of sixth order corrections to the electron anomaly and guess what will be done in next future. The 72 sixth order vertex graphs can be divided into three classes : light-light, vacuum polarization and three-photon exchange graphs.

i) Light-light graphs. They have been thoroughly investigated. No insurmountable difficulties were found, an enormous amount of work is still to be done - and at present there are no results.

ii) Vacuum polarization insertions. The analytic evaluation is by now completed[10] both for the electron and the muon anomalies. For the muon one has to consider also an interesting class of so-called mass-dependent graphs, containing vacuum polarization insertions due to electron loops. They originate terms in $\lg(m_\mu/m_e)$, whose structure is related to the renormalization group of QED[21].

iii) Three photon exchange graphs. They are drawn in fig.8), in order of (presumably) increasing difficulty.

Fig. 8). Three photon exchange graphs (mirror graphs, if any, to be added).

Each graph of fig.8) corresponds to 5 vertex graphs of usual type - graph c) for instance, corresponds to the graphs of fig.6). Graphs a,b) have been worked out, analytically, by Levine and Roskies, using hyperspherical coordinates[22,23], the results for c) are eqs. 28, 29). A straightforward extension of the method we described is expected to work for d), which is being investigated[24], as well for e). It is hoped that in the course of the work hints for the remaining graphs will be found.

REFERENCES AND FOOTNOTES

1. B. Lautrup, this volume.
2. J. Schwinger, Phys.Rev. 73, 416(1948).
3. A. Peterman, Helv.Phys.Acta 30, 407(1957).
   C.M. Sommerfield, Ann.Phys.(N.Y.) 5, 26(1958).
4. M. Levine and J. Wright, Phys.Rev. 8, 3171(1973).
5. P. Cvitanovic and T. Kinoshita, Phys.Rev. D10, 4007(1974).
6. P.T. Olsen and E.R. Williams, Proceedings of the AMCO-5 Conference, Paris 2-6 June 1975.
7. CODATA. Bulletin n.11, december 1973.
8. J.C. Wesley and A. Rich, Phys.Rev. A4, 1341 (1971).
9. A. Rich, Proceedings of the AMCO-5 Conference, Paris 2-6 June 1975.

10. R. Barbieri and E. Remiddi, Nucl.Phys. B 90, 233(1975).
11. R. Barbieri, M. Caffo and E. Remiddi, Physics Letters 57B, 460 (1975) and in preparation.
12. H. Strubbe, CERN internal report DD/74/5(1974).
13. H. Nielsen, Nova Acta Naturae Curiosorum 90, p.125, Halle 1909.
14. K.S. Kölbig, J.A. Mignaco and E. Remiddi, B.I.T. 10, 38(1970). R. Barbieri, J.A. Mignaco and E. Remiddi, Nuovo Cimento 11A, 824 and 865 (1972).
15. G. 't Hooft and M. Veltman, Nucl.Phys. B44, 189 (1972); Diagrammar, CERN 73-9 (1973); see also R. Gastmans, this volume.
16. See for instance R. Gastmans and R. Meuldermans, Nucl.Phys. B63, 277 (1973). The author acknowledges a fruitful discussion on the subject with R. Gastmans.
17. As matter of fact $V_1(u)$, $S_1(u)$ of eqs. 7,10) do acquire I.R. divergences when renormalized by subtractions at $u = m^2$. Similarly, "trivial" IR divergences can be present in graphs of figs. 3 a,c) and 4) as a consequence of the renormalization of the inserted second order vertex graph.
18. D.R. Yennie, S.C. Frautschi and H. Suura, Ann.Phys.(NY) 13, 379 (1961).
19. The separate results for graphs c_2, c_3, c_4 have been obtained immediately after the Cargèse 1975 Summer Institute and are reported here for completeness. To the author knowledge, numerically identical results have been independently obtained in semi-analytic form by M. Levine and R. Roskies (private communication, June 1975).
20. R. Barbieri and E. Remiddi, unpublished.
21. B. Lautrup and E. de Rafael, Nucl.Phys. B70, 317 (1974); R. Barbieri and E. Remiddi, Phys.Letters 57B, 273 (1975).
22. M. Levine and R. Roskies, Phys.Rev.Letters 30, 772(1973) and Phys.Rev. D9, 421 (1974).
23. The graphs of class a) have also been evaluated by K.A. Milton, W.Y. Tsai and L. De Raad, Jr., Phys.Rev. D9, 1809 (1974).
24. D. Oury and E. Remiddi, in progress.

GAUGE THEORIES : AN INTRODUCTION

R. GASTMANS *

Institute of Theoretical Physics

University of Leuven, Leuven, Belgium

(*) Bevoegdverklaard navorser, N.F.W.O., Belgium.

I. INTRODUCTION

Why all the excitement about unified gauge theories of weak and electromagnetic interactions ? Basically, because there is enough in them to get most physicists interested.

First of all, they show that weak processes and electromagnetic phenomena are intimately linked together and are merely different aspects of a unifying theory. Whenever a theoretician is able to unify, he feels he has made progress !

Also, our theoretician is happy because, after so many years of futile attempts, he finally succeeded in writing down a renormalizable theory of weak interactions. He is now able to calculate any process to any order, knowing that at the end he will come up with a finite answer. In other words, he now has more predictive power than he had before.

Technically, he gained more insight in quantum field theory, as he was obliged to study and understand better things like gauge fields, ghosts, spontaneous symmetry breaking, renormalization, Ward identities, etc. This new knowledge already proved quite valuable in other fields of elementary particle physics, like hadron physics and quantum gravity.

Of course, it also made the experimentalist happy, since he was provided with new particles to look for and new interactions to detect. This proved to be quite stimulating as physics with high energy neutrinos became feasible.

How far do we stand now ? First of all, there seems to be no significant contradiction between experiment and predictions from the gauge theories. What is more, weak neutral currents are established now and are easily interpreted in these models. Maybe, the recently discovered particles with very long lifetimes are connected with the charm quantum number which could be useful to suppress the neutral strangeness-changing currents. The near future will no doubt tell us in how far the gauge theories are correct ... or incorrect.

By now, many excellent introductions[1-8] have been written on the subject, and these notes certainly do not claim much originality in the presentation of the main features. They are merely a combination of what is in the literature with small variations here and there to cope with the estimated needs of the students in theoretical and experimental physics at this Summer Institute.

The lectures are thus organized as follows : Sec.II reviews what is wrong with the V-A theory and the intermediate vector boson model. This leads us to study generalized gauge invariance (Sec. III). Only after the introduction of spontaneous symmetry breaking, can we fully exploit this invariance and construct a "realistic" model for the leptons. We chose the Weinberg-Salam model and present its main features in Sec. IV. To be able to carry out closed-loop calculations, we study the dimensional regularization procedure in Sec. V. All this knowledge is then put to work in a few applications in Sec. VI such as elastic (anti-) neutrino-electron scattering, $e^+e^- \to \mu^+\mu^-$, the self-charge of the neutrino, and the anomalous magnetic moment of the muon.

II. UNRENORMALIZABILITY OF NON-GAUGE THEORIES

1. V-A Theory[9]

Processes involving only leptons, like μ-decay, are very well described by the phenomenological Lagrangian

$$L_{int} = -\frac{G}{\sqrt{2}} J^+_\mu(x)\, J^\mu(x), \tag{1}$$

where the lepton current

$$J_\mu = \bar{\mu}\gamma_5(1-\gamma_5)\nu_\mu + \bar{e}\gamma_\mu(1-\gamma_5)\nu_e ,$$

and the Fermi coupling constant

$$G = (1.0262 \pm 0.0001) \times 10^{-5}/m_p^2 ,$$

with m_p, the proton mass.

When supplemented with the Cabibbo current, also the semi-leptonic processes are given a good description[10,11]. For the non-leptonic processes, the situation is not so satisfactory, but this too can perhaps be remedied.

This theory is called the V-A theory, as only vector and axial vector couplings occur.

Let us take this Lagrangian seriously, and suppose we want to calculate a process like $\nu_\mu e \to \nu_\mu e$. This process is possible in second order, via the Feynman diagram of Fig. 1. Its matrix element is given by

$$M = i^2\, \frac{G^2}{2} \int \frac{d^4k}{(2\pi)^4}\; \bar{\nu}_\mu(p_3)\gamma_\mu(1-\gamma_5)\frac{i}{\not{p}_1+\not{K}-m}\,\gamma_\nu(1-\gamma_5)\nu_\mu(p_1)$$
$$\bar{e}\,(p_4)\gamma^\mu(1-\gamma_5)\frac{i}{\not{p}_2-\not{K}}\,\gamma^\nu(1-\gamma_5)e(p_2).$$

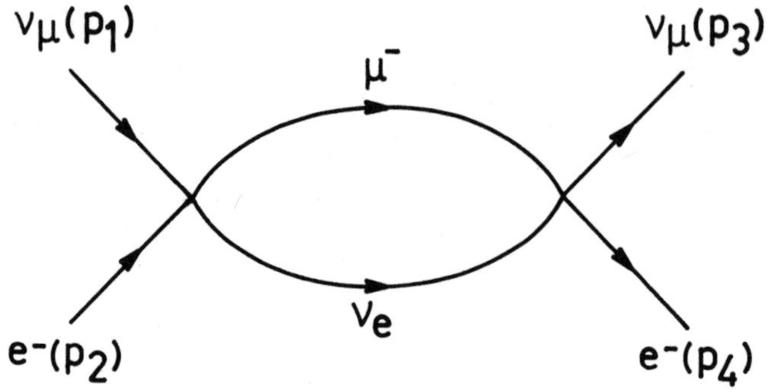

Fig. 1. Feynman diagram for $\nu_\mu e \to \nu_\mu e$ in V-A theory.

The integral over the virtual momenta diverges quadratically for high k values :

$$M \sim \int d^4k \ \frac{1}{k^2} \sim \Lambda^2 \quad .$$

This situation is general : for any process, higher order weak corrections produce infinities which cannot be eliminated through the renormalization process, i.e., through the redefinition of masses and coupling constants. It is thus a complete mystery why the Lagrangian (1) would be so successful at the phenomenological level, since it has to be supplemented with all sorts of counter-terms with arbitrary coupling constants to absorb the divergences.

Also, radiative corrections due to electromagnetism give problems. They are necessary because of the precision of the experiments, in the comparison of the coupling constants for μ-decay and β-decay, e.g.

For μ-decay, they happen to be finite[12], which can easily be understood following the arguments of Ref. 13. For $\mu^- \to e^-\bar{\nu}_e\nu_\mu$, we have in V-A (see fig. 2)

$$M_{V-A} = -i \ \frac{G}{\sqrt{2}} \ \bar{\nu}_\mu\gamma_\mu(1-\gamma_5)\mu \ \bar{e}\gamma^\mu(1-\gamma_5)\nu_e \ .$$

A Fierz transformation[14], allows one to write

$$M_{V-A} = -i \ \frac{G}{\sqrt{2}} \ \bar{\nu}_\mu\gamma_\mu(1-\gamma_5)\nu_e \ \ \bar{e}\gamma^\mu(1-\gamma_5)\mu \ .$$

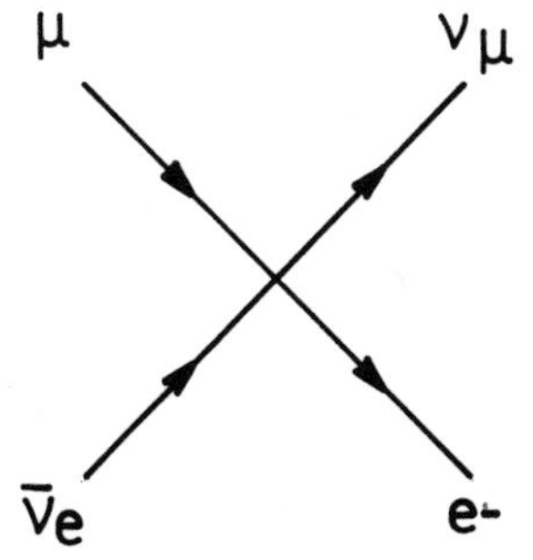

Fig. 2. Feynman diagram for $\mu^- \to e^- \bar{\nu}_e \nu_\mu$ in V-A theory.

The $\nu\bar{\nu}$-pair acts very much as if it were coupled to a neutral current in which the mass of the charged leptons changes at the interaction point. Since the divergences at the one-loop level in such theories are independent of the masses, the radiative corrections are very much like in QED, and can thus be handled.

However, $\nu_e e^- \to \nu_e e^-$ has electromagnetic divergences of order α^2, which cannot be eliminated ;[15] and so does β-decay in order α [12,13,16]. Here no such Fierz transformation can be done, which would leave one with neutral currents only.

2. The IVB-Model

Since the Lagrangian (1) was originally introduced in analogy with QED, why not postulate a vector particle in analogy with the photon? This new particle is traditionally called the intermediate vector boson (IVB). To respect the short range character of the weak interactions, it must be a heavy vector boson. Thus,

$$L_{int} = g\, J_\mu W^{+\mu} + h.c., \qquad (2)$$

with W_μ^+ the field operator which annihilates a negatively charged spin-1 boson. Notice that g, the semiweak coupling constant, is now dimensionless, like the charge e in QED.

Do we reproduce the successes of Lagrangian (1) ? Let us again look at μ-decay. In the IVB-model we have (see fig. 3)

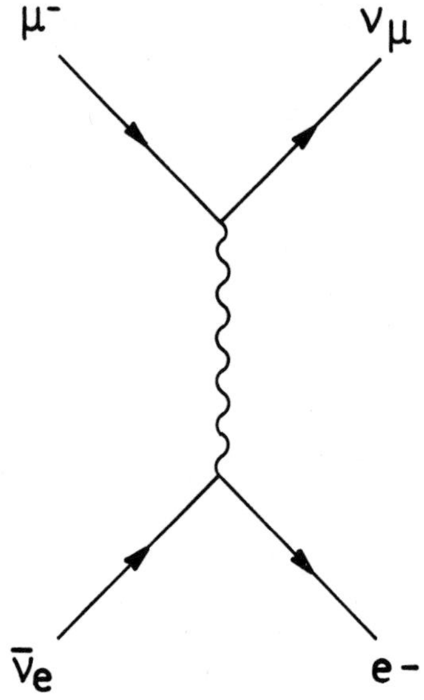

Fig. 3. Feynman diagram for $\mu^- \to e^- \bar{\nu}_e \nu_\mu$ in the IVB-model.

$$M_{IVB} = -g^2\ \bar{\nu}_\mu \gamma^\mu (1-\gamma_5)\mu(-i)\ \frac{g_{\mu\nu} - q_\mu q_\nu / M^2}{q^2 - M^2}\ \bar{e}\gamma^\nu (1-\gamma_5)\nu_e\ .$$

If $M^2 \gg q^2$ and $M^2 \gg m_e^2$, we can neglect the $q_\mu q_\nu$ term, and

$$M_{IVB} \simeq -\, i\, \frac{g^2}{M^2}\ \bar{\nu}_\mu \gamma^\mu (1-\gamma_5)\,\mu\, \bar{e}\gamma_\mu (1-\gamma_5)\nu_e$$

$$= M_{V-A}\ ,$$

provided we identify

$$\frac{g^2}{M^2} = \frac{G}{\sqrt{2}}\quad .$$

Until recently, only processes involving momentum transfers of the order of lepton or hadron masses were of importance (mainly decay processes). Now, however, with the advent of high energy neutrino beams large momentum transfers are obtained in deep inelastic scattering processes, and the mass of the IVB must be larger than ∿20 GeV to explain the experimental results[17].

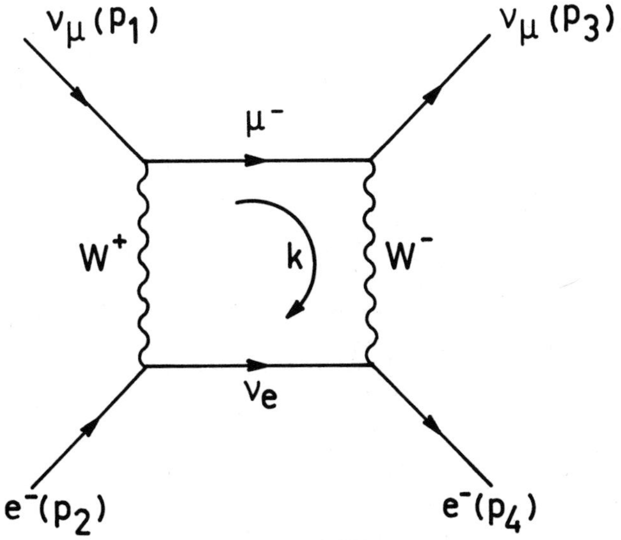

Fig. 4. Feynman diagram for $\nu_\mu e \to \nu_\mu e$ in the IVB-model .

Do we now have a renormalizable theory of weak interactions? To find out, we again look at $\nu_\mu e \to \nu_\mu e$. The relevant Feynman diagram is given in Fig. 4 and

$$M = i^4 g^4 \int \frac{d^4k}{(2\pi)^4} \; \bar{\nu}_\mu(p_3)\gamma_\mu(1-\gamma_5) \frac{i}{\not{p}_1+\not{K}-m} \gamma_\nu(1-\gamma_5)\nu_\mu(p_1)$$

$$\bar{e}(p_4)\gamma_\alpha \; (1-\gamma_5) \frac{i}{\not{p}_2-\not{K}} \gamma_\beta(1-\gamma_5)e(p_2)$$

$$(-i) \frac{g_{\mu\alpha}-(p_1-p_3+k)_\mu(p_1-p_3+k)_\alpha/M^2}{(p_1-p_3+k)^2-M^2} \; (-i) \frac{g_{\nu\beta}-k_\nu k_\beta/M^2}{k^2-M^2} \; .$$

For high k-values, the longitudinal parts of the W-propagators dominate, and

$$M \sim \int d^4k \, \frac{1}{K\!\!\!/}\frac{1}{K\!\!\!/} \, \frac{k_\mu k_\alpha}{k^2} \, \frac{k_\nu k_\beta}{k^2} \sim \int \frac{d^4k}{k^2} \sim \Lambda^2 \quad .$$

The matrix element is again quadratically divergent, therefore excluding the possibility that (2) might be a renormalizable interaction. So, comparing with V-A theory, we do not seem to have made much progress, but at least we did not make things worse and we managed to introduce a dimensionless coupling constant.

Consideration of higher order electromagnetic corrections to weak processes or weak virtual corrections to electromagnetic processes in the IVB-theory also does not lead to a definite conclusion as to its superiority over the V-A theory. True, the contribution of weak corrections to the anomalous magnetic moment of the muon can be made finite by giving the IVB itself a specific intrinsic moment[18)], in contradistinction to the V-A case[19)]. But, the introduction of a charged vector boson leads to difficulties of its own : its dynamic anomalous magnetic moment diverges at the one-loop level unless one introduces still more new particles[20)].

3. High Energy Behavior

It is clear from the above example that the difficulties with renormalizability stem from the longitudinal parts of the W-propagators.

Intuitively, one believes that the high energy behavior of Born terms is related to the renormalizability of the theory. The argument roughly goes as follows : [21)] the higher order fourpoint amplitude is related through a fixed-t dispersion integral in s to the lower order amplitudes. When the absorptive part does not fall off with s, subtractions in the theory, which cannot be absorbed in redefinitions of the original parameters, must be introduced.

Since it is somewhat simpler to study Born amplitudes rather than closed-loop diagrams, I propose to study what the high energy behaviour of Born amplitudes is for the production of longitudinally polarized W's.

Take $\nu\bar{\nu} \to W^+W^-$ [22)] (see fig. 5), its matrix element being

$$M_t = i^2 g^2 \; \varepsilon^-_\mu(\bar{k}) \; \varepsilon^+_\nu(k) \; \bar{v}(\bar{p})\gamma^\mu(1-\gamma_5) \, \frac{i(p\!\!\!/-K\!\!\!/)}{(p-k)^2} \, \gamma^\nu(1-\gamma_5)u(p) \; .$$

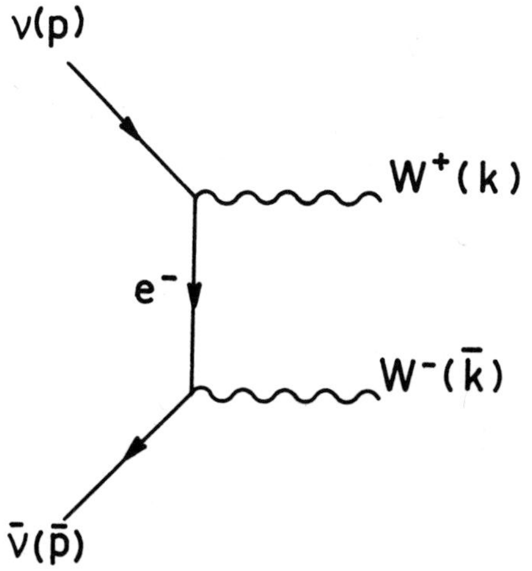

Fig. 5. Feynman diagram for $\nu\bar{\nu} \to W^+W^-$ in lowest order.

We neglected the electron mass, which is irrelevant for these considerations. Summing over all polarizations, we have

$$\sum|M_t|^2 = \frac{g^4}{(p-k)^4} \mathrm{Tr}[\gamma^\mu(1-\gamma_5)(\not{p}-\not{K})\gamma^\nu(1-\gamma_5)\not{p}\gamma^\rho(1-\gamma_5)(\not{p}-\not{K})\gamma^\sigma(1-\gamma_5)\bar{\not{p}}]$$

$$[-g^{\mu\sigma}+\bar{k}^\mu\bar{k}^\sigma/M^2]\,[-g^{\nu\rho}+k^\nu k^\rho/M^2] \;.$$

We used the relation that

$$\sum_{pol} \varepsilon_\mu(k)\varepsilon^*_\nu(k) = -g_{\mu\nu} + k_\mu k_\nu/M^2 \;.$$

To study the high energy behavior, we concentrate on the longitudinal terms, which will dominate :

$$\sum|M_t|^2 = \frac{g^4 8}{M^4(p-k)^4} \mathrm{Tr}\,[\bar{\not{K}}(1-\gamma_5)(\not{p}-\not{K})\not{K}\,\not{p}\,\not{K}\,(\not{p}-\not{K})\bar{\not{K}}\,\bar{\not{p}}] \;.$$

Neglecting terms like $k^2 = M^2$ compared to $p \cdot k$, etc., we get

$$\sum |M_t|^2 = \frac{2g^4}{M^4(p\cdot k)^2} \mathrm{Tr}\,[\bar{\not{K}}\ \not{p}\ \not{K}\ \not{p}\ \not{K}\ \not{p}\ \bar{\not{K}}\ \bar{\not{p}}]$$

$$= \frac{4g^4}{M^4(p\cdot k)} \mathrm{Tr}\,[\bar{\not{K}}\ \not{p}\ \not{K}\ \not{p}\ \bar{\not{K}}\ \bar{\not{p}}]$$

$$= \frac{8g^4}{M^4} \mathrm{Tr}\,[\bar{\not{K}}\ \not{p}\ \bar{\not{K}}\ \bar{\not{p}}]$$

$$= 64\ \frac{g^4}{M^4}\ (p\cdot\bar{k})(\bar{p}\cdot\bar{k})\ .$$

In the c.m.s., $\bar{p}\cdot\bar{k} \simeq E^2(1-\cos\theta)$, where θ is the scattering angle and E the energy. Thus

$$\sum |M_t|^2 = 64\ \frac{g^4}{M^4}\ E^4(1-\cos^2\theta) = 32\ G^2E^4\sin^2\theta\ .$$

With the chosen normalization,

$$\frac{d\sigma}{d\Omega} = \frac{32\ G^2E^4\sin^2\theta}{64\ \pi^2 s} = \frac{G^2E^2\sin^2\theta}{8\pi^2}\ ,$$

and the scattering amplitude becomes

$$f(E,\theta) = \frac{GE\ \sin\ \theta}{2\sqrt{2}\ \pi}\ (E \to \infty)\ .$$

This amplitude is in fact the helicity amplitude $f_{0,0;\ \frac{1}{2},-\frac{1}{2}}$ as only helicities 0,0 contribute to the $k_\mu k_\nu$ terms of the W polarization sum. Also, for neutrinos the helicity is always +1/2 and -1/2 for antineutrinos.

We can now make a partial wave expansion à la Jacob and Wick[23)]

$$f_{0,0;\frac{1}{2},\ -\frac{1}{2}} = \sum_J (2J+1)\ F^J_{0,0;\ \frac{1}{2},\ -\frac{1}{2}}(E)\ e^{i\phi} d^J_{10}(\theta)\ ,$$

where we have introduced the Wigner rotation functions, $d^J_{\lambda\mu}$. As $d^1_{10}(\theta) = -\frac{1}{\sqrt{2}}\sin\theta$, we get, neglecting phases

$$F^J_{0,0;\frac{1}{2},\ -\frac{1}{2}}(E) = \frac{GE}{6\pi}\ \delta_{J,1}\ . \qquad (3)$$

Now, the total cross section is given by

$$\sigma = \int d\Omega |f(\theta)|^2 = \sum_{J,J'} (2J+1)(2J'+1)F^J F^{J'*} \int d\Omega\, d^J_{10}\, d^{J'}_{10}$$

$$= 4\pi \sum_J (2J+1)|F^J|^2 \quad . \qquad (4)$$

On the other hand,

$$\sigma = \frac{4\pi}{E^2} \sum_J (2J+1)\sin^2\delta_J \ ,$$

where δ_J are the phase shifts. Unitarity requires

$$|\sin \delta_J| \leq 1 \ ,$$

and from Eqs.(3) and (4)

$$|F^J| \leq \frac{1}{E} \ ,$$

$$GE^2 \leq 6\pi \quad .$$

So, as soon as the energy becomes too large, the longitudinal W's begin to violate unitarity ($E \sim 10^3$ GeV).

We are thus led to introduce new exchanges to eliminate the linear growth with E in the amplitude of Eq. (3).

Exchanging another electron in the t-channel only gives constructive interference. So, we must exchange something in the s-channel (see fig. 6) and/or the u-channel (see fig. 7).

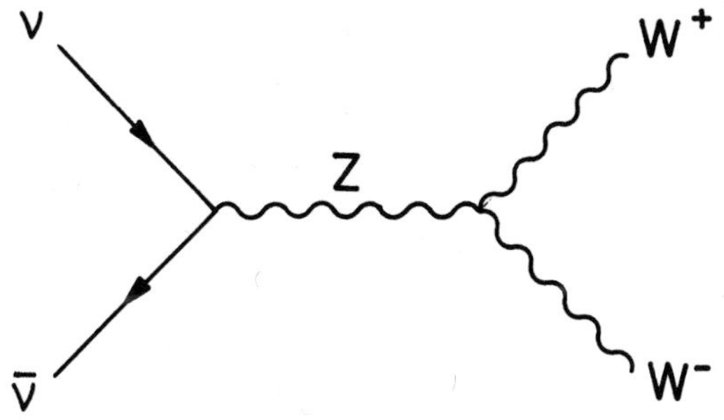

Fig. 6. A s-channel Feynman diagram for $\nu\bar{\nu} \to W^+W^-$.

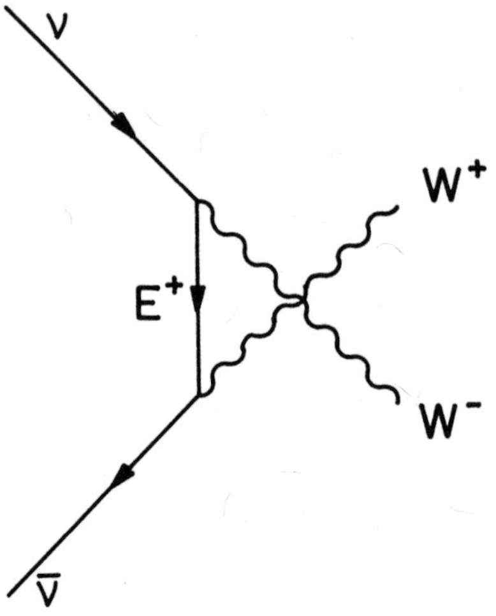

Fig.7 : A u-channel Feynman diagram for $\nu\bar{\nu} \to W^+W^-$.

Let us examine more in detail the u-channel possibility. To conserve lepton number and electric charge, it must necessarily be a positive lepton, which we shall call E^+. The u-channel amplitude is

$$M_u = i^2 g'^2 \; \varepsilon_\mu^-(\bar{k})\varepsilon_\nu^+(k) \; \bar{v}(\bar{p})\gamma^\nu(1-\gamma_5) \frac{i(\not{p}-\not{K})}{(p-\bar{k})^2} \gamma^\mu(1-\gamma_5)u(p) \; ,$$

where in leading order we neglected the E^+ mass.

For longitudinally polarized W's, we have

$$\varepsilon_\mu(k) = \frac{k_\mu}{M} + \mathcal{O}(\frac{M}{k_0}) \; ,$$

as can be seen from boosting the four-vector $\varepsilon_\mu = (0,0,0,1)$ along the z-axis.

Then,

$$M_u \simeq -2i \frac{g'^2}{M^2} \bar{v}(\bar{p}) \not{K} \frac{\not{p}-\bar{\not{K}}}{(p-\bar{k})^2} \bar{\not{K}} (1-\gamma_5)u(p)$$

$$= i \frac{g'^2}{M^2} \frac{1}{p\cdot\bar{k}} \bar{v}(\bar{p}) \not{K}\not{p}\bar{\not{K}} (1-\gamma_5)u(p) \; ,$$

$$= 2i \frac{g'^2}{M^2} \bar{v}(\bar{p}) \not{K} (1-\gamma_5)u(p), \qquad (5)$$

where use has been made of the Dirac equation

$$\not{p}\ u(p) = 0 \ .$$

The same manipulations of the s-channel amplitude lead to

$$M_s \simeq i \frac{g^2}{M^2} \frac{1}{p\cdot k} \ \bar{v}(\bar{p})\ \bar{\not{k}}\ \not{p}\ \not{k}\ (1-\gamma_5)u(p)$$

$$= 2i \frac{g^2}{M^2} \bar{v}(\bar{p})\ \bar{\not{k}}\ (1-\gamma_5)u(p) \qquad (6)$$

$$= -2i \frac{g^2}{M^2} \bar{v}(\bar{p})\ \not{k}\ (1-\gamma_5)u(p) ,$$

using also momentum conservation in this case.

Comparing Eqs.(5) and (6), we see that the leading terms are cancelled provided we choose $g'^2 = g^2$. In the derivation, we neglected terms of order m_e^2/E^2, m_E^2/E^2 and M^2/E^2; our amplitude F^J therefore now behaves as

$$F^J \sim \frac{G \times \text{mass}^2}{E} \quad ,$$

and there is no fear of violating unitarity any longer.

Several authors[21,24] have examined many other cases besides $\nu\bar{\nu} \to W^+W^-$, and were able to establish the following results :

1) to realize the cancellation of the terms of leading order for all Born processes, the vector particles must have a Yang-Mills structure, i.e., a structure based on generalized gauge invariance involving charged vector fields;
2) one either introduces weak neutral currents, or new leptons (or both);
3) for the cancellation of the non-leading terms, at least one extra scalar particle is required.

All these considerations turned out to be very enlightening in our understanding of the mechanisms involved. But, as stated in the beginning of this paragraph, they are based on an intuitive argument linking the high energy behavior of Born amplitudes to the renormalizability of the theory. It has recently been shown by Komen[25] that this argument is not necessarily true in at least one renormalizable model. Therefore, we shall no longer pursue these lines, but concentrate on the role of generalized gauge invariance in all this.

III. GENERALIZED GAUGE INVARIANCE

1. Noether's Theorem[26)]

A weak form of Noether's theorem can be phrased as follows : if, without the use of the field equations, the Lagrangian density L is invariant under some internal symmetry transformation, i.e. a local transformation on the fields of the type

$$\phi_i(x) \to \phi_i'(x) = \phi_i(x) - i\varepsilon\lambda_{ij}\phi_j(x) \qquad (i,j=1,2,\ldots,n),$$

then a conserved current exists. The λ_{ij}'s are independent of x, and ε is an infinitesimal parameter.

Indeed, for an infinitesimal transformation, we can make a Taylor series expansion of ϕ_i' around ϕ_i, and to first order

$$0 = L(\phi_i', \partial_\mu \phi_i') - L(\phi_i, \partial_\mu \phi_i)$$

$$= \sum_i \left[\frac{\partial L}{\partial \phi_i} \delta\phi_i + \frac{\partial L}{\partial(\partial_\mu \phi_i)} \delta(\partial_\mu \phi_i)\right] .$$

Using the field equations, we find

$$0 = \sum_i \left[\partial_\mu \left(\frac{\partial L}{\partial(\partial_\mu \phi_i)}\right) \delta\phi_i + \frac{\partial L}{\partial(\partial_\mu\phi_i)} \partial_\mu \delta \phi_i \right]$$

$$= \partial_\mu \left[\sum_i \frac{\partial L}{\partial(\partial_\mu\phi_i)} \delta\phi_i\right] .$$

The four-vector

$$j^\mu = -i \frac{\partial L}{\partial(\partial_\mu \phi_i)} \lambda_{ij}\phi_j$$

is thus a conserved vector :

$$\partial^\mu j_\mu = 0 .$$

2. QED : an Abelian theory[3,4)]

The Lagrangian for a free Dirac spinor,

$$L = \frac{i}{2} [\bar\psi\gamma_\mu\partial^\mu\psi - \partial^\mu\bar\psi\gamma_\mu\psi] - m \bar\psi \psi ,$$

is invariant under the space-time independent (global) phase transformation

$$\psi \to e^{-i\alpha}\psi,$$

$$\bar{\psi} \to e^{+i\alpha}\,\bar{\psi}\ .$$

This phase transformation is called a gauge transformation of the first kind and leads to the conserved current

$$j_\mu = \bar{\psi}\,\gamma_\mu\psi\ .$$

We now want to extend this symmetry to space-time dependent (local) phase transformations :

$$\psi(x) \to e^{-i\alpha(x)}\psi(x)\ , \tag{7}$$

$$\bar{\psi}(x) \to e^{+i\alpha(x)}\,\bar{\psi}(x),$$

where $\alpha(x)$ is a continuous function of the coordinates. This transformation is a gauge transformation of the second kind.

Clearly, L is not invariant as

$$L \to L + \bar{\psi}\,\gamma_\mu\psi\,\partial^\mu\alpha(x)\ .$$

To cancel this term, we add an extra term to the Lagrangian, involving a vector field, A_μ, which interacts with the spinor field and which also undergoes a gauge transformation, i.e.,

$$L_{int} = -e\,\bar{\psi}\,\gamma^\mu\psi\,A_\mu$$

and $$A_\mu(x) \to A_\mu(x) + \frac{1}{e}\,\partial_\mu\alpha(x)\ . \tag{8}$$

Indeed,

$$L_{int} \to L_{int} - \bar{\psi}\,\gamma^\mu\,\psi\partial_\mu\alpha(x),$$

and $L' = L + L_{int}$ is invariant under the simultaneous gauge transformations (7-8). If we now add the gauge invariant kinetic energy term for the field A_μ itself to L_{int}, we get

$$L_{QED} = \frac{i}{2}\,[\bar{\psi}\,\gamma_\mu\partial^\mu\psi - \partial^\mu\bar{\psi}\,\gamma_\mu\psi] - m\bar{\psi}\,\psi$$

$$- \frac{1}{4}\,(\partial_\mu A_\nu - \partial_\nu A_\mu)^2 - e\,\bar{\psi}\,\gamma^\mu\psi\,A_\mu,$$

which is the highly successful Lagrangian of QED.

From all this it follows that if we insist on gauge invariance of the second kind, the photon must necessarily exist and be massless (a mass term $-1/2\,\lambda^2 A_\mu A^\mu$ would not be invariant).

Therefore, the photon field is called a gauge field.

It is well known that L_{int} can be obtained from the Dirac Lagrangian by the minimal substitution rule

$$\partial_\mu \psi \to D_\mu \psi = [\partial_\mu + i e A_\mu] \psi .$$

The quantity D_μ is called the covariant derivative.

Obviously, QED is an Abelian gauge theory as

$$\psi \to e^{i\alpha}\psi \to e^{i\alpha} e^{i\beta}\psi = e^{i\beta}e^{i\alpha}\psi \leftarrow e^{i\beta}\psi \leftarrow \psi .$$

Finally, it should be noted that the coupling strength can be different for different charged fields. Indeed, let ψ' be another charged Dirac field, then

$$L_{QED} + \frac{i}{2}[\bar{\psi}'\gamma_\mu\partial^\mu \psi' - \partial^\mu\bar{\psi}'\gamma_\mu\psi'] - m'\bar{\psi}'\psi' + eQ\,\bar{\psi}'\gamma^\mu \psi' A_\mu$$

is invariant under the gauge transformations

$$\psi \to e^{-i\alpha(x)}\psi,$$

$$\psi' \to e^{-iQ\alpha(x)}\psi' ,$$

$$A_\mu \to A_\mu + \frac{1}{e}\partial_\mu \Lambda ,$$

as can be easily verified.

In the proof of the renormalizability of QED, the concept of gauge invariance is very important. Since the photon couples to a conserved current, any longitudinal component in the photon field is without effect in the calculation of S-matrix elements. In particular, the term in the photon propagator depending on a gauge fixing parameter has no effect. It should by now be clear that we would like to realize a similar situation for the IVB we have been considering.

3. Non-Abelian Gauge symmetry[3,4,27)]

The weak currents we encountered are charged currents involving two different particles (the e and ν_e, e.g.). If the idea of gauge symmetry is going to be relevant, we should consider gauge transformations which mix particles of different charges : something like "electronic isospin" [28)].

We therefore introduce the SU(2) spinor

$$\psi = \begin{pmatrix} \nu_e \\ e^- \end{pmatrix} ,$$

the components of which we shall give a common mass, m, temporarily. Then,

$$L_m = \frac{i}{2} [\bar{\psi} \gamma^\mu \partial_\mu \psi - \partial_\mu \bar{\psi} \gamma^\mu \psi] - m \bar{\psi} \psi$$

is invariant under the global transformation

$$\psi \rightarrow e^{-i \frac{\vec{\tau}}{2} \vec{\Lambda}} \psi ,$$

$$\bar{\psi} \rightarrow \bar{\psi} e^{+i \frac{\vec{\tau}}{2} \vec{\Lambda}} ,$$

where $\vec{\Lambda}$ is a constant isovector. By Noether's theorem, the isospin currents

$$\vec{J}_\mu = \bar{\psi} \gamma_\mu \frac{\vec{\tau}}{2} \psi$$

are conserved currents.

Yang and Mills[29)] originally thought of extending this type of invariance to local transformations. Let us see what happens when $\vec{\Lambda} = \vec{\Lambda}(x)$. To simplify things, we shall take $\vec{\Lambda}$ infinitesimally small. Our original Lagrangian is no longer invariant, but

$$L_m \rightarrow L_m + \frac{i}{2} \bar{\psi} \gamma^\mu (-i \frac{\vec{\tau}}{2} \cdot \partial_\mu \vec{\Lambda}) \psi - \bar{\psi} (+i \frac{\vec{\tau}}{2} \cdot \partial_\mu \vec{\Lambda}) \gamma^\mu \psi$$

$$= L_m + \bar{\psi} \gamma^\mu \frac{\vec{\tau}}{2} \psi \, \partial_\mu \vec{\Lambda} . \qquad (9)$$

We already learned that in such a case we have to introduce extra vector fields (gauge fields). We do this via the generalized minimum substitution rule

$$\partial_\mu \psi \rightarrow D_\mu \psi = \partial_\mu \psi + i g \frac{\vec{\tau}}{2} \vec{W}_\mu \psi ,$$

where we have introduced an isotriplet of vector fields $\vec{W}_\mu$. Applying this substitution to L_m, we introduce an interaction term

$$L_{int} = -g \bar{\psi} \gamma^\mu \frac{\vec{\tau}}{2} \psi W_\mu .$$

The next thing to do is to find the transformation properties of the W's, which would leave $L_m + L_{int}$ invariant. The QED example suggests we try

$$\vec{W}_\mu \to \vec{W}_\mu + \frac{1}{g}\,\partial_\mu \vec{\Lambda} \ . \tag{10}$$

We find

$$\begin{aligned} L_{int} \to L_{int} &- ig\,\bar{\psi}\,(\frac{\vec{\tau}}{2}\cdot\vec{\Lambda})\gamma^\mu(\frac{\vec{\tau}}{2}\cdot\vec{W}_\mu)\psi \\ &+ ig\,\bar{\psi}\,\gamma^\mu(\frac{\vec{\tau}}{2}\cdot\vec{W}_\mu)(\frac{\vec{\tau}}{2}\cdot\vec{\Lambda})\psi \\ &- \bar{\psi}\,\gamma^\mu\,\frac{\vec{\tau}}{2}\,\psi\,\partial_\mu\vec{\Lambda} \ . \end{aligned} \tag{11}$$

The last term in (11) cancels the extra term in (9), but we are still left with an extra piece

$$-ig\,\bar{\psi}\,\gamma^\mu[\frac{\tau_a}{2},\frac{\tau_b}{2}]\psi\,\Lambda^a\,W^b_\mu \ . \tag{12}$$

So, we add another term to Eq.(10) :

$$W^i_\mu \to W^i_\mu + \frac{1}{g}\,\partial_\mu\Lambda^i - i\lambda^i_{ab}\,\Lambda^a W^b_\mu \ , \tag{13}$$

and try to determine λ^i_{ab} . The last term in (13) gives

$$L_{int} \to L_{int} + ig\,\bar{\psi}\gamma^\mu\,\frac{\tau^i}{2}\,\psi\,\lambda^i_{ab}\,\Lambda^a W^b_\mu \ , \tag{14}$$

and the compensation between (12) and (14) will take place provided

$$[\frac{\tau_a}{2},\frac{\tau_b}{2}] = \lambda^i_{ab}\,\frac{\tau^i}{2} \ .$$

It follows that the λ^i_{ab} are the structure constants of the algebra. In our example , the algebra is SU(2) and

$$\lambda^c_{ab} = i\,\varepsilon_{abc} \ ,$$

where ε_{abc} is the totally antisymmetric tensor.

So, we succeeded in writing down a Lagrangian

$$\begin{aligned} L_m + L_{int} &= \frac{i}{2}\,[\bar{\psi}\,\gamma^\mu D_\mu\psi - (\overline{D_\mu\psi})\gamma^\mu\,\psi\,] - m\,\bar{\psi}\,\psi \\ &= \frac{i}{2}\,[\bar{\psi}\,\gamma^\mu\partial_\mu\psi - \partial_\mu\bar{\psi}\,\gamma^\mu\psi\,] - m\,\bar{\psi}\,\psi - g\bar{\psi}\,\gamma^\mu\,\frac{\vec{\tau}}{2}\,\psi\,\vec{W}_\mu \ , \end{aligned}$$

which is invariant under the local gauge transformations

$$\psi \to \psi - i\vec{\Lambda}(x)\cdot\frac{\vec{\tau}}{2}\psi ,$$

$$\vec{W}_\mu \to \vec{W}_\mu + \frac{1}{g}\partial_\mu\vec{\Lambda}(x) + \vec{\Lambda}(x) \times \vec{W}_\mu . \tag{15}$$

Again, we found that the existence of the isovector fields $\vec{W}_\mu$ was essential to realize this invariance. They are, like the photon, gauge fields : the Yang-Mills gauge fields.

We still have to construct a "free" Lagrangian for the $\vec{W}_\mu$ fields which does not spoil the gauge invariance. We shall follow a similar procedure, i.e. start with something obvious like

$$L' = -\frac{1}{4}\vec{G}_{\mu\nu}\vec{G}^{\mu\nu}, \quad \vec{G}_{\mu\nu} = \partial_\mu\vec{W}_\nu - \partial_\nu\vec{W}_\mu,$$

examine its transformation properties, and cancel the extra terms when we encounter them. Well, under (15)

$$\vec{G}_{\mu\nu} \to \vec{G}_{\mu\nu} + \partial_\mu\vec{\Lambda}\times\vec{W}_\nu - \partial_\nu\vec{\Lambda}\times\vec{W}_\mu + \vec{\Lambda}\times\vec{G}_{\mu\nu} . \tag{16}$$

So, we must add a term to $\vec{G}_{\mu\nu}$. We have another isovector $\vec{W}_\mu\times\vec{W}_\nu$, so let us try :

$$\vec{W}_\mu \times \vec{W}_\nu \to \vec{W}_\mu \times \vec{W}_\nu + \frac{1}{g}\partial_\mu\vec{\Lambda} \times \vec{W}_\nu + \frac{1}{g}\vec{W}_\mu \times \partial_\nu\vec{\Lambda}$$

$$+ (\vec{\Lambda} \times \vec{W}_\mu) \times \vec{W}_\nu + \vec{W}_\mu \times (\vec{\Lambda} \times \vec{W}_\nu)$$

$$= \vec{W}_\mu \times \vec{W}_\nu + \frac{1}{g}\partial_\mu\vec{\Lambda} \times \vec{W}_\nu - \frac{1}{g}\partial_\nu\vec{\Lambda}\times\vec{W}_\mu + \vec{\Lambda}\times(\vec{W}_\mu \times \vec{W}_\nu).$$

To cancel the $\partial_\mu\vec{\Lambda}$ terms we introduce

$$\vec{F}_{\mu\nu} = \vec{G}_{\mu\nu} - g\vec{W}_\mu \times W_\nu$$

$$= \partial_\mu\vec{W}_\nu - \partial_\nu\vec{W}_\mu - g\vec{W}_\mu \times \vec{W}_\nu,$$

then

$$\vec{F}_{\mu\nu} \to \vec{F}_{\mu\nu} + \vec{\Lambda} \times \vec{G}_{\mu\nu} - g\vec{\Lambda} \times(\vec{W}_\mu \times \vec{W}_\nu)$$

$$= \vec{F}_{\mu\nu} + \vec{\Lambda} \times \vec{F}_{\mu\nu},$$

and

$$L_{YM} = -\frac{1}{4}\vec{F}_{\mu\nu}\cdot\vec{F}^{\mu\nu}$$

$$\rightarrow L_{YM} - \frac{1}{4}(\vec{\Lambda}\times\vec{F}_{\mu\nu})\cdot\vec{F}^{\mu\nu} - \frac{1}{4}\vec{F}_{\mu\nu}\cdot(\vec{\Lambda}\times\vec{F}^{\mu\nu})$$

$$= L_{YM}$$

is the desired Yang-Mills Lagrangian. So, the total Lagrangian $L_m + L_{YM} + L_{int}$ is invariant under local gauge transformations, which again do not allow a mass term of the form $M^2\vec{W}_\mu\cdot\vec{W}^\mu$.

Invariance under the local gauge transformations (15) with a constant vector leads, via Noether's theorem, to the conserved isospin currents

$$\vec{J}_\mu = \bar{\psi}\gamma_\mu\frac{\vec{\tau}}{2}\psi + \vec{F}_{\mu\nu}\times\vec{W}^\nu .$$

Notice the appearance of a term due to the $\vec{W}$'s themselves in this current. Obviously, this originates in the fact that $\vec{W}$'s also carry isospin. They interact with themselves and produce 3- and 4-W vertices in Feynman diagrams. All this is quite different from QED, as the photon only couples to charged particles, and consequently not to itself.

One easily convinces oneself that, in contrast to the Abelian case, the Yang-Mills fields could not possibly couple to another spinor doublet with a different coupling constant while preserving the local gauge invariance.

Indeed, let there be another

$$L'_{int} = Q\,g\,\bar{\psi}'\gamma^\mu\frac{\vec{\tau}}{2}\psi'\vec{W}_\mu ,$$

with ψ' transforming as

$$\psi' \rightarrow e^{-iQ\frac{\vec{\tau}}{2}\cdot\vec{\Lambda}}\psi' ,$$

then reexamining the arguments leading to the commutation rules among the τ's , one finds they are only satisfied if $Q^2 = Q$, so either $Q = 1$ or $Q = 0$.

This fact has important consequences for making models of weak interactions. On the one hand, it says something about the universality of the interaction, but, on the other hand, it makes life more complicated for the model maker who can no longer eliminate unwanted processes by reducing some coupling constant, as that would destroy the renormalizability of his theory.

4. So what?

The reader may have found the preceding paragraph quite interesting, but he probably realized that all this does not give him a theory of weak interactions. True, we have a renormalizable theory[30] involving charged vector mesons, but they are massless, and we know that M_W must be larger than $\sim$ 20 GeV ! We also know that introducing a mass term in the Lagrangian destroys the renormalizability.

We also stated in the preceding chapter, while examining the high energy behavior of Born amplitudes, that a scalar meson was necessary to cancel residual terms in the $E \to \infty$ limit. In § 3, no mention was made of scalar particles.

Finally, weak interactions involve vector and axial vector currents, and we only talked about vector currents. To have conserved axial currents, however, we must have massless leptons, as can easily be seen by considering the transformation of L_m under

$$\psi \to \psi - i \frac{\vec{\tau}}{2} \vec{\Lambda} \gamma_5 \psi,$$

$$\bar{\psi} \to \bar{\psi} - i \bar{\psi} \gamma_5 \frac{\vec{\tau}}{2} \vec{\Lambda} \quad .$$

It follows that

$$\bar{\psi}\psi \to \bar{\psi}\psi - 2i \bar{\psi} \gamma_5 \frac{\vec{\tau}}{2} \vec{\Lambda} \psi ,$$

so, unless m = 0, we have no invariance. We are thus led to consider a world of massless fermions as well.

To make contact with the real world, we must find a way to introduce masses without spoiling the renormalizability of the theory. This "miracle" is achieved by the wonderful mechanism of spontaneous symmetry breaking.

IV. SPONTANEOUS SYMMETRY BREAKING

1. The Goldstone theorem[31]

Spontaneous symmetry breaking occurs when the Lagrangian describing the system is invariant under some transformation, but the ground state of the system is not. A classical example is the

Heisenberg ferromagnet described by an infinite array of spin -1/2 particles, interacting with their nearest neighbours. The Lagrangian is invariant under spatial rotations, but in the ground state, all spins align leading to a macroscopic magnetization in some arbitrary direction.

Coleman calls this a hidden symmetry[32)], since in such a world it would be very hard to detect the exact symmetry of the Lagrangian, as there would always be some preferred direction around to corrupt the experiments.

Let us first look at a very simple example[32)], a real self-interacting scalar field :

$$L = \frac{1}{2} (\partial_\mu \phi)^2 - V(\phi). \tag{17}$$

To look for the vacuum state, we determine the lowest energy state of the Hamiltonian :

$$\begin{aligned} \mathcal{H} &= \frac{\partial L}{\partial(\partial_o \phi)} \partial_o \phi - L \\ &= (\partial_o \phi)^2 - \frac{1}{2} (\partial_o \phi)^2 + \frac{1}{2} (\partial_k \phi)^2 + V(\phi) \\ &= \frac{1}{2} (\partial_o \phi)^2 + \frac{1}{2} (\partial_k \phi)^2 + V(\phi) . \end{aligned}$$

So, the state of lowest energy will be the one for which the vacuum expectation value $<0|\phi|0>$ minimizes the potential, V, while being a constant. If this minimum does not occur for $<0|\phi|0> = 0$, we should rewrite (17) in terms of the shifted field,

$$\chi = \phi - <0|\phi|0> ,$$

to find out what the spectrum of small oscillations about the vacuum is, as those oscillations are identified in quantum field theory with particle masses. Indeed, only in terms of χ are one-particle states and the vacuum orthogonal.

Let us take, e.g.,

$$V(\phi) = -a\phi^2 + b\phi^4, \quad a>0, \ b>0 ,$$

then $\frac{\partial V}{\partial \phi} = -2a\phi + 4b\phi^3$.

The minimum of V is given by

$$\phi = \left(\frac{a}{2b}\right)^{1/2} \equiv \lambda \tag{18}$$

in the classical theory, and in quantum field theory (ignoring quantum corrections)

$$<0|\phi|0> = \lambda \ .$$

Rewriting (17) in terms of χ to discover the physics, we find,

$$L = \frac{1}{2}(\partial_\mu \chi)^2 + a\chi^2 + 2a\lambda\chi + a\lambda^2$$

$$- b\chi^4 - 4b\lambda\chi^3 - 6b\lambda^2\chi^2 - 4b\lambda^3\chi - b\lambda^4$$

$$= \frac{1}{2}(\partial_\mu \chi)^2 + (a - 6b\lambda^2)\chi^2 - 4b\lambda\chi^3 - b\chi^4 + a\lambda^2 - b\lambda^4$$

$$= \frac{1}{2}(\partial_\mu \chi)^2 - 2a\chi^2 - 4b\lambda\chi^3 - b\chi^4 + b\lambda^4 \ . \qquad (19)$$

Eq.(18) makes us drop the term linear in χ . Also, the constant term, $b\lambda^4$, can be dropped as it merely shifts the overall energy, clearly without any physical consequences.

More interesting is the $-4b\lambda^2\chi^2$ term which introduces a mass $m = 2\sqrt{a}>0$ for the χ-field. In the original Lagrangian (17), the term $+a\phi^2$ could not be identified with a mass term, as it had the wrong sign.

Finally, the term χ^3 gives a self-interaction which spoils the symmetry $\phi \to -\phi$ of the original Lagrangian. This term is responsible for the "spontaneous symmetry breaking".

As we ultimately want to apply this mechanism to the case of gauge symmetries, let us see what happens when we have a continuous symmetry.

Take a theory with two scalar fields ϕ_1 and ϕ_2 and a potential[32)]

$$V(\phi_1,\phi_2) = \alpha\,[\phi_1^2 + \phi_2^2 - a^2]^2 \ , \qquad \alpha>0 \ .$$

The symmetry in this case is the invariance under rotations of the fields (see fig. 8)

$$\phi_1 \to \phi_1 \cos\theta + \phi_2 \sin\theta \ ,$$

$$\phi_2 \to -\phi_1 \sin\theta + \phi_2 \cos\theta \ .$$

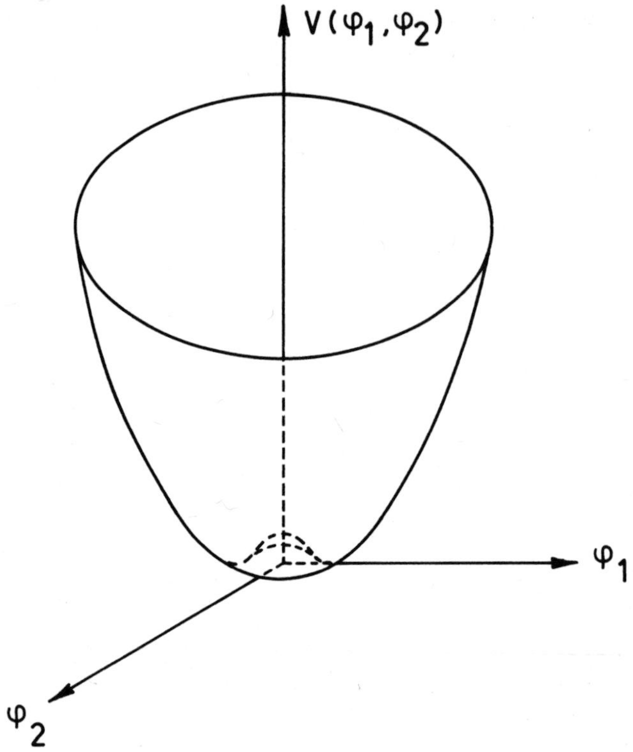

Fig. 8. The potential $V(\phi_1,\phi_2)$.

The potential is minimized on the circle

$$\phi_1^2 + \phi_2^2 = a^2 .$$

This relation defines many vacuum states, so let us choose

$$\langle\phi_1\rangle = a ,$$
$$\langle\phi_2\rangle = 0 .$$

Again, we shift the fields :

$$\chi_1 = \phi_1 - a ,$$

$$\chi_2 = \phi_2 ,$$

and the Lagrangian becomes

$$L = \frac{1}{2} (\partial_\mu \chi_1)^2 + \frac{1}{2} (\partial_\mu \chi_2)^2 - \alpha[\chi_1^2 + \chi_2^2 + 2a \chi_1]^2 .$$

So, in reality, our theory describes a massless field, χ_2, and a massive one, $\chi_1 (m = \sqrt{8}\, a\alpha)$. Obviously, we have lost the manifest symmetry between the two fields ϕ_1 and ϕ_2 . The massless spin-0 field χ_2 is the "Goldstone boson" associated with the spontaneous breakdown of the continuous SO(2) symmetry of the original Lagrangian.

Following the arguments of ref. 3, one can intuitively understand why massless particles must exist when the vacuum is degenerate. It is defined as a state with zero energy and momentum, so different vacua can only differ if they contain different numbers of quanta with zero energy and momentum and vacuum quantum numbers. These quanta are precisely the Goldstone bosons. Also the number of Goldstone bosons can be understood by counting the degrees of freedom in the problem. In the above example, there was only one degree of freedom (motion along a circle) associated with the transformation of one vacuum into another, hence the existence of only one Goldstone boson.

All these observations can be put together in the following theorem[33] : given a field theory with the "usual" axioms, if the Lagrangian is invariant under a continuous group G_n with n parameters, and if the solutions exhibit the symmetry $G_m \subset G_n$, then (n-m) massless spin-0 zero bosons are necessarily contained in the theory.

At the end of the preceding chapter we promised that spontaneous symmetry breaking would help us introducing masses into the theory without spoiling the renormalizability, and now we succeeded in proving that this leads to the existence of massless spinless particles which are nowhere found in nature. The way out is provided by the fact that the gauge theories do not obey the "usual" axioms of field theory. E.g., in QED, we cannot maintain Lorentz covariance and avoid the introduction of states with negative norm. This is the case in the Gupta-Bleuler formalism[34], where the gauge fixing condition reads

$$\frac{\partial A_\mu^+}{\partial x_\mu} |\phi\rangle = 0 ,$$

(the subscript + refers to annihilation operators only), but where the norm of a state vector $|\phi\rangle$ in the Hilbert space is redefined by

$$N(\phi) = \langle\phi|\eta|\phi\rangle ,$$

where η is some Hermitian operator. The operator η is connected with the number of scalar photons in $|\phi\rangle$, leading to a norm which is no longer positive definite.

2. The Brout-Englert-Guralnik-Hagen-Higgs-Kibble Mechanism[35)]

We shall consider what happens to the Goldstone bosons in the presence of a gauge field, and we take the abelian case first, as it considerably simplifies the notations. So, we take a charged scalar field with self-interactions to allow for spontaneous symmetry breaking and couple it to our abelian gauge field, the photon:

$$L = -\frac{1}{4} F_{\mu\nu}F^{\mu\nu} + [(\partial^\mu - ie A^\mu)\phi^+][(\partial_\mu + ie A_\mu)\phi] + a\phi^+\phi - b(\phi^+\phi)^2 ,$$

$$a>0,\ b>0 . \tag{20}$$

The gauge invariance is established by transforming

$$\phi \to e^{-i\alpha(x)} \phi,$$

$$A_\mu \to A_\mu + \frac{1}{e} \partial_\mu \alpha(x) \tag{21}$$

When the electromagnetic field is turned off, we have the Goldstone situation of the preceding paragraph, where the scalar field has a nonvanishing vacuum expectation value. Here too, the gauge invariance of the model allows us to choose a vacuum for which this value is real

$$\lambda = (a/2b)^{1/2} \equiv \frac{v}{\sqrt{2}} .$$

Introducing two real fields ϕ_1 and ϕ_2 by the equation

$$\phi = \lambda + \frac{1}{\sqrt{2}} (\phi_1 + i \phi_2) ,$$

we have

$$\langle 0|\phi_1|0\rangle = \langle 0|\phi_2|0\rangle = 0 .$$

Following the example of ref.1, it is however more advantageous in this case to introduce two other real fields, χ and θ , through

$$\phi = \frac{1}{\sqrt{2}} (v + \chi) e^{i\theta} .$$

The gauge transformation (21) now reads

$$\chi \to \chi ,$$

$$\theta \to \theta - \alpha .$$

Obviously, the choice $\alpha = \theta$ transforms θ into zero. Also,

$$A_\mu \to B_\mu = A_\mu + \frac{1}{e} \partial_\mu \theta ,$$

$$F_{\mu\nu} = \partial_\mu A_\nu - \partial_\nu A_\mu \to G_{\mu\nu} = \partial_\mu B_\nu - \partial_\nu B_\mu .$$

Notice that the field B_μ is gauge invariant. Rewriting the Lagrangian (20) in terms of the fields χ and B_μ, we find that the field θ has disappeared from the Lagrangian altogether.

Indeed,

$$\begin{aligned} L &= -\frac{1}{4} G_{\mu\nu} G^{\mu\nu} + \frac{1}{2} [\partial_\mu \chi - iev\, B_\mu - ie\chi\, B_\mu] [\partial^\mu \chi + iev\, B^\mu + ie\chi B^\mu] \\ &\quad + \frac{1}{2} a (v^2 + 2v\chi + \chi^2) - \frac{1}{4} b (v^2 + 2v\chi + \chi^2)^2 \\ &= -\frac{1}{4} G_{\mu\nu} G^{\mu\nu} + \frac{1}{2} e^2 v^2 B_\mu^2 \\ &\quad + \frac{1}{2} (\partial_\mu \chi)^2 - \frac{1}{2} (-a + 3v^2 b) \chi^2 \qquad (22) \\ &\quad - bv\chi^3 - \frac{1}{4} b \chi^4 + \frac{1}{2} e^2 B_\mu^2 (\chi^2 + 2v\chi) \\ &\quad + v\chi(a - v^2 b) + \frac{1}{2} av^2 - \frac{1}{4} bv^4 . \end{aligned}$$

Again, we can drop the irrelevant constant, as well as the linear term in χ as $v^2 = a/b$.

In the course of these manipulations, several miracles have happened. First, our field B_μ has become massive ($M = ev$); secondly, our field χ has also developed a mass ($m = \sqrt{2a}$); and, last but not least, we have no Goldstone boson left ! The phase field, θ, has been "eaten up" by the massless gauge field and by giving it a third degree of freedom, made the vector field massive.

In the jargon, this is sometimes described by the phrase : the would-be Goldstone has been gauged away.

The equivalence of Lagrangians (20) and (22) is at best indicated by the derivation of (22) from (20). Indeed, all the transformations we made were done as if we were dealing with c-numbers, whereas in fact they are operator transformations. The renormalizability of Lagrangian (20) therefore does not imply the renormalizability of Lagrangian (22), but suggests it. Indeed, we are now dealing again with massive vector fields which lead to unrenormalizable theories by simple power counting, because of their naughty propagators. But, one can verify at the one-loop level that Lagrangian (22) "remembers" enough of its original from (20) through the particular coupling constants and masses it has that cancellations of divergences do occur. However, a more powerful and more formal proof has to be given that the renormalizability to all orders can be established after spontaneous symmetry breaking. As well for the Abelian case as for the Yang-Mills case, these proofs have been given[36-37], in the early seventies, and they provided a renewed interest in the study of the Weinberg-Salam model[38], which had already been proposed in 1967.

3. The Weinberg-Salam model[38,4]

We now want to construct a renormalizable model of the weak and electromagnetic interactions of the leptons. We immediately have the choice of either introducing new leptons or additional bosons. Let us arbitrarily take the second choice: then the isotriplet $\vec{W}_\mu$ shall be given a Yang-Mills structure and the isosinglet B_μ will be an Abelian gauge field. Our symmetry group is thus $SU(2) \times U(1)$.

We have two leptons, μ and ν (later we add e and ν_e), which lead to weak charged currents $\bar{\mu}\,\gamma_\mu(1-\gamma_5)\nu$, which selects the "left handed" projection of the fields. We thus introduce a left handed doublet

$$L = \frac{1-\gamma_5}{2} \binom{\nu}{\mu} .$$

With this notation,

$$J_\mu = \bar{\mu}\gamma_\mu(1-\gamma_5)\nu = 2(\bar{\nu},\bar{\mu}) \frac{1+\gamma_5}{2} \gamma_\mu \begin{pmatrix} 0 & 0 \\ 1 & 0 \end{pmatrix} \frac{1-\gamma_5}{2} \binom{\nu}{\mu}$$

$$= 2\,\bar{L}\gamma_\mu\tau_-\,L .$$

Of course, $J^+_\mu = 2\,\bar{L}\,\gamma_\mu\,\tau_+\,L$.

The electromagnetic interaction selects muons and

$$J_\mu^{em} = \bar{\mu}\,\gamma_\mu \mu = \bar{\mu}\,\frac{1+\gamma_5}{2}\,\gamma_\mu\,\frac{1-\gamma_5}{2}\,\mu + \bar{\mu}\,\frac{1-\gamma_5}{2}\,\gamma_\mu\,\frac{1+\gamma_5}{2}\,\mu$$

$$= (\bar{\nu},\bar{\mu})\,\frac{1+\gamma_5}{2}\,\gamma_\mu \begin{pmatrix} 0 & 0 \\ 0 & 1 \end{pmatrix} \frac{1-\gamma_5}{2} \binom{\nu}{\mu}$$

$$+ \bar{\mu}\,\frac{1-\gamma_5}{2}\,\gamma_\mu\,\frac{1+\gamma_5}{2}\,\mu$$

$$= \bar{L}\,\gamma_\mu\,\frac{1-\tau_3}{2}\,L + \bar{R}\,\gamma_\mu\,R\,,$$

where we have introduced the right handed singlet

$$R = \frac{1+\gamma_5}{2}\,(\mu)\,.$$

We can thus hope to have a reasonable, symmetric theory, if we choose the interaction

$$\mathcal{L}_{int} = -g\,\bar{L}\,\gamma_\mu\,\frac{\vec{\tau}}{2}\,L\,\vec{W}^\mu - g'\,\left(\frac{1}{2}\,\bar{L}\,\gamma_\mu L + \bar{R}\gamma_\mu\,R\right)B^\mu\,.$$

At least in this way we have all the lepton currents we need in the interaction.

Our gauge invariant, renormalizable Lagrangian is then

$$\mathcal{L} = -\frac{1}{4}\,G_{\mu\nu}G^{\mu\nu} - \frac{1}{4}\,B_{\mu\nu}B^{\mu\nu}$$

$$+ \frac{i}{2}\,[\bar{L}\,\gamma^\mu D_\mu L - (\overline{D_\mu L})\gamma^\mu L] + \frac{i}{2}\,[\bar{R}\,\gamma^\mu D_\mu R - (\overline{D_\mu R})\gamma^\mu R\,]\,,$$

with

$$D_\mu L = [\partial_\mu + i\,g\,\frac{\vec{\tau}}{2}\cdot\vec{W}_\mu + i\,g'\,\frac{1}{2}\,B_\mu]L\,,$$

$$D_\mu R = [\partial_\mu + i\,g'\,B_\mu]R\,.$$

We already know that the charged W's give us the IVB model of the weak interactions. What type of neutral current interaction have we generated ? The photon field is determined by the fact that it couples to the electromagnetic current. Let us therefore introduce

$$\begin{pmatrix} A_\mu \\ Z_\mu \end{pmatrix} = \begin{pmatrix} \cos\theta & -\sin\theta \\ \sin\theta & \cos\theta \end{pmatrix} \begin{pmatrix} B_\mu \\ W^3_\mu \end{pmatrix}, \quad \begin{pmatrix} B_\mu \\ W^3_\mu \end{pmatrix} = \begin{pmatrix} \cos\theta & \sin\theta \\ -\sin\theta & \cos\theta \end{pmatrix} \begin{pmatrix} A_\mu \\ Z_\mu \end{pmatrix},$$

where θ is a mixing angle (the Weinberg angle), giving

$$\begin{aligned} L_{\text{neutral}} &= -g\,\bar{L}\gamma^\mu \frac{\tau_3}{2} L\,(-\sin\theta\,A_\mu + \cos\theta\,Z_\mu) \\ &\quad - \frac{1}{2} g'\,\bar{L}\,\gamma^\mu L\,(\cos\theta\,A_\mu + \sin\theta\,Z_\mu) \\ &\quad - g'\,\bar{R}\,\gamma^\mu R\,(\cos\theta\,A_\mu + \sin\theta\,Z_\mu)\,. \end{aligned}$$

If we put

$$g'\cos\theta = g\sin\theta = e\,, \quad \mathrm{tg}\theta = g'/g\,,$$

then

$$\begin{aligned} L_{\text{neutral}} &= -e[\,\bar{L}\,\gamma_\mu \frac{1-\tau_3}{2} L + \bar{R}\,\gamma_\mu R]A^\mu \\ &\quad - [\,g\cos\theta\,\bar{L}\gamma_\mu \frac{\tau_3}{2} L + \frac{1}{2} g'\sin\theta\,\bar{L}\gamma_\mu L + g'\sin\theta\,\bar{R}\gamma_\mu R]Z^\mu \\ &= -e\,J^{em}_\mu A^\mu \\ &\quad -[\frac{g}{2}\cos\ (\bar{\nu},\bar{\mu}) \begin{pmatrix} 1 & 0 \\ 0 & -1 \end{pmatrix} \gamma_\mu \frac{1-\gamma_5}{2} \begin{pmatrix} \nu \\ \mu \end{pmatrix} \\ &\quad + \frac{g'}{2}\sin\ (\bar{\nu},\bar{\mu})\gamma_\mu \frac{1-\gamma_5}{2} \begin{pmatrix} \nu \\ \mu \end{pmatrix} + g'\sin\theta\,\bar{\mu}\gamma_\mu \frac{1+\gamma_5}{2}\mu]\,Z^\mu \\ &= -e\,J^{em}_\mu A^\mu + [\bar{\mu}\,\gamma_\mu(g_V - g_A\gamma_5)\mu - g_A\,\bar{\nu}\gamma_\mu(1-\gamma_5)\nu]\,Z^\mu\,, \end{aligned}$$

with

$$g_V = \frac{1}{4} g\cos\theta - \frac{3}{4} g'\sin\theta = \frac{1}{4} g\cos\theta\,(1 - 3\mathrm{tg}^2\theta)\,,$$

$$g_A = \frac{1}{4} g\cos\theta + \frac{1}{4} g'\sin\theta = \frac{1}{4} g\sec\theta\,.$$

The coupling to the neutral vector boson Z_μ thus leads to a departure from pure V-A couplings for the muon.

We now generate masses for the particles through the Brout et al. mechanism. We have to make three vector bosons massive, i.e. three would-be Goldstone bosons are necessary. The simplest way to do this is by introducing four scalar fields, e.g. an isodoublet of complex fields :

$$\phi = \begin{pmatrix} \phi_+ \\ \phi_o \end{pmatrix} ,$$

which means adding a Lagrangian

$$L_\phi = (D^\mu\phi)^+ D_\mu\phi + a\phi^+\phi - b(\phi^+\phi)^2$$

with

$$D_\mu\phi = [\,\partial_\mu + ig\,\frac{\vec{\tau}}{2}\cdot\vec{W}_\mu - ig'\,\frac{1}{2}\,B_\mu]\phi ,$$

to allow for spontaneous symmetry breaking. The term in g is completely fixed by the non-abelian gauge structure, whereas the term g' is chosen so that ϕ_o does not couple to the photon field A_μ. Indeed, the neutral interaction is given by

$$(ig\,\frac{1}{2}\,\tau_3 W_3 - ig'\,\frac{1}{2}\,B_\mu)\begin{pmatrix} \phi_+ \\ \phi_o \end{pmatrix}$$

$$= \frac{1}{2}\,i\begin{pmatrix} [\,g(-\sin\theta\,A_\mu + \cos\theta\,Z_\mu) - g'(\cos\theta\,A_\mu + \sin\theta\,Z_\mu)]\,\phi_+ \\ \\ [-g(-\sin\theta\,A_\mu + \cos\theta\,Z_\mu) - g'(\cos\theta\,A_\mu + \sin\theta\,Z_\mu)]\,\phi_o \end{pmatrix}$$

$$= \frac{1}{2}\,i\begin{pmatrix} [\,-2e\,A_\mu + g\,\cos\,\theta(1 - tg^2\theta)Z_\mu]\,\phi_+ \\ \\ -\,g\,\sec\theta\,Z_\mu\phi_o \end{pmatrix} .$$

To implement the spontaneous symmetry breaking, we make the Higgs transformation

$$\phi = e^{i\,\vec{\tau}\cdot\vec{\theta}} \begin{pmatrix} 0 \\ \frac{\lambda+\chi}{\sqrt{2}} \end{pmatrix}$$

with $\lambda = (a/b)^{1/2}$, the non-zero vacuum expectation value, and χ the nontrivial field superimposed on the vacuum. Using the gauge invariance of the theory, we choose a vacuum for which the θ-fields disappear from the Lagrangian. Thus,

$$\phi \to \begin{pmatrix} 0 \\ \frac{\lambda+\chi}{\sqrt{2}} \end{pmatrix} , \tag{23}$$

and
$$L_\phi \to \left| \left[\partial_\mu - \frac{i}{2} g \sec\theta\, Z_\mu + i\,\frac{g}{2} \begin{pmatrix} 0 & W_1 - iW_2 \\ W_1 + iW_2 & 0 \end{pmatrix} \right] \begin{pmatrix} 0 \\ \frac{\lambda+\chi}{\sqrt{2}} \end{pmatrix} \right|^2$$
$$+ a\,\frac{1}{2}(\lambda+\chi)^2 - b\,\frac{1}{4}(\lambda+\chi)^4$$

$$= \frac{1}{2}(\partial_\mu\chi)^2 + \frac{1}{8} g^2 \sec^2\theta\, Z_\mu^2 (\lambda+\chi)^2 + \frac{1}{8} g^2 (W_1^2 + W_2^2)(\lambda+\chi)^2$$
$$+ \frac{1}{2}\chi^2 (a - 3b\lambda^2) - b\lambda\chi^3 - \frac{1}{4} b\,\chi^4 \ (+\ \text{constant})$$

$$= \frac{1}{2}(\partial_\mu\chi)^2 - \frac{1}{2}(2a)\,\chi^2$$
$$+ \frac{1}{2}\left(\frac{1}{2} g\lambda \sec\theta\right)^2 Z_\mu^2 + \frac{1}{2}\left(\frac{1}{2} g\,\lambda\right)^2 \left(|W_\mu^+|^2 + |W_\mu^-|^2\right)$$
$$+\ \text{interaction terms.}$$

Our theory therefore describes, after spontaneous symmetry breaking,

i) a massive neutral scalar, $m = \sqrt{2a}$;
ii) a massive neutral vector, $M_Z = 1/2\ g\lambda|\sec\theta|$;
iii) two massive charged vectors, $M_W = 1/2\ g\lambda$;
iv) a massless neutral vector, $M_\gamma = 0$.

It is immediately seen that $M_Z > M_W$. Because of the relation

$$\frac{G}{\sqrt{2}} = \frac{1}{M_W^2} \left(\frac{g}{2\sqrt{2}}\right)^2 ,$$

we have

$$\lambda^2 = \frac{1}{\sqrt{2}G} \rightarrow \lambda = 246 \text{ GeV},$$

a value very close to the unitarity cut-off. Finally,

$$M_W = \frac{1}{2} g\lambda = \frac{1}{2} e\lambda \frac{1}{|\sin\theta|} \geqslant \frac{1}{2} e\lambda = 37.292 \text{ GeV}$$

$$M_Z = \frac{1}{2} g\lambda \; |\sec\theta| = e\lambda \left|\frac{1}{\sin 2\theta}\right| \geqslant e\lambda = 74.584 \text{ GeV} .$$

Notice that both lower limits cannot be obtained simultaneously.

Finally, the mass of the χ boson is completely undetermined.

We still have to give a mass to the muon. Since direct mass terms break the gauge invariance, we try to generate the mass via the same mechanism of spontaneous symmetry breaking, i.e. we introduce

$$\mathcal{L}' = -f(\bar{R}\,\phi^+ L + \bar{L}\,\phi\, R),$$

which is a gauge invariant and renormalizable interaction. The replacement (23) then gives

$$\mathcal{L}' \rightarrow \frac{-f}{\sqrt{2}} \Big[\bar{\mu} \frac{1-\gamma_5}{2} (0, \lambda+\chi) \frac{1-\gamma_5}{2} \binom{\nu}{\mu} + (\bar{\nu}, \bar{\mu}) \frac{1+\gamma_5}{2} \begin{pmatrix} 0 \\ \lambda+\chi \end{pmatrix} \frac{1+\gamma_5}{2} \mu \Big]$$

$$= \frac{-f}{2\sqrt{2}} [\lambda\, \bar{\mu}\, (1-\gamma_5)\mu + \lambda\bar{\mu}(1+\gamma_5)\mu$$

$$+ \chi\bar{\mu}\, (1-\gamma_5)\mu + \chi\, \bar{\mu}(1+\gamma_5)\mu\,]$$

$$= \frac{-f}{\sqrt{2}} [\lambda\, \bar{\mu}\, \mu + \chi\, \bar{\mu}\, \mu] \quad .$$

So, identifying

$$\frac{f\lambda}{\sqrt{2}} = m_\mu \ ,$$

we succeeded in giving the muon a mass, while keeping the neutrino massless.

Also,

$$\frac{f}{\sqrt{2}} = \frac{m_\mu}{\lambda} = 4.3 \times 10^{-4} \quad ,$$

which is a very small number.

To introduce e and ν_e, we can follow exactly the same arguments, except that we must take a new parameter f' to generate the electron mass.

Recapitulating, we can say that we introduced the constants g,g', λ, a, f and f' which we fixed by equating certain combinations of them to e, G, m_e,m_μ , which means we have 2 arbitrary constants, e.g.,θ , the mixing angle, and m_χ , the mass of the Higgs scalar.

For the sake of completeness, it should be mentioned that the Weinberg-Salam model is by no means the only renormalizable model describing unified weak and electromagnetic interactions of leptons and vector bosons. The main feature of this model is that it contains no new leptons, whereas other models do. For a classification of these models, the interested reader is referred to Ref.39.

V. DIMENSIONAL REGULARIZATION

Before we start exploring the physical consequences of the Weinberg-Salam theory, we still have to learn how to calculate amplitudes when closed loops are present. Indeed, at intermediate stages of the calculations, infinities will appear, and one has to give formal definitions of the quantities one is calculating. This operation is called regularization.

In the proof of renormalizability of the theory, essential use is made of the so-called Ward-Takahashi identities which relate different Green's functions in the theory to one another (e.g., vertex functions and self-energies). These Ward-Takahashi identities are consequences of the gauge invariance of the Lagrangian.

It is therefore essential that any regularization procedure preserves the symmetries of the theory, e.g. one should be careful when introducing auxiliary masses when gauge invariance requires a vanishing mass. One regularization scheme which satisfies this criterium and which is also very well suited for practical calculations is the dimensional regularization method [37,40].

Let us first look at an example where only scalar particles are involved. We calculate the self-energy of the scalar in a $\lambda\phi^3$ theory, for which the Feynman diagram of fig.9 is relevant. The integral we have to evaluate is

$$I_4 = \int d^4k \frac{1}{[k^2-m^2][k^2-2\, p\cdot k]} ,$$

with $p^2 = m^2$. Let us calculate I_4 as if all virtual momenta were n-dimensional. Then,

$$I_n = \int d^nk \frac{1}{[k^2-m^2][k^2-2p\cdot k]} .$$

Of course, p_μ is a physical four-vector with only 4 components. In the restframe for p_μ, we have

$$I_n = \int dk_o \frac{\omega^{n-2} d\omega\, d\Omega_{n-1}}{[k_o^2-\omega^2-m^2][k_o^2-2mk_o-\omega^2]} ,$$

where $\omega = (k_1^2 + k_2^2 + \ldots + k_{n-1}^2)^{1/2}$. The integral over the angular coordinates can easily be done :

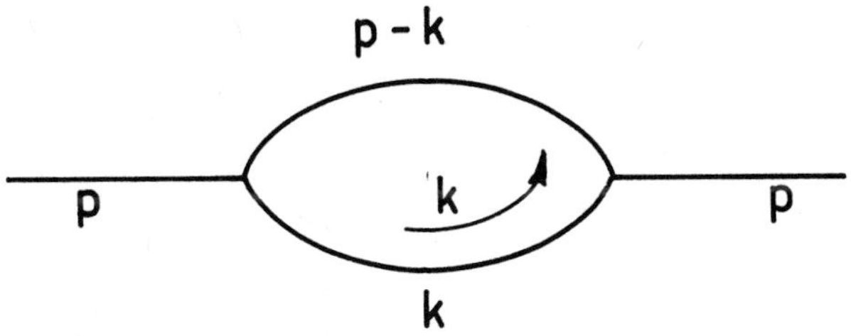

Fig. 9 : Self-energy diagram in a ϕ^3 theory.

$$\int d\Omega_{n-1} = \int_0^{2\pi} d\theta_1 \int_0^{\pi} d\theta_2 \sin\theta_2 \ldots \int_0^{\pi} d\theta_{n-2} \sin^{n-3}\theta_{n-2}$$

$$= 2\pi\,(\sqrt{\pi})^{n-3}\,\frac{\Gamma(1)\,\Gamma(\frac{3}{2})\,\ldots\,\Gamma(\frac{n-2}{2})}{\Gamma(\frac{3}{2})\,\Gamma(2)\,\ldots\,\Gamma(\frac{n-1}{2})} = \frac{2\pi^{\frac{n-1}{2}}}{\Gamma(\frac{n-1}{2})} ,$$

and

$$I_n = \frac{2\pi^{\frac{n-1}{2}}}{\Gamma(\frac{n-1}{2})} \int_{-\infty}^{\infty} dk_o \int_0^{\infty} d\omega \frac{\omega^{n-2}}{[k_o^2-\omega^2-m^2][k_o^2-2mk_o-\omega^2]}$$

When $n \geqslant 4$, our integral diverges as

$$I_n \sim \int^{\infty} \frac{dk\; k^{n-1}}{k^4} = \int^{\infty} dk\; k^{n-5} .$$

Now, we also find that near $\omega=0$, we have divergences for $n\leqslant 1$. But, if $1 < n < 4$, we have a well defined integral which we now proceed to calculate.

$$I_n = \int d^n k \int_0^1 dx \frac{1}{[k^2-2p\cdot kx-m^2(1-x)]^2}$$

$$= \int d^n \ell\; dx \frac{1}{[\ell^2-m^2(1-x+x^2)]^2}$$

$$= i\,\frac{2\pi^{n/2}}{\Gamma(\frac{n}{2})} \int_0^{\infty} \frac{\ell^{n-1}\; d\ell dx}{[\ell^2+m^2(1-x+x^2)]^2}$$

$$= i\,\frac{2\pi^{n/2}}{\Gamma(\frac{n}{2})}\,\frac{\Gamma(\frac{n}{2})\,\Gamma(2-\frac{n}{2})}{2\Gamma(2)} \int_0^1 dx[m^2(1-x+x^2)]^{\frac{n}{2}-2}$$

$$= i\pi^{n/2}\,\Gamma(2-\frac{n}{2}) \int dx\,[m^2(1-x+x^2)]^{n/2-2}$$

This expression has a pole at $n = 4$, which is a consequence of the divergence of the integral in four dimensions. The procedure is now to take this expression as the regularized definition of the integral I_4, keeping n, the dimensionality of space-time, as a new parameter in the theory, and letting $n \to 4$ continuously at the end of the calculations.

The basic formula in the dimensional regularization scheme is

$$\int d^n k \; \frac{1}{[k^2 - L + i\varepsilon]^m} = i\,(-)^m \, \pi^{\frac{n}{2}} \; \frac{\Gamma(m - \frac{n}{2})}{\Gamma(m)} \, L^{\frac{n}{2} - m} \quad . \qquad (24)$$

Ultimately, we are only interested in the $n \to 4$ behavior of I_n. We, therefore, expand in a Laurent series around $n = 4$:

$$\Gamma(2 - \tfrac{n}{2}) = \frac{1}{2 - \frac{n}{2}} \; \Gamma(3 - \tfrac{n}{2}) = \frac{-2}{n-4} \quad \Gamma(3 - \tfrac{n}{2})$$

$$= \frac{-2}{n - 4} \; \Gamma(1) + \Gamma'(1) + \mathcal{O}(n-4) \; ,$$

$$a^{(n-4)/2} = 1 + \frac{n - 4}{2} \; \ln a + \; \mathcal{O}((n-4)^2) \; ,$$

$$\to \quad I_n \simeq i\,\pi^2 \{ \frac{-2}{n-4} + \Gamma'(1) - \ln \pi - \int dx \, \ln \; [m^2(1-x+x^2)] \} \; ,$$

which is the result we were after.

The main advantages of the procedure are :

i) the form of the integrand is retained, which is important for establishing Ward identities;

ii) the procedure is manifestly covariant, unitary, and causal in the limit $n \to 4$;

iii) all formal manipulations, like shifting, symmetrizing, are allowed, provided one also imposes the rule $g^{\mu\nu} g_{\mu\nu} = n$.

To establish this last rule, consider the integral[41,42]

$$J = \int d^n k \; \frac{k_\mu}{[k^2 - m^2]^2} \quad ,$$

which is zero for reasons of covariance. Shifting gives

$$J = \int d^nk \frac{(k+p)_\mu}{[(k+p)^2-m^2]^2}$$

$$= \int d^nk \{\frac{k_\mu}{[k^2-m^2]^2} + \frac{p_\mu}{[k^2-m^2]^2} - \frac{4\ k\cdot p\ k_\mu}{[k^2-m^2]^3}\} + \mathcal{O}(p^2) .$$

If we take $k_\mu \to 0$,

$$k_\mu k_\nu \to \frac{1}{4} k^2 g_{\mu\nu},$$

to symmetrize the integrand, we get

$$J \simeq \int d^nk \{ \frac{1}{[k^2-m^2]^2} - \frac{k^2}{[k^2-m^2]^3}\} p_\mu$$

$$= -m^2 p_\mu \int d^nk \frac{1}{[k^2-m^2]^3}$$

$$= i\ m^2 p_\mu\ \pi^{n/2}\ \frac{1}{2}\ \Gamma(3-\frac{n}{2})\ m^{n-6} \underset{n\to 4}{\to} i\ p_\mu\ \pi^2 \frac{1}{2} \neq 0 .$$

If, however, we take

$$k_\mu \to 0 ,$$
$$k_\mu k_\nu \to \frac{1}{n} k^2 g_{\mu\nu}, \tag{25}$$

we obtain

$$J \simeq \int d^nk \{ \frac{1}{[k^2-m^2]^2} - \frac{4k^2/n}{[k^2-m^2]^3} \} p_\mu$$

$$= \int d^nk \{ \frac{1-4/n}{[k^2-m^2]^2} - \frac{4m^2/n}{[k^2-m^2]^3} \} p_\mu$$

$$= i\pi^{n/2} \{(1-\frac{4}{n})\ \Gamma(2-\frac{n}{2})\ m^{n-4} + \frac{4}{n}\ m^2\ \frac{1}{2}\ \Gamma(3-\frac{n}{2})\ m^{n-6}\} p_\mu$$

$$= i\pi^{n/2}\ m^{n-4}\ p_\mu\ [-\frac{2}{n}(2-\frac{n}{2})\ \Gamma(2-\frac{n}{2}) + \frac{2}{n}\Gamma(3-\frac{n}{2})]$$

$$= 0 .$$

The substitution rule (25) is equivalent with $g^{\mu\nu} g_{\mu\nu} = n$, as can be seen from contracting both sides with $g^{\mu\nu}$.

Now we turn to fermions. How do we define $\not{k} = k^{\mu}\gamma_{\mu}$? Bardeen [43] suggests to retain the four γ-matrices and not to introduce n-4 extra ones. Thus, fermion propagators depend only on the first 4 components of the virtual momenta. The γ_5 then anticommutes as usual, but closed loops involving fermions only are not regularized. These fermion loop diagrams form a separate class of diagrams on their own, and have to be regularized by some other method[44].

This asymmetric treatment of boson and fermion lines leads to computational complications as

$$\mathrm{Tr}[\gamma^{\mu}\gamma^{\nu} + \gamma^{\nu}\gamma^{\mu}] = 2g^{\mu\nu} ,$$

and thus

$$\mathrm{Tr}[\gamma^{\mu}\gamma_{\mu}] = 16 = 4g^{\mu}{}_{\mu} \rightarrow g^{\mu}{}_{\mu} = 4 .$$

This shows that in order to assign a value to $g^{\mu}{}_{\mu}$, one has to remember whether the indices μ run from 0 to 4 or from 0 to n.

There are cases, however, where one can consistently extend the Diracology to n dimensions. One such example is the case of normal parity spinor loops[7], or the calculation of static quantities (g-2, etc.) which do not involve pseudoscalar terms[20]. Then, to satisfy the relevant Ward-Takahashi identities, one can extend the number of γ-matrices to n and make γ_5 anticommute with all the n other γ's. In this case, $g^{\mu}{}_{\mu} = n$ can be taken to hold all the time.

The Diracology in n dimensions then leads to the following formulae :

$$\gamma^{\mu}\gamma_{\mu} = n ,$$

$$\gamma^{\mu} \not{a} \gamma_{\mu} = -\not{a}\gamma^{\mu}\gamma_{\mu} + 2 a_{\mu}\gamma^{\mu} = (2-n)\not{a} ,$$

$$\gamma^{\mu} \not{a} \not{b} \gamma_{\mu} = -\not{a}\gamma^{\mu}\not{b}\gamma_{\mu} + 2 \not{b} \not{a}$$

$$= 4\, a\cdot b + (n-4) \not{a} \not{b} ,$$

$$\gamma^{\mu} \not{a} \not{b} \not{c} \gamma_{\mu} = -2 \not{c} \not{b} \not{a} + (4-n) \not{a} \not{b} \not{c} .$$

A few remarks are in order. Sometimes, application of the basic formula (24) leads to strange, unexpected, but perfectly consistent results. E.g., for the badly divergent integrals

$$\int d^n k (k^2)^m = 0 \,, \qquad m \geqslant 0 \;.$$

Also

$$\int d^n k \frac{k^2}{[k^2-m^2]^2} = \int d^n k \left\{ \frac{1}{k^2-m^2} + \frac{m^2}{[k^2-m^2]^2} \right\}$$

$$= i\,\pi^{n/2} \left\{ -\Gamma(1-\tfrac{n}{2}) m^{n-2} + m^2\,\Gamma(2-\tfrac{n}{2}) m^{n-4} \right\}$$

$$= -i\,\pi^{n/2}\,\frac{n}{2}\,\Gamma(1-\tfrac{n}{2}) m^{n-2}$$

$$= \frac{n}{2} \int \frac{d^n k}{k^2-m^2} \,,$$

although intuitively one expects to have the same leading divergence for both integrals, independently of n.

VI. APPLICATIONS

First, we consider a few consequences of the Weinberg-Salam theory where only tree diagrams are relevant for the discussion, and then, a few examples of one-loop calculations.

1. $\nu_\mu(\bar{\nu}_\mu) + e^- \to \nu_\mu(\bar{\nu}_\mu) + e^-$

The experimental evidence for the existence of neutral currents in neutrino-induced semi-leptonic interactions is by now quite overwhelming[45]. This strongly suggests that the neutral currents should also be observable in purely leptonic processes as predicted by the Weinberg-Salam theory. Three candidates for the reaction $\bar{\nu}_\mu e^- \to \bar{\nu}_\mu e^-$ have been found in the Gargamelle experiment[46,47].

For the reaction $\nu_\mu e^- \to \nu_\mu e^-$, the theory predicts that it can proceed in lowest order via Z-exchange,[48] leading to the Feynman diagram of fig. 10. Its matrix element is given by

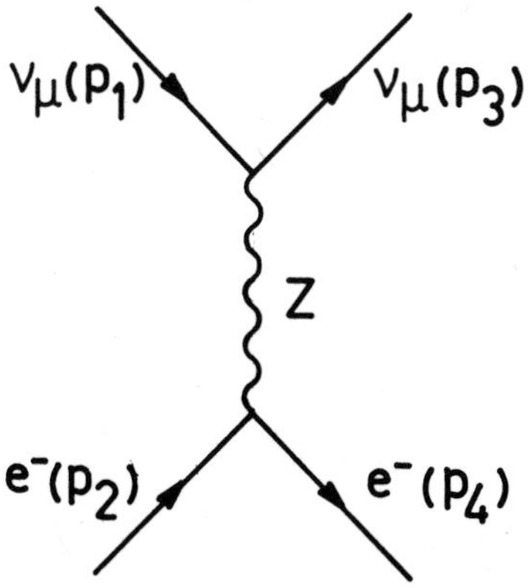

Fig. 10. Feynman diagram for $\nu_\mu e^- \rightarrow \nu_\mu e^-$ in the Weinberg-Salam model.

$$M = i^2(-g_A)\ \bar{\nu}(p_3)\gamma^\mu(1-\gamma_5)\nu(p_1)\ \bar{e}(p_4)\gamma^\nu(g_V-g_A\ \gamma_5)e(p_2)$$

$$(-i)\ \frac{g_{\mu\nu}-(p_1-p_3)_\mu\ (p_1-p_3)_\nu/M_Z^2}{(p_1-p_3)^2 - M_Z^2}$$

$$\simeq i\ \frac{g_A}{M_Z^2}\ \bar{\nu}(p_3)\gamma^\mu(1-\gamma_5)\nu(p_1)\bar{e}(p_4)\gamma_\mu(g_V-g_A\gamma_5)e(p_2),$$

for $(p_1-p_3)^2 \ll M_Z^2$. Conventionally, this amplitude is written as

$$M = -i\ \frac{G}{\sqrt{2}}\ \bar{\nu}(p_3)\gamma^\mu(1-\gamma_5)\nu(p_1)\bar{e}(p_4)\gamma_\mu(C_V-C_A\gamma_5)e(p_2)\ .$$

A simple calculation then shows that

$$C_V = -\frac{1}{2} + 2\sin^2\theta\ , \qquad C_A = -\frac{1}{2}\ ,$$

where θ is the Weinberg angle. With this matrix element one can now calculate the total cross section, which will be a function of

$\sin^2\theta$. Similarly, one can calculate the cross section for $\bar{\nu}_\mu e^- \to \bar{\nu}_\mu e^-$ (see fig. 11).

With the experimental upper limit (90% confidence level)[47]

$$\sigma(\bar{\nu}_\mu e^- \to \bar{\nu}_\mu e^-) \lesssim 0.17 \times 10^{-41} E_{\bar{\nu}} \ \mathrm{cm}^2/\mathrm{electron},$$

where $E_{\bar{\nu}}$ is the lab energy of the incident anti-neutrino, one finds the limit (90% C.L.)

$$\sin^2\theta \lesssim 0.45.$$

For the neutrino induced reaction, the limit is[49]

$$\sigma(\nu_\mu e^- \to \nu_\mu e^-) \lesssim 0.7 \times 10^{-41} E_\nu \ \mathrm{cm}^2/\mathrm{electron},$$

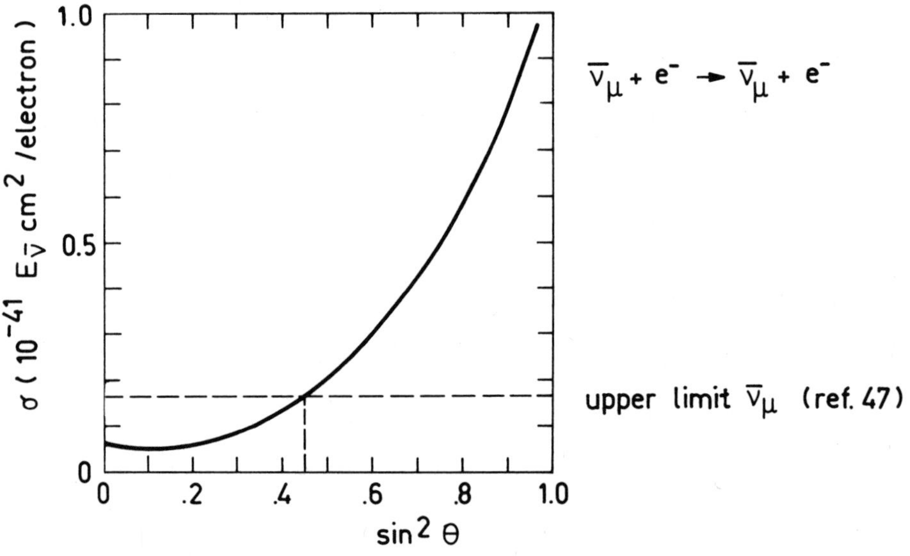

Fig. 11. The total cross section for $\bar{\nu}_\mu + e^- \to \bar{\nu}_\mu + e^-$ as a function of the mixing parameter in the Weinberg-Salam model.

which implies

$$0.1 \lesssim \sin^2\theta \lesssim 0.65.$$

2. $\bar{\nu}_e + e^- \rightarrow \bar{\nu}_e + e^-$

The previous reactions were typical examples of processes which are absent in lowest order V-A theory. The reaction $\bar{\nu}_e + e^- \rightarrow \bar{\nu}_e + e^-$ is present in V-A theory, but is predicted to be different in the Weinberg-Salam model. Indeed, there are now two Feynman diagrams for this process (see fig. 12), whereas in V-A, only the W-exchange would contribute. The matrix element is given by

$$M \simeq i^2(-g_A)\ \bar{\nu}(p_1)\gamma^\mu(1-\gamma_5)\nu(p_3)\ \bar{e}(p_4)\gamma_\mu(g_V-g_A\gamma_5)e(p_2)\ \frac{(-i)}{(-M_Z^2)}$$

$$+\ i^2\ \left(\frac{g}{2\sqrt{2}}\right)^2\ \bar{\nu}(p_1)\gamma^\mu(1-\gamma_5)e(p_2)\bar{e}(p_4)\gamma_\mu(1-\gamma_5)\ \nu(p_3)\ \frac{(-i)}{(-M_W^2)}$$

../..

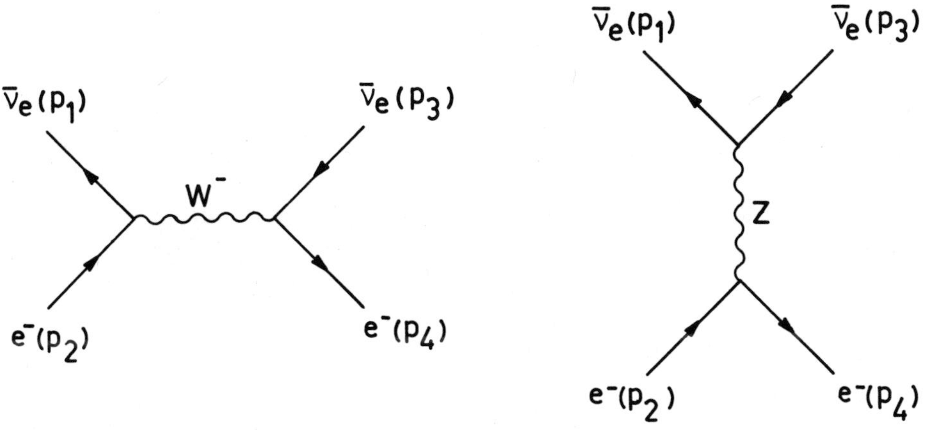

Fig. 12. Feynman diagram for the reaction $\bar{\nu}_e + e^- \rightarrow \bar{\nu}_e + e^-$ in the Weinberg-Salam model.

$$= i \frac{g_A}{M_Z^2} \bar{\nu}(p_1)\gamma^\mu(1-\gamma_5)\nu(p_3)\ \bar{e}(p_4)\gamma_\mu(g_V-g_A\gamma_5)\ e(p_2)$$

$$- i \frac{g^2}{8M_W^2} \bar{\nu}(p_1)\ (1-\gamma_5)\nu(p_3)\ \bar{e}(p_4)\gamma_\mu(1-\gamma_5)e(p_2)$$

$$= - i \frac{G}{\sqrt{2}} \bar{\nu}(p_1)\gamma^\mu(1-\gamma_5)\nu(p_3)\ \bar{e}(p_4)\gamma_\mu(C'_V-C'_A\gamma_5)\ e(p_2)\ ,$$

with

$$C'_V = \frac{1}{2} + 2 \sin^2\theta, \quad C'_A = \frac{1}{2}\ .$$

Using antineutrinos from a reactor, Gurr, Reines, and Sobel[50] were able to set an upper limit for this reaction (90% C.L.);

$$\sigma(\bar{\nu}_e\ e^- \rightarrow \bar{\nu}_e\ e^-) \lesssim 1.6 \times 10^{-41}\ E_{\bar{\nu}}\ cm^2/\text{electron}.$$

To translate this in a limit on $\sin\theta$, one has to take into account the detection efficiency, which is also a function of $\sin\theta$. This was done by Baltay[51] with the result (see fig. 13)

$$\sin^2\theta \lesssim 0.35 \qquad (90\%\ \text{C.L.})$$

The fact that also semileptonic processes are compatible with a value of $\sin^2\theta \sim 0.3\text{-}0.5$ is quite encouraging for the Weinberg-Salam model [52].

3. $e^+e^- \rightarrow \mu^+\mu^-$

This process, involving colliding electron-positron beams, is traditionally seen as a classic test of QED at high energies. It receives in lowest order its contribution from one photon exchange (see fig. 14). In the Weinberg-Salam model, also Z- and χ-exchange will contribute, and the question is whether this upsets the agreement between QED and experiment[53].

We can readily forget the χ-contribution because of its very small coupling constants with e and μ . For γ and Z, we have[54-56]

$$M_\gamma = i^2(-e)^2\ \bar{\mu}(p_3)\gamma^\mu\mu(p_4)\ \bar{e}(p_2)\gamma_\mu e(p_1)\ \frac{(-i)}{s}\ ,$$

$$M_Z \simeq i^2\ \bar{\mu}(p_3)\gamma^\mu(g_V-g_A\gamma_5)\ \mu(p_4)\bar{e}(p_2)\gamma_\mu(g_V-g_A\gamma_5)e(p_1)\frac{(-i)}{(-M_Z^2)}\ ,$$

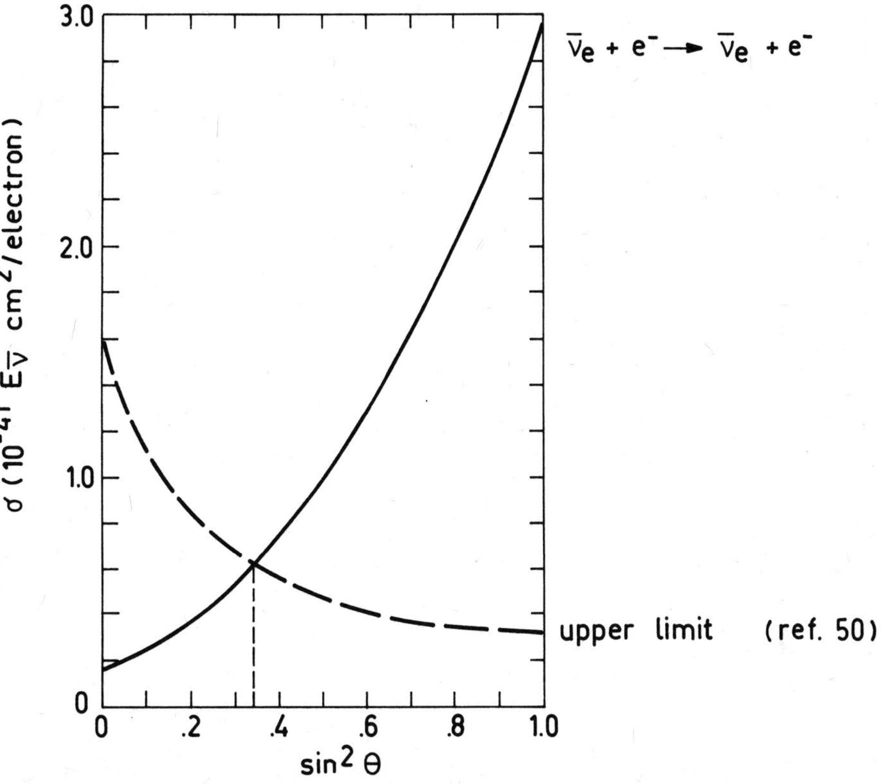

Fig. 13. The total cross-section for $\bar{\nu}_e + e^- \rightarrow \bar{\nu}_e + e^-$ as a function of the mixing parameter in the Weinberg-Salam model.

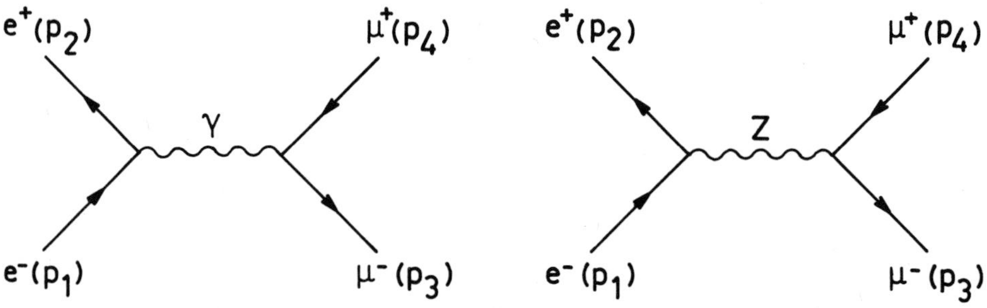

Fig. 14. Feynman diagrams for the reaction $e^+e^- \to \mu^+\mu^-$ in the Weinberg-Salam model.

where $s = (p_1 + p_2)^2$.

The matrix element M_γ leads to a cross section

$$\frac{d\sigma^\circ}{d\Omega} = \frac{\alpha^2}{4s} (1 + \cos^2\theta_s) ,$$

where θ_s stands for the scattering angle. The interference between the two graphs now leads to a corrected cross section

$$\frac{d\sigma}{d\Omega} = \frac{\alpha^2}{4s} [1 + \cos^2\theta_s - \frac{g_V^2}{2\pi\alpha} \frac{s}{M_Z^2} (1 + \cos^2 \theta_s) - \frac{g_A^2}{\pi\alpha} \frac{s}{M_Z^2} \cos \theta_s] .$$

In the derivation of this result, lepton masses have been neglected, and $M_Z^2 >> s$. Thus, the rate for this experiment is changed by an effect of the order of $s/M_Z^2 \sim 2$ % for the highest beam energies.

A quantity which may be easier to measure is the angular asymmetry

$$A(\theta) = \frac{d\sigma(\theta) - d\sigma(\pi-\theta)}{d\sigma(\theta) + d\sigma(\pi-\theta)} \simeq - \frac{g_A^2}{\pi\alpha} \frac{s}{M_Z^2} \frac{\cos\theta_s}{1+\cos^2\theta_s}$$

$$= - \frac{\sqrt{2}}{4\pi\alpha} G s \frac{\cos \theta_s}{1 + \cos^2\theta_s} .$$

Thus, interference between weak and electromagnetic processes leads to a forward-backward asymmetry of a few percent.

Unfortunately, pure QED corrections of order α^3 also lead to such an asymmetry, and make a detailed discussion dependent on the experiment. These corrections are however known in principle and have indeed been calculated [55,57]. They also turn out to be of the same order of magnitude as the weak contribution. As the weak effects grow with s, however, they may turn out to be quite important for the furture generation of colliding e^+e^- beam facilities.

Just as for $e^+e^- \to \mu^+\mu^-$, these interference effects are also there for $e^+e^- \to e^+e^-$ [56] and $e^-e^- \to e^-e^-$ [58]. Unfortunately, the purely electromagnetic term is enhanced in these processes due to the pole at t=0.

4. The charge of the neutrino[20]

In lowest order, the neutrino has no direct coupling to the photon : it is electrically neutral. But, in higher order, the neutrino does couple to the photon. Because of CP and γ_5 invariance, the neutrino has only one electromagnetic form factor, $F(q^2)$, on mass shell, and the current matrix element can be written as

$$M_\mu = ie\,F(q^2)\,\bar{\nu}\gamma_\mu(1-\gamma_5)\nu\ .$$

The introduction of a counter-term in the Lagrangian to force the neutrino charge to be zero would spoil the renormalizability of the theory, hence, to all orders, we must have $F(0) = 0$.

At the one-loop level, there are two Feynman diagrams (see fig. 15) which could give the neutrino a charge. Before calculating these diagrams, we have to derive the γ-W vertex function. The relevant piece of the Lagrangian is

$$L_{WWW} = +\frac{g}{2}(\partial^\mu\vec{W}^\nu - \partial^\nu\vec{W}^\mu)\cdot(\vec{W}_\mu \times \vec{W}_\nu)$$

$$= g\,\partial^\mu W^{1\nu}(W^2_\mu W^3_\nu - W^3_\mu W^2_\nu)$$

$$+ g\,\partial^\mu W^{2\nu}(W^3_\mu W^1_\nu - W^1_\mu W^3_\nu)$$

$$+ g\,\partial^\mu W^{3\nu}(W^1_\mu W^2_\nu - W^2_\mu W^1_\nu)\ .$$

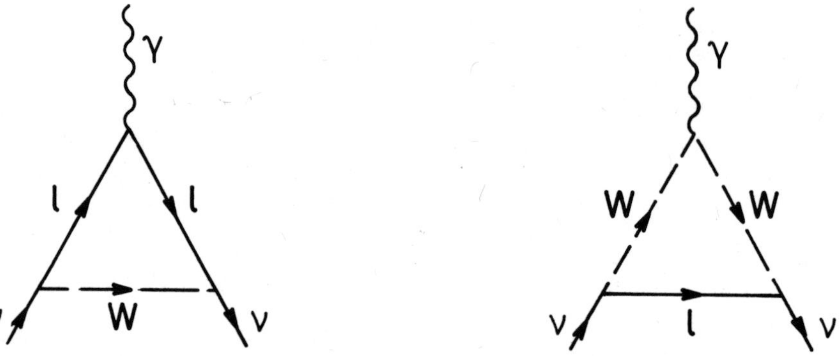

Fig.15. Feynman diagrams contributing to the self-charge of the neutrino.

Substituting

$$W^1_\mu = \frac{W^+_\mu + W^-_\mu}{\sqrt{2}} \quad , \qquad W^2_\mu = -i \; \frac{W^+_\mu - W^-_\mu}{\sqrt{2}} \quad ,$$

we find

$$\begin{aligned} L_{WWW} = & -i \frac{g}{2} (\partial^\mu W^{+\nu} + \partial^\mu W^{-\nu}) \, [(W^+_\mu - W^-_\mu) \, W^3_\nu - W^3_\mu (W^+_\nu - W^-_\nu)] \\ & -i \frac{g}{2} (\partial^\mu W^{+\nu} - \partial^\mu W^{-\nu}) \, [\, W^3_\mu (W^+_\nu + W^-_\nu) - (W^+_\mu + W^-_\mu) W^3_\nu \,] \\ & -i \frac{g}{2} \partial^\mu W^{3\nu} \, [(W^+_\mu + W^-_\mu)(W^+_\nu - W^-_\nu) - (W^+_\mu - W^-_\mu)(W^+_\nu + W^-_\nu)] \\ = & -ig \; \partial^\mu \, W^{+\nu} \, (W^-_\nu W^3_\mu - W^-_\mu W^3_\nu) \\ & -ig \; \partial^\mu \, W^{-\nu} \, (W^+_\mu W^3_\nu - W^+_\nu W^3_\mu) \\ & -ig \; \partial^\mu \, W^{3\nu} \, (W^-_\mu W^+_\nu - W^+_\mu W^-_\nu) \; . \end{aligned}$$

Substituting $W^3_\mu \to -\sin\theta\, A_\mu$, we obtain

$$L_{\gamma WW} = ie\, \partial^\mu W^{+\nu}(W^-_\nu A_\mu - W^-_\mu A_\nu) + ie\, \partial^\mu W^{-\nu}(W^+_\mu A_\nu - W^+_\nu A_\mu)$$

$$+ie\, \partial^\mu A^\nu (W^-_\mu W^+_\nu - W^+_\mu W^-_\nu) \ .$$

For the labelling of fig. 16, we get the vertex function

$$V_{\alpha\beta\mu} = ie\,[\, i(p+Q)_\mu g_{\alpha\beta} - i(p+Q)_\beta\, g_{\alpha\mu} - i(p-Q)_\alpha g_{\beta\mu} + i(p-Q)_\mu g_{\alpha\beta}$$

$$-i2Q_\beta\, g_{\alpha\mu} + i2Q_\alpha g_{\beta\mu}]$$

$$= -e\,[\, 2p_\mu\, g_{\alpha\beta} - (p-3Q)_\alpha\, g_{\beta\mu} - (p+3Q)_\beta g_{\alpha\mu}] \,.$$

It is useful for later calculations to notice that

$$(p+Q)^\alpha (p-Q)^\beta V_{\alpha\beta\mu} = -e\,[\, 2p_\mu (p^2-Q^2)$$

$$-(p^2-2p\cdot Q-3Q^2)(p-Q)_\mu - (p^2+2p\cdot Q-3Q^2)(p+Q)_\mu]$$

$$= -e\,[\, p_\mu (2p^2-2Q^2-p^2 + 2p\cdot Q + 3Q^2-p^2-2p\cdot Q+3Q^2) \qquad (26)$$

$$+Q_\mu (p^2-2p\cdot Q-3Q^2-p^2-2p\cdot Q+3Q^2)]$$

$$= -4e\,[\, Q^2 p_\mu - p\cdot Q\, Q_\mu\,] \,.$$

We can now proceed with the calculation of the neutrino self-charge. We evaluate the Feynman diagrams of fig.15 in the limit $Q_\mu \to 0$, and, consequently, we can put $p_\mu = 0$. The first diagram then yields

$$M_1 = i^3 \left(\frac{-g}{2\sqrt{2}}\right)^2 (-e) \int \frac{d^4k}{(2\pi)^4}\; \bar{\nu}\, \gamma_\alpha (1-\gamma_5)\, \frac{i}{\not{k}}\, \not{\epsilon}\, \frac{i}{\not{k}}\, \gamma_\beta (1-\gamma_5)\nu$$

$$(-i)\, \frac{g^{\alpha\beta} - k^\alpha k^\beta / M_W^2}{k^2 - M_W^2} \ .$$

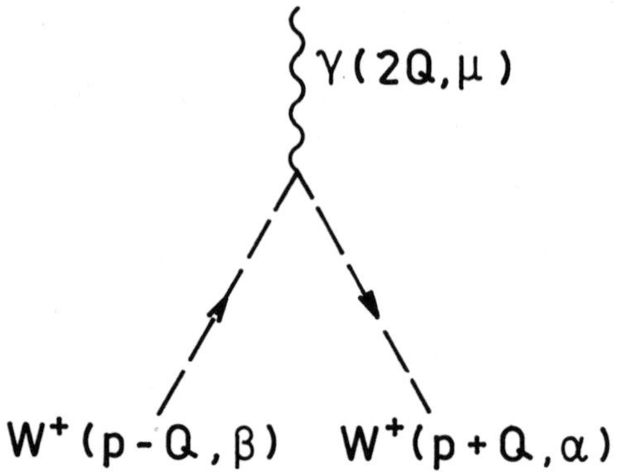

Fig. 16. The γWW vertex.

To simplify the calculations, we shall put the lepton mass to zero, although the proof can be given including the terms in m_ℓ.

As in this case we have to evaluate the normal and the abnormal parity contributions, we shall take 4-dimensional γ-matrices, which can be done as the n-dimensional momentum of the W-propagator regularizes the quadratically divergent integral. We then have

$$M_1 = -\frac{1}{4} eg^2 \int \frac{d^nk}{(2\pi)^4} \bar{\nu}(\gamma_\alpha \not{k} \not{\varepsilon} \not{k} \gamma^\alpha - \frac{1}{M_W^2} \tilde{k}^4 \not{\varepsilon})(1-\gamma_5)\nu \, \frac{1}{\tilde{k}^4(k^2-M_W^2)} ,$$

where now

$$k^2 = k_0^2 - \ldots - k_{n-1}^2 , \quad \tilde{k}^2 = k_0^2 - k_1^2 - k_2^2 - k_3^2 .$$

We proceed :

$$M_1 = -\frac{1}{4} e\, g^2 \int \frac{d^nk}{(2\pi)^4} \bar{\nu} (\tilde{k}^2 - \tilde{k}^4/M_W^2) \not{\varepsilon} (1-\gamma_5)\nu \, \frac{1}{\tilde{k}^4(k^2-M_W^2)} ,$$

and

$$F_1(0) = \frac{1}{4} i g^2 \int \frac{d^n k}{(2\pi)^4} \left[\frac{1}{\tilde{k}^2(k^2-M_W^2)} - \frac{1}{M_W^2(k^2-M_W^2)} \right] .$$

The second term can readily be evaluated using Eq.(24). It yields

$$-\frac{1}{4} ig^2 \frac{1}{M_W^2} (-i) \pi^{n/2} \Gamma(1-\frac{n}{2}) M_W^{n-2} / (2\pi)^4$$

$$= -\frac{1}{4} g^2 \pi^{n/2} M_W^{n-4} \Gamma(1-\frac{n}{2})/(2\pi)^4 .$$

In the first term, we begin by integrating over the angular variables and the length of the extra n-4 components :

$$\int dk_4 \ldots dk_{n-1} \frac{1}{k^2-M_W^2} = \int d\omega\, \omega^{n-5} \, d\Omega_{n-4} \frac{1}{\tilde{k}^2-\omega^2-M_W^2}$$

$$= \frac{2\pi^{(n-4)/2}}{\Gamma(\frac{n-4}{2})} \int d\omega \frac{\omega^{n-5}}{\tilde{k}^2-\omega^2-M_W^2}$$

$$= -\pi^{(n-4)/2} \Gamma(3-\frac{n}{2}) (M_W^2-\tilde{k}^2)^{n/2-3} .$$

The first term then gives

$$-\frac{1}{4} ig^2 \pi^{(n-4)/2} \Gamma(3-\frac{n}{2}) \int \frac{d^4\tilde{k}}{(2\pi)^4} \frac{(M_W^2-\tilde{k}^2)^{n/2-3}}{\tilde{k}^2}$$

$$= -\frac{1}{4} g^2 \pi^{(n-4)/2} \Gamma(3-\frac{n}{2}) \int \frac{d^4\tilde{\ell}}{(2\pi)^4} \frac{(M_W^2+\tilde{\ell}^2)^{n/2-3}}{\tilde{\ell}^2}$$

$$= -\frac{1}{2} g^2 \pi^{n/2} \Gamma(3-\frac{n}{2}) \int_0^\infty \frac{\ell^3 d\ell}{(2\pi)^4} \frac{(M_W^2+\ell^2)^{n/2-3}}{\ell^2}$$

$$= -\frac{1}{2} g^2 \pi^{n/2} \Gamma(3-\frac{n}{2}) \int_0^\infty d\ell \; \ell (M_W^2+\ell^2)^{n/2-3}/(2\pi)^4$$

$$= -\frac{1}{4} g^2 \pi^{n/2} \Gamma(3-\frac{n}{2}) \frac{\Gamma(1)\Gamma(2-n/2)}{\Gamma(3-\frac{n}{2})} M_W^{n-4}/(2\pi)^4$$

$$= -\frac{1}{4} g^2 \pi^{n/2} M_W^{n-4} \Gamma(2-\frac{n}{2})/(2\pi)^4 .$$

Repeated use has been made of the angular integration formula on page 144 and of the relation

$$\int_0^\infty dx \frac{x^\beta}{(x^2+M^2)^\alpha} = \frac{1}{2} \frac{\Gamma(\frac{\beta+1}{2}) \Gamma(\alpha-\frac{\beta+1}{2})}{\Gamma(\alpha)(M^2)^{\alpha-(\beta+1)/2}} .$$

Putting these two results together, we have

$$F_1(0) = -\frac{1}{4} g^2 \pi^{n/2} M_W^{n-4} [\Gamma(1-\frac{n}{2}) + \Gamma(2-\frac{n}{2})] /(2\pi)^4$$

$$= \frac{1}{8} g^2 \pi^{n/2} M_W^{n-4} (n-4) \Gamma(1-\frac{n}{2})/(2\pi)^4 . \qquad (27)$$

For the second Feynman diagram, we have

$$M_2 = i^3 (\frac{-g}{2\sqrt{2}})^2 (-e) \int \frac{d^4k}{(2\pi)^4} \bar{\nu}\gamma_\alpha(1-\gamma_5) \frac{i}{(-K)}\gamma_\beta (1-\gamma_5)\nu$$

$$[2 k_\mu g_{\rho\sigma} - k_\rho g_{\sigma\mu} - k_\sigma g_{\rho\mu}] \varepsilon^\mu$$

$$(-i)^2 \frac{[g^{\alpha\rho}-k^\alpha k^\rho/M_W^2] [g^{\beta\sigma}-k^\beta k^\sigma/M_W^2]}{[k^2-M_W^2]^2}$$

$$= -\frac{1}{4} eg^2 \int \frac{d^nk}{(2\pi)^4} \bar{\nu} \gamma_\alpha K \gamma_\beta(1-\gamma_5)\nu \varepsilon^\mu[2k_\mu g_{\rho\sigma}-k_\rho g_{\sigma\mu}-k_\sigma g_{\rho\mu}]$$

$$\frac{[g^{\alpha\rho}g^{\beta\sigma} - k^\alpha k^\rho g^{\beta\sigma}/M_W^2-k^\beta k^\sigma g^{\alpha\rho}/M_W^2]}{\tilde{k}^2(k^2-M_W^2)^2} ,$$

as the two longitudinal parts of the W-propagators give no contribution .

$$M_2 = -\frac{1}{4} eg^2 \int \frac{d^n k}{(2\pi)^4} \bar{\nu} \{-4k_\mu K - \tilde{k}^2\gamma_\mu - \tilde{k}^2\gamma_\mu$$

$$- 2[2\tilde{k}^2 k_\mu K - k^2 \tilde{k}^2\gamma_\mu - \tilde{k}^2 k_\mu K]/M_W^2 \}(1-\gamma_5)\nu\, \varepsilon^\mu$$

$$\frac{1}{\tilde{k}^2(k^2-M_W^2)^2} .$$

$$F_2(0) = \frac{1}{4} ig^2 \int \frac{d^n k}{(2\pi)^4} \frac{-\tilde{k}^2-2\tilde{k}^2-2\,(\frac{1}{4}\tilde{k}^4 - k^2\tilde{k}^2)/M_W^2}{\tilde{k}^2(k^2-M_W^2)^2}$$

$$= \frac{1}{4} ig^2 \int \frac{d^n k}{(2\pi)^4} \frac{-3-2\,(\frac{1}{4}\tilde{k}^2-k^2)/M_W^2}{(k^2-M_W^2)^2}$$

$$= \frac{1}{4} ig^2 \int \frac{d^n k}{(2\pi)^4} \frac{-3-2\,(\frac{1}{n}-1)k^2/M_W^2}{(k^2-M_W^2)^2}$$

$$= \frac{1}{4} ig^2 \int \frac{d^n k}{(2\pi)^4} \left[\frac{-2(1-n)/nM_W^2}{k^2-M_W^2} + \frac{-3-2(1-n)/n}{(k^2-M_W^2)^2}\right]$$

$$= -\frac{1}{4} g^2 \frac{\pi^{n/2}}{(2\pi)^4} \left[2\,\frac{1-n}{n}\,\frac{1}{M_W^2}\,\Gamma(1-\frac{n}{2})M_W^{n-2} - (3+\frac{2(1-n)}{n})\,\Gamma(2-\frac{n}{2})M_W^{n-4}\right]$$

$$= -\frac{1}{4} g^2 \pi^{n/2} M_W^{n-4} \left[2\,\frac{1-n}{n}\,\Gamma(1-\frac{n}{2}) - \frac{2+n}{n}\,\Gamma(2-\frac{n}{2})\right]/(2\pi)^4$$

$$= -\frac{1}{8} g^2 \pi^{n/2} M_W^{n-4} (n-4)\,\Gamma(1-\frac{n}{2})/(2\pi)^4 . \qquad (28)$$

It follows from Eqs.(27,28) that

$$F(0) = F_1(0) + F_2(0) = 0 ,$$

as must be the case in a consistent regularization scheme.

One may wonder what the model has to say about the charge radius of the neutrino, defined as

$$\langle r^2 \rangle = 6 \left. \frac{\partial F(q^2)}{\partial q^2} \right|_{q^2=0} .$$

Notice first of all that the charge radius is not a static quantity: to measure it, one has to penetrate the neutrino, i.e., we have to make a scattering experiment. But if we perform elastic $e\nu$ scattering, say, we also have to consider the competing processes like two Z or two W exchange, etc. Indeed, in the Weinberg-Salam model, all particles which couple to the photon, also couple to the Z.

Now, the theory only guarantees finite total S-matrix elements, and not finite form factors. Indeed, explicit calculations show that F'(0) is infinite, and the neutrino charge radius is not a physical quantity in this theory.

5. The g-2 of the muon[59,20].

The anomalous magnetic moment of the muon is a static quantity which provides a classic test for QED. Experimentally[60], its value is

$$a_\mu^{exp} = (1165895 \pm 27) \times 10^{-9} ,$$

and, theoretically,

$$a_\mu^{th} = (1165908 \pm 10) \times 10^{-9} ,$$

This number includes the contribution from QED up to sixth order, as well as the hadronic contribution. Future experiments promise to bring down the uncertainty to $\sim 12 \times 10^{-9}$.

As one can see, the agreement between theory and experiment is good :

$$a_\mu^{exp} - a_\mu^{th} = -(13 \pm 37) \times 10^{-9} .$$

It is thus reasonable to expect that the contributions due to the weak interactions will lie in the range

$$-50 \times 10^{-9} \leqslant a_\mu^{weak} \leqslant 24 \times 10^{-9} \ .$$

Let us see if this is the case. The most general electron-photon vertex compatible with current conservation and CP-invariance is

$$M_\mu = -ie\,\bar{u}\,[F_1(q^2)\gamma_\mu + F_2(q^2)\, i\sigma_{\mu\nu}\, q^\nu/2m$$

$$+ F_3(q^2)\gamma_5(\gamma_\mu q^2 - \not{q}q_\mu)/4M_W^2]\ u\ ,$$

and the anomaly is given by

$$a_\mu = F_2(0)\ .$$

Apart from the pure QED contributions, the Weinberg-Salam model also predicts weak contributions to a_μ at the one-loop level arising from the Feynman diagrams of fig. 17. Suppose we calculate the contribution due to the W-exchange graph :

$$M_\mu^W = i^3 \left(\frac{-g}{2\sqrt{2}}\right)^2 (-e) \int \frac{d^4k}{(2\pi)^4}\ \bar{u}(p+Q)\gamma^\rho(1-\gamma_5)\ \frac{i}{\not{p}-\not{K}}\ \gamma^\sigma(1-\gamma_5)u(p-Q)$$

$$(-i)^2\ \frac{g_{\rho\beta}-(k+Q)_\rho(k+Q)_\beta/M_W^2}{[(k+Q)^2-M_W^2]}\ \ \frac{g_{\sigma\alpha}-(k-Q)_\sigma(k-Q)_\alpha/M_W^2}{[(k-Q)^2-M_W^2]}$$

../..

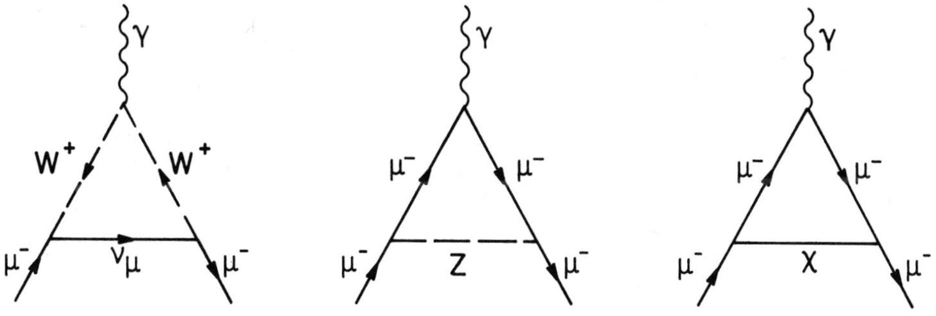

Fig. 17. Feynman diagrams contributing to the anomalous magnetic moment of the muon.

$$[\ -2k_\mu\ g^{\alpha\beta} + (k+3Q)^\alpha\ g^\beta{}_\mu + (k-3Q)^\beta\ g^\alpha{}_\mu].$$

Superficially, this matrix element is quartically divergent, but we already know (eq.(26)) that the two longitudinal parts of the W-propagator give a contribution quadratic in Q, which does not contribute to a_μ . In the same limit,

$$\frac{1}{[(k+Q)^2-M_W^2]}\ \frac{1}{[(k-Q)^2-M_W^2]} \to \frac{1}{[k^2-M_W^2]^2} \quad .$$

Also, as we are interested in F_2 only, which is a normal parity part of the diagram, we can do all the Diracology in n dimensions. Thus

$$M_\mu^W \to \frac{eg^2}{4} \int \frac{d^nk}{(2\pi)^4}\ \bar{u}\ \gamma^\rho(\not{p}-\not{K})\gamma^\sigma(1-\gamma_5)u$$

$$[g_{\rho\beta}g_{\sigma\alpha} - (g_{\rho\beta}(k-Q)_\sigma(k-Q)_\alpha + g_{\sigma\alpha}(k+Q)_\rho(k+Q)_\beta)/M_W^2\ ]$$

$$\frac{[-2k_\mu g^{\alpha\beta} + (k+3Q)^\alpha g^\beta{}_\mu + (k-3Q)^\beta g^\alpha{}_\mu]}{[k^2-M_W^2]^2\ [k^2-2p\cdot k + m^2]}$$

$$\to \frac{eg^2}{4} \int \frac{d^nk}{(2\pi)^4}\ \bar{u}\ \ \{\gamma^\rho(\not{p}-\not{K})\gamma^\sigma\ [-2k_\mu g_{\rho\sigma} + (k+3Q)_\sigma g_{\rho\mu} + (k-3Q)_\rho g_{\sigma\mu}]$$

$$- \frac{1}{M_W^2}\ [\gamma^\rho(\not{p}-\not{K})(\ \not{K}-\not{Q})(-2k_\mu(k-Q)_\rho + (k^2+2k\cdot Q)g_{\rho\mu} + k_\mu(k-3Q)_\rho)$$

$$+ (\not{K}+\not{Q})(\not{p}-\not{K})\gamma^\sigma(-2k_\mu(k+Q)_\sigma + k_\mu(k+3Q)_\sigma + (k^2-2k\cdot Q)g_{\sigma\mu})\ ]\}u$$

$$\times \frac{1}{[k^2-M_W^2]^2\ \ [k^2-2p\cdot k+m^2]}$$

$$\rightarrow \frac{eg^2}{4} \int \frac{d^nk}{(2\pi)^4}\, \bar{u}\, \{2(n-2)k_\mu(m-\not{K})+\gamma_\mu(\not{p}-\not{K})(\not{K}+3\not{Q})+(\not{K}-3\not{Q})(\not{p}-\not{K})\gamma_\mu$$

$$-\frac{1}{M_W^2}\,[\gamma^\rho(\not{p}-\not{K})(\not{K}-\not{Q})(-k_\mu(k+Q)_\rho+(k^2+2k\cdot Q)g_{\rho\mu})$$

$$+\,(\not{K}+\not{Q})(\not{p}-\not{K})\gamma^\sigma(-k_\mu(k-Q)_\sigma+(k^2-2k\cdot Q)g_{\sigma\mu})]\}\; u$$

$$\times\,\frac{1}{[k^2-M_W^2]^2\;[k^2-2p\cdot k+m^2]}$$

$$\rightarrow \frac{eg^2}{4} \int \frac{d^nk}{(2\pi)^4}\, \bar{u}\, \{2(n-2)k_\mu(m-\not{K})+\gamma_\mu(\not{p}\not{K}-3m\not{Q}-3\not{K}\not{Q})$$

$$+\,(\not{K}\not{p}+3m\not{Q}+3\not{Q}\not{K})\gamma_\mu\;-\;[-2k_\mu(\not{K}+\not{Q})(-k^2+2k\cdot p-m^2+m\not{p}-m\not{K})$$

$$+\,(k^2+2k\cdot Q)\gamma_\mu(-k^2+2k\cdot p-m^2+m\not{p}-m\not{K})$$

$$+\,(k^2-2k\cdot Q)(-k^2+2k\cdot p-m^2+m\not{p}-m\not{K})\gamma_\mu\;]\;/M_W^2\}\; u$$

$$\frac{1}{[k^2-M_W^2]^2\;[k^2-2k\cdot p+m^2]}$$

$$\rightarrow \frac{eg^2}{2} \int \frac{d^n\ell}{(2\pi)^4}\;\frac{dx(1-x)}{[\ell^2-M_W^2(1-x)+m^2x(1-x)]^3}$$

$$\bar{u}\,\{2(n-2)m\,p_\mu\,x(1-x)-3m(1-x)\,[\gamma_\mu,\not{Q}]$$

$$-\,[-2m^2\,(\ell_\mu+p_\mu x)(-\not{\ell}+m(1-x))+(\ell^2+m^2x^2)m[\gamma_\mu,\not{Q}\,]\,(1-x)$$

$$-\,2m\,p\cdot\ell\;x[\gamma_\mu,\not{\ell}]\;-2m\;\ell\cdot Q\;[\gamma_\mu,\not{\ell}]]\;/M_W^2\;\}u$$

$$\rightarrow \frac{eg^2 m}{2} \int \frac{d^n \ell}{(2\pi)^4} \frac{dx(1-x)}{[\ell^2-(1-x)(M_W^2-m^2x)]^3}$$

$$\bar{u}\,\{2(n-2)\,x\,(1-x)p_\mu + 3(1-x)\,[\not{Q},\gamma_\mu] - [-2m^3x(1-x)p_\mu$$

$$-(1-x)(\ell^2+m^2x^2)\,[\not{Q},\gamma_\mu] + \frac{2}{n}\,\ell^2(1+x)\,[\not{Q},\gamma_\mu]]\;/M_W^2\,\}u\ .$$

The Gordon decomposition,

$$\bar{u}(p+Q)\gamma_\mu u(p-Q) = \bar{u}(p+Q)\,[\,p_\mu/m + i\sigma_{\mu\nu}Q^\nu/m]\;u(p-Q)\ ,$$

allows us to substitute

$$p_\mu \rightarrow -im/e\ , \quad [\not{Q},\gamma_\mu] \rightarrow 2im/e$$

in the calculation of a_μ :

$$a_\mu^W = i\,\frac{g^2m^2}{2} \int \frac{d^n\ell}{(2\pi)^4} \frac{dx(1-x)}{[\ell^2-(1-x)(M_W^2-m^2x)]^3}$$

$$\{2(2-n)x(1-x) + 6(1-x) -[\ell^2\ (-2(1-x) + 4(1+x)/n)$$

$$+ m^2(2x(1-x) - 2x^2(1-x))]\,/M_W^2\ \}.$$

As

$$\int d^n\ell\,\frac{\ell^2}{[\ell^2-C]^m} = \int d^n\ell\,\{\frac{1}{[\ell^2-C]^{m-1}} + \frac{C}{[\ell^2-C]^m}\}$$

$$= i(-)^{m-1}\,\pi^{n/2}\,[\,C^{n/2+1-m}\,\Gamma(m-1-\tfrac{n}{2})/\Gamma(m-1) - C^{n/2+1-m}\,\Gamma(m-\tfrac{n}{2})/\Gamma(m)]$$

$$= i(-)^{m-1}\,\pi^{n/2}\,C^{n/2+1-m}\,[\,m-1-(m-1-\tfrac{n}{2})]\;\Gamma(m-1-\tfrac{n}{2})/\Gamma(m)$$

$$= i(-)^{m-1}\,\pi^{n/2}\,C^{n/2+1-m}\,\frac{n}{2}\;\Gamma(m-1-\tfrac{n}{2})/\Gamma(m),$$

we have

$$a_\mu^W = - \frac{g^2 m^2}{2} \frac{\pi^{n/2}}{(2\pi)^4} \int dx(1-x)$$

$$\{- \frac{1}{2} \Gamma(3 - \frac{n}{2}) [(1-x)(M_W^2-m^2x)]^{n/2-3} (1-x)(6+4x-2nx)$$

$$- \frac{1}{2} \Gamma(2 - \frac{n}{2}) [(1-x)(M_W^2-m^2x)]^{n/2-2} (-n(1-x)+2+2x)/M_W^2$$

$$+ \frac{1}{2} \Gamma(3 - \frac{n}{2}) [(1-x)(M_W^2-m^2x)]^{n/2-3} 2x(1-x)^2 m^2/M_W^2\} .$$

Expanding around n=4 and dropping the $\Gamma'(1)$ and $\ln\pi$ terms, we find

$$a_\mu^W = - \frac{g^2 m^2}{2} \frac{\pi^2}{(2\pi)^4} \int dx(1-x)\{- \frac{3-2x}{M_W^2-m^2x} + \frac{x(1-x)m^2/M_W^2}{M_W^2-m^2x}$$

$$+ \frac{-2+6x}{(n-4)M_W^2} + (-1+3x)\ln [(1-x)(M_W^2-m^2x)] /M_W^2$$

$$-(1-x)/M_W^2\} .$$

The pole term gives no contribution as

$$\int_0^1 dx(1-x)(-2+6x) = \int_0^1 dx\ x\ (4-6x) = 0 .$$

In the limit $m^2 << M_W^2$, we finally have

$$a_\mu^W = \frac{g^2 m^2}{32\pi^2 M_W^2} \int dx(1-x) [(3-2x)-(-1+3x)\ln(1-x)+(1-x)]$$

$$= \frac{g^2 m^2}{32\pi^2 M_W^2} \int dx\ x(1+3x-(2-3x)\ \ln x)$$

$$= \frac{g^2 m^2}{32\pi^2 M_W^2} [\frac{1}{2} + 1 + \frac{2}{4} - \frac{3}{9}]$$

.../..

$$= \frac{G\ m^2}{8\pi^2\ \sqrt{2}} \quad [\frac{10}{3}]$$

$$\simeq 3.3 \times 10^{-9} \quad ,$$

for the muon case. The Z -exchange graph gives a contribution dependent on $\sin\theta$, but of the same order of magnitude. For $\sin^2\theta \sim 0.4$, $a_\mu^Z \simeq -1.5 \times 10^{-9}$. The χ-exchange contribution can be neglected for $m_\chi >> m$, but for $m_\chi \sim m$, $a_\mu^\chi \sim 10^{-9}$.

It is clear, from the mere size of these contributions, that the agreement between pure QED calculations and experiment is not upset by these weak effects. Conversely, no limit on m_χ can be deduced from the present or planned experiments on the muon anomaly.

REFERENCES

1. B. Zumino, lectures at the Cargèse Summer Institute, July 1972, CERN TH 1550 (1972).
2. J.C. Taylor, An introduction to gauge fields and renormalizable theories of charged vector mesons, Rutherford Laboratory preprint, RPP/T/29 (1972).
3. L.M. Sehgal, Unified theories of weak and electromagnetic interactions : an elementary introduction, Aachen preprint, PITHA-68 (1973).
4. A. De Rújula, An elementary introduction to the unified theories of weak and electromagnetic interactions, Harvard preprint (1973).
5. Riazuddin, An introduction to spontaneously broken gauge theories of weak and electromagnetic interactions, Daresbury preprint (1973).
6. C.H. Llewellyn Smith, An introduction to renormalizable models of weak and electromagnetic interactions and their experimental consequences, CERN TH 1710 (1973).
7. K.T. Mahanthappa, Lectures on spontaneously broken gauge theories of weak and electromagnetic interactions, University of Texas at Austin preprint (1973).
8. E.S. Abers and B.W. Lee, Phys.Reports 9, 1(1973).
9. E.C.G. Sudarshan and R.E. Marshak, Phys.Rev. 109, 1860(1958); R.P. Feynman and M. Gell-Mann, Phys.Rev. 109, 193(1958); J.J. Sakurai, Nuovo Cimento 7, 649 (1958).
10. See e.g. R.E. Marshak, Riazuddin and C.P. Ryan, Theory of weak interactions in particle physics (Wiley-Interscience, New York, 1969).
11. For a more recent review, see M.K. Gaillard, lectures at the 1975 Cargèse Summer Institute, July 1975.

12. S.M. Berman, Phys.Rev. 112, 267 (1958); T. Kinoshita and A. Sirlin, Phys.Rev. 113, 1652 (1959); L. Durand, III, L.F. Landovitz and R.B. Marr, Phys.Rev. 130, 1188 (1963).
13. S.M. Berman and A. Sirlin, Ann.Phys. (N.Y.) 20, 20(1962).
14. M. Fierz, Z. Physik 104, 553 (1937). See also e.g., ref. 10, pp.82-84.
15. S.L. Adler, Phys.Rev. 177, 2426 (1969); J. Bell and R. Jackiw, Nuovo Cimento 60, 47 (1969).
16. G. Källén, Nucl.Phys. B1, 225 (1967).
17. C. Franzinetti, Proceedings 6th International Symposium on electron and photon interactions at high energies, ed. by H. Rollnik and W. Pfeil (North-Holland Publ.Co., Amsterdam 1974), p.366.
18. S.J. Brodsky and J.D. Sullivan, Phys.Rev. 156, 1644 (1967); T. Burnett and M.J. Levine, Phys.Rev. 24B, 467 (1967).
19. R.N. Amado and L. Holloway, Nuovo Cimento 30, 1083 (1963); 30, 1572 (1963).
20. W.A. Bardeen, R. Gastmans and B. Lautrup, Nucl.Phys. B46, 319 (1972).
21. C.H. Llewellyn Smith, Phys.Letters 46B, 233 (1973).
22. M. Gell-Mann, M.L. Goldberger, N. Kroll and F.E. Low, Phys.Rev. 179, 1518 (1969).
23. M. Jacob and G.C. Wick, Ann.Phys. (N.Y.) 1, 404 (1959).
24. J.M. Cornwall, D.N.Levin and G. Tiktopoulos, Phys.Rev.Lett. 30, 1268 (1973).
25. G.J. Komen, Phys.Lett. 55B, 389 (1975).
26. J.D. Bjorken and S.D. Drell, Relativistic Quantum Fields, (McGraw Hill, New York 1965), pp.19-20.
27. J. Bernstein, Rev.Mod.Phys. 46, 7(1974).
28. S. Weinberg, Phys.Rev.Letters 19, 1264 (1967); 27, 1688(1971).
29. C.N. Yang and R.C. Mills, Phys.Rev. 96, 191(1954).
30. R.P. Feynman, Acta Phys.Polon. 26, 697(1963); B.S. deWitt, Phys.Rev. 162, 1195, 1239 (1967); L.D. Faddeev and V.N. Popov, Phys.Letters 25B, 29(1967); V.N. Popov and L.D. Faddeev, Kiev JTP, Report JTP 67-36, translation NAL TH 4-57; L.D. Faddeev, Theor. and Math.Phys. 1, 1(1970); S. Mandelstam, Phys.Rev. 175, 1580 (1968); E.S. Fradkin and I.V. Tuytin, Phys.Letters 30B, 562(1969) and Phys.Rev. D2, 2841 (1970); M. Veltman, Nucl.Phys. B21, 288(1970).
31. J. Goldstone, Nuovo Cimento 19, 154(1961); Y. Nambu and G. Jona-Lasinio, Phys.Rev. 122, 345 (1961); 124,246 (1961).
32. S. Coleman, "Secret Symmetry : an introduction to spontaneous symmetry breakdown and gauge fields", proceedings of the 1973 International Summer School of Physics, "Ettore Majorana", (to be published).
33. J. Goldstone, A. Salam and S. Weinberg, Phys.Rev. 127, 965(1962); R.F. Streater, Proc.Roy.Soc.(London) A287, 510(1965); D. Kastler, D.W. Robinson, and A. Swieka, Comm.Math.Phys.2, 108(1966).
34. See e.g., F. Mandl, Introduction to quantum field theory (Inter-

science, New York,1966) pp.57-71.
35. F. Englert and R. Brout, Phys.Rev.Letters 13, 321(1964); P.W. Higgs, Phys.Letters 12, 132 (1964); Phys.Rev.Lett. 13, 508 (1964); Phys.Rev. 145, 1156 (1966); G.S. Guralnik, C.R. Hagen and T.W. Kibble, Phys.Rev.Letters 13, 585 (1964); T.W. Kibble, Phys.Rev. 155, 1554 (1967).
36. G. 't Hooft, Nucl.Phys. B33, 173 (1971); B35, 167(1971); G. 't Hooft, and M. Veltman, Nucl.Phys. B50, 318(1972); B.W. Lee, Phys.Rev. D5, 823(1972); B.W. Lee and J. Zinn-Justin, Phys.Rev. D5, 3121(1972), 5, 3137 (1972); 5, 3155 (1972); 7, 1049 (1973).
37. G. 't Hooft and M. Veltman, Nucl.Phys. B44, 189 (1972).
38. S. Weinberg, Phys.Rev.Letters 19, 1264 (1967); 27, 1688 (1971); A. Salam in Elementary Particle Physics, ed. by N. Svartholm (Almquist and Wiksells, Stockholm, 1968), p.367.
39. J.D. Bjorken and C.H. Llewellyn Smith, Phys.Rev. D7, 887 (1973).
40. C.G. Bollini and J.J. Giambiagi, Phys.Lett. 40B, 566 (1972); J.F. Ashmore, Nuovo Cimento Lett., 4, 289 (1972); G.M. Cicuta and E. Montaldi, Nuovo Cimento Lett. 4, 329 (1972).
41. This example was suggested by B. Lautrup.
42. G. 't Hooft and M. Veltman, Diagrammar, CERN 73-9, yellow report (1973).
43. W.A. Bardeen, in Proceedings of the 16th High energy physics conference, NAL, Batavia, Ill., September 1972, ed. by J.D. Jackson and A. Roberts (NAL, 1973), vol.2, pp.295-298.
44. W.A. Bardeen, Phys.Rev. 184, 1849 (1969); R.W. Brown, C.C. Shih and B.L. Young, Phys.Rev. 186, 1491 (1969).
45. F.J. Hasert et al., Phys.Letters 46B, 138 (1973); A. Benvenuti et al., Phys.Rev.Letters 32, 800 (1974); B. Aubert et al., Phys.Rev.Letters 32, 1454, 1457 (1974); S.J. Barish et al., Phys.Rev.Letters 33, 448 (1974); B.C. Barish et al., Phys.Rev. Letters 34, 538 (1975).
46. F.J. Hasert et al., Phys.Letters 46B, 121 (1973).
47. Gh. Bertrand-Coremans et al. (Gargamelle Neutrino-Freon Collaboration), contributed paper at the Colloque International sur la physique de neutrino à haute énergie, Paris, 1975 .
48. G. 't Hooft, Phys.Letters 37B, 195 (1971); H.H. Chen and B.W. Lee, Phys.Rev. D5, 1874 (1972).
49. V. Brisson, Search for the processes ($\nu_\mu + e^- \to \nu_\mu + e^-$) and ($\bar{\nu}_\mu + e^- \to \bar{\nu}_\mu + e^-$), Proceedings of the XVI International Conference on High energy physics, National Accelerator Laboratory, Batavia, Ill., September 1972, ed. by J.D. Jackson and A. Roberts, (National Accelerator Laboratory), Vol. 2, p. 195.
50. H.S. Gurr, F. Reines and H.W. Sobel, Phys.Rev.Letters 28, 1406 (1972).
51. C. Baltay, Review of the experimental situation on neutral currents, proceedings neutrino '72 Europhysics conference, Balatonfüred, Hungary, June 1972, ed. by A. Frenkel and G. Marx (OMKDK Technoinform, Budapest) Vol.1, p.199 .

52. D.C. Cundy, Plenary report on neutrino physics, proceedings of the XVII International Conference on high energy physics, London, 1974, ed. by J.R. Smith (Rutherford Laboratory, Chilton, 1974), p. IV-131.
53. B.L. Beron et al., Phys.Rev.Letters 33, 663 (1974); J.E. Augustin et al., Phys.Rev.Letters 34, 233 (1975).
54. J. Godine and A. Hankey, Phys.Rev. D6, 3301 (1972); A. Love, Nuovo Cimento Lett. 5, 113 (1972); G.V. Grigoryan and A. Khoze, Yerevan Phys.Institute report (unpublished); V.K. Cung, A.K. Mann, and E.A. Paschos, Phys.Lett. 41B, 355 (1972).
55. I.B.Khriplovich, Sovj.J.Nucl.Phys. 17, 298(1973).
56. R. Budny, Phys.Lett. 45B, 340 (1973).
57. D.A. Dicus, Phys.Rev. D8, 890(1973); R.W. Brown, V.K. Cung, K.O. Mikaelian and E.A. Paschos, Phys.Lett. 43B, 403 (1973); F.A. Berends, K.J.F. Gaemers and R. Gastmans, Nucl.Phys. B57, 381 (1973); B63, 381 (1973); A.B. Kraemer and B. Lautrup, preprint NBI-HE-75-4 (to be published).
58. R. Gastmans and Y. Van Ham, Phys.Rev. D10, 3629 (1974).
59. R. Jackiw and S. Weinberg, Phys.Rev. D5, 2396 (1972); I. Bars and M. Yoshimura, Phys.Rev. D6, 374 (1972).
60. J. Bailey et al., Phys.Lett. 55B, 420 (1975).

RENORMALIZATION OF GAUGE FIELDS MODELS

C. BECCHI

University of Genova and

I.N.F.N. (Sezione di Genova)

INTRODUCTION

The aim of these lectures is to introduce a new, hopefully regularization independent, approach to the renormalization of gauge field models.

The relevance of gauge field models and in particular of the models involving Yang-Mills fields associated with non abelian transformations to the construction of renormalizable theories of the weak and electromagnetic interactions has been exhaustively discussed in this school[1]. Their renormalizability is widely analysed in the existing literature[2].

We shall describe here a new approach to renormalizable gauge field models[3]. This approach is essentially based on the extensive use of a renormalized version of the Schwinger action principle[4] which has been recently proved[5] in the framework of the Bogoliubov-Parasiuk-Hepp-Zimmermann (B.P.H.Z.) renormalization scheme[6].

Our study is also based on a newly discovered algebraic property of the Faddeev-Popov[7] gauge field lagrangians; namely their invariance under a set of non linear transformations (Slavnov[8] transformations) which explicitly involve the Faddeev-Popov Fermi scalar ghost fields.

We shall limit our analysis to the SU(2) Higgs-Kibble model[9]. This is a typical example to which our results can be systematically applied since the Lie algebra which characterizes the gauge transformations is semisimple and the theory does not involve massless fields.

In the first section we summarize in terms of the Schwinger action principle some formal results of lagrangian field theory and we state their renormalized version.

In section two we discuss at the classical level the algebraic properties of the SU(2) Higgs-Kibble-Faddeev-Popov lagrangian and we exhibit its invariance under Slavnov transformations.

The extension of these results to the full quantum level is discussed in the third section.

SECTION I

This is an introductory section on some results of renormalized lagrangian field theory. The main reason to include it in these lectures is to make clear some aspects of the language which are not (yet) of common use in quantum field theory.

For the sake of simplicity we shall only discuss the case of a scalar field which is supposed to be massive. The generalization of the main results to the case of non scalar fields is straightforward.

For systems with a finite number of degrees of freedom the Schwinger action principle can be derived from the Schrödinger equation[4].

Let $q = (q_1,\dots,q_n)$, $p = (p_1,\dots,p_n)$ be the canonical coordinates of a system whose dynamics is described by the hamiltonian $H\,(p,q,t)$; $\{\lambda\}$ stands for a set of parameters.

We indicate by $|\varphi,T;t,\lambda>$ the Schrödinger state at time t coming from the state $|\varphi>$ at time T and by $q(t)$, $p(t)$, $H_\lambda(t)$ the Heisenberg canonical operators and hamiltonian, $F_\lambda(t)$ an arbitrary operator.

Introducing the lagrangian : $L_\lambda(t) = L_\lambda(q(t),\dot{q}(t),t) = p(t)\dot{q}(t) - H_\lambda$ we get :

$$\frac{\partial}{\partial\lambda_i} <\psi,T';0,\lambda|F_\lambda(t)|\,\varphi,T;0\,\lambda> =$$

$$\frac{i}{\hbar}\int_T^{T'} dt'\{<\psi,T';0,\lambda|\;\left(F_\lambda(t)\frac{\partial}{\partial\lambda_i}L_\lambda(t')\right)_+|\,\varphi,T;0\lambda> + <\psi,T';0,\lambda|$$

$$\frac{\partial}{\partial\lambda_i}F_\lambda(t)\,|\varphi,T;0,\lambda>\} \qquad (1)$$

where $(\;)_+$ is the time-ordering

Let us now consider a field theory lagrangian density :

$$\mathcal{L}(x) = \mathcal{L}_0(x) + \mathcal{L}_I(x) + \mathcal{L}_S(x) = \varphi_{(x)}(K\varphi)(x) + eP_I(x) + J(x)\varphi(x) \tag{2}$$

where $\mathcal{L}_0$ is the free part and K(x,y) the kernel of the wave operator; the interaction part $\mathcal{L}_I$ is supposed to be a polynomial eP_x in the field and J(x) is a c-number source. Indicating by $|\Omega>$ the vacuum and by $|\Omega,in>$ and $|\Omega,out>$ the states $|\Omega,-\infty;0,e,J(x)>$ and $|\Omega,+\infty;0,e,J(x)>$ respectively we define the functional :

$$Z[e,J] = <\Omega,out|\ \Omega\ in> \tag{3}$$

By repeated use of eq.(1) we see that Z[e,J] is the functional generating the vacuum expectation values of the time-ordered products of fields:

$$\left(\frac{\hbar}{i}\right)^n \prod_{i=1}^{m} \frac{\delta}{\delta J(x_i)} Z[e,J] = <\Omega,out|\left(\prod_{i=1}^{n} \varphi(x_i)\right)_+ |\Omega,in> \tag{4}$$

we have also

$$\frac{\partial}{\partial e} Z[e,J] = \frac{i}{\hbar} <\Omega,out| \int dx\ \frac{\partial}{\partial e} \mathcal{L}(x)\ |\Omega,in> \tag{5}$$

Since according to the definition (eq.(3)) Z is trivially invariant under the local (canonical) infinitesimal field transformation :

$$\varphi(x) \rightarrow \varphi(x) + \lambda(x)\ \phi[\varphi](x) \tag{5}$$

where ϕ is a polynomial in the field and $\lambda(x)$ an infinitesimal c-number local parameter, we have :

$$<\Omega,out|\ \delta\int dx\ [\mathcal{L}_0 + \mathcal{L}_I + \mathcal{L}_S](x)|\ \Omega,\ in > \tag{7}$$

$$\equiv <\Omega\ ,out|\ \int dx\ \lambda(x)\ [J(x)\ \phi[\varphi](x) + \int dy\ \frac{\delta}{\delta\lambda(x)} (\mathcal{L}_0 + \mathcal{L}_I)(y)]\ |\Omega,in> = 0$$

In the framework of the renormalized perturbation theory the functional Z is defined through the Gell-Mann-Low expression :

$$Z[e,J] = \frac{<\Omega^{(o)}|\,(e^{\frac{i}{\hbar}\int dx\,(\mathcal{L}_I^{(o)} + \mathcal{L}_S^{(o)})(x)})_+|\Omega^{(0)}>}{<\Omega^o\,|(e^{\frac{i}{\hbar}\int dx\,\mathcal{L}_I^{(o)}(x)})_+\,|\Omega^{(o)}>} \equiv <e^{\frac{i}{\hbar}\int dx\,\mathcal{L}_S(x)}> \qquad (8)$$

(the superscript $^{(o)}$ meaning the interaction picture), as a formal power series in the source $J(x)$. The time-ordering symbol is defined by the subtraction procedure. The coefficient of the nth order term in $J(x)$ is the formal sum of all the renormalized Feynman amplitudes with n external legs.

The functional Z_c defined by

$$Z[e,J] = e^{\frac{i}{\hbar} Z_c [e,J]} \qquad (9)$$

is another formal series in $J(x)$. The coefficient of the nth order term is now the formal sum of all the connected Feynman amplitudes with n external legs. Since a connected Feynman amplitude with L loops is proportional to $\hbar^n$ (*) the functional Z_c can be considered as a double formal series in $J(x)$ and $\hbar$ whose term of lowest order in $\hbar$ represents the tree approximation (classical) limit.

The renormalized Feynman amplitudes have typical power counting properties which are simply expressed in the case of proper (amputated one-particle irreducible) amplitudes, i.e. those which cannot be broken into two parts by cutting only one propagator.

The Fourier transformed proper n-field amplitude :

$$<\varphi(0) \prod_{1}^{n-1}{}_i \;\tilde{\varphi}(p_i) >_+^{PROP} = \Gamma^{(n)}_{(p_i)} \qquad (10)$$

satisfies the condition :

$$\lim_{\lambda\to\infty} \lambda^{n-4-\varepsilon} \;\Gamma^{(m)}(\lambda p_i) = 0 \qquad (11)$$

for any positive ε if the momenta are euclidean (imaginary energies) and non exceptional (no partial sum vanishes).

(*) The Feynman rules prescribe a factor $\hbar^{-1}$ for each vertex, $\hbar$ for each internal line and the numbers of loops L, vertices v, and internal lines l are connected by : L = l-v+I.

To state the renormalized version of eq.(5) and eq.(7) we have first to define a suitable class of renormalized local operators. For any positive integer d we shall define O_d a local operator of class d if it satisfies for any integer n and positive ε :

$$\lim_{\lambda\to\infty} \lambda^{n-d-\varepsilon} \langle O_d(0) \prod_1^n{}_i \tilde{\varphi}(\lambda p_i)\rangle_+^{PROP} = 0 \tag{12}$$

if the momenta are euclidean and non exceptional.

Now a renormalized local operator must be specified prescribing a suitable system of normalization conditions. After Zimmerman[6] we can state the following theorem. A local operator P_d of class d is uniquely specified by the Taylor series to order d-n about the origin of the proper amplitudes :

$$t^{d-n}_{\{p_i\}} \langle P_d(0) \prod_1^n{}_i \tilde{\varphi}(p_i)\rangle_+^{PROP} \tag{13}$$

for all the integers n up to d. Hence, given a basis $\{p_i\}$ of polynomials of dimension non exceeding d in the fields and their derivatives (⁎) we can define a basis of local operators of class d $\{N_d(P_i)\}$ by the normalization condition :

$$t^{d-n}_{\{p_i\}} \langle N_d(P_i)(0) \prod_d^n{}_i \tilde{\varphi}(p_i)\rangle_+^{PROP} = \langle N_d(P_i)(0) \prod_1^n{}_i \tilde{\varphi}(p_i)\rangle_+^{TRIVIAL} \tag{14}$$

where by trivial we mean the tree approximation contribution.

This is a simplified definition of the Zimmerman N-product operators.

We are now in condition to state the renormalized versions of eq.(5) and eq.(7) which have been proved in the framework of the B.P.H.Z. renormalization scheme by Gomes, Lowenstein and Lam[5].

If λ is a parameter of the theory there exists a local polynomial Λ_λ in the fields and their derivatives whose dimension is lower than five and such that :

$$\partial_\lambda Z = \frac{i}{\hbar} \langle \int dx\, N_4[\Lambda_\lambda]\,(x)\, e^{\frac{i}{\hbar}\int dy\, \mathcal{L}_S(y)} \rangle_+ \tag{15}$$

Comparing with the tree approximation we see that in the classical limit ($\hbar = 0$), Λ_λ coincides with $\partial_\lambda \mathcal{L}$. Hence :

(⁎) A scalar field has dimension one and each derivative increases the dimension by one.

$$\Lambda_\lambda = \partial_\lambda \mathcal{L} + \hbar Q_\lambda \tag{16}$$

for some Q_λ.

Given a formal infinitesimal field transformation $\varphi(x) \to \varphi(x) + \lambda(x) N_d [\phi] (x)$ where $\lambda(x)$ is a c-number function and ϕ is a local field polynomial, there exists a local polynomial Λ_ϕ in the fields and their derivatives whose dimension is lower than 4+d-I and such that :

$$< \int dx\, \lambda(x) \,[J N_d [\phi] + N_{4+d-1}[\Lambda_\phi]]\,(x)\, e^{\frac{i}{\hbar}\int dy\, \mathcal{L}_S(y)} >_+ = 0 \tag{17}$$

Again comparing with the tree approximation yields :

$$\Lambda_\phi = \frac{\delta}{\delta\, \lambda(x)} \int dy\, \mathcal{L}(y) + \hbar Q_\phi \quad . \tag{18}$$

This is a weak formulation of the renormalized quantum action principle. Indeed eq.(15) and eq.(17) simply amount to a power counting prescription, which, however, will be sufficient to prove rather sophisticated results on the renormalizable gauge field models.

To conclude this introductory section let us recall the connection between the scattering operator S and the functional Z. According to the L.S.Z. asymptotic theory, S is given in the perturbative sense in terms of Z and of the canonically quantized asymptotic field φ^{in} by :

$$S = :\exp\int dx dy \varphi^{in}(x) K^{as}(x,y) \frac{\delta}{\delta J(y)} : Z \,\Big|_{J=0} = \Sigma\, Z \,\Big|_{J=0} \tag{19}$$

where K^{as} is the asymptotic wave operator derived from the renormalized propagator.

SECTION 2.

Since the Higgs-Kibble mechanism and the structure of the Yang-Mills lagrangian have been discussed in the first week of this School, we shall not examine here the reasons upon which the choice of the Faddeev-Popov lagrangian is based, but simply describe the structure of the SU(2) Higgs-Kibble model and its algebraic properties.

Let $\{\varphi_1\} \equiv (\{\pi_\alpha\},\sigma)$ with $i = 1,\ldots,4$, $\alpha = 1,2,3$ be a multiplet of scalar fields and a^μ_α $(\alpha = 1,2,3)$ be a multiplet of Yang-Mills photons. An infinitesimal gauge transformation of parameter $\delta\omega_\alpha(x)$ is defined by :

$$\delta\pi_\alpha = \frac{e}{2}[\ \varepsilon^{\beta\gamma}_\alpha \delta\omega_\beta \pi_\gamma + (\sigma+F)\delta\omega_\alpha\ ] \equiv \frac{\delta\pi_\alpha}{\delta\omega_\beta}\delta\omega_\beta$$

$$\delta\sigma = -\frac{e}{2}\pi^\alpha\delta\omega_\alpha = \frac{\delta\sigma}{\delta\omega_\alpha}\delta\omega_\alpha \tag{20}$$

$$\delta a^\mu_\alpha = \partial^\mu\delta\omega_\alpha + e\varepsilon^{\beta\gamma}_\alpha \delta\omega_\beta\ a^\mu_\gamma \equiv \frac{\delta a_\alpha}{\delta\omega_\beta}\delta\omega_\beta$$

The parameter F plays the role of the σ field vacuum expectation value. A lagrangian invariant under such ted in terms of the antisymmetric covariant field tensor $G^{\mu\nu} = \partial^\mu a^\nu_\alpha - \partial^\nu a^\mu_\alpha - e\varepsilon^{\beta\gamma}_\alpha a^\mu_\beta a^\nu_\gamma$ and of the covariant derivatives : $\{D^\mu\varphi_i\} \equiv (\{\partial^\mu\pi_\alpha - \frac{e}{2}[\varepsilon^{\beta\gamma}_\alpha a^\mu_\beta\pi_\gamma + (\sigma+F)a^\mu_\alpha]\}, \partial^\mu\sigma + \frac{e}{2}(\pi^\mu a^\mu_\alpha))$, it has the form :

$$\mathcal{L}_{inv} = -\frac{1}{4}(C^\alpha_{\mu\nu})^2 + \frac{1}{2}(D_\mu\varphi_i)^2 - H(\underline{\varphi + F}) \tag{21}$$

Here H is a polynomial in $\pi_\alpha\pi^\alpha + (\sigma+F)^2$ invariant under the transformation (20) and satisfying the condition :

$$\partial_{\varphi_i} H(\underline{\varphi+F})\Big|_{\varphi=0} = 0 \tag{22}$$

which ensures that the classical vacuum corresponds to the configuration $\{\varphi_i\} = 0$.

From eq. (22) combined with the invariance condition for H it turns out that the mass matrix of the scalar fields $\|\partial_{\varphi_i}\partial_{\varphi_j}H\||_{\varphi=0}$ has three null eigenvalues corresponding to the π_α modes and a positive eigenvalue corresponding to the σ field. This is a direct consequence of the Goldstone theorem[10].

Writing the bilinear (free) part of the lagrangian one can study the wave operator of the fields. Looking in particular at the coupled photon $-\pi$ channel the free lagrangian is

$$\mathcal{L}_{0,inv} = -\frac{1}{4}(\partial^\mu a^\nu_\alpha - \partial^\nu a^\mu_\alpha)^2 + \frac{e^2}{8}F^2(a^\mu_\alpha)^2 + \frac{1}{2}(\partial^\mu\pi_\alpha)^2 + \frac{eF}{2}\partial_\mu\pi^\alpha a^\mu_\alpha \tag{23}$$

The Fourier transformed wave operator factorizes in two sub-matrices, the first corresponding to the transverse modes of the photon, the second to the coupled longitudinal photon and π channels. They are respectively :

$$K^{T\ \alpha,\beta}_{\mu\nu}(k) = -\delta^{\alpha\beta}(g^{\mu\nu} - \frac{k^{\mu}k^{\nu}}{k^2})(k^2 - \frac{e^2F^2}{4})$$

$$((K^{\alpha\beta}_{a_2,\pi})) \equiv \delta^{\alpha\beta} \begin{pmatrix} \frac{e^2F^2}{4} & -ik\frac{eF}{2} \\ ik\frac{eF}{2} & k^2 \end{pmatrix} \qquad (24)$$

The first equation exibits the Higgs phenomenon (the photons are massive). The determinant of the matrix shown in eq.(24) is identically zero. It follows that our invariant theory is not directly quantizable.

One of the ways out of this problem is to eliminate the degrees of freedom with degenerate wave operator by means of a field dependent gauge transformation[9, 11]. This reduces the degrees of freedom of the model to a minimal number, the transverse modes of the photons and the σ field to which "physical" particles are associated. However, the resulting lagrangian is not of the renormalizable type. This is an essential argument in favour of the Faddeev-Popov lagrangian which is the natural extension to the case of non abelian gauge fields of the usual lagrangian of Q.E.D.

The Faddeev-Popov lagrangian involves, beyond the scalar and vector fields already considered, two conjugate multiplets of scalar fields $\{c_\alpha\}$, $\{\bar{c}_\alpha\}$, $(\alpha = 1,2,3)$ labelled by the adjoint representation of SU(2). These fields are quantized according to the Fermi statistics in a Fock space with indefinite metrics. The lagrangian is :

$$\mathcal{L} \equiv \mathcal{L}(\{\varphi_i\},\{a^{\mu}_{\alpha}\},\{c_{\alpha}\},\{\bar{c}_{\alpha}\}) = \mathcal{L}_{inv}(\{\varphi_i\},\{a^{\mu}\})$$

$$- \frac{1}{K}\left[\frac{(g^{\alpha})^2}{2} - c_{\alpha}(x)\, \mathcal{M}^{\alpha\beta}(x,y)\, \bar{c}_{\beta}(y)\right] \qquad (25)$$

with $g^{\alpha} = \partial_{\mu} a^{\mu,\alpha} + \rho\pi^{\alpha}$ and $\mathcal{M}^{\alpha,\beta}(x,y) = \frac{\delta g^{\alpha}(x)}{\delta\omega_{\beta}(y)}$.

This modified version of the Higgs-Kibble model can be staightforwardly quantized in a Fock space with indefinite metrics. The corresponding perturbation theory is renormalizable by power counting. However we have to show that the resulting S operator restricted to the above defined physical subspace of the Fock space is unitary in the perturbative sense and independent from the parameters specifying the Faddeev-Popov term of the lagrangian. At least in the tree approximation this is equivalent to show that the matrix elements of the gauge operator g_α within physical states vanish. This is nothing but the well known supplementary condition of Q.E.D.

The algebraic property of $\mathcal{L}$ which we have referred to in the introduction is its invariance under the following system of infinitesimal transformations that we call Slavnov transformations :

$$\begin{aligned}
\delta\pi_\alpha &= \delta\lambda\, [-\tfrac{e}{2} E_\alpha^{\beta\gamma}\pi_\beta \bar{c}_\gamma + \tfrac{e}{2}(\sigma+F)\bar{c}_\alpha] \equiv \delta\lambda\, P_\alpha \\
\delta\sigma &= \delta\lambda\, [-\tfrac{e}{2}\pi^\alpha \varepsilon_\alpha] \equiv \delta\lambda\, P_o \qquad\qquad (26)\\
\delta a^\mu_\alpha &= \delta\lambda [\partial^\mu \bar{c}_\alpha - e\varepsilon_\alpha^{\beta\gamma} a^\mu_\beta \bar{c}_\gamma] \equiv \delta\lambda\, P^\mu_\alpha \\
\delta\bar{c}_\alpha &= \delta\lambda\, [\tfrac{e}{2}\varepsilon_\alpha^{\beta\gamma}\bar{c}_\beta \bar{c}_\gamma] \equiv \delta\lambda\, \bar{P}_\alpha \\
\delta c_\alpha &= \delta\lambda\, [\partial_\mu a^\mu_\alpha + \rho\pi_\alpha] \equiv \delta\lambda\, g_\alpha
\end{aligned}$$

$\delta\lambda$ is a space-time independent infinitesimal parameter which commutes with $\{\varphi_i\}$ and $\{a^\mu_\alpha\}$ but anticommutes with $\{c_\alpha\}$ and $\{\bar{c}_\alpha\}$ and for two transformations labelled by $\delta\lambda_1$ and $\delta\lambda_2$, $\delta\lambda_1$ and $\delta\lambda_2$ anticommute.

The invariance of the action corresponding to $\mathcal{L}$ can be checked immediately using the composition law of the gauge transformations :

$$[\frac{\delta}{\delta\omega_\alpha(x)}, \frac{\delta}{\delta\omega_\beta(y)}] = \varepsilon_\gamma^{\alpha\beta}\ \delta(x-y)\ \frac{\delta}{\delta\omega_\Gamma(x)} \qquad (27)$$

To express the Slavnov invariance of the lagrangian as a functional differential equation for Z_c we add to $\mathcal{L}$ a system of source terms introducing the external fields $\{\gamma_\alpha, \gamma^\mu_\alpha, \gamma_o, \zeta_\alpha\}$ coupled to $\{P_\alpha, P^\mu_\alpha, P_o, \bar{P}_\alpha\}$:

$$\mathcal{L}^{(\eta)} \equiv \mathcal{L}(\{\varphi_i\},\{a^\mu_\alpha\},\{c_\alpha\},\{\bar{c}_\alpha\}) + \gamma_\alpha P^\alpha + \gamma^\alpha_\mu P^\mu_\alpha + \gamma_o P_o + \zeta^\alpha \bar{P}_\alpha \tag{28}$$

It is a remarkable property of the Slavnov transformations that they leave invariant this modified lagrangian. (Because the polynomials $\{P_\alpha, P^\mu_\alpha, P_o, \bar{P}_\alpha\}$ are invariant).

Upon introducing the sources $\{J_\alpha, J_o, j^\mu_\alpha, \bar{\xi}_\alpha, \xi_\alpha\}$ for the fields $\{\pi_\alpha, \sigma, a^\mu_\alpha, c_\alpha, \bar{c}_\alpha\}$ eq. (7) yields

$$\mathcal{S} Z \equiv \int dx \, [J_\alpha \frac{\delta}{\delta\gamma_\alpha} + J_o \frac{\delta}{\delta\gamma_o} + j^\mu_\alpha \frac{\delta}{\delta\gamma^\mu_\alpha} - \xi^\alpha \frac{\delta}{\delta\zeta\alpha}$$

$$- \bar{\xi}^\alpha (\partial_\mu \frac{\delta}{\delta\gamma^\alpha_\mu} + \rho \frac{\delta}{\delta\gamma^\alpha})] \quad (x) \; Z =$$

$$\frac{i}{\hbar} \langle\Omega,\text{out} \mid \int dx \; [\; J_\alpha P^\alpha + J_o P_o + j^\alpha_\mu P^\mu_\alpha - \xi^\alpha \bar{P}_\alpha - \bar{\xi}^\alpha g_\alpha](x) \; |\Omega,\text{in}\rangle \tag{29}$$

whenever the lagrangian is Slavnov invariant, that is at least at the tree level. Multiplying eq.(29) by Σ_{phys}, the restriction to the physical degrees of freedom of the operator Σ defined in eq. (19), and suppressing the external sources we get the desired supplementary condition :

$$- \frac{\hbar}{i} \frac{\delta}{\delta\xi(0)} \sum_{phys} \mathcal{S} Z \Big|_{J=0} = i\hbar \, (\partial_\mu \frac{\delta}{\delta j^\alpha_\mu} + \rho \frac{\delta}{\delta j^\alpha}) \; (0) \sum_{phys} Z\Big|_J$$

$$= \sum_{phys} \langle\Omega,\text{out}| \; g^\alpha_{(x)} \; |\Omega,\text{in}\rangle = 0 \tag{30}$$

Hence we can consider a good renormalization program for our model to try to preserve the Slavnov identity (eq.(29)) to any order of the perturbation series. However this attitude is legitimate only if the Faddeev-Popov lagrangian is uniquely determined (up to the addition of a divergence and a multiplicative renormalization of the fields) by the Slavnov invariance condition. The dimensional constraints insuring the renormalizability of the theory are understood.

In the cases in which the gauge group is semisimple this uniqueness condition is verified. In the abelian case there may arise new terms. For instance in Q.E.D. in the Stueckelberg gauge the photon mass term $\int [\, a_\mu a^\mu/2 + c\,\bar{c}\,](x)\,dx$ is Slavnov invariant.

If the external fields are assigned dimension two the most general dimension four lagrangian carrying no Faddeev-Popov charge $Q^{\phi\pi}$

$$Q^{\phi\pi} J^\alpha = 2\, J^\alpha$$

$$Q^{\phi\pi} c_\alpha = c_\alpha \;;\quad Q^{\phi\pi}\gamma_\alpha = \gamma_\alpha \;;\quad Q^{\phi\pi}\gamma_o = \gamma_o \;;\quad Q^{\phi\pi}\gamma^\alpha_\mu = \gamma^\alpha_\mu$$

$$Q^{\phi\pi}\pi_\alpha = Q^{\phi\pi}\sigma = Q^{\phi\pi} a^\alpha_\mu = 0 \tag{31}$$

$$Q^{\phi\pi}\bar{c}^\alpha = -\,\bar{c}^\alpha$$

is :

$$\mathcal{L}(\phi) + \gamma_\alpha \mathcal{P}^\alpha + \gamma_o \mathcal{P}_o + \gamma^\alpha_\mu \mathcal{P}^\mu_\alpha + J^\alpha \bar{\mathcal{P}}_\alpha \equiv \bar{\mathcal{L}}^{(\eta)}$$

where ϕ stands for the quantized fields and the polynomials $\{\mathcal{P}^\alpha\ \mathcal{P}^\mu_o, \mathcal{P}_\alpha, \bar{\mathcal{P}}_\alpha\}$ have dimension non exceeding two.

The Slavnov invariance condition writes in the functional differential form :

$$\int dx \Big[\mathcal{P}_\alpha \frac{\delta}{\delta\pi_\alpha} + \mathcal{P}_o \frac{\delta}{\delta\sigma} + \mathcal{P}_\alpha \frac{\delta}{\delta a^\mu_\alpha} + \bar{\mathcal{P}}_\alpha \frac{\delta}{\delta c_\alpha} + g_\alpha \frac{\delta}{\delta c_\alpha}\Big] \bar{\mathcal{L}}^{(\eta)} = \mathcal{S}\bar{\mathcal{L}}^{(\eta)} = 0 \tag{32}$$

Looking in particular at the coefficients of the external fields we get :

$$\mathcal{S}\mathcal{P}_\alpha = \mathcal{S}\mathcal{P}_o = \mathcal{S}\mathcal{P}^\mu_\alpha = \mathcal{S}\bar{\mathcal{P}}_\alpha = 0 \tag{33}$$

from which the structure of the Slavnov transformation given in eq.(26) can be reconstructed. Indeed, for example, under the stated conditions $\bar{\mathcal{P}}_\alpha$ has the form :

$$\bar{\mathcal{P}}_\alpha = \frac{1}{2}\,\Gamma^{\beta\gamma}_\alpha\, \bar{c}_\beta\, \bar{c}_\gamma \tag{34}$$

and eq. (33) writes :

$$\mathcal{S}\bar{\mathcal{P}}_\alpha = \frac{1}{2}\,\Gamma^{\beta\lambda}_\alpha\,\Gamma^{\gamma\delta}_\lambda\, \bar{c}_\beta \bar{c}_\gamma \bar{c}_\delta = 0 \tag{35}$$

which is nothing but the Jacobi identity :

$$\Gamma^{\beta\lambda}_{\alpha}\,\Gamma^{\gamma\delta}_{\lambda} + \Gamma^{\gamma\lambda}_{\alpha}\Gamma^{\delta\beta}_{\lambda} + \Gamma^{\delta\lambda}_{\alpha}\,\Gamma^{\beta\gamma}_{\lambda} = 0 \tag{36}$$

Let us now come back to the external field independent part of the lagrangian $\bar{\mathcal{L}}(\phi)$. If $\bar{\mathcal{L}}(\phi)$ is Slavnov invariant a fortiori:

$$\mathcal{S}^2\bar{\mathcal{L}}(\phi) \equiv \int dx(\partial_\mu \mathcal{J}^\mu_\alpha + \rho\mathcal{P}_\alpha)(x)\, \frac{\delta}{\delta c_\alpha(x)}\bar{\mathcal{L}}(\phi) =$$

$$\int dx(\mathcal{M}\bar{c})_\alpha\,(x)\, \frac{\delta}{\delta c_\alpha(x)}\, \bar{\mathcal{L}}(\phi) = 0 \tag{37}$$

Now $\bar{\mathcal{L}}(\phi)$ can be written inthe form :

$$\bar{\mathcal{L}}(\phi) = \bar{\mathcal{L}}_{inv}(\{\pi_\alpha\},\sigma,\{a^\mu_\alpha\}) + \Delta\mathcal{L}(\{\pi_\alpha\},\sigma,\{a^\mu_\alpha\})$$

$$+ L^{\alpha\beta\gamma\delta}c_\alpha c_\beta \bar{c}_\gamma c_\delta + c_\alpha(K\bar{c})^\alpha \tag{38}$$

where $\bar{\mathcal{L}}_{inv}$, invariant under Slavnov transformations, is itself a solution of eq.(32). By eq.(37) it is obvious that :

$$L^{\alpha\beta,\gamma\delta} = 0 \tag{39}$$

and that :

$$\int dx(\mathcal{M}\bar{c})_\alpha\,(x)\,(K\,\bar{c})^\alpha\,(x) = 0 \tag{40}$$

Writing out the general form of K yields if the gauge group is semisimple :

$$c_\alpha(K\,\bar{c})^\alpha = c_\alpha\,\Gamma^{\alpha\alpha'}\,(\mathcal{M}\bar{c})_{\alpha'} \tag{41}$$

where $\Gamma^{\alpha\alpha'}$ is a numerical symmetrical matrix. Going back to eq.(32) yields :

$$\bar{\mathcal{L}}(\phi) = \mathcal{L}_{inv}(\{\pi_\alpha\},\sigma,\{a^\mu_\alpha\}) + c_\alpha\,\Gamma^{\alpha\alpha'}(\mathcal{M}\bar{c})_{\alpha'} - \frac{1}{2}\,g_\alpha\Gamma^{\alpha\alpha'}g_{\alpha'} \tag{42}$$

which is identical with eq.(25) modulo a redefinition of g_α and c_α.

SECTION 3

We now discuss the extension of the Slavnov identity (eq.(29)) to the quantum level. Since the time at our disposal is restricted we shall only sketch out the complete proof which can be found in the referred literature. In particular we shall translate the Slavnov transformations to the quantum level in the form :

$$
\begin{aligned}
\delta\pi_\alpha &= \delta\lambda\, N_2 \left[-\frac{e}{2}\,\varepsilon_\alpha^{\beta\gamma}\,\pi_\beta\,\bar{c}_\gamma + \frac{e}{2}\,(\sigma+F)\,\bar{c}_\alpha \right] \equiv \delta\lambda\, N_2[P_\alpha] \\
\delta\sigma &= -\delta\lambda\, N_2 \left[\frac{e}{2}\,\pi^\alpha\,\bar{c}_\alpha \right] \equiv \delta\, N_2[P_o] \\
\delta a_\alpha^\mu &= \delta\lambda \left[\partial^\mu \bar{c}_\alpha - N_2 \left[e\,\varepsilon_\alpha^{\beta\gamma}\, a_\beta^\mu\, \bar{c}_\gamma \right] \right] \equiv \delta\lambda\, N_2[P_\alpha^\mu] \qquad (43) \\
\delta\bar{c}_\alpha &= \delta\lambda\, N_2 \left[\frac{e}{2}\,\varepsilon_\alpha^{\beta\gamma}\,\bar{c}_\beta\bar{c}_\gamma \right] \equiv \delta\lambda\, N_2[\bar{P}_\alpha] \\
\delta c_\alpha &= \delta\lambda \left[\partial_\mu a_\alpha^\mu + \rho\pi_\alpha \right] = \delta\lambda\, \mathcal{E}_\alpha
\end{aligned}
$$

thus forgetting the possibility of a finite renormalization of the polynomials ($P_\alpha, P_o, P_\alpha^\mu, \bar{P}_\alpha$). This avoids a lot of algebra without spoiling the meaning of the proof.

Let us then consider a quantum extension of the model corresponding through some renormalization procedure to the lagrangian $\mathcal{L}_{eff} = \mathcal{L} + \hbar\mathcal{L}'$. This is a formal power series in $\hbar$ whose lowest order term is the classical lagrangian $\mathcal{L}$.

After an infinitesimal Slavnov transformation the renormalized action principle yields :

$$
\begin{aligned}
&\frac{i}{\hbar} < \int dx \left[J_\alpha N_2 [P^\alpha] + J_o N_2[P_o] + J_\alpha^\mu N_2[P_\mu^\alpha] - \xi^\alpha N_2[\bar{P}_\alpha] - \bar{\xi}^\alpha g_\alpha \right. \\
&\qquad \left. + N_5(\Lambda) \right] \times - e^{\frac{i}{\hbar}\int dy \mathcal{L}_S(y)} >_+^{CONN} = \qquad (44) \\
&= \mathcal{S} Z_c + \frac{i}{\hbar} < \int dx\, N_5(\Lambda)(x)\, e^{\frac{i}{\hbar}\int dy\, \mathcal{L}_S(y)} >_+^{CONN} = 0
\end{aligned}
$$

with :

$$
\Lambda = \mathcal{S}\mathcal{L}_{eff} + \hbar Q \qquad (45)
$$

vanishing for $\hbar = 0$. In eq.(45) Q sums up the radiative corrections. In the above mentioned simplifying hypothesis Λ does not depend on the external fields.

We want to show that, up to higher order terms in $\hbar$,

$$\int dx \, \Lambda \equiv \bar{\Lambda} = \mathcal{S}\Omega \tag{46}$$

for some dimension four functional Ω. If it is so comparing with eq.(45) we see that the equation :

$$\bar{\Lambda} = 0 \tag{47}$$

can be solved in terms of $\mathcal{L}_{eff}$ order by order in $\hbar$.

Let us now introduce a new external field carrying $Q^{\phi\pi} = 1$ coupled to the vertex $N_5[\Lambda](x)$. In terms of the β-dependent connected Green functional $Z_c^{(\beta)}$ eq.(44) writes :

$$\mathcal{S}Z_c^{(\beta)} = \int dx \frac{\delta}{\delta\beta(x)} Z_c^{(\beta)} + \int dx \, \beta(x) < N_6 [\Lambda'] (x) \, e^{\frac{i}{\hbar}\int dy \mathcal{L}_S(y)} >_{+}^{CONN} \tag{48}$$

The second term in the right hand side of eq.(48) comes from the quantum variation of the β coupling. Hence :

$$\Lambda'(x) = \mathcal{S}\Lambda(x) + O(\hbar\Lambda) \tag{49}$$

Now we compute :

$$\begin{aligned} \mathcal{S}^2 Z_c &= - \int dx \left[\bar{\xi}^{\alpha}\left(\partial_{\mu}\frac{\delta}{\delta\gamma_{\mu}^{\alpha}} + \rho \frac{\delta}{\delta\gamma^{\alpha}}\right)\right](x) \; Z_c = \\ &= \mathcal{S}\int \frac{\delta}{\delta\beta(x)} dx \, Z_c^{(\beta)}\Big|_{\beta=0} \\ &= - \int dx < N_6[\Lambda'] (x) \, e^{\frac{i}{\hbar}\int dy \mathcal{L}_S(y)} >_{+}^{CONN} \end{aligned} \tag{50}$$

since, due to the anticommutativity of β, $\int dx \, dy \frac{\delta}{\delta\beta(x)} \frac{\delta}{\delta\beta(y)} = 0$
Now eq.(50) leads to a consistency condition for Λ.
Indeed eq.(50) is a perturbed version of the equation :

$$\int dx \, [\bar{\xi}^{\alpha} \, (\partial_{\mu} \frac{\delta}{\delta\gamma^{\alpha}_{\mu}} + \rho \frac{\delta}{\delta\gamma^{\alpha}})] \quad (\lambda) \; Z_{c} = 0 \tag{51}$$

and we can show that in the class of models in which eq.(51) is verified :

$$\bar{\xi}^{\alpha}(x) = L^{\alpha\alpha'} \, (\partial_{\mu} \frac{\delta}{\delta\gamma^{\alpha}_{\mu}} + \rho \frac{\delta}{\delta\gamma^{\alpha}}) \; (x) \quad Z_{c} \tag{52}$$

for some symmetrical matrix L. This is indeed a solution of eq. (51) which in the tree approximation coincides with the general solution. Hence eq.(52) is the general solution of eq.(51) in the sense of formal power series in $\hbar$.

Comparing eq.(51) with eq.(50) leads to :

$$\begin{aligned}\bar{\xi}^{\alpha} (x) = \; & L^{\alpha\alpha'} (\partial_{\mu} \frac{\delta}{\delta\gamma^{\alpha}_{\mu}} + \rho \frac{\delta}{\delta\gamma^{\alpha}}) (x) \; Z_{c} \\ & + <N_{3} \, [\frac{\delta}{\delta c^{\alpha}} R^{(\Lambda)}] \; (x) \; e^{\frac{i}{\hbar} \int dy \mathcal{L}_{S}(y)} >^{CONN}_{+}\end{aligned} \tag{53}$$

for some dimension 4 functional $R^{(\Lambda)}$ of order Λ . Substituting eq.(53) into eq.(50) yields :

$$\begin{aligned}& \int dx \; <N_{3}[\frac{\delta}{\delta c_{\alpha}} R^{\Lambda}] (x) \quad e^{\frac{i}{\hbar} \int dy \mathcal{L}_{S}} >^{CONN}_{+} \\ & <[\partial_{\mu} N_{2} \; [P^{\alpha}_{\mu}] + \rho N_{2} \; [P_{\alpha}]] \; (x) \; e^{\frac{i}{\hbar} \int dy \mathcal{L}_{S}} >^{CONN}_{+} \\ & = \int dx \; <N_{6} \; [\Lambda'] \; (x) \; e^{\frac{i}{\hbar} \int dy \mathcal{L}_{S}} >^{CONN}_{+}\end{aligned} \tag{54}$$

which to the minimal order in $\hbar$ for which Λ is non vanishing implies the relation :

$$\int dx \; (\mathcal{M} \bar{c})_{\alpha} \; \frac{\delta}{\delta c_{\alpha}} \; R^{(\Lambda)} \equiv \mathcal{S}^{2} R^{(\Lambda)} = \bar{\Lambda}' = \mathcal{S} \bar{\Lambda} \tag{55}$$

where the radiative corrections to Λ have been forgotten and $\bar{\Lambda}' = \int dx \; \Lambda'(x)$.

Now eq.(55) is solved by

$$\bar{\Lambda} = \mathcal{S} R^{(\Lambda)} + \bar{\Lambda}_{\natural} \tag{56}$$

for some $\bar{\Lambda}_{\natural}$ satisfying :

$$\mathcal{S}\, \bar{\Lambda}_{\natural} = 0 \tag{57}$$

It remains to show that eq.(57) implies that

$$\bar{\Lambda}_{\natural} = \mathcal{S}\, \hat{\bar{\Lambda}}_{\natural} \tag{58}$$

for some $\hat{\bar{\Lambda}}_{\natural}$.

We use for $\bar{\Lambda}_{\natural_\alpha}$ the expansion :

$$\bar{\Lambda}_{\natural} = \int dx\, \Lambda(x)\, \bar{c}_\alpha(x) + \int dx\, dy\, dz\, \Lambda^{\alpha\beta\gamma}(x,y,z)\, c_\alpha(x)\bar{c}_\beta(y)\bar{c}_\gamma(z) +$$

$$\int dx\, \Lambda^{\alpha\beta,\gamma\delta\eta}\, (c_\alpha c_\beta \bar{c}_\gamma \bar{c}_\delta \bar{c}_\eta)(x) \tag{59}$$

Since eq.(57) implies :

$$\mathcal{S}^2\, \bar{\Lambda}_{\natural} = 0$$

a discussion similar to that found at the end of section 2 shows that

$$\Lambda^{\alpha\beta\gamma}(x,y,z) = \Lambda^{\alpha\beta,\gamma\delta\eta} = 0 \tag{60}$$

Using now eq.(57) yields :

$$\frac{\delta}{\delta\omega_\alpha(x)}\, \Lambda^\beta(y) - \frac{\delta}{\delta\omega_\beta(y)}\, \Lambda^\alpha(x) - e\, \varepsilon^{\alpha\beta}_{\ \ \gamma}\, \delta(x-y)\Lambda^\gamma(x) = 0 \tag{61}$$

which is nothing else than the Wess-Zumino[12)] consistency condition.

One can show that the general solution of eq.(61) is in our case

$$\Lambda^\alpha(x) = \frac{\delta}{\delta\omega^\alpha(x)}\, \hat{\bar{\Lambda}}_{\natural} \tag{62}$$

from which eq.(58) immediately follows thus completing the proof of the Slavnov identity.

For models involving fermion fields and gauge groups admitting a totally antisymmetric invariant tensor $D^{\alpha\beta\gamma}$ with indices in the adjoint representation the general solution of eq.(61) is no more eq.(62) but

$$\Lambda^{\alpha}_{(x)} = \frac{\delta}{\delta\omega^{\alpha}(x)} \hat{\bar{\Lambda}}_{\natural} + h_{\alpha}(x)$$

where $h_{\alpha}(x)$ is the Adler-Bardeen anomaly [13].

CONCLUSION

In section 3 we have discussed the existence of a renormalized quantum extension of the SU(2) Higgs-Kibble classical model.

It remains to investigate how the parameters defining our model can be fixed by means of a suitable system of normalization conditions in order the theory can be interpreted as an operator theory in the Fock space of the free fields.

Then we should show the unitarity and independence from the parameters which define the gauge function of the restriction of the scattering operator S to the physical subspace of the Fock space.

For lack of time we do not examine these problems which are solved in the referred literature[3].

We should also mention that very recent technical improvements[14] allow to extend our method to the analysis of models involving massless particles. There is also some progress in the study of models whose gauge group is not semisimple.

We can thus hope that in some more or less near future we shall be able to discuss models really relevant to the physics of the weak interactions.

ACKNOWLEDGEMENTS

These lectures contain a reduced version of the analysis of the renormalization of gauge field models due to A. Rouet and R. Stora in collaboration with the author. The author wishes to thank in particular R. Stora for his interest in the draft of this paper.

REFERENCES

1. B. Lautrup; elsewhere in this volume.
2. A complete bibliography about the theory of gauge fields can be found for instance in :
E.S. Abers, B.W.Lee : Phys.Reports 9C n°1 (nov.1973)
M. Veltman : invited talk presented at the International Symposium on electrons and photons at high energies, Bonn 27-31 August 1973.
J. Zinn-Justin : Lectures given at the International Summer Institute for Theoretical Physics, Bonn 1974.
3. C. Becchi, A. Rouet, R. Stora, Comm.Math.Phys. 42, 127(1975).
C. Becchi, A. Rouet, R. Stora, Marseille preprints, 74/P639 (october 1974) and 75/P723 (april 1975).
4. P. Schwinger, Phys.Rev. 82, 914 (1951);
C.S. Lam, Nuovo Cimento 38, 1754 (1965).
K. Symanzik, Lectures on lagrangian quantum field theory, DESY T-71/1 (february 1971).
5. J.H. Lowenstein, Comm.Math.Phys. 24, 1(1971) and Seminars on renormalization theory, Vol.II, Univ. of Maryland Technical Report n° 31 -068 (december 1972).
M. Gomes, J.H. Lowenstein, Pittsburg report (august 1972)
Y-M.P. Lam, Phys.Rev. D6, 2145 (1972)
P. Breitenlohner, D. Maison : dimensional renormalization and the action principle, MPI-PAE/PTh 25/74 (may 1975).
6. W. Zimmermann, Ann.Phys. 77, 536-570 (1973).
7. L.D. Faddeev, V.N. Popov, Phys.Letters 25B, 29 (1967).
8. A.A. Slavnov, Teor.i Mat.Fiz., 10, 153 (1972).
J.C. Taylor, Nucl.Phys. B33, 436 (1971).
9. P. Higgs, Phys.Lett. 12, 132 (1964); Phys.Rev. 145, 1156(1966);
T.W.B. Kibble, Phys.Rev. 155, 1554 (1967).
10. J. Goldstone, Nuovo Cimento, 19, 154(1961).
11. S. Coleman, J. Wess, B. Zumino, Phys.Rev. 177, 2239 (1969).
12. J. Wess, B. Zumino, Phys.Letters 37B, 95 (1971).
13. W.A Bardeen, Phys.Rev. 184, 1848 (1969).
14. J.H. Lowenstein, W. Zimmerman, Nucl.Phys. B86, 77(1975) and in the Max-Planck-Institute Münich) preprints : MPI-PAE/PTh5 and 6/75 (march 1975).
J.H. Lowenstein, Auxiliary mass formulation of the pure Yang-Mills model. NYU/TR3/75.
C. Becchi, in preparation.
A. Rouet, T. Clark, renormalization of gauge theories with massless propagators. Max-Planck Institute, preprint (Münich).

GAUGE FIELDS AND STRONG INTERACTIONS

Thomas APPELQUIST*

Department of Physics, Yale University

New Haven, Connecticut 06520

ABSTRACT

These three lectures deal with several aspects of the Yang-Mills quark gluon theory of strong interactions. A dispassionate overview is followed by a plunge into the physics of electron positron annihilation into hadrons. Looming large in this discussion are the newly discovered, long-lived mesons. The role of these new heavy hadrons in the quark gluon model is examined. It is suggested that they are heavy quark antiquark bound states and that their properties could provide a rather clear and simple experimental handle on the underlying field theory.

NOTE

At the time the lectures were being given and written up (July through September 1975), the experimental situation with respect to the new particles was developing and changing rapidly. At the risk of giving the published version of the lectures a somewhat acausal character, I have incorporated a discussion of some of these developments.

---

* Alfred P. Sloan Foundation Fellow

LECTURE 1

1. INTRODUCTION

There now exists an attractive and viable candidate (really a class of candidates) for a local quantum field theory of the strong interactions. The elementary fields in the strong interaction Lagrangian are quarks and vector gluons. Very little is known about most features of this theory. For example, it is not at all clear how to calculate the ground state properties of hadrons or even to ennumerate the spectrum of physical states. Nevertheless, its short distance behavior is well understood and its long range structure is very tantalizing. This structure suggests the possibility that the theory could produce long range forces that might permanently confine quarks and gluons to the interior of physical hadrons.

The discovery[1),2)] of heavy, long lived $J^P = 1^-$ hadrons within the last year is an important development for the quark gluon theory of strong interactions. It means that new, heavy quarks must be included in the model[3)]. Furthermore, the new particles should provide us with an important new experimental handle on the dynamics of the model. By this I mean that their features should reflect the properties of the underlying field theory more directly than the other hadrons. This is due to the large mass of the new quarks and their subsequent nonrelativistic bound state motion.

These three lectures are intended to be a survey of the Yang-Mills quark gluon theory of strong interactions with special emphasis on the role of the new hadrons. I will occasionally disgress into related matters such as weak and electromagnetic interactions but the main thrust will be strong interaction dynamics. In the first lecture, I will describe the model and discuss renormalization, the renormalization group and asymptotic freedom. Lecture II will begin with a discussion of quark mass renormalization. The remainder of this lecture will be devoted largely to electron positron annihilation into hadrons. The computation of the total cross section behavior in terms of the underlying quark gluon field theory is discussed and sources of possible perturbation theory breakdown identified. In Lecture III, particular attention is paid to the breakdown in the vicinity of a heavy quark-antiquark threshold. The possible role of asymptotic freedom in explaining the narrowness of the new hadrons is examined. I will review the status of the heavy quark-antiquark bound state as a nonrelativistic system and look at the spin dependent forces and the role of the Bethe-Salpeter equation. I will conclude with a few remarks about infrared behavior and quark confinement.

2. A DESCRIPTION OF THE MODEL

I will assume that the strong interactions are described by a local, renormalizable quantum field theory of quarks and gluons. The Lagrangian is

$$= -\frac{1}{4} F^a_{\mu\nu} F^{a\mu\nu} + \bar{\psi}(i\not{D} - m_0)\psi \qquad (1)$$

where ψ is a set of quark fields coming in several flavors, u,d,s along with one or more heavy quarks. Each flavor comes in three colors[4),5)] and color is taken to be an exact SU(3) gauge symmetry. Thus each quark color multiplet has a single mass and the colored vector mesons remain massless. $F^a_{\mu\nu}$ is the gauge covariant curl and D_μ is the covariant derivative

$$(D_\mu\psi)_n = \partial_\mu\psi_n - \frac{1}{2} g A^a_\mu(\lambda^a)_{nm} \psi_m \qquad (2)$$

where ψ_m is one of the color triplets. The symmetries of the theory are determined by the bare mass matrix m_0. Each flavor has its own electrical charge and the colored gluons are electrically neutral. I will always assume the existence of at least one heavy quark c. There may be others, but in the discussion of e^+e^- annihilation, I will espouse the view that only the c quark is operative at present energies.

Many of the properties of the colored quark model have been discussed by Fritzsch, Gell-Mann and Leutwyler[5)]. The emphasis in these lectures will be on the short distance structure of the model and I will devote most of the remainder of this lecture to introducing the essential ideas. Reference 5 is a good introductory overview of the model for those of you unacquainted with it.

3. RENORMALIZATION

The quantization of Yang-Mills theories has been discussed by many people. A recent lucid treatment is that of Lee and Zinn-Justin[6)] who develop the Feynman rules and discuss regularization and Ward-Slavnov identities. The Feynman rules are shown here in fig.1. Higher order computations (one or more closed loops) are best done using the dimensional continuation scheme[6)] as a regulator. I will occasionally be looking at higher order diagrams but only for the purpose of examining general features. It will not be necessary to do explicit computations requiring the explicit use of dimensional continuation.

By power counting, the theory is renormalizable so that a finite number of counterterms is sufficient to define ultraviolet

	Vertices	Bare Vertices
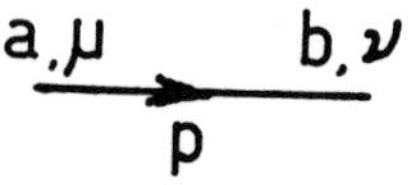	$-i\delta^{ab}\Delta_{\mu\nu}(p)$	$-i\delta^{ab}[(g_{\mu\nu}-p_\mu p_\nu/p^2)/p^2+ap_\mu p_\nu(p^2)^{-2}]$
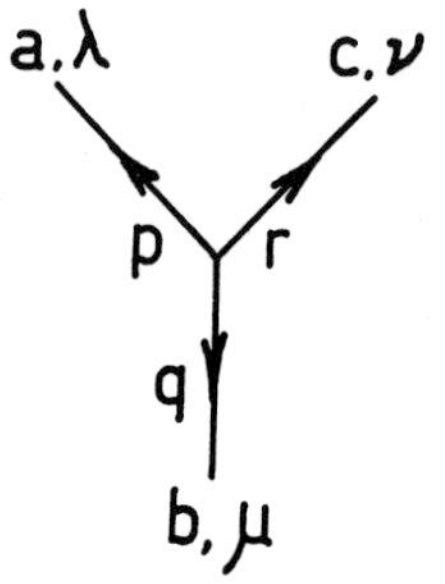	$i\Gamma^{abc}_{\lambda\mu\nu}(p,q,r)$	$f^{abc}[(p-q)_\nu g_{\lambda\mu}+(q-r)_\lambda g_{\mu\nu}+(r-p)_\mu g_{\nu\lambda}]$
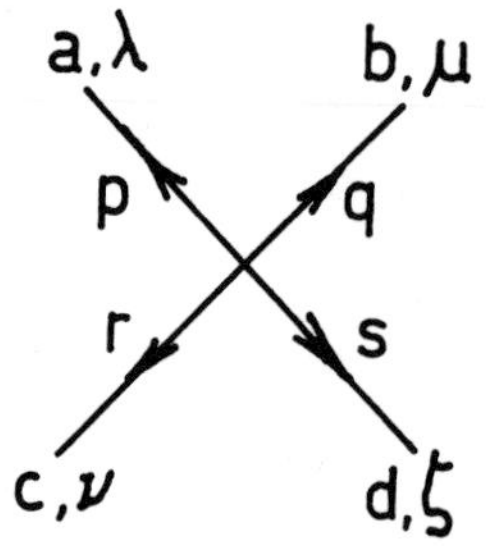	$i\Gamma^{abcd}_{\lambda\mu\nu\zeta}(p,q,r,s)$	$-if^{abf}f^{cdf}(g_{\lambda\nu}g_{\mu\zeta}-g_{\lambda\zeta}g_{\mu\nu})$ $-if^{acf}f^{bdf}(g_{\lambda\mu}g_{\nu\zeta}-g_{\lambda\zeta}g_{\mu\nu})$ $-if^{adf}f^{cbf}(g_{\lambda\nu}g_{\mu\zeta}-g_{\lambda\mu}g_{\zeta\nu})$
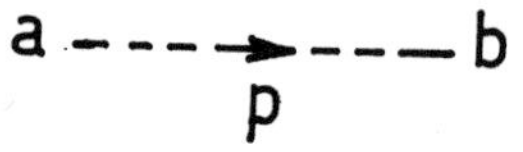	$\delta^{ab}iG(p)$	$i\delta^{ab}1/p^2$ F-P propagator
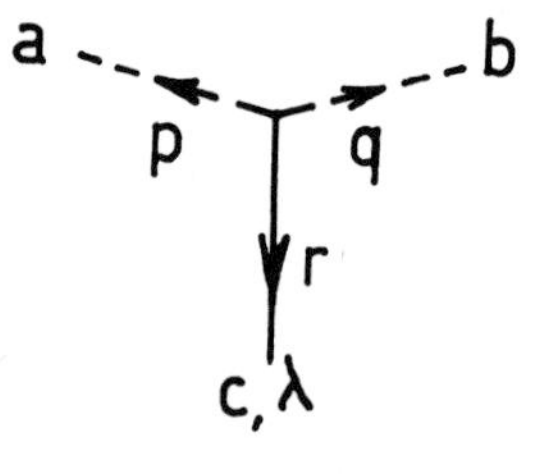	$i\gamma^{abc}_\lambda(p,q;r)$	$f^{abc}p_\lambda$

Fig. 1. Feynman rules for the pure Yang-Mills theory. The conventions are those of Bjorken and Drell. The inclusion of the quark propagator and the quark-quark-gluon vertex is straightforward.

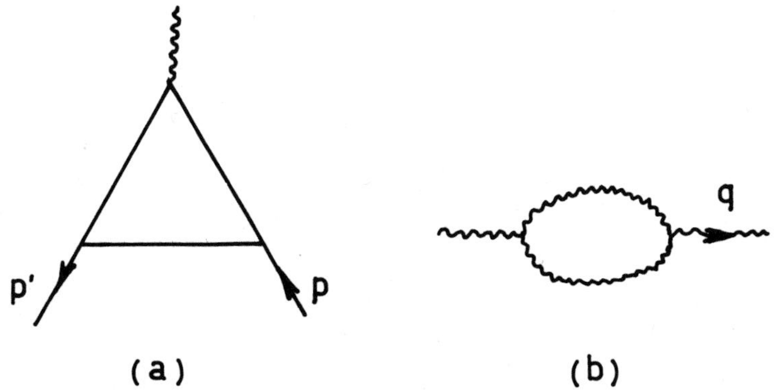

Fig. 2. One loop diagrams with infrared divergences on mass shell

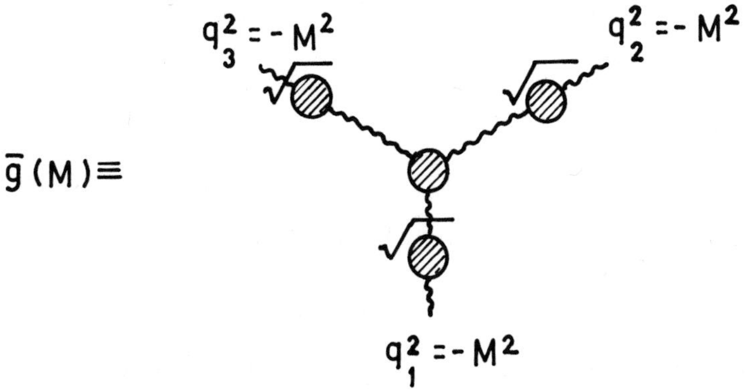

Fig. 3. A definition of the renormalized (running) coupling constant in a Yang-Mills theory.

finite Green's functions to all orders in perturbation theory. We thus have a perfectly satisfactory theory of Green's functions, all of which are calculable and finite as long as the external momenta are kept away from mass shell. In that limit, which unfortunately is the relevant limit for exploring particle structure and constructing the S-matrix, the theory is plagued by infrared divergences. It is easy to see, for example, that the quark gluon vertex diagram of fig. 2a behaves like $\log(p^2-m^2)$ in the limit $p^2 = p'^2 \to m^2$. Similarly, the gluon self-energy graph of fig. 2b behaves like $(q_{\mu\nu}q^2 - q_\mu q_\nu)\log q^2 + q_\mu q_\nu$ terms in the limit $q^2 \to 0$. On the nth loop level, there will be n log factors and unlike QED, the true infrared structure of the theory is unknown. In the next lecture, I will take a quick look at what little is known about the infrared structure and its speculated connection to quark confinement and the physical particle spectrum.

Because of the infrared divergences, the renormalized Green's functions must be defined by subtracting away from mass shell. A convenient way of defining the renormalized coupling constant $\bar{g}(M)$ is in terms of the gluon propagator and the 1PI three gluon vertex at a symmetric Euclidean point. This is shown in fig.3. Each of the external line factors symbolizes the object $\sqrt{d(k^2)}$ where $d(k^2)$ appears in the transverse part of the complete gluon propagator $-i(g_{\mu\nu} - k_\mu k_\nu/k^2)\ \frac{1}{k^2}\ d(k^2) - i\alpha\ k_\mu k_\nu/k^4$. The Ward-Slavnov identities[5] assure us that $\bar{g}(M)$ can equivalently be defined using say the quark-quark-gluon vertex.

I have so far considered only wave function and coupling constant renormalization. Mass renormalization and the question of how to define the renormalized quark masses is equally important but it is best to return tothat after a discussion of the renormalization group and asymptotic freedom.

4. THE RENORMALIZATION GROUP AND ASYMPTOTIC FREEDOM

The ideas of the renormalization group[7] and the property of asymptotic freedom[8] underlie much of what I will say about e^+e^- annihilation and the physics of $c\bar{c}$ bound states. The renormalization group is really the subject of Dr. Crewther's lectures and all I intend to do is to outline the notions that will be essential in lectures 2 and 3.

A renormalization mass M must be introduced to define the theory perturbatively. On the other hand, the physical content of the theory cannot depend on the choice of M and it is both possible and convenient to move it about. The existence of the renormalization group is the statement that a shift in M can be reabsorbed completely into multiplicative rescalings of field strengths and the coupling strength $\bar{g}(M)$. This feature of any renormalizable theory is most usefully expressed in terms of the partial differential equations of the renormalization group[7),9)].

These provide a framework for discussing both infrared and ultraviolet asymptotic behavior of the theory. This is in general a nontrivial problem precisely <u>because</u> of the essential presence of M. Even when the quark masses can be neglected the naive use of scale invariance is impossible. Perturbation theory is plagued by arbitrarily high powers of logarithms involving M which can sum up to modify the naive dimensional predictions. The differential equations of the renormalization group provide a framework for discussing these modifications, which depend on the properties of a given field theory. One such modification is asymptotic freedom.

A field theory is asymptotically free if $\bar{g}(M) \to 0$ as $M \to \infty$. The deciding one loop calculation reveals that the color SU(3) gauge theory is asymptotically free providing that $N \leq 16$. Recall that N is the number of quark flavors. As $M \to \infty$, $\bar{g}^2(M) \sim 1/\log M$, indicating that the short distance structure of the theory can be calculated perturbatively in terms of a small coupling constant. Since this is only an asymptotic theory, the question of when the small coupling regime is reached can only be answered experimentally. The experimental applications of asymptotic freedom are very limited since very few experimental quantities depend <u>only</u> on the short distance structure of the theory. The most direct experimental application is inelastic lepton scattering. The momentum transfer dependence of the structure function moments can be shown, by the use of the Wilson operator product expansion[10)], to depend only on short distance structure. Approximate Bjorken scaling of the structure functions can then be explained by asymptotic freedom providing that for $M > 1$ or 2 GeV, $\alpha_S(M) \equiv \bar{g}^2(M)/4\pi \ll 1$. This establishes the scale for the onset of the weak coupling region.

Some of the dynamical considerations in the next two lectures will involve some unconventional uses of asymptotic freedom. For this reason, it is important to understand in some detail the limitations on its use. Why is it difficult to experimentally isolate short distance behavior and thereby allow the use of perturbation theory ? The coupling constant $\bar{g}(M)$ has been defined in the deep Euclidean region ($M > 1$-2 GeV) so that it is small. The Green's functions of the theory can be calculated as a perturbation expansion in $\bar{g}(M)$ providing that this small coupling constant really is the appropriate expansion parameter. This will not be the case if large dynamical factors accompany each power of $\bar{g}(M)$. This can happen only if the object being calculated is sensitive to dimensional factors (such as a momentum or a mass) which are much smaller than M. The dynamical factors are typically powers of log(p/M) where p is the small momentum and if they accompany each power a $\bar{g}(M)$, then the true expansion parameter is not $\bar{g}(M)$ but instead of $\bar{g}$ appropriate to the smaller momentum scale. If this scale is less than a few hundred MeV, or, say one GeV to be conservative, $\bar{g}$ must become strong and the use of perturbation theory is impossible.

This then is the restriction on the use of the asymptotic freedom. With M taken greater than one or two GeV, $\bar{g}(M)$ is small but perturbation theory can be used only if small (< 1 GeV) dimensional parameters do not crucially enter the calculation. This is the case for example with deep Euclidean Green's functions and the leptoproduction structure function moments. In the later, the Wilson expansions must be used to disentangle the large dimensional parameters (momentum transfer and M) from the small ones (mass and binding energy of the target nucleon). The disentangling or elimination of small dimensional parameters will be of great concern in the next two lectures.

5. THE CHOICE OF FLAVORS

The fourth quark flavor was given a raison d'être long ago in the classic paper of Glashow, Iliopoulos and Maiani[11]. This is the celebrated GIM mechanism for the suppression of $\Delta S = 1$ weak neutral currents[12]. The fourth quark whose role in the strong interactions I will be discussing, could well be the GIM quark. That is, it could enter the weak currents as prescribed by GIM and have an electrical charge of 2/3. It is to be emphasized, however, that this is not necessary. No commitment to a particular theory of weak interactions need be made for the discussion of the color gauge theory of strong interactions.

There are by now many reasons for considering the possibility of even more than four quark flavors. In my opinion, none of these are yet as compelling as the theoretical (GIM) and experimental (new particles) evidence in favor of the fourth quark. Nevertheless, when one begins to contemplate the behavior of $\sigma_{TOT}(e^+e^- \to \text{hadrons})$, the strange and wonderful things being discovered in the high energy neutrino experiments and the arcane problems of triangle anomalies in weak and electromagnetic interactions, a natural if not profound questions emerges : Why not ? Indeed, a variety of models incorporating additional heavy quarks and/or heavy leptons have already been constructed. A lucid review of this work has recently been given by R.M. Barnett[13]. For the remainder of these lectures, I will assume that only the fourth quark c is playing a role in the current e^+e^- experiments. There is some evidence[14], however, that heavy leptons may also be being produced.

LECTURE II

6. MASS RENORMALIZATION

The strong interaction Lagrangian (1) has a bare quark mass matrix m_0. From the point of view of strong interaction phenomenology, it is reasonable to assume that it is a God given parameter in the strong interaction Lagrangian. The question of where it comes from is very interesting but very likely a deeper problem than strong interaction physics. Its origin is surely connected with the ultimate unification of the fundamental interactions and the breakdown of weak and electromagnetic symmetries[15]. Our strong interaction Lagrangian is an effective Lagrangian. It can be used in isolation up to energies Λ above which the weak and electromagnetic interactions become comparable to the strong [16]. If the effective weak and electromagnetic Lagrangian is renormalizable (a Higgs-gauge theory for example), Λ will be well above attainable laboratory energies[17]. I will assume this to be the case.

It is useful to think of m_0 as a renormalized mass matrix defined at the Euclidean momentum scale Λ. The effective mass matrix at laboratory energies will be related to $m_0(\Lambda)$ by renormalization effects due mainly to strong interactions since the weak and electromagnetic interactions stay small below Λ. The question of how to define these laboratory renormalized masses is a matter requiring some thought, especially since the quarks may be permanently confined inside color singlet hadrons. One can imagine defining a mass matrix $m(M)$ at a sliding Euclidean scale M [18], which approaches m_0 as $M \to \Lambda$, which stays nearly equal to m_0 through the weak coupling region down to one or two GeV, and which finally becomes some appropriate constituent quark model mass below one GeV. In the weak coupling regime, $m_0(M)$ may be experimentally accessible[19] but its connection to the constituent masses is obscured by strong coupling effects.

In the case of the charmed quark, it is possible to define a useful constituent mass in a precise field theoretic way. As we shall see, this relies on the fact that this mass sits in the weak coupling region and that binding energies are small. A convenient, but by no means unique, scheme is the following. A renormalized mass matrix m is obtained by the common rescaling $m_i = Zm_{0i}$. Then Z is adjusted so that for the c quark, m_c is the threshold of a cut in the propagator to any finite order of perturbation theory. Wave function renormalization must be done off shell as discussed in Lecture 1. Then as $p^2 \to m_c^2$, the c quarks propagator will behave like

$$\frac{1}{p^2-m_c^2}\ [\log(p^2-m_c^2)]^n$$

where n is the order of perturbation theory. This completely imitates quantum electrodynamics but, unlike that theory, the threshold behavior to all orders is unknown due to infrared instability. Nevertheless, we shall see that m_c, defined in this way, is experimentally accessible in a way that light quark masses are not. The experimental significance of the light quark masses as defined here is obscure but that isn't important for the heavy quark computations I will describe.

A crucial ingredient in some of our applications of asymptotic freedom is the rapid onset of the weak coupling regime. In particular, perturbation theory becomes possible well before the heavy c quark can be excited. For Euclidean momenta on the order of m_c, perturbation theory can be used although m_c will play a prominent role and the result will be far from scale invariant. This takes some gettting used to since the familiar short distance applications of the renormalization group are to an energy region well above all the mass parameters. The onset of the weak coupling region at about 1 GeV is determined by the light quark and gluon sector of the theory.

7. $\sigma_{TOT}(e^+e^- \to$ HADRONS) AND ASYMPTOTIC FREEDOM

The most direct application of asymptotic freedom to e^+e^- annihilation is a dispersive constraint on the total cross section. The hadronic vacuum polarization tensor $\Pi_{\mu\nu}(q) = (g_{\mu\nu}q^2 - q_\mu q_\nu) \times \Pi(q^2)$ can be calculated[20] perturbatively for space like q^2 less than -1 GeV^2. Since this is a deep Euclidean Green's function, the light quark masses can be scaled to zero and $\Pi(q^2)$ becomes a function of dimensionless ratios involving q^2, m_c and the renormalization mass M : $\Pi[q^2/M^2, m_c^2/M^2, \bar{g}(M)]$. The absence of sensitivity to the light quark masses (the small dimensional parameters) is insured by the mass singularity analysis of Kinoshita[21]. In effect, the external momentum provides an infrared shield insuring the existence of the limit $m_q \to 0$. $\Pi(q^2)$ is related to the total cross section by the dispersion relation

$$\Pi(q^2) = \frac{q^2}{\pi} \int_0^\infty \frac{ds\ R(s)}{s(s-q^2)} \tag{3}$$

where

$$R(s) \equiv \frac{3s}{4\pi\alpha^2}\ \sigma_{TOT}\ (e^+e^- \to \text{hadrons}) \tag{4}$$

and $s = E_{CM}^2$.

This can be used to put a bound on the total cross section behavior[22]. This is an important and useful fact but I want to go on to a more speculative use of asymptotic freedom.

R(s) appears to be flat from $E_{CM} \simeq 1$ GeV to $E_{CM} \simeq 3$ GeV and may again be flattening out above 5 to 6 GeV [23]. In the lwoer region, $R(s) \simeq 2$ so that it is given quite nicely by the parton model with the nine light quarks. The parton model calculation is zeroth order perturbation theory directly in the time like region as indicated in fig.4. Since quarks and gluons have not been seen and may never be seen as physical particles, can this be justified? In the language of the parton model, one simply says that once the probability for the creation of the $q\bar{q}$ pair has been correctly computed, the final state interactions that produce the physical hadrons occur with probability one. They do not affect the total cross section. In the remainder of this section and the next, I will use the property of asymptotic freedom to give some quantitative justification to this direct use of perturbation theory in the time like region. There are sources of perturbation theory breakdown however, and I will discuss them in Sec.9.

There are three energy regions to consider.

I. Low energies. Light quark masses negligible but the heavy c quark cannot be excited.

II. High energies. All quark masses negligible.

III. A transitional region.

I will first consider the use of asymptotic freedom in regions I and II and consider how the perturbation expansion could break down. Then I will look at the transitional region III. Here the perturbation expansion surely breaks down and we shall look in some detail at the way this happens.

To examine regions I or II in low orders of perturbation theory, we can set the masses of the operative quarks equal to zero. The zeroth order (parton) contribution is shown in fig.4 and we have $R^{(0)} = \sum_i Q_i^2$ where Q_i is the charge of the ith quark. This gives $R^{(0)}=2$ below charm threshold and $R^{(0)}=10/3$ in the 4 quark GIM model. To see whether this result is reliable, we examine the higher orders. The second order contribution is shown in fig.5. An explicit calculation for zero quark mass gives R(s) through second order.

$$R^{(2)}(s) = \sum Q_i^2 \; \{ \; + \frac{4}{3} \; \frac{3\,\alpha_s(M)}{4\pi} \} \qquad (5)$$

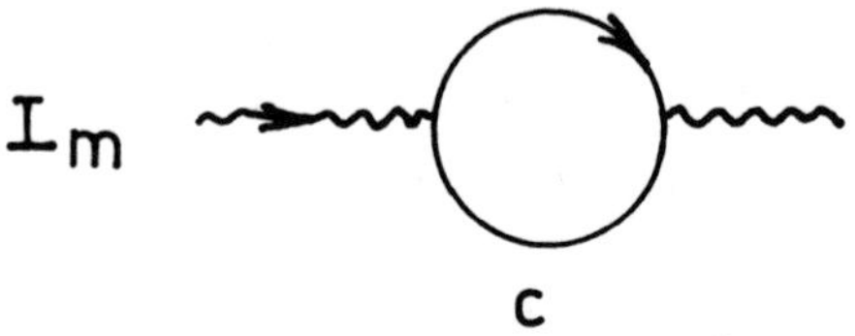

Fig. 4. The zeroth order, parton model contribution to R(s). It gives $R^{(0)} = 2$ in the colored triplet model.

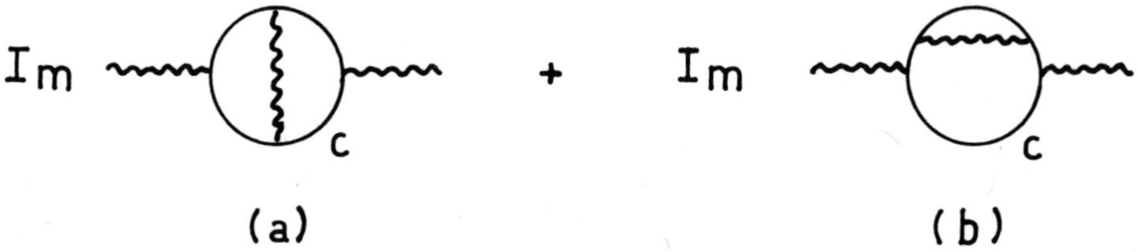

Fig. 5. The second order (order α_s) contributions to R(s).

This result is taken directly from electrodynamics[24)25)]. The only modification is the factor of 4/3 in second order which comes from an SU(3) color sum. Since $\alpha_s(M) \ll 1$ for $M > 1$ GeV, the second order term is a small correction.

8. MASS SINGULARITIES

It is important to understand the simplicity of the second order result before going on. In doing so, we shall encounter the essential property of the theory which underlies the use of perturbation theory at timelike as well as spacelike momenta. R(s) is dimensionless and could therefore be a function of dimensionless ratios of E_{CM}, quark masses m_q and the renormalization mass M. There is no M dependence through second order since no renormalization subtractions need be done. The overall divergence is absent since an imaginary part is being taken and the subgraph divergences in graphs (5a) and (5b) cancel through the electromagnetic Ward identity. Explicit M dependence will enter in the next order.

The absence of explicit quark mass dependence (the existence of the limit $m_q \to 0$) is insured by the same mass singularity theorem of Kinoshita[21,26)] that underlies perturbation theory, in the Euclidean region. Thus no small dimensional parameters enter the calculation to this order and R(s) remains constant. Perturbation theory seems to be converging.

The application of the Kinoshita theorem to $\sigma_{TOT}(e^+e^- \to \text{hadrons})$ is particularly straightforward. The idea is that any graph contributing to $\Pi(q)$ is finite when all internal masses are taken to zero since the external momentum provides an infrared cutoff. This is true for timelike as well as spacelike q^2 and in the time-like case, it applies to the absorptive as well as dispersive parts. The sum of the contributions to $\sigma_{TOT}(e^+e^- \to \text{hadrons})$ corresponding to the different Cutkosky cuts of a single Feynman diagram will be free of mass singularities. This is a generalization of the Block-Nordseick analysis in quantum electrodynamics to a situation with self-coupled massless fields. Individual contributions will contain mass singularities leading typically to terms like $\log E_{CM}/m_q$ or logarithmic singularities due to the masslessness of the gluons. However, they will cancel in the total cross section.

I can perhaps best put the mass singularity theorem in perspective by making a list of qualifying remarks.

1. It applies only to renormalizable theories. It has been proven[21)] for all except Yang-Mills theories and in that case I have checked in low orders that the theorem is true. I am assuming here that it works to all orders[27)].

2. It is applicable to any properly defined total transition probability. In general, however, there are mass singularities associated with the incoming lines and these must be treated carefully[21]. For e^+e^- annihilation the initial line is a single of shell photon and there are no mass singularities.

3. It says nothing about partial transition probabilities. In the case of e^+e^- annihilation into hadrons, it can be applied only to the total cross section. In particular, it completely leaves open the question of quark confinement. To answer this question, one would have to look at more than σ_{TOT}. A constraint would have to be put on the final state to force the macroscopic separation of the $q\bar{q}$ pair. A confining theory would inevitably lead to the presence of small dimensional parameters such as the inverse $q\bar{q}$ separation length or the energy resolution of the quark detector or perhaps mass shell singularities of the type discussed in Sec.6. Perturbation theory is not likely to be useful.

In fourth order and beyond, renormalization subtractions become necessary and R(s) picks up s dependence in the form of powers of $\log s/M^2$. They remain inocuous until s becomes much larger than M. It is then sensible to express the perturbation expansion in terms of a $\bar{g}(s)$ appropriate to the larger energy scale. As $s \to \infty$, $\bar{g}^2(s) \sim 1/\log s$ and the parton model result is approached from above. I will imagine keeping M fixed at around 3 GeV. Then for a sizable energy range, $\log s/M^2$ is of order one and it isn't necessary to introduce a sliding coupling constant $\bar{g}(s)$. The perturbation expansion will converge as long as there are no mass singularities and this is insured to this order and beyond by the Kinoshita analysis. It is important to be able to scale all the light quark masses to zero. It is also possible to scale m_c to zero in region III although this isn't always appropriate and certainly not necessary since $m_c > 1$ GeV.

The mass singularity theorem has apparently justified the use of perturbation theory for $\sigma_{TOT}(e^+e^- \to \text{hadrons})$. The parton model is the dominant, zeroth order contribution and higher order corrections can be calculated. The second order result (5) is in fact a good fit to the total cross section from one to three GeV since with $\alpha_s(M) \simeq 0.25$ [28], $R^{(2)}(s) \simeq 2.2$. Above the transitional region ($E_{CM} > 5.5$ GeV), R(s) is again flat and equal to about 5 [23]. This clearly indicates there is something going on in addition to, or in place of, the excitation of 12 GIM quarks. An attractive possibility from several points of view is the production of heavy leptons[14,29]. As far as the quark sector is concerned however, it seems both experimentally and theoretically that perturbation theory could be a reliable tool for the total cross section.

9. SOURCES OF BREAKDOWN

The theoretical situation isn't nearly that good. There is a rather obvious source of perturbation theory breakdown that appears in higher orders. In the Euclidean region, the Kinoshita analysis assures us that the internal masses any diagram can be scaled to zero. Since the light quark masses are assumed to be small (<<1GeV), this is an appropriate thing to do for momenta well above 1 GeV. In the physical region, however, the perturbation expansion contains branch points corresponding to nominal multiple quark thresholds. If quarks are confined, these thresholds disappear in favor of physical particle thresholds when the theory is saloed to all orders. Nevertheless they are there in finite orders and it is only appropriate to scale $m_q \to 0$ in a given diagram if E_{CM} is well above all the nominal quark thresholds of that diagram.

For any value of E_{CM}, diagrams exist (perhaps in very high order) which contain thresholds at or above E_{CM}. This is the source of a new small dimensional parameter -- the distance to nearby nominal quark thresholds. This will no doubt cause perturbation theory breakdown in high orders and it is probably connected to the formation of the physical particles. In ref. 13, some plausibility arguments are given that the high order effects average out to be a small contribution to R(s) but this has not been proven. In the next lecture, I will assume that because of the convergence of the perturbation expansion in low orders, R(s) can indeed be computed perturbatively in regions I and II.

There is a way to avoid the multiple threshold singularities and yet to improve upon the dispersive constraint (4). (q^2) can be calculated in the complex q^2 plane with Re $q^2 > 0$ provided that Im q^2 is taken non-zero and large enough to shield from the singularities of the physical region. With Im $q^2 > 1$ gGeV, there will be no sensitivity to small (<1 GeV) dimensional parameters and perturbation theory can be applied. The dispersion relation for (q^2) then leads to a prediction for R(s) averaged over a region of order 1 GeV [30]. It must be given by the parton model computation.

It remains a problem to understand why the parton model works on a much more local levels. The experiments average over energy intervals on the order of 1 MeV and the parton model works well from 1 to 3 GeV. The use of perturbation theory locally for R(s) is not fully justified but I will assume that it is possible unless an obvious breakdown appears in low orders.

LECTURE III

10. $c\bar{c}$ THRESHOLD BEHAVIOR

Recall from the last lecture that away from the transitional region, perturbation theory converges through low orders. This consequence of the Kinoshita analysis was used to give some justification to the use of perturbation for σ_{TOT}. In the transitional region, on the other hand, the breakdown of perturbation theory is immediate, coming already at second order.

The second order computation of R(s), (Eq.5) is easily generalized to include the heavy quark mass. The result is[24,25]

$$R^{(2)}(s) = \sum_{\substack{\text{light}\\\text{quarks}}} Q_i^2 \left\{1 + \frac{4}{3} \frac{3\,\alpha_s(M)}{4\pi}\right\}$$
$$+ \sum_{\substack{\text{heavy}\\\text{quarks}}} Q_i^2\, \theta(s-4m_c^2)\, v\, \frac{3-v^2}{2} \left\{1 + \frac{4}{3}\alpha_s(M) f(v)\right\}, \tag{6}$$

where v is the velocity of the heavy quark or antiquark, $v=\sqrt{1-4m_c^2/s}$. The overall heavy quark factor $v\,\frac{3-v^2}{2}$ is just s-wave phase space and it approaches 1 as $m_c \to 0$. The function f(v) is complicated, involving combinations of Spence functions, but it is well represented (to ± 1 percent) by an interpolating formula due to Schwinger[25] :

$$f(v) = \frac{\pi}{2v} - \frac{3+v}{4}\left(\frac{\pi}{2} - \frac{3}{4\pi}\right) \tag{7}$$

This formula is exact in the limits $v \to 0$ and $v \to 1$ and as $v \to 1$, $f(v) \to \frac{3}{4\pi}$ in agreement with the zero mass calculation.

As $v \to 0$, $f(v) \to \frac{\pi}{2v}$. This behavior comes from graph (5a) and is a consequence of a Coulomb-like final state interaction. In n^{th} order, n uncrossed gluon exchanges give n factors of 1/v and the perturbation expansion breakds down. In the limit $v \to 0$, these leading terms are just the expansion of the nonrelativistic Coulomb enhancement factor[25])

$$\frac{\frac{4}{3}\alpha_s \frac{\pi}{v}}{1 - \exp\left(-\frac{4}{3}\alpha_s \frac{\pi}{v}\right)} = \frac{|\psi_{COUL}(0)|^2}{|\psi_{COUL}(\infty)|^2} \tag{8}$$

representing the increased probability for $c\bar{c}$ production in an attractive Coulomb field. This breakdown of the perturbation expansion for small v is connected to a breakdown below $E_{CM} = 2m_c$. There, the sum of uncrossed gluon exchanges is responsible for the formation of positronium-like bound states. A typical contribution below threshold is shown in fig. 6.

If this were the only breakdown of perturbation theory, the uncrossed ladders could be summed as in QED and one could perturb about that result. There is, however, a more serious perturbation theory breakdown due to the Yang-Mills infrared structure. As threshold is approached from above or below, the typical momentum transfer flowing through exchanged gluons decreases. This means that the higher order corrections to these lines become more and more important, increasing the effective coupling strength. When the typical exchanged momentum becomes less than about one GeV, the theory becomes strongly coupled and the perturbation expansion breaks down.

I can estimate the width of the transitional region inside of which the Yang-Mills infrared structure leads to a breakdown of the perturbation expansion. For E_{CM} above $2\,m_c$, a measure of the typical momentum transfer $\langle k \rangle$ is $\langle k^2 \rangle \simeq \frac{1}{4}(E^2_{CM} - 4\,m_c^2)$. If $\langle k \rangle$ is to be less than one GeV, then $E^2_{CM} < 4\,(m_c^2 + 1\ \mathrm{GeV}^2)$. Below $E_{CM} = 2\,m_c$, the typical momentum transfer is a graph like the one in fig. 6 is $\langle k \rangle \simeq \sqrt{m_c^2 - E^2_{CM}/4}$. In a Coulombic theory, this is just the Bohr momentum at each bound state. Thus if we take $E_{CM} = 2m_c - \frac{1}{4}(\frac{3}{3}\alpha_s)^2\, m_c \frac{1}{n^2}$, then $\langle k \rangle = \frac{1}{2}(\frac{4}{3}\alpha_s) m_c \frac{1}{n}$.
The entire range in E^2_{CM} inside of which the typical momentum transfer is less than one GeV is thus

$$\Delta(E^2_{CM}) \simeq 2E_{CM}\,\Delta E_{CM} \simeq 8\ \mathrm{GeV}^2 \quad . \tag{9}$$

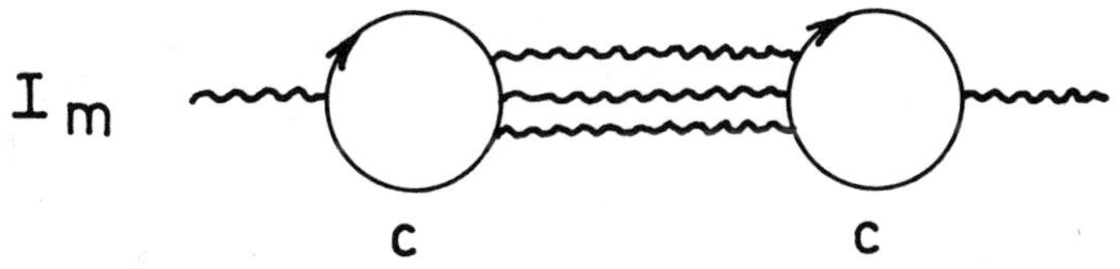

Fig. 6. A contribution to σ_{TOT} below threshold. Positronium-like resonances arise from uncrossed ladder exchanges in the c quark loops.

With $E_{CM} \simeq$ 3-4 GeV, $\Delta E_{CM} \gtrsim$ 1 GeV. In the absence of some careful higher order calculations, this can be taken only as an order of magnitude estimate. The experimental curve for R(s)[23] suggests the transitional region to be perhaps a factor of two larger than this. It extends from $E_{CM} \simeq$ 3 GeV to $E_{CM} \simeq$ 5 GeV.

Beyond $E_{CM} \simeq$ 5 GeV, perturbation theory should again be applicable. R(s) will be given, through second order, by eq.6 with α_s replaced by the running coupling constant $\bar{\alpha}(s)$:

$$\bar{\alpha}(s) = \frac{\alpha_s}{1 + \frac{25}{12\pi}\alpha_s \log s/M^2} . \qquad (10)$$

If the GIM model were correct, the approach to 10/3 would be quite rapid. At E_{CM} = 6 GeV for example, we find R(s) = 3.5 ± 0.2 where the estimated error comes from the uncertainty in the value of α_s and from the uncalculated higher order terms. Beyond E_{CM} = 6 GeV, the rate of decrease is very slow and in fact R(s) should remain nearly constant through $E_{CM} \simeq$ 9 GeV. This results from an interplay between the slowly rising two particle phase space factor and the slowly decreasing $\bar{\alpha}(s)$. The experiments clearly indicate that the GIM model is not the entire description of e^+e^- physics above E_{CM} = 3 GeV.

11. $c\bar{c}$ BOUND STATES

If one or more of the $c\bar{c}$ bound states lies outside the strong coupling transitional region, then it will be essentially Coulombic. A simple way to see if this is the case for the J(3.1) is to estimate the Bohr momentum $k_B = \frac{1}{2}(\frac{4}{3}\alpha_s)m_c$. For $\alpha_s \lesssim 0.25$ and $m_c \lesssim 2$GeV, $k_B \lesssim 300$ MeV and with momentum transfers this small, the effective coupling strength associated with the binding will be large. Thus the J(3.1) cannot be Coulombic. With distances as large as about one Fermi being important, the long-range confining forces are already playing a role. Computations in terms of the underlying field theory must be replaced by more phenomenological considerations.

It is important to point out that a heavy quark-antiquark system becomes more Coulombic as the quark mass increases. The Bohr momentum is determined by the equation

$$k_B = \frac{1}{2}\left(\frac{4}{3}\alpha_s(k_B)\right)m_Q \qquad (11)$$

and if m_Q is large enough so that $k_B > 1$ GeV, $\alpha_s(k_B)$ will be small ($\lesssim 0.25$) and the $Q\bar{Q}$ ground state will be Coulombic.

For $m_Q \gtrsim$ 6-7 GeV, this is the case and all the ground state properties (in particular the hyperfine splitting) are computable. What better reason could there be for building the next generation e^+e^- machines than to look for a truely Coulombic hadron ?

A great deal of phenomenological work has been done on the $c\bar{c}$ system[31,32,33]. The essential ingredient of this work is non-relativistic motion. Despite the fact that the $c\bar{c}$ bound state is non-Coulombic, there is good evidence that it remains non-relativistic and loosely bound. The first generation of this work[32,33] used a spin independent Schroedinger equation formalism with an effective long range $c\bar{c}$ potential. A popular choice for this effective potential includes a Coulombic short distance piece, a linearly rising long range piece and a constant V_0 to represent intermediate range and spin-dependent forces.

$$V(r) = V_0 + ar - \frac{4/3\ \alpha_s}{r} \quad . \tag{12}$$

The linear long range growth has been suggested by several theoretical considerations[34] but it has in no way been <u>derived</u> from the underlying Yang-Mills field theory.

Two important things have emerged from this work. Firstly there is the consistency check that the system is indeed nonrelativistic. From fits to the leptonic width of the J(3.1) (the wave function at the origin) and the J(3.1)-ψ(3.7) splitting, all groups agree that $a/m_c^2 \ll 1$. This qualitative feature is almost certainly independent of the specific form of the confining potential.

Secondly, there are quantitative predictions. In addition to the J(3.1) and ψ(3.7), other levels are predicted :

1. Additional radial recurrence , the recurrence at 4.2 GeV being the first. The 4.2 is already quite broad, sitting well above the threshold for the production of a pair of "charmed" mesons, $(c\bar{q})$ and $(\bar{c}q)$.

2. Pseudoscalar partners of the $J^{PC} = 1^{--}$ vector mesons, the $\eta_c, \eta'_c, \ldots$.[12] They can be reached by magnetic dipole transition from the vector states.

3. Four intermediate P-wave states centered at around 3.5 GeV. Their quantum numbers are $J^{PC} = 0^{++}$, 1^{++}, 2^{++} and 1^{+-} with the even C states reachable by electric dipole transitions from the ψ(3.7).

4. D-wave states in the neighborhood of 4.1 - 4.2 GeV. These can mix with the ψ(4.2) and this might offer an explanation for the structure emerging in this region[23]. Fine structure involves the spin dependent forces and is more difficult to predict. Of particular interest are the J(3.1) -η_c hyperfine splitting and the

splittings among the P-states. I will return to a discussion of spin dependent forces in Section 13.

The recent discoveries of intermediate levels around 3.4 and 3.5 GeV and a possible level at around 2.8 GeV offer a great deal of support for this general picture[35,36]. The level at 2.8 GeV could well be the η_c and the intermediate levels could be one or more of the P-states and/or the η_c'. A great deal of work will be necessary to complete the correspondence between the predicted and experimentally discovered levels.

Electric and magnetic dipole transitions have been estimated[31] or computed[32,37] by several groups. Most of the predicted transitions are quite sensitive to the fine structure splittings. For example, the transition $J(3.1) \rightarrow \eta_c \gamma$ varies as $(\Delta M)^3$ where ΔM is the $J(3.1) - \eta_c$ hyperfine splitting. A rather simple line of reasoning (see section 13) suggested that this splitting might be on the order of 100 MeV[37,38]. In this case, the magnetic dipole width is on the order of 1 KeV. In the newly discovered level at around 2.8 GeV[35] is indeed the η_c, then the width will be substantially larger. Scaling up by $(\Delta M)^3$ will probably be an overestimate because of corrections to the dipole approximation and spin dependence in the overlap integral. It can be no more than about 10 keV[39].

The physics of the $\psi(3.7)$ is likely to be much more complicated than a $c\bar{c}$ pair moving in some effective potential. It sits very close to the threshold for the production of a $(c\bar{q})$ and $(\bar{c}q)$ pair. Coupling to these decay channels can affect the position and decays of the $\psi(3.7)$. The radiative widths to the P-states seem to be about one order of magnitude smaller than the original estimates which neglected this effect as well as fine structre. Some recent work has begun to take this into account[40,41].

12. TOTAL HADRONIC WIDTHS

Surely the outstanding problem for the $c\bar{c}$ model of the new hadrons is the narrow width of the J(3.1). The Okubo-Iizuka-Zweig rule[42] which describes the large enhancement of $\phi \rightarrow K\bar{K}$ over $\phi \rightarrow \pi\pi\pi$ seems to be much more strongly operative here. The hadronic width of the J(3.1) is about 50 KeV which is perhaps a factor 10^{-4} below a "typical" width for a hadron of this mass. A mechanism was suggested by Politzer and myself[3] to explain this fact making use of the asymptotic freedom of the model and I want to review the argument behind this rather unusual use of asymptotic freedom. It is far from air tight but I would at least like to convince you that some analysis has gone into it and that it remains a viable possibility. It makes one rather striking experimental prediction which I will discuss at the end of this section.

I suggested in the last lecture that if the perturbation expansion converges in low orders, perturbation theory can be applied locally to the computation of R(s). This is manifestly not the case at the position of a resonance but it may still be possible to understand the widths of the J(3.1) and η_c by the use of perturbation theory. The widths involve both a large mass scale ($E_{CM} \simeq$ 3 GeV) and a small one (momentum transfer ≃ 300 MeV), and if perturbation theory is to be useful, then these two dependences must be disentangled. The part depending only on the large energy scale can then be computed in perturbation theory.

First of all, recall that the leptonic width of the J is given by the expression

$$\Gamma(J \rightarrow \bar{\ell}\ell) = |\psi(0)|^2 16\ (Q_c \alpha)^2\ \frac{1}{M_J^2}\ . \qquad (13)$$

$\psi(0)$ is the non-relativistic wave function at the origin and Q_c is the charge of the c quark in units of e. $|\psi(0)|^2$ is of course not calculable in perturbation theory but the leptonic width is a measure of this probability factor.

What we suggested[3)] is that the hadronic width could also be expressed as the product of $|\psi(0)|^2$ and a calculable matrix element. The annihilation of the $c\bar{c}$ state into some final state consisting of light quarks and gluons is shown in fig.7. The B_n amplitude is defined to be two particle irreducible in the $c\bar{c}$ channel. The A amplitude contains the forces that produce the bound state and it contains mass singularities since the c and $\bar{c}$ lines sit very close to $p^2 = m_c^2$. (Recall from the phenomenological work that the $c\bar{c}$ system is non-relativistic and weakly bound). It cannot be calculated pertubatively and it is responsible for the factor $|\psi(0)|^2$ when the subsequent annihilation is local.

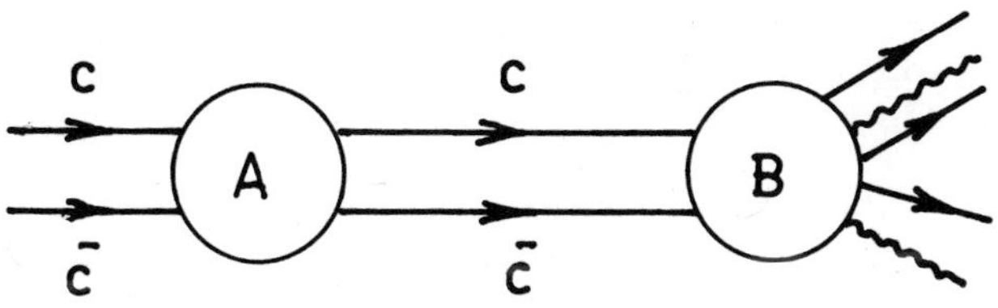

Fig. 7. The transition of a $c\bar{c}$ pair into a state consisting of gluons and light quarks

If the B_n amplitude can be shown to be insensitive to the small dimensional factors, it can be calculated perturbatively. What we suggested is that $|B_n|^2$ summed over all final states $\sum_n |B_n|^2$ depends only on the large factor E_{CM} and M (the renormalization mass) and can be computed in perturbation theory. An analysis through low orders of perturbation theory leads to the following conclusions :

1. $\sum_n |B_n|^2$ is free of mass singularities associated with the initial lines by virtue of its two particle irreducibility in the $c\bar{c}$ channel. This can be seen most easily for η_c decay where the Born term is the annihilation into two gluons. It is straight forward to check that the two particle irreducibility does indeed shield out mass singularities and therefore eliminate dependence on small dimensional parameters like $p^2 - m_c^2$. I recommend verifying this for a few simple graphs.

2. $\sum_n |B_n|^2$ is free of final state mass singularities because of the inclusive sum. This application of the Kinoshita theorem is analogous to the computation of R(s) off resonance. The analysis has only been carried to order α_s^6 and I am currently trying to generalize it[43].

If this analysis is correct to all orders, then the dominant contribution to J(3.1) and η_c decay is the annihilation into 3 gluons and 2 gluons respectively. As in the parton model, these minimal constituent states then evolve into ordinary hadrons with probability one. The expressions for the widths can be taken over from the corresponding expressions for ortho and para positronium decay.

$$\Gamma_{Had}(J) = |\psi(0)|^2 \frac{16}{9\pi} (\pi^2 - 9) \frac{5}{18} \alpha_s^3 \frac{1}{m_c^2} \tag{14}$$

$$\Gamma_{Had}(\eta_c) = |\psi(0)|^2 \frac{8}{3} \alpha_s^2 \frac{1}{m_c^2} . \tag{15}$$

These are zero binding approximations which are subject to relativistic and binding corrections. We have estimated that these corrections could be as much as 20 % [3].

The J(3.1) width can be fit by choosing $\alpha_s(3.1) \simeq 0.2$. One is then led to the rather striking prediction that the hadronic width of the η_c should be a few MeV. If the $\eta_c \to \gamma\gamma$ width is a few KeV[44], then the $\gamma\gamma$ branching ratio will be on the order of 10^{-3}. There is so far one experimental number relevant to this prediction. The DESY group has reported[37]

$$\frac{\Gamma(J \to \gamma\eta_c)}{\Gamma(J \to \text{all})} \times \frac{\Gamma(\eta_c \to \gamma\gamma)}{\Gamma(\eta_c \to \text{all})} \simeq 2 \times 10^{-4} . \tag{16}$$

In the last section, it was suggested that the $J \to \gamma\eta_c$ branching ratio might well be on the order of 10 %. If this is the case, then the $\gamma\gamma$ branching ratio of might well be 10^{-3}. If this branching ratio is much larger, then the very small magnetic dipole transition becomes hard to understand.

If the minimal gluon mechanism is wrong, then it is hard to see why the η_c width should be much larger than the J width. Although no sign of breakdown appears in low orders, it could be that this is misleading and that the expansion breaks down in high orders[45]. If this is the case, then the decay is described by some other mechanism, for example charmed meson intermediate states, $\psi \to D\bar{D} \to$ non charmed hadrons[46]. As long as the minimal gluon mechanism is at least an important part of the whole story, the η_c should be broad. A large width for this state would be an important piece of support for the quark gluon theory.

The application of weak coupling methods to the light hadrons is much more speculative[47]. The direct channel energies are small and the bound states are relativistic. It is nevertheless an interesting possibility that these ideas have something to do with the classic examples of the O-I-Z rule[42].

13. CONCLUDING REMARKS

Since the $c\bar{c}$ system appears to be non relativistic, it should be easier to deal with than any other hadronic system. The motion of the $c\bar{c}$ pair is governed by an effective Hamiltonian corresponding to some effective Bethe-Salpeter kernel. If this kernel has some simple structure in terms of Greens functions, then a connection can be established between the bound state properties and the color gauge theory which is much more direct than for the old hadrons. Whether the dominant part of the kernel is simple enough to do this is not yet clear.

For a Coulombic system, the dominant kernel is single photon exchange $[\gamma^\mu]_1 \; 1/k^2 [\gamma_\mu]_2$ and this leads to the Breit-Fermi Hamiltonian[48]. The appropriate kernel for the $c\bar{c}$ system is not a priori evident. An analysis of diagrams indicates that any contribution to the kernel <u>without internal $c\bar{c}$ loops</u> is equally important. This is because every additional factor of $\alpha_s(M_J)$ is accompanied by a logarithm of momentum transfer or binding energy. This product must be taken to be O(1).

In order to make some numerical estimates, several groups assumed a simple form for the $c\bar{c}$ kernel. Although surely over simplified, it seemed to me at least to be a sensible order of magnitude guess. The guess was $[\gamma_\mu]_1 \, D(k^2) \, [\gamma^\mu]_2$ with $D(k^2)$ chosen to give the spin independent potential discussed in Section II. [37,38] This structure looks like single gluon exchange with a dressed gluon propagator. It then must be assumed, for example, that $D(k^2) \sim 1/k^4$ as $k \to 0$.

This assumption fixes the spin dependent forces and the analog of the Breit-Fermi Hamiltonian can be written down. Numerical computations are still underway and it looks as though sensible fine structure emerges. The spin-spin interaction goes like $1/r \; \vec{\sigma}_1 \cdot \vec{\sigma}_2$ at large distances and the hyperfine splitting comes out around 70-90 MeV. The experimental number may be larger by a factor of three or four[35].

The true structure of the kernel is surely more complicated than this. Certainly there is no reason to assume a tensor structure of the form $[\gamma_\mu]_1 [\gamma^\mu]_2$. Even if a dressed single glyon exchange is dominant, the vertex corrections each have a Pauli momentum piece in addition to the Dirac piece. The truth is probably still more complicated[49] and $c\bar{c}$ fine structure measurements are extremely important to get a handle on this structure.

The fundamental problems remain. Are the colored quarks and gluons permanently confined? Can $c\bar{c}$ properties, much less the properties of ordinary hadrons be computed starting from the color gauge theory? Absolute confinement is an attractive and widely discussed idea[50]. The necessary strongly attractive long range forces could well be a consequence of the infrared instability of the asymptotically free gauge theory. Some quantitative support for this idea has come from the lattice gauge work of Witson and Kogut and Susskind[51] but the way it comes about in a continuum theory is far from clear.

Finally, the existence of charmed hadrons made of a heavy quark and one or more light quarks is an inescapable consequence of the color gauge theory. The experimental situation with respect to these states is confused at the moment[52]. They may or may not have been seen in several experiments. Since they decay weakly, the experimental signature is of course dependent on how the heavy quark enters the weak current. Nothing could be more welcome at the moment than some strong evidence for the existence of charmed hadrons.

REFERENCES AND FOOTNOTES

1. J.J. Aubert et al., Phys.Rev.Lett. 33, 1404 (1974).
2. J.-E. Augustin et al., Phys.Rev.Lett. 33, 1406 (1974).
3. Much of what I will say in these lectures about the physics of heavy quarks is based on work by myself and H.D. Politzer. Most of it is published in two papers. T. Appelquist and H.D. Politzer, Phys.Rev.Lett. 34, 43 (1975); T. Appelquist and H.D. Politzer, Institute for Advanced Study preprint , February 1975, to be published in the Physical Review.
4. O.W. Greenberg, Phys.Rev.Lett. 13, 598 (1964).
5. For a modern treatment and a discussion of the many experimental advantages of the colored quark model, see H. Fritzsch, M.-Gell-Mann and H. Leutwyler, Phys.Lett. 47B, 365 (1973).
6. B.W. Lee and J. Zinn Justin, Phys.Rev. D5, 3121 (1972).
7. F. Low and M. Gell-Mann, Phys.Rev. 95, 1300 (1954); N.N. Bogoliubov and D.V. Shirkov, Introduction to the Theory of Quantized Fields, New York, Interscience Publ. 1959.
8. H.D. Politzer, Phys.Rev.Lett. 30, 1346 (1973); D.J. Gross and F. Wilczek, Phys.Rev.Lett. 30, 1343 (1973); A recent review of asymptotic freedom and some of its applications has been given by H.D. Politzer, Phys.Reports 14C, 130 (1974).
9. C. Callan, Phys.Rev. D2, 1541 (1970). K. Symanzik, Comm.Math. Phys. 18, 227 (1970).
10. K.G. Wilson, Phys.Rev. 179, 1499 (1969), and unpublished work. See also W. Zimmerman in "Lectures on Elementary Particles and Quantum Field Theory",edited by S. Deser et al., MIT Press, Cambridge, Mass., 1970.
11. S.L. Glashow, I. Iliopoulos and L. Maiani, Phys.Rev. D2, 1285 (1970).
12. For a recent review of the GIM mechanism and its implications for weak phenomenology see M.K. Gaillard, B.W. Lee and J.Rosner, Rev.Mod.Phys. 47, 277 (1975).
13. R.M. Barnett, Invited talk, 6th Hawaii Topical Conference in Particle Physics, University of Hawaii, Honolulu, August 1975, and FERMILAB-Conf-75/71-THY, September 1975.
14. G. Feldman, apporteurs talks at the 1975 International Symposium on Lepton and Photon Interactions at High Energies, Stanford Calif., M. Perl, Invited talk at the 1975 meeting of the Division of Particles and Fields of the American Physical Society, Seattle.
15. M. Weinstein, Phys.Rev. D1, 1854 (1973), J.C. Pati and A. Salam, Phys.Rev. D8, 1240 (1973) and Phys.Rev.Lett. 31, 661 (1973). H. Georgi and S.L. Glashow, Phys.Rev.Lett. 32, 438 (1974).
16. K. Wilson, Phys.Rev. D3, 1818 (1971). H. Georgi, H. Quinn and S. Weinberg, Phys.Rev.Lett. 33, 451 (1974).
17. If the effective weak Lagrangian is non renormalizable (the Fermi theory for example) then Λ might be large but ultimately attainable ($E_{CM} \simeq 10^5$ GeV in the Fermi theory for example).

18. This corresponds to performing the mass renormalization subtraction at the same large Euclidean momentum at which the wave function subtraction is carried out.
19. R.L. Jaffe and C.H. Llewellyn Smith, Phys.Rev. D7, 2506 (1973). H. Leutwyler, Phys.Lett. 48B, 431 (1974). M. Testa, Physics Letters 56B, 53 (1975).
20. T. Appelquist and H. Georgi, Phys.Rev. D8, 4000 (1973); A. Zee, Phys.Rev. D8, 4038 (1973).
21. T. Kinoshita, I. Math.Phys. 3, 650 (1962).
22. S.L. Adler, Phys.Rev. D10, 3714 (1974). This analysis for the 9 colored Gell-Mann Zweig quarks can easily be extended to include one or more heavy quarks.
23. R. Schwitters, Rapporteurs talk at the 1975 International Symposium on Lepton and Photon Interactions at High Energies, Stanford, California.
24. R. Jost and J.M. Luttinger, Helv.Phys.Acta 23, 201 (1950). G. Kallen and A. Sabry, Dansk.Vidensk.Selsk. 29, no.17 (1955).
25. J. Schwinger, "Particles, Sources and Fields, Vol.II", Chapter 5-4, Addison-Wesley, New York 1973.
26. T.D. Lee and M. Nauenberg, Phys.Rev. 133, B1544 (1954).
27. The theorem has been checked to low orders (two loops) by myself and J. Carazzone. I have been informed by Professor T. Kinoshit that such a check has also been performed by Dr. A. Ukawa.
28. This order of magnitude for $\alpha_s(M)$ (with M equal to a few GeV) has been suggested by analyses of Bjorken scaling in electroproduction. See the third paper of ref.8.
29. H. Harari, Rapporteurs talk at the 1975 International Symposium on Lepton and Photon Interactions at High Energies, Stanford, California. The heavy lepton should decay semi-leptonically at least 60 percent or 70 percent of the time so that it makes a sizable contribution to R(s).
30. H.R. Quinn, private communication and E.Poggio, H.R. Quinn and S. Weinberg, manuscriptin preparation.
31. T. Appelquist, A. DeRujula, H.D. Politzer and S.L. Glashow, Phys.Rev.Lett. 34, 365 (1974).
32. E. Eichten, K. Gottfreid, T. Kinoshita, J. Kogut, K.D. Lane and T.M. Yan, Phys.Rev.Lett. 34, 369 (1975).
33. B. Harrington, S.Y. Park and A. Yildiz, Phys.Rev.Lett. 34, 168 (1975). H. Schnitzer and J.S. Kang, Brandeis University, preprint, 1975. K. Jhung, K. Chung and R.S. Willey, University of Pittsburgh preprint (1975).
34. The Coulomb potential in one spatial dimension is proportion to r so that this long range growth is a natural feature of any scheme in which hadronic physics is essentially one dimensional at large distances.
35. Bjorn Wiik, Rapporteurs talks at the 1975 International Symposium on Lepton and Photon Interactions at High Energies, Stanford, California.
36. G.J. Feldman et al., Phys.Rev.Lett. 35, 821 (1975).

37. J. Borenstein and R. Shankar, Phys.Rev.Lett. 34, 619 (1975). J. Borenstein, Harvard Univ. Ph.D. thesis, May 1975, unpublished. J. Pumplin, W. Repko and A. Sato, Michigan State Preprint, July 1975.
38. H.J. Schnitzer, Brandeis University Preprint, June 1975 and July 1975.
39. The total width of the J(3.1) is about 70 keV. By adding up the observed partial widths (ref.23) and estimating isospin related neutral channels, no more than 10 keV remains unaccounted for.
40. J. Kogut and L. Susskind, Phys.Rev.Lett. 34, 767 (1975) and Cornell preprint CLNS-303(1975). R. Barbieri et al. CERN preprint TH2026 (1975).
41. E. Eichten et al., Cornell preprint CLNS-316 (1975).
42. S. Okubo, Phys.Letters 5, 165 (1963); J. Iizuka, K. Okada and O. Shito, Prog.Theo.Phys. 35, 1061 (1966); G. Zweig, 1964, unpublished.
43. Details of this analysis will be presented in a future publication.
44. This value is predicted if the decay is described by the Born graph for $c\bar{c}$ into $\gamma\gamma$(ref.3). This is on even less firm ground than the other predictions since two real photons are being emitted. The perturbation expansion can break down due to vector dominance-like corrections on each photon line.
45. A small momentum dependence could enter from a sharing of the momentum with multiple gluon emission. See ref.3 for a discussion of this.
46. G. Cohen-Tannoudji, C. Gilain, G. Girardi, U. Maor and A. Morel, I would like to thank Uri Maor for several informative discussions.
47. A. DeRujula and S.L. Glashow, Phys.Rev.Lett. 34, 46 (1975).
48. J. Schwinger, Particles, Sources and Fields (Addison-Wesley, Reading, 1973), Vol.2, 326-346, 420; A. Akhiezer and V. Beretski, QED (Interscience, New York, 1965), 514-530; G. Kallen, QED (Springer-Verlag, New York, 1972), 182-188; H. Bethe and E. Salpeter, Quantum Mechanics of One and Two Electron Atoms (Academic Press, New York, 1957), 178-193.
49. The structure of the $c\bar{c}$ kernel has been looked at in some detail by Ivan Muzinich and myself. If it is assumed that the potential (12) is due to dressed single gluon exchange, then the fact that $a/m_c^2 \ll 1$ can be used to show that this is a consistent assumption. The high order terms in the kernel skeleton expansion are supressed by powers of a/m_c^2. The main problem with dressed ladder dominance is gauge invariance. It has not been demonstrated.
50. See for example R. Dashen, Rapporteurs talk at the 1975 International Symposium on Lepton and Photon Interactions at High Energies, Stanford, California.

51. K. Wilson, Phys.Rev. D10, 2445 (1974); J. Kogut and L. Susskind, Phys.Rev. D11, 395 (1975).
52. See Rapporteurs talks by G. Feldman, C. Rubbia, B. Barish and N. Samios at the 1975 International Symposium on Lepton and Photon Interactions at High Energies, Stanford, California.

PHENOMENOLOGY OF ELECTRON POSITRON ANNIHILATION INTO HADRONS

Otto Nachtmann

Institute for Theoretical Physics

University of Bern

TABLE OF CONTENTS

1. INTRODUCTION

Electron positron annihilation into hadrons certainly has been one of the most exciting topics of high energy physics over the last two years or so. I think it is fair to say that the experimental findings culminating in the discovery of the narrow $\psi(J)$ resonances [1)-3)] - of which the $\psi 3,1$ was discovered simultaneously in proton beryllium scattering[4)] - have surpassed the imagination of most theoreticians. Indeed, the popular view was that the total cross section would decrease smoothly as 1/s for center of mass energies $\sqrt{s}$ above the prominent low mass resonances ρ, ω, ϕ. What experimentalists found is shown schematically in fig.1, where we have plotted the famous ratio

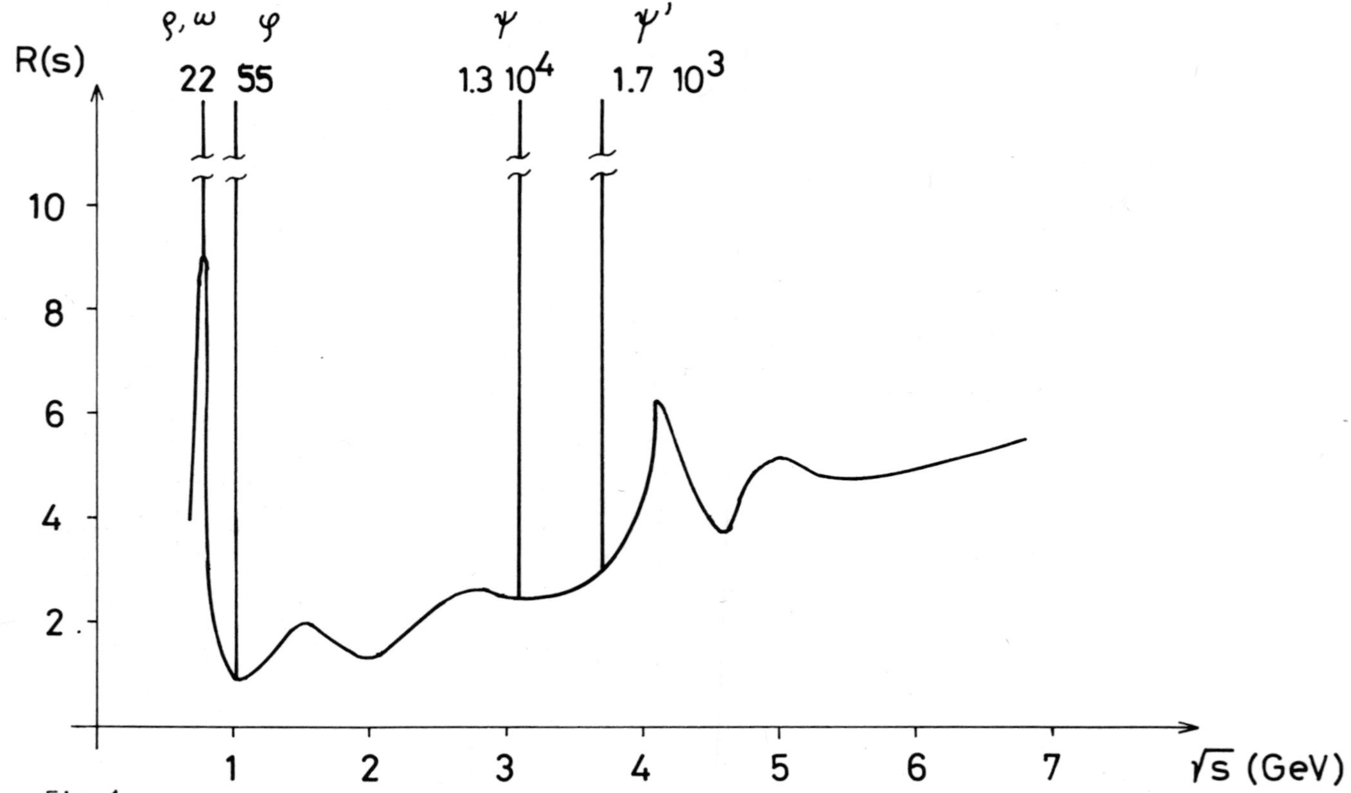

Fig. 1
The ratios $R(s) = \sigma(e^+e^- \to \text{hadrons})/\sigma_{QED}(e^+e^- \to \mu^+\mu^-)$ as function of the c.m. energy $\sqrt{s}$.

$$R(s) = \sigma(e^+e^- \to \text{hadrons})/\sigma_{QED}(e^+e^- \to \mu^+\mu^-)$$

$$\equiv \sigma(e^+e^- \to \text{hadrons})/(\frac{4\pi\alpha^2}{3}\ \frac{1}{s})$$

R(s) instead of being constant as implied by a 1/s decrease of the cross section shows very remarkable structure in the energy range explored to date.

In section 2 we will discuss these findings. Our guide will be the quark model. Resonances are understood as quark antiquark bound states. The asymptotic behaviour of R(s) for $s \to \infty$ should give information on the number and charges of the quarks.

In section 3 we propose to have a look at final states. The study of the decays of resonances is essential in order to clarify their nature i.e. their quantum numbers. At very high energies scaling phenomena are expected to play an important role.

2. THE TOTAL CROSS SECTION

2.1 Kinematics and general ideas

We want to discuss electron positron annihilation into hadrons via one photon exchange (fig. 2).

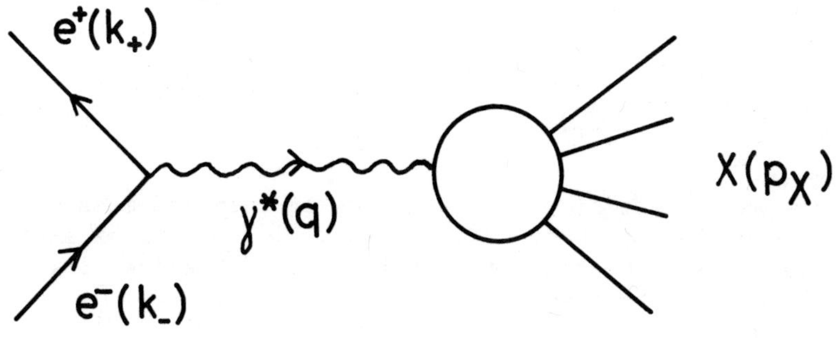

Fig. 2.
Electron positron annihilation into hadrons via one photon exchange. $q = k_+ + k_-, s = q^2$.

$$e^+(k_+) + e^-(k_-) \to \gamma^*(q) \to \text{hadrons}$$

$$q = k_+ + k_- \tag{2.1}$$

The production of the virtual photon through the annihilation of an electron and a positron is described by standard quantum electrodynamics and can be calculated explicitly. The unknown and therefore interesting quantity is the probability for the conversion of the virtual photon into hadrons which tells us about strong interactions. Our theoretical ignorance about this process is represented by the blob in fig. 2.
Splitting off all explicitly calculable factors we can write the total cross section for the reaction eq.(2.1) as follows :

$$\sigma_{tot}(e^+e^- \to \text{hadrons}) = \frac{4\pi\alpha^2}{3}\frac{R(s)}{s}$$

$$s = (k_+ + k_-)^2 = q^2 \tag{2.2}$$

where α is the fine structure constant and $R(s)$ measures the rate of conversion of the virtual photon into hadrons

$$(-g_{\rho\lambda}q^2 + q_\rho q_\lambda)\frac{R(q^2)}{6\pi} = \sum_X (2\pi)^4\delta(q-p_X) <0|J_\rho|x> <x|J_\lambda|0>$$

$$= \int dx\, e^{iqx} <0|[J_\rho(x)J_\lambda(0)]|0> \tag{2.3}$$

J_λ is the hadronic part of the electromagnetic current and the sum in eq. (2.3) runs over all hadronic states.
It is instructive to calculate the total cross section for production of muon pairs (fig. 3). The result is for $s \gg m_\mu^2$:

$$\sigma(e^+e^- \to \mu^+\mu^-) = \frac{4\pi\alpha^2}{3}\frac{1}{s} \tag{2.4}$$

If hadrons were a set of point particles of spin 1/2 mass m_i and charge Q_i then the hadron production (fig. 2) could be as easily calculated as the muon production (fig. 3). Comparing eqs. (2.2) and (2.4) we would find for $s \gg m_i^2$

$$R(s) = \sum Q_i^2 \tag{2.5}$$

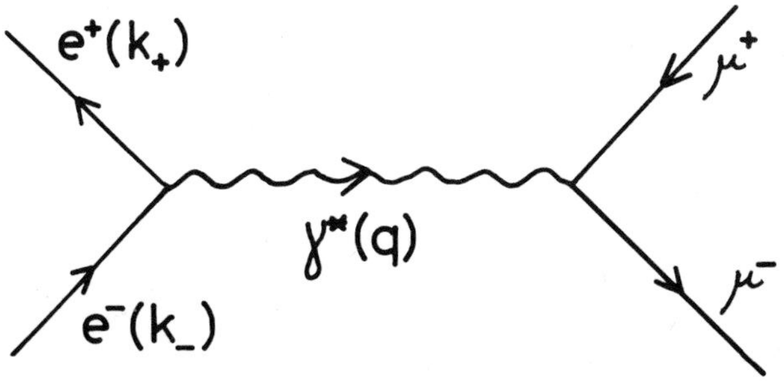

Fig. 3 : Muon pair production by electron positron annihilation

In strong interactions one is of course far from dealing with one or a few point particles. Nevertheless, if we look at fig.1, we find that the experimental value of R(s) is not far from our expectation for a few point particles except at the resonances where R(s) shoots off much higher. (This experimental finding is far from obvious. In fact, before experiments were started there were strong believes that nothing except the tails of the ρ,ω,ϕ mesons would be seen at high energies, implying $R(s) \sim s^{-2}$ for large s).

The simplest phenomenological model which tries to explain this fact is the parton model, with the partons identified as quarks. The idea is then, that the virtual photon creates at first a quark antiquark pair (fig. 4). If quark and antiquark interact only weakly at short distances and if they are not just in a resonating state, i.e. do not come back to the production point again, the transition rate should not depend on the ultimate fate of the $q\bar{q}$ pair, described by the blob in fig. 4 and R(s) should be given by the naive value eq. (2.5) obtained for point particles[5]. The sum in eq.(2.5) should run over all "active" quarks, i.e. those with $m_i^2 \ll s$.

For large separations the forces between quarks must of course become very strong because we do not see quarks coming out. They must be somehow confined permanently. These binding forces also

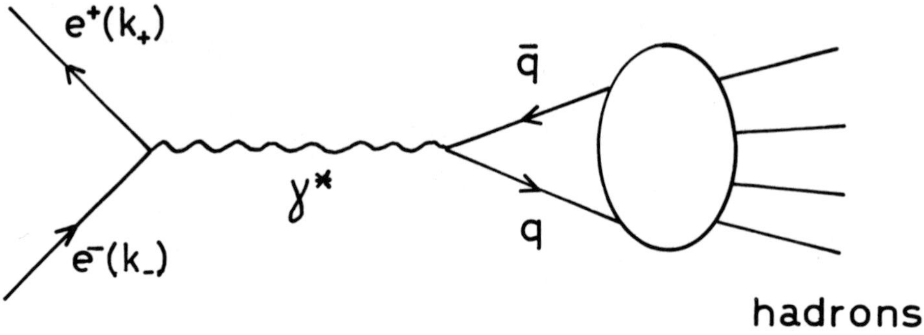

Fig. 4.

Electron positron annihilation into hadrons proceeding via the creation of a quark antiquark pair which then turns into hadrons

give rise to quark antiquark bound states which are identified with the observed resonances.

What will be the order of magnitude of R(s) at the peak of a resonance ? The photon creates initially a $q\bar{q}$ pair as in fig. 4. But now quark and antiquark stay together for a time Γ^{-1} where Γ is the total decay rate of the resonance. This enhances the wave function in the interaction region. Let L characterise the linear dimension of the interaction region (e.g. $\sim m_\pi^{-1}$). The density of quarks in the interaction region should be proportional to the time spent there. For free quarks moving with a velocity of order 1 that time should be of order L, for a resonating quark antiquark system it is Γ^{-1}, therefore the free quark result eq. (2.5) will be multiplied by an enhancement factor Γ^{-1}/L :

$$R(\text{at peak of resonance}) \sim \frac{Q^2}{\Gamma L} \tag{2.6}$$

where Q is the charge of the resonating quark. This gives indeed the correct order of magnitude of the observed values in fig. 1.

All this works rather nicely for the region $\sqrt{s} \lesssim 3$ GeV if we identify the quarks with the familiar u,d,s quarks, where each quark comes in three color varieties u_A, A=1,2,3 (e.g. red, green, blue) etc. (table 1).

	T	T_3	Q	S
u_A	1/2	1/2	2/3	0
d_A	1/2	-1/2	-1/3	0
s_A	0	0	-1/3	-1

Table 1 : quantum numbers of the usual quarks

As you know, an extra degree of freedom, usually called color is introduced[6) 7)] in order to solve the problem of quark statistics and to fix up the π_o decay constant as calculated from the Adler-Bell-Jackiw anomaly[8)]. All observed states at least below 3 GeV in mass are assumed to be color singlets.

The low mass resonances ρ,ω,ϕ are then $^3S_1$ bound states of the usual quarks with the quark content to a good approximation given by

$$\rho^o = \frac{1}{\sqrt{6}} \sum_A (u_A\bar{u}_A - d_A\bar{d}_A)$$

$$\omega = \frac{1}{\sqrt{6}} \sum_A (u_A\bar{u}_A + d_A\bar{d}_A)$$

$$\phi = \frac{1}{\sqrt{3}} \sum_A s_A\bar{s}_A \qquad (2.7)$$

The electromagnetic current is bilinear in quark fields q

$$J_\lambda = \bar{q}\, Q\gamma_\lambda q \qquad (2.8)$$

where Q is the quark charge matrix. In the colored quark model discussed above the quark spinor carries 2 indices

$$q = (q_{\alpha A}),\ \alpha = u,d,s; \quad A = 1,2,3$$

and the charge matrix is given by

$$Q = \bar{Q}_{\alpha\beta}\ \delta_{AB}$$

$$(\bar{Q}_{\alpha\beta}) = \begin{pmatrix} 2/3 & & \\ & -1/3 & \\ & & -1/3 \end{pmatrix} \qquad (2.9)$$

In the Han-Nambu version of the three triplet model[6)] the charge matrix Q has in fact an additional piece which transforms like an octet under the color group. Below color-threshold however only color singlet states contribute in the sum over intermediate states in eq.(2.3). Since the vacuum is also supposed to be a color singlet any non color singlet piece of the current eq. (2.8) will be inoperative and the effective charge matrix will be exactly as shown in eq. (2.9).

For values of $\sqrt{s}$ above the low energy resonances ρ, ω, ϕ and below a possible color threshold we would then expect

$$R = \mathrm{tr}\ Q^2 = 2 \qquad (2.10)$$

Eq.(2.10) is in rough agreement with the data up to $\sqrt{s} \sim 3$ GeV (fig.1.).

The excitement starts at 3.1 GeV. If we stick to the quark picture then the $\psi_{3,1}$ resonance must be some new phenomenon, either a state involving new quarks - the prototype of such theories is the charm theory[9) 10)] - or a state where the color degree of freedom is excited[11) 12)].

As you know one of the main challenges theory has to confront is to explain the small width of the $\psi_{3,1}$ and $\psi_{3,7}$ resonances (table 2).

V	Γ_V(total) (MeV)
ρ(770)	150 ± 10
ω(783)	10,0 ± 0,4
ϕ(1019)	4,2 ± 0,2
ψ(3100)	0,069 ± 0,015
ψ(3700)	0,225 ± 0,056
ψ(4200)?	∿300

Table 2 : Total width of vector mesons. Data are from refs. 13), 14).

Two approaches have been taken concerning this problem. Either the small width of $\psi_{3.1}$ and $\psi_{3.7}$ is the consequence of a dynamical effect (this is the explanation given e.g. in the charm theory) or it is the consequence of the ψ-particles carrying explicitly a new quantum number conserved under strong interactions. This would be analogous to the explanation of the long lifetime of the K mesons by the introduction of the conserved quantum number strangeness. This latter possibility is explored e.g. in the color models of refs. 11), 12).

At the moment of writing these lectures it is entirely unclear which one if any of the models proposed so far will survive the demolishing influence of experimental facts. We feel that it is useless to give a complete listing of all these models here. Our policy will be instead to take as starting point the three triplet colored quark model for low lying hadrons. By choosing rather arbitrarily two of the proposed models we will then explore the generalizations in the two obvious directions : adding other color triplets or thawing the color degree of freedom. But before discussing models we want to have a look at some general properties of ψ-particles.

2.2 Production and decay of ψ-particles

We will assume throughout that ψ-particles are vector mesons with negative parity and charge conjugation which couple to the photon in a manner similar to ρ,ω,ϕ. This is supported by experiment[14]. The ψ-photon coupling constant F_ψ is defined by

$$\langle 0|J_\mu|\psi(p,\varepsilon)\rangle = F_\psi \frac{m_\psi \cdot \varepsilon_\mu}{\sqrt{2Vp_0}} \tag{2.11}$$

where V is the normalisation volume and our states are normalized to 1.

The γ-ψ coupling governs both the contribution of ψ-production to the total cross section or R(s) eq.(2.3) and the leptonic width of ψ.

From fig. 5 we find easily

$$R(s) = 12\pi|F_\psi|^2 \quad \frac{m_\psi \Gamma(\psi \to \text{hadrons})}{(s-m_\psi^2)^2 + m_\psi^2\,\Gamma_\psi^2} \tag{2.12}$$

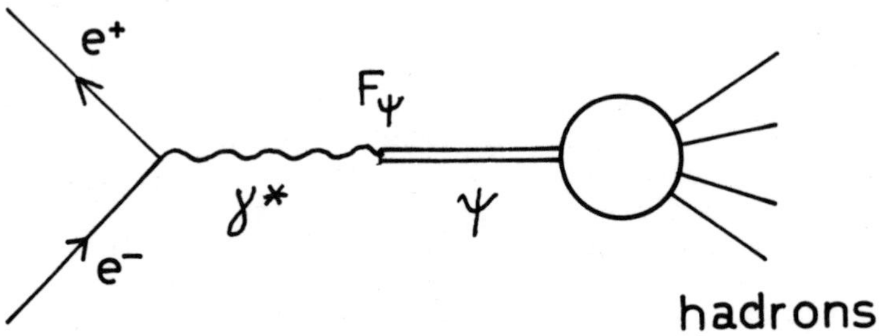

(a)

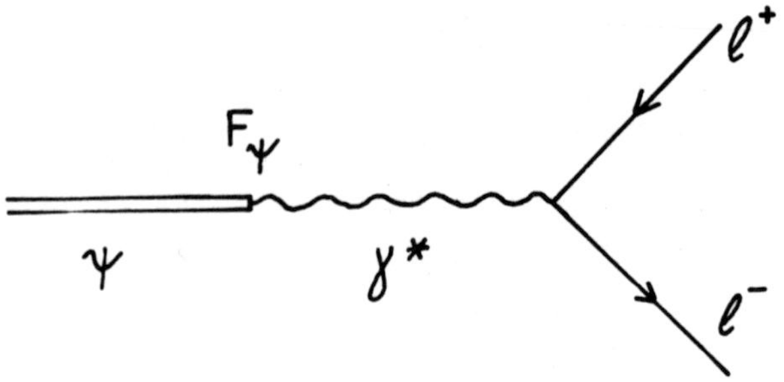

(b)

Fig. 5.

(a) contribution of ψ production to the total cross section eq.(2.2).
(b) leptonic decay of a ψ; ℓ stands for electron or muon.

$$\Gamma(\psi \rightarrow \ell^+\ell^-) = \frac{4\pi\alpha^2}{3} \frac{|F_\psi|^2}{m_\psi} \tag{2.13}$$

where Γ_ψ is the total width of the ψ-particle and ℓ stands for electron or muon. The lepton mass has been neglected. Experimentally F_ψ is quite similar to the coupling constants of the ρ, ω, ϕ mesons to the photon (table 3).

V	$\Gamma(V \rightarrow e^+e^-)$ (keV)	$F_V/m_{\pi+}$
$\rho(770)$	6,4 ± 0,86	1,06 ± 0,06
$\omega(783)$	0,76 ± 0,17	0,37 ± 0,04
$\phi(1019)$	1,34 ± 0,10	0,56 ± 0,02
$\psi(3100)$	4,8 ± 0,6	1,85 ± 0,12
$\psi(3700)$	2,2 ± 0,3	1,37 ± 0,09
$\psi(4200)$	4 ± 2	2,0 ± 0,5

Table 3 : Experimental values for electronic decay widths of vector mesons[13) 14)] and photon-vector mesons coupling constants.

One also finds that μ-e universality holds for leptonic decays of ψ-particles[14)]. Therefore it seems that the new vector mesons and the old ones couple to the photon in quite a similar way.

For their decay the ψ-particles have various options (to order α in the matrix element). They can decay directly into hadrons (fig. 6a), they can decay into hadrons via a photon (fig. 6b) or they can decay into leptons (fig. 5b). The total decay rate is therefore written as a sum :

$$\Gamma_{tot} = \Gamma_h + \Gamma_{ee} + \Gamma_{\mu\mu} + \Gamma_{\gamma h} + \Gamma_{int} \tag{2.14}$$

where Γ_{ee}, $\Gamma_{\mu\mu}$ are the leptonic widths, Γ_h, $\Gamma_{\gamma h}$ the widths due to the decays into hadrons corresponding to fig. 6(a) and (b) respectively and Γ_{int} is the interference term between the processes of fig. 6(a) and (b).

Up to now we have discussed the production of ψ-particles by electron positron annihilation to lowest order in α. In fact higher order vacuum polarisation effects are very big in the region of the narrow resonances $\psi_{3,1}$ and $\psi_{3,7}$. They are however easily

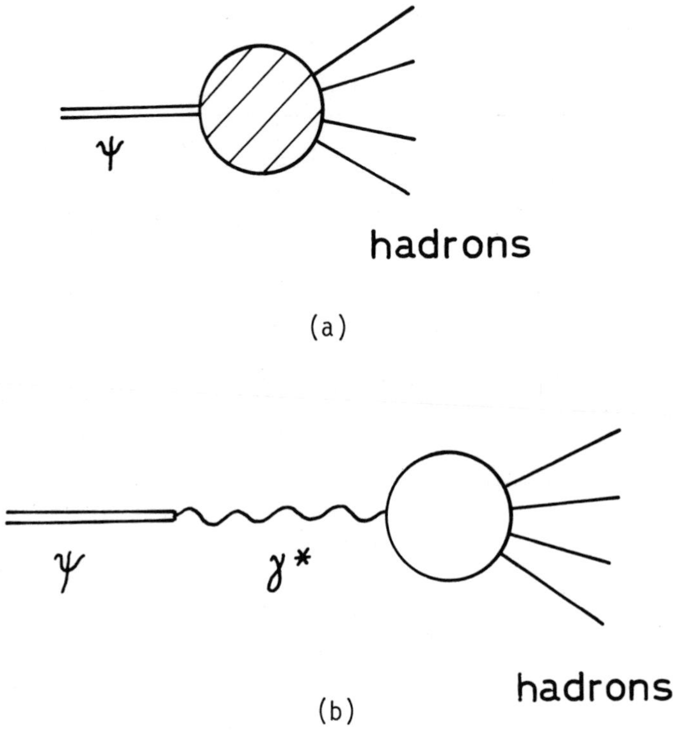

Fig. 6.
(a) Direct and (b) photon mediated decay of ψ into hadrons.

summed up and seen to produce just a shift in the mass and width of the resonance. This is discussed in the appendix where also some useful formulae are collected.

The hadronic final states found in the decay of the $\psi_{3,1}$ seem to have mostly odd G-parity. The photonic process fig. 6(b) is however a source of even G-parity states and the experimental finding suggests that it is the only source[14]. One would then be led to ascribe odd G-parity to the $\psi_{3,1}$ particle. But this conclusion is not inescapable as will be seen in section 2.4.

We will now turn to a discussion of specific models.

2.3 Addition of further quarks

The simplest extension of the three-triplet scheme evidently is a four -triplet scheme. A very specific and detailed scheme of this type has been proposed in refs. 9-10). A "charmed" quark-triplet is added to the three "known" ones, charm being a conserved quantum number very much like strangeness (table 4).

	T	T_3	Q	S	C
u_A	1/2	1/2	2/3	0	0
d_A	1/2	-1/2	-1/3	0	0
s_A	0	0	-1/3	-1	0
c_A	0	0	2/3	0	1

Table 4 : Quantum numbers of quarks in the charm model[9][10].
A = 1,2,3 is the color index.

Mesons are built in the usual way as $q\bar{q}$ bound states. All observable states are supposed to be color singlets. The familiar pseudoscalar octet + singlet under SU(3) is extended to a quindeciment + singlet under SU(4) and similarly for vector mesons. The quantum numbers and the quark content of these mesons are listed in table 5 [15]. One expects mixing among the three states with C=0 and T=0 in analogy to η,η' and ω,ϕ mixing.

The electromagnetic current is neutral with respect to electric charge, strangeness and charm :

$$J_\lambda = \frac{2}{3}\bar{u}\gamma_\lambda u - \frac{1}{3}\bar{d}\gamma_\lambda d - \frac{1}{3}\bar{s}\gamma_\lambda s + \frac{2}{3}\bar{c}\gamma_\lambda c \qquad (2.15)$$

C	T	Meson 0^-	Meson 1^-	Quark content
0	1	π^+	ρ^+	$u\bar{d}$
		π^o	ρ^o	$\frac{1}{\sqrt{2}}(u\bar{u} - d\bar{d})$
		π^-	ρ^-	$d\bar{u}$
0	1/2	K^+	K^{*+}	$u\bar{s}$
		K^o	K^{*o}	$d\bar{s}$
		K^-	K^{*-}	$s\bar{u}$
		$\bar{K}^o$	$\bar{K}^{*o}$	$s\bar{d}$
0	0	η	ω	$\frac{1}{\sqrt{2}}(u\bar{u} + d\bar{d})_m$
		η'	ϕ	$(s\bar{s})_m$
		η_c	ϕ_c	$(c\bar{c})_m$
1	1/2	D^+	D^{*+}	$c\bar{d}$
		D^o	D^{*o}	$c\bar{u}$
1	0	F^+	F^{*+}	$c\bar{s}$
-1	1/2	D^-	D^{*-}	$d\bar{c}$
		$\bar{D}^o$	$\bar{D}^{*o}$	$u\bar{c}$
-1	0	F^-	F^{*-}	$s\bar{c}$

Table 5 : Quark content of pseudoscalar and vector mesons in the charm model. C = charm, T = isospin. The meson states with C=T=0 will be mixtures of the quark states listed. This is indicated by m.

therefore only four of the vector mesons listed in table 5 can couple directly to the photon : $\rho^o,\omega,\phi,\phi_c$. It is tempting to identify ϕ_c with one of the new vector mesons and to assume in analogy with "ideal" mixing in the ω-ϕ case that ϕ_c is an almost pure $c\bar{c}$ state.

$$\psi(3100) \equiv \phi_c \sim c\bar{c} \qquad (2.16)$$

This is in fact what is proposed in ref. 10).

The trouble comes from two sources :
(i) Experimentalists found another narrow resonance $\psi_{3,7}$ and a broad structure, possibly a resonance at $\sqrt{s} \sim 4,2$ GeV.
(ii) ϕ_c with the quark content as in eq.(2.16) does not carry a new quantum number explicitly. No quantum number conservation prevents its decay into ordinary hadrons. This makes it hard to understand the very small width of the $\psi_{3,1}$ and $\psi_{3,7}$ (table 2).

Both of these puzzles are supposedly solved in an ingeniously simple and appealing picture : the charmonium model[16]. Quarks and antiquarks are thought to interact among each other by exchanging an octet of colored gluons in much the same way as electrons and positrons interact by exchanging photons. As the Coulomb potential between electron and positron produces a bound state : positronium so does the strong potential $V_{st}(r)$ it produces "charmonia", bound states of a charmed quark and antiquark. In such a potential model there will be a ground state - identified with $\psi_{3,1}$ - there may be excited states - perhaps the $\psi_{3,7}$ is one of them - and there may be something like an ionisation threshold - perhaps the structure around $\sqrt{s} = 4,2$ GeV is connected with this threshold.

This model of quarks interacting with colored gluons in a way invariant under non abelian gauge transformations produces another interesting effect : asymptotic freedom[17] which means essentially that strong interactions become weak at short distances and can then be treated in a perturbative way very much like quantum electrodynamics. This weakening of the force at short distances is made responsible for the small width of the $\psi_{3,1}$ and $\psi_{3,7}$.

The basic idea is an extension of the argument used to explain the suppression of the three pion decay of ϕ relative to the corresponding decay of the ω .

$$\frac{\Gamma(\phi \to 3\pi)}{\Gamma(\omega \to 3\pi)} \simeq 0,07 .$$

The ϕ meson consists of a good approximation of strange quarks only, $\phi \sim s\bar{s}$ and does not like to decay into pions which contain only non-strange quarks. Instead it couples easily to K-mesons which contain stange quarks (Zweig's rule). Similarly it is assumed that the $\psi_{3,1}$ is a very pure state of charmed quarks $\psi_{3,1} \sim c\bar{c}$ and prefers to couple to charmed mesons, i.e. mesons containing charmed quarks. An "allowed" decay would then be e.g.

$$\psi(3100) \to D^+D^-$$

corresponding to an "allowed" quark diagram (fig. 7) where no one of the quark lines starts and ends in the same meson. However these "allowed" decays may be energetically forbidden if the mesons

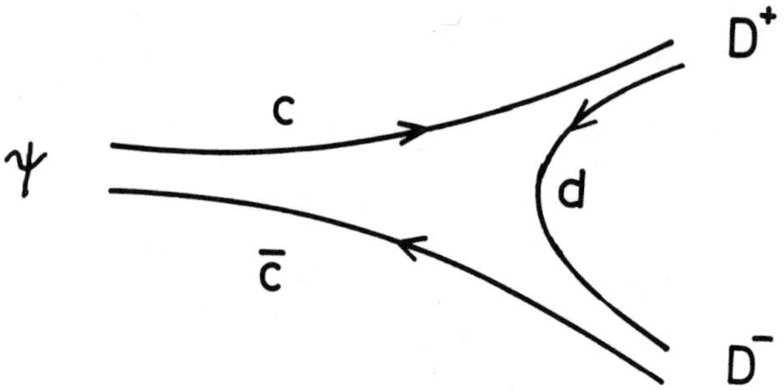

Fig. 7 Quark diagram for an "allowed" coupling $\psi D^+ D^-$

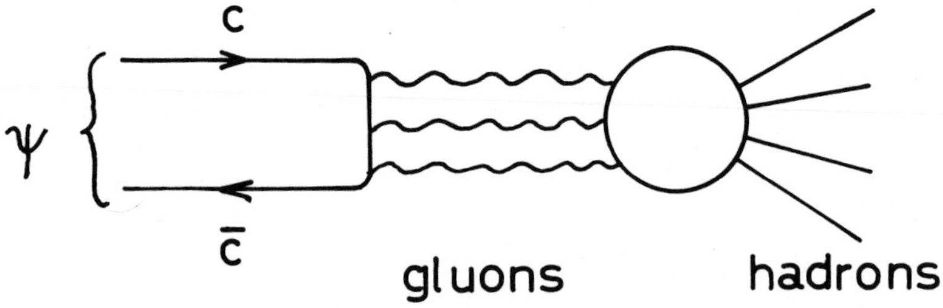

Fig. 8
A charmed quark and antiquark annihilating through colored gluons and producing ordinary hadrons

carrying charm explicitly like D mesons are heavy enough. In this case the $\psi_{3.1}$ meson can only decay into normal hadrons not containing a c quark. This could happen through the annihilation of the c and $\bar{c}$ quarks into gluons which then turn into normal hadrons. From charge conjugation invariance the process can only proceed via an odd number of gluons. A single gluon is ruled out by color symmetry therefore the simplest possibility is three gluon annihilation (fig. 8). Due to the large c-quark mass in the Fermion-propagators of fig. 8 this annihilation can only occur when the c and $\bar{c}$ quark come close together. Since the coupling at short distance becomes progressively weaker in the model under consideration, this decay is a slow process.

This detailed picture is supposed to follow from an underlying quark gluon lagrangian

$$L = \bar{q}\,(i\gamma^{\mu}D_{\mu} - m)q - \frac{1}{4}\,G^{a}_{\mu\nu}G^{a\mu\nu} \qquad (2.17)$$

where $q = (q_{\alpha A})$ is the quark spinor carrying an SU(4) index $\alpha = u,d,s,c$ and a color index $A = 1,2,3$. $G^{a}_{\mu\nu}$ $a = 1,\ldots,8$ is the gluon field strength, D_{μ} the covariant derivative. In terms of the gluon field A^{a}_{μ} :

$$G^{a}_{\mu\nu} = \partial_{\mu}A^{a}_{\nu} - \partial_{\nu}A^{a}_{\mu} + g\, f_{abc}\, A^{b}_{\mu}A^{c}_{\nu}$$

$$D_{\mu} = \delta_{\alpha\beta}\delta_{AB}\partial_{\mu} - ig\, A^{a}_{\mu}\,\delta_{\alpha\beta}\,\frac{1}{2}\,\lambda^{a}_{AB} \qquad (2.18)$$

g is the quark gluon coupling constant.

This may or may not be the correct starting point to solve all the puzzles of strong interactions. In any case it seems worthwhile to pursue the idea of a quark antiquark system moving in a potential well and producing bound states a little further[16]. Motivated by the lagrangian eq.(2.17) which produces asymptotic freedom or better by simplicity we start by considering a Coulomb potential and assume that the quarks move non-relativistically. The Schrödinger equation for the space part of the wave function of the $c\bar{c}$ system is then

$$\{-\frac{1}{m_c}\,\Delta - \frac{4\alpha_s}{3|\vec{x}|}\}\,\psi(\vec{x}) = E_B\psi(\vec{x})$$

$$\alpha_s = \frac{g^2}{4\pi} \qquad (2.19)$$

where $\vec{x}$ is the relative distance between c and $\bar{c}$ quark, E_B the binding energy and α_s characterises the strength of the potential. Everybody has calculated the energy levels following from eq.(2.19) in the beginners course on quantum mechanics. The spins of the quarks can be combined anti-parallel to give singlets and parallel to give triplets. The lowest lying energy levels are as shown in table 6.

Charge conjugation and parity of the state $n^{2s+1}L_j$ are

$$C = (-1)^{\ell+s} \qquad p = (-1)^{\ell+1} \qquad (2.20)$$

Therefore the only states listed in table 5 corresponding to vector mesons with spin-parity-charge conjugation 1^{--} are the $1\,^3S_1$ and $2\,^3S_1$ states. This suggests the identification

$$\psi(3100) \sim 1\,^3S_1$$
$$\psi(3700) \sim 2\,^3S_1 \tag{2.21}$$

$n^{2s+1}L_j$	E_B
$1\,^1S_o,\ 1\,^3S_1$	$-\frac{m_c}{4}(\frac{4}{3}\alpha_s)^2$
$2\,^1S_o,\ 2\,^3S_1$; $2\,^1P_1$; $2\,^3P_o,\ 2\,^3P_1,\ 2\,^3P_2$	$-\frac{m_c}{16}(\frac{4}{3}\alpha_s)^2$
-----	----

Table 6 : Lowest lying energy levels of the "charmonium" system.

Another quantity of interest which can be calculated easily in the potential model is the leptonic decay rate of the two states eq. (2.21). From the diagram fig. 9 we find in general

$$\Gamma(^3S_1 \to e^+e^-) = \frac{16\pi\alpha^2}{9\,m_c^2}\,|\psi(0)|^2 \tag{2.22}$$

and inserting the Coulomb wave functions

$$\Gamma(1\,^3S_1 \to e^+e^-) = \frac{2}{9}\,\alpha^2\,(\frac{4}{3}\,\alpha_s)^3\,m_c$$

$$\Gamma(2\,^3S_1 \to e^+e^-) = \frac{1}{36}\,\alpha^2(\frac{4}{3}\,\alpha_s)^3\,m_c \tag{2.23}$$

From the mass and leptonic width of the $\psi_{3,1}$ we can therefore deduce values for m_c and α_s .

$$2m_c - (\frac{4}{3}\,\alpha_s)^2\,\frac{m_c}{4} \simeq 3100 \text{ MeV}$$
$$\tag{2.24}$$
$$\frac{2}{9}\,\alpha^2\,(\frac{4}{3}\,\alpha_s)^3\,m_c \simeq 5 \text{ keV} \qquad m_c \simeq 1{,}61 \text{ GeV},\ \alpha_s \simeq 0{,}48$$

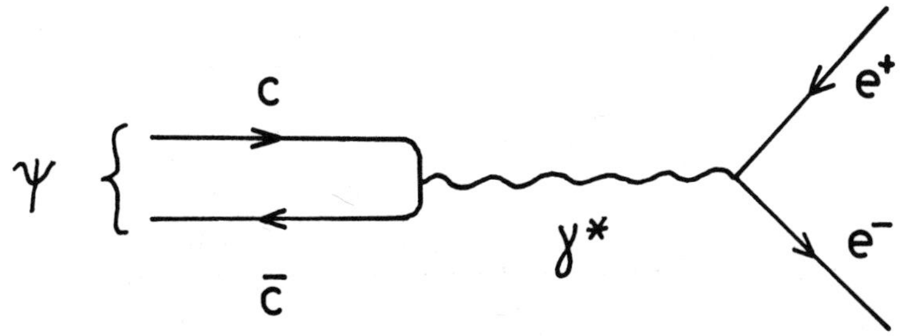

Fig. 9 Leptonic decay of a ψ meson viewed as a $c\bar{c}$ bound state.

From the virial theorem we find for the mean values of the kinetic and potential energies T and V

$$\langle T\rangle + \langle V\rangle = E_B$$

$$2\langle T\rangle = - \langle V\rangle \tag{2.25}$$

and therefore for the ground state

$$\langle T\rangle = (\tfrac{4}{3}\,\alpha_s)^2 \; \frac{m_c}{4} \simeq 0{,}16 \text{ GeV} \ll m_c$$

which justifies the non-relativistic treatment.

Of course the Coulomb potential cannot be the whole story or else we would see quarks coming out. Somehow the potential well must become infinitely high at large distances thereby confining quarks (fig. 10). Such a modification of the Coulomb potential will of course also affect the energy levels. Since the 2 S wave is more extended than the 2 P wave the effect will be larger for 2 S waves and the level ordering will be as shown in fig. 11. An immediate consequence of such a situation is the possibility of γ ray transition as also indicated in fig. 11. This is indeed what is obtained from explicit calculations[16] where the level spacing 2S - 2P is found to be of order ∿100 - 200 MeV and the radiative width of the 2 $^3S_1$ state identified with $\psi_{3,7}$ of the order of 200 keV.

An extensive search for monochromatic γ rays was undertaken by experimentalists but so far nothing with energy $E_\gamma > 60$ MeV has been found[14]. One way out would be to assume that the force between a

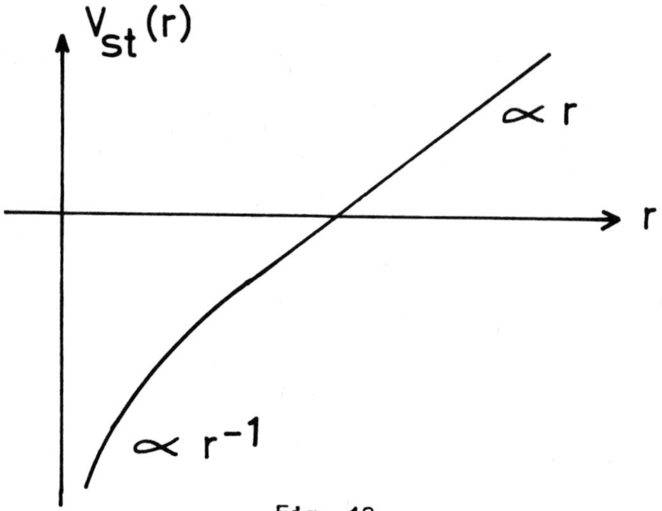

Fig. 10.

Possible confining potential between a quark and antiquark $\propto 1/r$ for $r \to 0$ and $\propto r$ for $r \to \infty$.

c and $\bar{c}$ quark is more complicated than shown in fig. 10. We could e.g. change the short distance part of the potential. Something more singular at the origin than a Coulomb potential would lower the 2 S level relative to the 2 P level. Of course this would mean giving up asymptotic freedom. As a curiosity we note that in any theory with anomalous dimensions the potential generated by one gluon exchange must be more singular at the origin than the Coulomb potential, if positivity arguments can be applied to the gluon propagator.

Other ad hoc modifications of the Schrödinger equation (2.19) can be imagined in order to raise the 2P level e.g. introducing a term $\propto \vec{L}^2$ where $\vec{L}$ is the angular momentum operator. But we feel that this road would not be very productive.

More trouble is encountered by the simple charm model with only four quark triplets (table 4) as we move to higher energy. The model predicts a threshold for production of charmed mesons (D, F in table 5) and an associated rise in the cross section eq.(2.

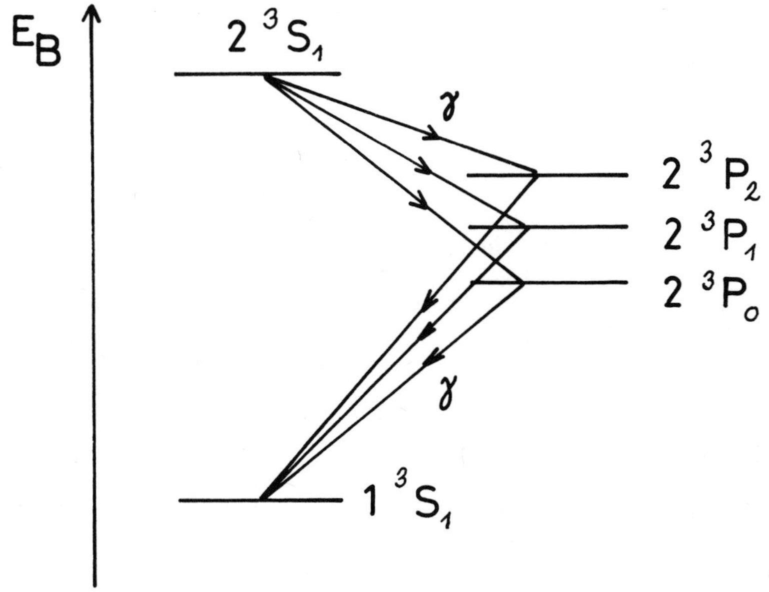

Fig. 11.
Schematic level ordering in a confining potential as shown in fig. 10.

Above the threshold the ratio R(s) should fall, perhaps with some wiggles[16] and ultimately approach a constant from above. Asymptotic freedom allows one to make a definite prediction[18]. For $s \to \infty$ the model lagrangian eq. (2.17) predicts

$$R(s) = \frac{10}{3} \left[1 + \frac{16}{25 \ln s/s_0} + \ldots \right] \tag{2.26}$$

where s_0 is an unknown scale parameter.

Experimentally structure around $\sqrt{s}$ = 4 GeV is observed which could be interpreted as a threshold (fig. 1). However nothing unusual was found up to now in the final state, especially no charmed mesons[14]. Preliminary data at higher energies[14] are also in contradiction with eq. (2.26). The ratio R(s) is both too big, around 5 to 6 and the general tendency is a slow rise with energy instead of the slow decrease predicted by eq. (2.26). Of course there is the unknown scale parameter s_0 in eq. (2.26) and one might try to push the asymptotic regime where eq. (2.26) is supposed to be valid higher up in energy. However if the border of the famous land asymptopia is pushed too far beyond the horizon it risks to have its name changed to utopia.

As a side remark we note that some time ago there has been a speculation that a ratio R(s) approaching a constant from below could be expected in a class of models where strong interactions are described by an infrared stable fixpoint of the renormalization group[19].

To end this section on charm theories we draw the following tentative conclusions :
(i) the absence of monochromatic γ rays from the decay of the $\psi_{3,7}$ suggests that this is not just an excited state of the $\psi_{3,1}$ but built of different quarks or the same quarks in a different combination;
(ii) the high value of R(s) for $\sqrt{s} > 5$ GeV also suggests that more than four quark triplets are needed. A model of this kind is e.g. proposed in ref. 20;
(iii) the apparent absence of new particles in the final state for $\sqrt{s} \gtrsim 4$ GeV is a puzzle for all theories which introduce new quarks and explain the rise in R(s) around $\sqrt{s} \sim 4$ GeV precisely by the production of new particles.

2.4 Thawing the color degree of freedom.

The general idea of color models is to stick to nine quarks as in the usual three triplet-three color model but to drop the restriction that all observable states must be color singlets. Among the many different versions of color models we propose to examine the model of ref. 12 in more detail.

The starting point is the Han-Nambu 9 quark model[6]. There are nine basic quarks $q_{\alpha A}$ α = u,d,s, A = 1,2,3. Strong interactions are thought to be approximately invariant under a group SU(3)'xSU(3)" which transforms the basic quarks according to a $(3,\bar{3})$ representation

$$q_{\alpha A} \rightarrow U'_{\alpha\beta}\, U''^{*}_{AB}\, q_{\beta B} \qquad (2.27)$$

Low lying hadrons are color singlets, i.e. states invariant under SU(3)". For them the observable symmetry group reduces to SU(3)', the usual SU(3) group of strong interactions.

The charges of the quarks are chosen as shown in table 7.

A \ α	u	d	s
1	0	-1	-1
2	1	0	0
3	1	0	0

Table 7 : Charges of quarks in the Han-Nambu scheme.

The electromagnetic current is written in the usual way

$$J_\mu = \bar{q}\gamma_\mu Q q \tag{2.28}$$

where the charge matrix Q as given in table 7 is conveniently split into two pieces Q' and Q" which transform according to representations (8,1) and (1,8) under SU(3)' x SU(3)".

$$Q = Q' + Q''$$

$$Q' = (\bar{Q}_{\alpha\beta}\ \delta_{AB})$$

$$Q'' = -(\delta_{\alpha\beta}\bar{Q}_{AB})$$

$$Q = \begin{pmatrix} 2/3 & & \\ & -1/3 & \\ & & -1/3 \end{pmatrix} \tag{2.29}$$

There is a generalized Gell-Mann Nijishima relation for both Q' and Q"

$$Q' = T_3' + \frac{1}{2} Y'$$

$$Q'' = T_3'' + \frac{1}{2} Y'' \tag{2.30}$$

where T_3', Y', T_3'', Y'' are the third components of isospin and the hypercharges of SU(3)' and SU(3)".

Below the color threshold the color octet piece of the current never contributes and we only see the color average of the charge matrix

$$Q_{eff} = Q'$$

which gives effective charges 2/3, -1/3, -1/3 to u, d and s quarks as is immediately seen from table 7. Therefore below color threshold we expect for the ratio R(s) eq.(2.2)

$$R(s) = \mathrm{tr}\, Q^2_{eff} = 2$$

Once the color degree of freedom is excited we expect many new $q\bar{q}$ meson states. The familiar pseudoscalar and vector meson nonets are extended to families of 81 mesons groupingin representations

$$(1,1),\ (8,1),\ (1,8),\ (8,8) \qquad (2.31)$$

of SU(3)' x SU(3)". These mesons are conveniently labeled by symbols like

$$(\pi,\pi),\ (\pi,\eta),\ (\rho,\omega)\ \ldots$$

where the first (second) symbol characterises the position of the particle in the SU(3)' (SU(3)") multiplet.

How many vector mesons can couple to the photon in the manner of eq. (2.11)? Suppose first that SU(3)" is exact. The current eq. (2.28) has pieces transforming as members of the (8,1) and (1,8) representations. From eq.(2.30) it is clear that only vector mesons with the quantum numbers of T_3', Y', T_3'', Y" can couple. But SU(3) is broken and the eighth component of an SU(3)' octet will mix with the corresponding SU(3)' singlet. Taking this into account we find easily that the following vector mesons can couple to the photon :

in (1,1) + (8,1) :

$$(\rho_o,\omega_o),(\omega,\omega_o),(\phi,\omega_o) \qquad (2.32)$$

in (1,8) + (8,8) :

$$(\omega,\rho_o),(\omega,\omega_8)$$

$$(\phi,\rho_o),(\phi,\omega_8)$$

where the vector mesons (ω,ρ_o), (ω,ω_8) are degenerate in mass and so are (ϕ,ρ_o), (ϕ,ω_8).

The singlet-octet mixing angles in (1,1) + (8,1) and (1,8) + (8,8) need not be identical.

Electromagnetic and weak effects (by inducing tadpole terms) ar expected to break the degeneracy within the two doublets of eq.(2.33

Therefore one would expect to see doublets of new vector mesons spaced by a few MeV. Such a structure has not been observed. A way out is to break SU(3)" e.g. by introducing in analogy to the usual SU(3) breaking a term transforming as (1,Y") in the strong hamiltonian[12]. This leaves color isospin T" and color hypercharge Y" as exactly conserved quantum numbers as far as strong interactions are concerned. Now a total of eight vector mesons can couple to the photon. They are listed in table 8.

T'	Y'	T"	Y"	(a,b)	exp
1	0	0	0	$(\rho_o,\omega_o)_m$	ρ(770)
0	0	0	0	$(\omega,\omega_o)_m$	ω(783)
0	0	0	0	$(\phi,\omega_o)_m$	ϕ(1019)
0	0	1	0	(ω,ρ_o)	ψ(3100)
0	0	1	0	(ϕ,ρ_o)	ψ(3700)
0	0	0	0	$(\omega,\omega_8)_m$	ψ(4200) ?
1	0	0	0	$(\rho_o,\omega_8)_m$	
0	0	0	0	$(\phi,\omega_8)_m$	ψ(4900) ?

Table 8 : quantum numbers and a possible assignment of vector mesons in a broken color scheme. States with the same values for T', Y', T" and Y" can mix among each other. The subscript m indicates that the states are not pure color singlets or octets but mixed.

The color singlets get a slight admixture of color octet states and are identified with the usual ρ,ω,ϕ . The states (ω,ρ_o) and (ϕ,ρ_o) carry explicitly a new quantum number, color isospin which is conserved under strong interactions. They cannot decay strongly into normal hadrons due to this selection rule. This is the basic explanation of the small width of $\psi_{3,1}$ and $\psi_{3,7}$ in this model.

The states $(\omega,\omega_8)_m$, $(\rho_o,\omega_8)_m$ and $(\phi,\omega_8)_m$ on the other hand have an admixture of color singlet states. Their decay into normal hadrons is therefore not inhibited, they should show up as broad resonances. In the model of ref. 12 it is assumed that (ω,ω_8) and $(\rho_o,\omega_8)_m$ are degenerate in mass as are the usual ρ and ω and that they form together the big bump observed at 4,2 GeV (fig. 1). A further broad vector meson is predicted : $(\phi,\omega_8)_m$ and with some imagination one can identify it with an apparent structure seen in the data around $\sqrt{s}$ = 4,9 GeV. We note as a positive point for this model that no new particles are expected from the decay of the (4200) in agreement with experiment.

For larger energies R(s) should approach a value 4 which is at least closer to the observed value 5 - 6 (fig.1) than the charm predication 10/3.

Next we turn to decays of the $\psi_{3,1}$. In the model under consideration this meson carries color isospin T" = 1. If decays into possibly lower lying states with T" = 1 like

$$\psi_{3,1} = (\omega,\rho_o) \rightarrow (\pi_o,\pi_o) + \pi_+ + \pi_- \tag{2.34}$$

are forbidden energetically then $\psi_{3,1}$ will be stable under strong interactions. Its decay should then be electromagnetic since electromagnetism violates conservation of both normal and color isospin. The naive expectation would therefore be to see many radiative decays of the $\psi_{3,1}$. This is not born out by experiment[14] and one has to invent various ad hoc mechanism to explain this and the fact that $\psi_{3,1}$ likes to decay into states with odd G-parity.

The mechanism proposed in ref.12) in order to perform these miracles is similar to the one which was invented for η-decay. Remember that η has even G-parity but ∿54% of its decays lead to 3 pions. First order radiative decays with emission of a single γ account for only ∿5% of the decays. A model which accounts for these facts is the tadpole model of ref. 21). A piece transforming like the third component of isospin $(T_3', 1)$ is introduced in the hamiltonian. Decays via this tadpole which lead to odd G-parity states are found to dominate first order radiative decays by a factor 10.

For the decay of the $\psi_{3,1}$ a color tadpole is introduced transforming as $(1, T_3'')$. Decays of the $\psi_{3,1}$ via this new tadpole will again lead to odd G-parity states as observed experimentally. First order radiative decays remain a problem but are hard to estimate. A typical such decay would be

$$\psi_{3,1} \rightarrow \eta + \gamma \tag{2.35}$$

where the photon absorbes the color quantum number of the $\psi_{3,1}$. This decay is therefore similar to the decay

$$\omega \rightarrow \pi^\circ + \gamma \tag{2.36}$$

The rates are however very dissimilar[13)14)]

$$\Gamma(\omega \rightarrow \pi^\circ\gamma) = (870 \pm 60) \text{ keV}$$

$$\Gamma(\psi_{3,1} \rightarrow \eta\gamma) \simeq 0{,}35 \text{ keV} \tag{2.37}$$

A possible explanation is found in observing that η and γ from the decay of $\psi_{3,1}$ are much more energetic than π^{o} and γ from the decay of the ω. Therefore the wave functions of η and γ will oscillate rapidly over the extension of the $\psi_{3,1}$ particle and suppress the decay. In other words : heavy objects do not like to decay into two body channels. Quantitative estimates seem to be able to produce the correct order of magnitude for the suppression factor[12)].

It is even harder to estimate decays of $\psi_{3,1}$ into multihadrons plus γ. However the experimental evidence for or against such decays is also lacking.

Decays of the $\psi_{3,7}$ into $\psi_{3,1}$ plus normal hadrons are strong decays in the color model under consideration but could be suppressed by a dynamical rule like Zweig's rule since the relation of $\psi_{3,7}$ to $\psi_{3,1}$ is like the relation of ϕ to ω (see table 8). Decays of the $\psi_{3,7}$ will be discussed further in section 3.

We conclude the discussion of the color model of ref. 12) by nothing the following points :
(i) the great stability of the $\psi_{3,1}$ and $\psi_{3,7}$ is explained by the conservation of color isospin in strong interactions;
(ii) charged partners of $\psi_{3,1}$ and $\psi_{3,7}$ are predicted;
(iii) the decay of $\psi_{3,1}$ must be assumed to proceed mainly via a tadpole. Suppression of radiative decays is a problem.
(iv) the decay $\psi_{3,7}$ to $\psi_{3,1}$ plus hadrons is a strong decay but possibly hindered by the normal Zweig rule.

3. A LOOK AT THE FINAL STATE

In this section we want to discuss the final state observed in electron positron annihilation. Whereas a simple and rather successful picture exists for the total cross section as discussed in section (2.1) a complete description of the final state does not exist. In the quark language we have the situation as shown in fig. 4. The creation of the quark antiquark pair by the virtual photon is supposedly understood. How the quarks turn into ordinary hadrons however is quite mysterious and magic words like confinement etc. have to be pronounced. These mystical effects are all supposed to occur within the big blob in fig. 4.

We will first make some remarks on the decays of $\psi_{3,1}$ and $\psi_{3,7}$ and then discuss scaling phenomena.

3.1 Decays of $\psi_{3,1}$

The main features of the decay of the $\psi_{3,1}$ resonance have already been mentioned in section 2.2. Here we will concentrate on purely hadronic decays (fig. 6a). The final states seem to be predominantly multiparticle states with odd G-parity. This does not necessarily imply odd G-parity for the $\psi_{3,1}$ (see section 2.4).

It should be of interest to determine the SU(3) transformation properties of the final state in the decay of the $\psi_{3,1}$. This could be used to constrain the phantasy of model builders. A model where $\psi_{3,1}$ decays into hadrons through gluon annihilation (fig.8) would e.g. predict an SU(3) singlet final state.

In the color model of section 2.4 on the other hand the final state should have the transformation properties of a singlet octet mixture. The mixing angle can be determined from the electronic decay widths of the $\psi_{3,1}$ and $\psi_{3,7}$. This model predicts for the decay widths of the $\psi_{3,1}$ and singlet channels

$$\frac{\Gamma(\psi_{3,1} \to \text{octet})}{\Gamma(\psi_{3,1} \to \text{singlet})} = \frac{\Gamma(\psi_{3,7} \to e^+e^-)}{\Gamma(\psi_{3,1} \to e^+e^-)} \simeq \frac{1}{2} \tag{3.1}$$

where the numerical value is taken from table 3.

Let M_a $a = 1,\ldots,8$ be a meson octet of arbitrary spin. The operation of charge conjugation induces an orthogonal transformation among the M_a's.

$$\begin{aligned} C\,M_a &= \bar{M}_a = c_{ab}M_b \\ c_{ab}c_{a'b} &= \delta_{aa'} \end{aligned} \tag{3.2}$$

It follows that the SU(3) singlet state formed out of two such mesons has positive charge conjugation :

$$C : M_aM_a \to \bar{M}_a\bar{M}_a = M_aM_a \tag{3.3}$$

The $\psi_{3,1}$ with negative charge conjugation cannot decay into such a state if C is conserved.

If therefore the $\psi_{3,1}$ is a SU(3) singlet all decays into two mesons of the same octet are forbidden.

As an example consider the decay $\psi_{3,1} \to K^+K^-$. If $\psi_{3,1}$ is a pure singlet the decay is forbidden, if it has an octet component, the decay is allowed.

Experimentally[14)] this decay is indeed very rare :

$$\frac{\Gamma(\psi_{3,1} \to K^+K^-)}{\Gamma_{tot}} \lesssim 7.10^{-4} \qquad (3.4)$$

This suggests singlet but heavy particles do not like to decay into two light particles anyway (see section 2.4)so the argument is not conclusive.

Decays of $\psi_{3,1}$ into $p\bar{p}$ and $\Lambda\bar{\Lambda}$ have been observed[14)]. This offers another testing ground for the SU(3) properties of the $\psi_{3,1}$. If it is a singlet we would expect equal branching ratios for all decays $\psi_{3,1} \to B\bar{B}$ where B is any member of the baryon octet $B = p, n, \Lambda, \Sigma, \Xi$. However the baryons do not have equal masses and just this kinematic SU(3) breaking effect can alter theequality of the rates by a substantial factor. In special circumstances even by an order of magnitude.

3.2 Decays of $\psi_{3,7}$

A large fraction of the time the $\psi_{3,7}$ resonance decays to $\psi_{3,1}$ plus normal hadrons[14)]

$$\frac{\Gamma(\psi_{3,7} \to \psi_{3,1} + X)}{\Gamma(\psi_{3,7} \to all)} = 0,57 \pm 0,08 \qquad (3.5)$$

These cascade decays are further split into the charged pion and neutral decay modes

$$\frac{\Gamma(\psi_{3,7} \to \psi_{3,1}\ \pi^+\pi^-)}{\Gamma(\psi_{3,7} \to \psi_{3,1} + X)} = 0,56 \pm 0,10$$

$$\frac{\Gamma(\psi_{3,7} \to \psi_{3,1}\ \text{neutrals})}{\Gamma(\psi_{3,7} \to \psi_{3,1} + X)} = 0,44 \pm 0,03 \qquad (3.6)$$

Let us first look at the two pion cascade decay (fig. 12) :

$$\psi_{3,7} \to \psi_{3,1} + 2\pi \qquad (3.7)$$

The two pions in the final state can have isospin T = 0,1,2 and by simple Clebsch Gordan algebra this leads to the following predictions for neutral versus charged pion rates :

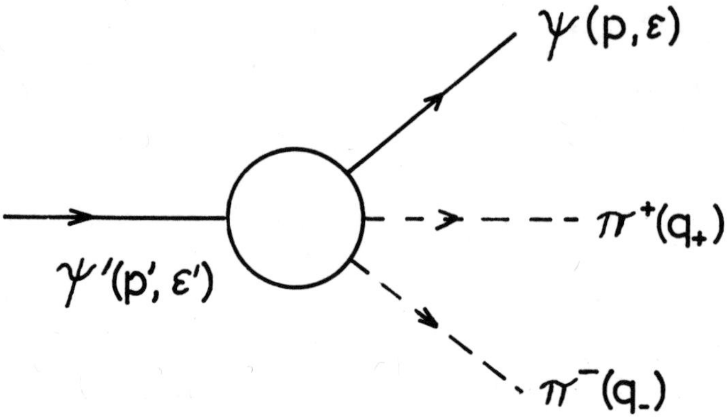

Fig. 12.

Decay $\psi_{3,7} \rightarrow \psi_{3,1} + \pi^+ + \pi^-$. The polarisation vectors and four momenta of the particles are indicated in brackets.

T	$\Gamma(\psi_{3,7} \rightarrow \psi_{3,1}\pi^\circ\pi^\circ)/\ \Gamma(\psi_{3,7} \rightarrow \psi_{3,1}\ \pi^+\pi^-)$
0	1/2
1	0
2	2

Comparing with eq.(3.6) and keeping in mind that "neutrals" stands mostly for $\pi^\circ\pi^\circ$ but includes also some η mesons, we see that the data excludes pure isospin T=1 or 2 but is quite compatible with pure isospin T=0.

The two pions in the reaction eq.(3.7) are of very low energy since the total energy available in the decay is only ~600 MeV. This is an ideal situation for applying PCAC ideas[22]. A few assumptions are made which eventually have to be checked by experiment :

(i) the two pions are emitted in a relative S-wave. This is suggested by centrifugal barrier considerations;

(ii) the two pions are in a state of pure isospin T=0;

(iii) the coupling $\psi_{3,1}\psi_{3,1}\pi$ and $\psi_{3,7}\psi_{3,7}\pi$ vanish. This is assured if the $\psi_{3,1}$ and $\psi_{3,7}$ are eigenstates of G-parity. In both the charm model of section 2.3 and the color model of section

2.4 the ψ mesons have this property.

Now one can apply the usual machinery of PCAC[23),24)]. The outcome is essentially the following. In the limit of zero pion mass the coupling of the pions to the ψ mesons is of the derivative type :

$$L' = -g\psi_\mu \psi'^\mu \, \partial^\lambda \vec{\pi} \partial_\lambda \vec{\pi} \qquad (3.8)$$

where $\psi_\mu (\psi'_\mu)$ is the field of the $\psi_{3,1}(\psi_{3,7})$ meson. This leads to a 2π mass spectrum

$$\frac{d\Gamma}{dt}(\psi_{3,7} \to \psi_{3,1}\pi^+\pi^-) = g^2 t^2 \frac{|\vec{p}_\psi|}{(2\pi)^3 . 16 m^2_{\psi'}} \left(1 - \frac{4m^2_\pi}{t}\right)^{1/2} \left(1 + \frac{|p_\psi|^2}{3m^2_\psi}\right)$$

$$t = (m_{\pi\pi})^2 = (q_+ + q_-)^2$$

$$|\vec{p}_\psi| = \frac{1}{2m_{\psi'}} (m^4_{\psi'} + m^4_\psi + t^2 - 2m^2_{\psi'} m^2_\psi - 2m_{\psi'}{}^2 t - 2m^2_\psi t)^{1/2} \qquad (3.9)$$

and a decay rate

$$\Gamma(\psi_{3,7} \to \psi_{3,1}\pi^+\pi^-) = g^2 I$$

$$I = 0{,}5 \, . \, 10^{-7} \text{ GeV}^5 \qquad (3.10)$$

The 2π mass spectrum is shown in fig. 13. The suppression at low t is due to the vanishing of the coupling eq.(3.8) whenever one of the pions has zero fourmomentum (Adler's zeros[24)]).

From the experimental value for the total decay rate[14)]

$$\Gamma(\psi_{3,7} \to \psi_{3,1}\pi^+\pi^-) \simeq 70 \text{ keV}$$

we find

$$g \simeq 0{,}65 \, f_\pi^{-2} \qquad (3.11)$$

where $f_\pi = 0{,}93 \, m_\pi$ is the pion decay constant. This value for g is quite similar to the value of the low energy 4π -coupling constant. Indeed in the chiral symmetry limit[23)] we have

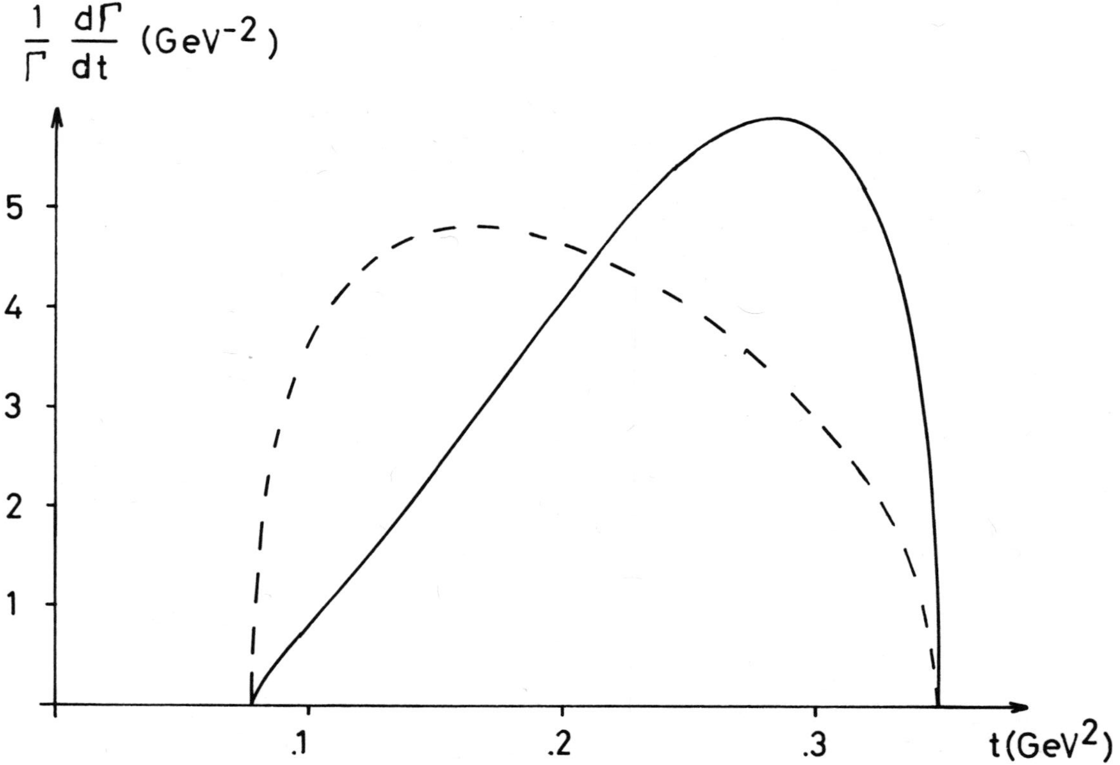

Fig. 13.
The $\pi^+\pi^-$ mass spectrum for the reaction $\psi_{3.7} \rightarrow \psi_{3.1} + \pi^+ + \pi^-$.
Dotted line : phase space distribution. Full line : the distribution predicted by the model discussed in the text (eq.(3.9)). Preliminary experimental results favour the full line [14),22)].

$$L_{4\pi} = -\frac{1}{2f_{\pi\pi}^2}(\vec{\pi}\vec{\pi})(\partial^{\lambda}\vec{\pi}\,\partial_{\lambda}\vec{\pi}) \tag{3.12}$$

We conclude that the decay $\psi_{3,7} \to \psi_{3,1} + 2\pi$ is a strong decay and not suppressed.

In a careful analysis of this reaction one should consider rescattering effects of the two pions. In ref. 22) it is found that these effects are small and that eq.(3.8) gives a good representation of the amplitude throughout the physical region. This casts serious doubts on the validity of any model where the two pions are emitted by the intermediary of an ε-meson since this would imply strong rescattering effects.

Another interesting decay is

$$\psi_{3,7}(p',\varepsilon') \to \psi_{3,1}(p,\varepsilon) + \eta(q) \tag{3.13}$$

Parity conservation requires the η to be emitted in a P-wave. The effective coupling for this process (fig. 14) is given by

$$L' = f_{\psi'\psi\eta}\,\varepsilon^{\lambda\mu\rho\nu}\,\partial_{\lambda}\psi'_{\mu}\,\partial_{\rho}\psi_{\nu}\eta \tag{3.14}$$

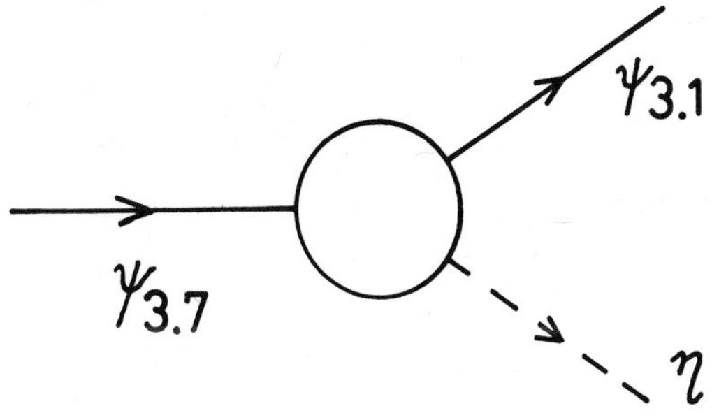

Fig. 14. Decay $\psi_{3,7} \to \psi_{3,1} + \eta$.

where ψ', ψ and η are the fields for the $\psi_{3,7}, \psi_{3,1}$ and η mesons. The decay rate for the reaction eq.(3.13) is easily calculated:

$$\Gamma(\psi_{3,7} \to \psi_{3,1}\eta) = \frac{f^2_{\psi'\psi\eta}}{4\pi} \frac{1}{3} |\vec{q}_{cm}|^3 \tag{3.15}$$

where $|\vec{q}_{cm}|$ is the momentum of the decay products in the c.m. system. From experiment we find[14]

$$\Gamma(\psi_{3,7} \to \psi_{3,1}\eta) \simeq 10 \text{ keV}$$

$$\frac{f^2_{\psi'\psi\eta}}{4\pi} = 3,7 \cdot 10^{-3} \text{ GeV}^{-2} \tag{3.16}$$

This value for the coupling constant can be compared with the coupling constant as obtained from the decay $\omega \to \pi^\circ\gamma$ in the Gell-Mann Sharp Wagner model[25] (fig. 15).

$$\frac{f^2_{\omega\rho\pi}}{4\pi} \simeq 16,4 \text{ GeV}^{-2} \tag{3.17}$$

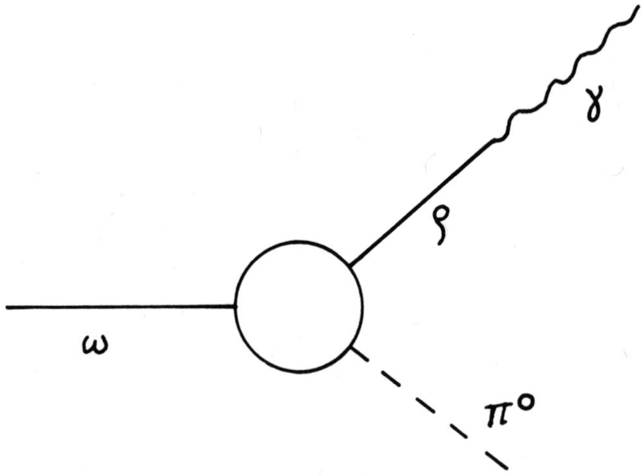

Fig. 15. Decay $\omega \to \pi^\circ\gamma$ in the Gell-Mann Sharp Wagner model.

We find that the $\psi'\psi\eta$ coupling is enormously suppressed. Some suppression can be understood if the $\psi_{3,1}$ and $\psi_{3,7}$ are both SU(3) singlets. Then only the SU(3) singlet component of the η can couple in eq.(3.14), but the η has only a small singlet admixture of order 17% [26]. Even taking this into account the $\psi'\psi\eta$ coupling seems to be suppressed relative to the $\omega\rho\pi$ coupling.

Other hadronic decay modes of the $\psi_{3,7}$ resonance have not been observed to date [14]. In particular it seems that the $\psi_{3,7}$ does not like to decay into many of the channels observed for the $\psi_{3,1}$.

3.3 Scaling in inclusive spectra.

We now turn to electron positron annihilation into hadrons with detection of one hadron in the final state (fig. 16)

$$e^+(k_+) + e^-(k_-) \to h(p) + X \tag{3.18}$$

where h stands for the detected hadron (e.g. a pion or a proton).

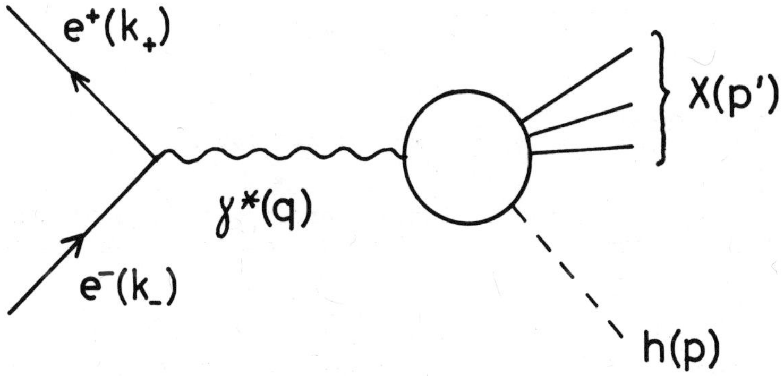

Fig. 16.
Electron positron annihilation into hadrons with detection of one hadron in the final state.

A useful set of variables are the energy E of h in the c.m. system, the angle θ of emission relative to the positron momentum and the scaling variable x defined below.

$$E = \frac{p\cdot q}{\sqrt{s}}\;, \quad q = k_+ + k_-\;, \quad s = q^2$$

$$\cos\theta = \left.\frac{\vec{p}\cdot\vec{k}_+}{|\vec{p}|\,|\vec{k}_+|}\right|_{c.m.}$$

$$0 \leq \theta \leq \pi$$

$$x = \frac{2E}{\sqrt{s}}$$

$$0 \leq x \leq 1 \tag{3.19}$$

The cross section for the reaction eq.(3.18) can be written as product of some explicitly calculable QED factors and an unknown hadronic tensor :

$$\bar{W}_{\rho\lambda}(p,q) = \sum_X (2\pi)^3\, 2\, Vp_0\, \delta(p'+p-q)$$

$$\langle 0|J_\rho|X(p')h(p)\rangle \quad \langle h(p)X(p')|J_\lambda|0\rangle \tag{3.20}$$

where J_λ is the electromagnetic current.

From gauge invariance, Lorentz and parity invariance we find the decomposition :

$$\bar{W}_{\rho\lambda}(p,q) = (-g_{\rho\lambda} + \frac{q_\rho q_\lambda}{q^2})\, \bar{W}_1(pq,q^2)$$

$$+ (p_\rho - \frac{(pq)}{q^2} q_\rho)\,(p_\lambda - \frac{(pq)}{q^2} q_\lambda)\, \bar{W}_2(pq,q^2) \tag{3.21}$$

The structure functions $\bar{W}_{1,2}(pq,q^2)$ are the analogs of the famous structure functions $W_{1,2}$ of deep inelastic electron nucleon scattering.

It is more convenient to use a helicity basis. Let us consider the c.m. system with the z axis chosen in the direction of flight of the hadron h and define polarization vectors for the virtual photon as shown in eq.(3.22).

$$q^\mu = (\sqrt{s},\ 0,\ 0,\ 0)$$

$$p^\mu = (E,\ 0,\ 0,\ (E^2-m^2)^{1/2})$$

$$\varepsilon_L^\mu = (0,\ 0,\ 0,\ 1) \qquad (3.22)$$

$$\varepsilon_\pm^\mu = \mp \frac{1}{\sqrt{2}} (0,\ 1, \pm\ i,\ 0)$$

The structure functions

$$\bar{W}_L(pq,q^2) = \varepsilon_L^{*\rho}\ \bar{W}_{\rho\lambda}\varepsilon_L^\lambda = \bar{W}_1 + (E^2-m^2)\bar{W}_2$$

$$\bar{W}_T(pq,q^2) = \varepsilon_\pm^{*\rho}\ \bar{W}_{\rho\lambda}\varepsilon_\pm^\lambda = \bar{W}_1 \qquad (3.23)$$

are then a measure of the inclusive decay probability of the virtual photon with longitudinal or transverse polarization relative to the direction of flight of the detected hadron.

The cross section for the reaction eq.(3.18) is now easily written down

$$\frac{p_o d\sigma}{d^3p} = \frac{\alpha^2}{2s^2} \{\bar{W}_T(1 + \cos^2\theta) + \bar{W}_L(1 - \cos^2\theta)\} \qquad (3.24)$$

It is instructive to calculate the inclusive distribution functions for the two reactions $e^+e^- \to \mu^+\mu^-$ and $e^+e^- \to \pi^+\pi^-$. The result of this simple exercise is as follows

for $e^+e^- \to \mu^+\mu^-$ and $s >> m^2$

$$\bar{W}_T = \delta(2pq - q^2)\ 2q^2$$

$$\bar{W}_L = 0 \qquad (3.25)$$

for $e^+e^- \to \pi^+\pi^-$ and $s >> m_\pi^2$

$$\bar{W}_T = 0$$

$$\bar{W}_L = \delta(2pq - q^2)\ s|F_\pi(s)|^2 \qquad (3.26)$$

where $F_\pi(s)$ is the pion form factor. Inserting in eq.(3.24) we see that the distribution for muons is proportional to $(1+\cos^2\theta)$ for pions to $(1-\cos^2\theta)$. This is a simple consequence of helicity conservation and γ_5 invariance for massless spin 1/2 particles. If we neglect the electron mass electron and positron can only annihilate in the two states shown in fig. 17a where they have opposite helicities. Therefore the virtual photon has helicity $\pm$ 1 in the beam direction. However for the production of a pion at an angle θ only photons with helicity zero in the direction of flight of the pion can contribute. The distribution of the pions reflects therefore the amount of helicity zero (with respect to the direction of flight of the pion) in a photon state of helicity $\pm$ 1 (with respect to the beam direction). This is easily calculated and found to be proportional to $(1-\cos^2\theta)$. A similar argument can be given to explain the $(1+\cos^2\theta)$ distribution for the muons.

We will now outline a simple theory for inclusive hadron production[27]. We assume that we are at sufficiently high energy so that both quark masses and hadron masses can be neglected. According to the picture developed in section 2.1 the virtual photon creates at first a quark antiquark (fig. 4), each particle carrying half of the energy available, i.e. $\sqrt{s}/2$. These quarks are then supposed to fragment into hadrons. Two basic assumptions are made :

(i) the hadrons come out with small transverse momentum relative to the parent quark. As a first approximation the transverse momentum will be neglected entirely;

(ii) the probability of finding a hadron h with energy E depends only on the nature of the parent quark and the ratio of the hadron energy E and the energy of the parent quark $\sqrt{s}/2$. This probability is described by a fragmentation function $D_q^h(x)$, $x = 2E/\sqrt{s}$. The "wee" region $x \lesssim m/\sqrt{s}$ is excluded in this consideration.

Now it is easy to calculate the inclusive cross section for hadrons. The production rate for quarks is easily obtained from the expression for muon production eq.(3.25). This has to be multiplied with the probability for quarks fragmenting into hadrons given by $D_q^h(x)$. The result is

$$d\sigma = \sum_i \frac{\alpha^2}{2s^2} \, d\Omega \int_0^\infty d|\vec{p}_i||\vec{p}_i| \; 2\, s\delta(2p_i q - q^2)$$
$$(1 + \cos^2\theta) Q_i^2 \, D_{q_i}^h(x) dx \qquad (3.27)$$

where the sum runs over all "active" quarks and antiquarks and Q_i is the charge of the i-th quark.

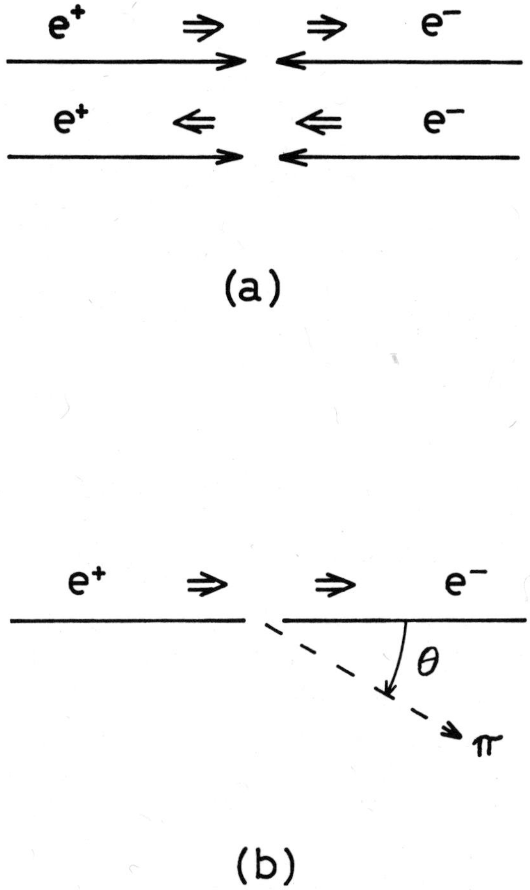

Fig. 17:
(a) Annihilation of zero mass electrons and positrons
(b) Production of a pion π at an angle θ to the beam direction.

The integration in eq.(3.27) is trivially done and the result is

$$\frac{sd\sigma}{dxd\Omega} = \frac{\alpha^2}{4}(1 + \cos^2\theta) \sum_i Q_i^2 D_{q_i}^h(x)$$

$$\frac{sd\sigma}{dx} = \frac{4\pi\alpha^2}{3} \sum_i Q_i^2 D_{q_i}^h(x)$$

$$x \gtrsim \frac{m}{\sqrt{s}} \tag{3.28}$$

We find that the cross section exhibits scaling behaviour in this simple model i.e. the dimensionless quantity $sd\sigma/dx$ depends only on the dimensionless variable $x = 2E/\sqrt{s}$. This is the analog of the famous Bjorken scaling in deep inelastic electron nucleon scattering. The inclusive distribution has been studied for three energies[28]. The experimental finding is shown schematically in fig. 18. Scaling seems to hold for $x \gtrsim 0{,}5$. For $x \lesssim 0{,}5$ however there are large deviations from scaling. This may be due to non-asymptotic terms or to the opening up of new channels like charmed meson production.

To conclude this brief discussion of inclusive distributions we list a few predictions of the simple model outlined above.

(i) The inclusive distribution should scale, i.e. $\frac{sd\sigma}{dx}$ should be a function of the scaling variable $x = 2E/\sqrt{s}$ only.

(ii) The angular dependence of the one particle inclusive distribution should be proportional to $(1+\cos^2\theta)$ if quarks have spin 1/2.

(iii) Hadrons should be produced in jets.

(iv) If the quark fragmentation frunctions $D_q^h(x)$ behave similarly to the quark distribution functions of the nucleon we expect $D_q^h(x) \propto \frac{1}{x}$ for $x \to 0$ and as a consequence a mean multiplicity $\langle n^h \rangle \propto \ln s$.

(v) The fragmentation functions of quarks should be universal and therefore as well measurable in inclusive electro- and neutrino-production.

APPENDIX

In this appendix we collect some useful formulae which arise in connection with the production and decay of ψ-particles.

Consider the coupled system of the photon and the ψ-meson. The propagator of this coupled system is

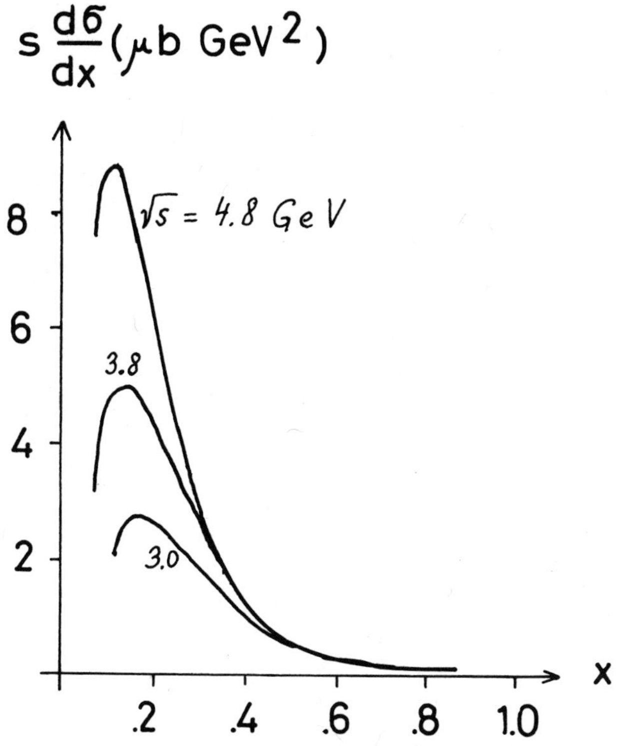

Fig. 18

Inclusive hadron distribution (from ref. 28)).

$$\frac{1}{i}\ \Delta^{ij}_{\mu\nu}(x) = \begin{pmatrix} \langle 0|T\{A_\mu(x)A_\nu(0)\}|0\rangle & \langle 0|T\{A_\mu(x)\psi_\nu(0)\}|0\rangle \\ \\ \langle 0|T\{\psi_\mu(x)A_\nu(0)\}|0\rangle & \langle 0|T\{\psi_\mu(x)\psi_\nu(0)\}|0\rangle \end{pmatrix}$$

$$= \int \frac{dq}{(2\pi)^4}\ e^{-iqx_i}\ \Delta^{ij}_{\mu\nu}\ (q) \tag{A.1}$$

where A_μ and ψ_μ are the photon and ψ-meson fields. We can decompose $\Delta^{ij}_{\mu\nu}(q)$ into a transverse and longitudinal part :

$$\Delta^{ij}_{\mu\nu}(q) = (-g_{\mu\nu} + \frac{q_\mu q_\nu}{q^2})\ \Delta^{ij}_T(q^2) + \frac{q_\mu q_\nu}{q^2}\ \Delta^{ij}_L\ (q^2) \tag{A.2}$$

In the following we will only consider the transverse part. We introduce the mass matrix by

$$\Delta_T^{-1}(q^2) = q^2 - M^2(q^2) \tag{A.3}$$

Time reversal invariance requires M^2 to be symmetric

$$\{M^2(q^2)\}^T = M^2(q^2) \tag{A.4}$$

when an upper T means transposed.

Let X be any final state and introduce the one particle irreducible (with respect to γ and ψ) transition amplitudes

$$a^\mu_X = \begin{pmatrix} \langle\gamma^\mu|T^\dagger|X\rangle \\ \\ \langle\psi^\mu|T^\dagger|X\rangle \end{pmatrix} = (-g^{\mu\nu} + \frac{q^\mu q^\nu}{q^2})\Delta_T^{-1} \begin{pmatrix} \langle 0|A_\nu|X\rangle \\ \\ \langle 0|\psi\ \ |X\rangle \end{pmatrix} \tag{A.5}$$

The unitarity equation can then be written as

$$(-g_{\mu\nu} + \frac{q_\mu q_\nu}{q^2})\ \frac{1}{2i}\ \{M^{2+}\ (q^2) - M^2(q^2)\}$$

$$= \frac{1}{2} \sum_X (2\pi)^4\ \delta(q-p_X)a_{X\mu}a^+_{X\nu} \tag{A.6}$$

Let m_o^2 and Γ_o be the mass and width of the ψ-particle in the absence of electromagnetism. We will approximate $M^2(q^2)$ by a constant in the vicinity of $q^2 = m_o^2$.

$$M^2(q^2) \simeq M^2(m_o^2) \equiv M^2 \tag{A.7}$$

Compute the eigenvectors and eigenvalues of M^2 remembering that M^2 is symmetric but not hermitian.

$$M^2\phi_i = \lambda_i\phi_i$$

$$\phi_i^T M^2 = \phi_i^T\lambda_i \qquad i = 1,2 \tag{A.8}$$

$$\phi_i^T\phi_j = \delta_{ij}$$

$$\lambda_1 = m^2 - im\Gamma$$

$$\lambda_2 = O(e^2)$$

$$N^2 = \phi_1^+ \phi_1 \qquad (\neq 1 \text{ in general})$$

$$\Delta_T(q^2) = \sum_{i=1}^{2} \frac{\phi_i\phi_i^T}{q^2 - \lambda_i} \tag{A.9}$$

Therefore the resonant part of the propagator is

$$\Delta_T^{res}(q^2) = \frac{\phi_1\phi_1^T}{q^2-m^2+im\Gamma} \tag{A.10}$$

A formula for the width Γ is most conveniently obtained by sandwiching the unitarity relation eq.(A.6) between ϕ_1^+ and ϕ_1.

$$\Gamma = \frac{1}{2m} \sum_X (2\pi)^4\delta(q-p_X)$$

$$(-\frac{1}{3}) \frac{1}{N^2} \cdot (\phi_1^+ a_X^\mu)(a_{X\mu}^+\phi_1) \tag{A.11}$$

Now we will make an expansion in e. The mass matrix is easily seen to be

$$M^2 = \begin{pmatrix} -e^2 m_o^2 \Pi_B & e\, m\, F \\ e\, m\, F & m_o^2 - i m_o \Gamma_o \end{pmatrix} \qquad (A.12)$$

where Π_B is the vacuum polarisation function of the photon excluding all graphs with a single intermediate ψ-meson and F is the γ-ψ coupling constant which will in general be a complex number. We find now easily :

$$m^2 - im\Gamma \simeq m_o^2 - i m_o \Gamma_o + e^2 F^2 - e^4 F^2 \Pi_B$$

$$\phi_1 \simeq \begin{pmatrix} \frac{eF}{m} \\ 1 \end{pmatrix}$$

$$N^2 \simeq 1 \qquad (A.13)$$

Let f be a set of final states. From eq.(A.11) we find the partial decay rate for the process $\psi \rightarrow f$:

$$\Gamma_f = \frac{1}{2m} \sum_{X \epsilon f} (2\pi)^4 \delta(q-p_X)$$

$$(-\frac{1}{3}) \quad \{<X|T|\gamma_\mu> \frac{eF}{m} + <X|T|\psi_\mu>\}^*$$

$$\{<X|T|\gamma^\mu> \frac{eF}{m} + <X|T|\psi^\mu>\} \qquad (A.14)$$

where the two terms in the brackets correspond to the processes of fig. 6(a) and (b) respectively. As an example we obtain for the leptonic width (ℓ = electron or muon)

$$\Gamma_{\ell^+\ell^-} = \frac{4\pi\alpha^2}{3} \frac{|F|^2}{m} \qquad (A.15)$$

Consider finally the resonant part of the cross section for e^+e^- annihilation into some final state f.

$$\sigma_f(s) = \sigma(e^+e^- \to "\psi" \to f) \tag{A.16}$$

If the summation over final states in f includes an integration over all momenta of the particles the cross section can be written in a simple form

$$\sigma_f(s) = 12\pi \frac{\Gamma_{e^+e^-} \Gamma_f}{(s-m^2)^2+(m\Gamma)^2}$$

$$\int d\sqrt{s}\ \sigma_f(s) = \frac{6\pi^2}{m^2} \frac{\Gamma_{e^+e^-} \Gamma_f}{\Gamma} \tag{A.17}$$

REFERENCES

1. J.E. Augustin et al., Phys.Rev.Letters 33, 1406 (1974).
2. C. Bacci et al., Phys.Rev.Letters 33, 1408 (1974).
3. G.S. Abrams et al., Phys.Rev.Letters 33, 1453 (1974).
4. J.J. Aubert et al., Phys.Rev.Letters 33, 1404 (1974).
5. N. Cabibbo, G. Parisi and M. Testa, Nuovo Cimento Lettere 4, 35 (1970).
6. M.Y. Han and Y. Nambu, Phys.Rev. 139, B1006 (1965), D10, 674 (1974); O.W. Greenberg, Phys.Rev.Letters 13, 598 (1964).
7. M. Gell-Mann, Acta Physica Austriaca Suppl. IX, 733 (1972).
8. S.L. Adler, Phys.Rev. 177, 2426 (1969).
   J.S. Bell and R. Jackiw, Nuovo Cimento 60A, 47 (1969).
9. J.D. Bjorken and S.L. Glashow, Phys.Letters 11, 255 (1964).
   S.L. Glashow, J. Iliopoulos and L. Maiani, Phys.Rev.D2,1285(1970).
10. S. Borchardt et al., Phys Rev.Letters 34, 38, 236 (1975).
    Th. Applequist and H.D. Politzer, Phys.Rev.Letters 34, 43 (1975).
    A. de Rujula and S.L. Glashow, Phys.Rev.Letters 34, 46(1975) and many others.
11. B.G. Kenny et al. Phys.Rev.Letters 34, 429 (1975).
    A.I. Sanda and H. Terazawa, Phys.Rev.Letters 34, 1403 (1975) and many others.
12. N. Marinescu and B. Stech, Heidelberg University Preprint (1975).
    G. Feldman and P. Matthews, Trieste Preprint (1975).
13. V. Chaloupka et al. Phys.Letters 50B, No.1 (1974).
14. All experimental information on the ψ-particles is taken from the talks of V. Lüth and G. Wolf at the International Conference on High Energy Physics, Palermo (Italy), June 1975.
15. M.K. Gaillard, B.W. Lee and J.L. Rosner, Rev.Mod.Phys. 47, 277 (1975).
16. Th.Applequist et al., Phys.Rev.Letters 34, 365 (1975).
    E. Eichten et al. Phys.Rev.Letters 34, 369 (1975).
    J. Kogut and L. Susskind, Phys.Rev.Letters 34, 767 (1975).

17. D. Gross and F. Wilczek, Phys.Rev.Letters 30, 1343 (1973). H.D. Politzer, Phys.Rev.Letters 30, 1346 (1973).
18. Th. Applequist and H. Georgi, Phys.Rev. D8, 4000 (1973). A. Zee, Phys.Rev. D8, 4038 (1973).
19. O. Nachtmann, Phys.Letters 51B, 469 (1974).
20. H. Harari, SLAC PUB 1568 (1975).
21. S. Coleman and S.L. Glashow, Phys.Rev. 134, B671 (1964).
22. D. Morgan and R. Pennington, Rutherford Laboratory Preprint RL-75-062; L.S. Brown and R.N. Cahn, Phys.Rev.Letters 35, 1 (1975).
23. S. Weinberg, Phys.Rev.Letters 17, 616 (1966).
24. S.L. Adler, Phys.Rev. 137, B1022 (1965).
25. M. Gell-Mann, D. Sharp and W.G. Wagner, Phys.Rev.Letters, 8, 261 (1962).
26. H. Pilkuhn et al., Nuclear Phys. B65, 460 (1973).
27. S.D.Drell, D.J. Levy and T.M. Yan, Phys.Rev. D1, 1617 (1970). R.P. Feynman, Photon Hadron Interactions (W.A. Benjamin 1972).
28. B. Richter in Proc. of the 18th International Conference on High Energy Physics, London (1974).

DYNAMICAL SYMMETRY BREAKDOWN

F. ENGLERT

Faculté des Sciences, Université Libre de Bruxelles

Bruxelles, Belgium

I. INTRODUCTION

Spontaneous breakdown of symmetry and hence degenerate vaccuum is a general feature of second order phase transition[1]; e.g. breakdown of rotational invariance in ferromagnetism, translational invariance in freezing, fermionic charge invariance in superconductivity. That such a mechanism could also be operative in relativistic quantum field theories was originally proposed by Nambu and Jona-Lasinio[2]. They showed that mass could arise from a "phase transition" where chiral charge conservation is spontaneously broken in the same way as the energy gap of a superconductor arises from the breakdown of fermionic charge conservation.

This suggested that more generally mass could originate from a type of phase transition in which a symmetry is spontaneously broken. In particular the possibility that Yang-Mills vector mesons acquire mass in such a way[3],[4] has led to a tentative unification of weak and electromagnetic interactions which may perhaps be extended to strong interactions[5]. In particular the possibility that such a mechanism may lead to quantized vortex lines[6] and monopoles[7] has revived the possibility of dealing with strong interactions along similar lines : indeed the connection between dual models and gauge fields now appears at least plausible.

However, in most models of vector meson mass generation the breakdown of the original symmetry is imputable to scalar fields acquiring a non vanishing vaccuum expectation value. Such inclusion of scalar mesons introduces a great amount of arbitrariness and moreover usually destroys the scale invariance of the problem ab initio[8]. This is unsatisfactory because a natural origin of the physical scale is a mass and therefore one expects the basic matter action to be scale invariant if mass finds its origin in spontaneous symmetry breakdown. Thus one is tempted to consider these scalar mesons as phenomenological quantities and to look for dynamical mass generation without recourse to them.

This is why one is led to restrict the basic Lagrangian to massless fermions which are minimally coupled to gauge fields. Such scale invariant Lagrangian are indeed derivable from a small set of symmetry principles and the scale invariance is a consequence of these other symmetries[8]. Moreover there are indications that such matter Lagrangians could play a very special role in building a consistent quantized theory of gravitation in which no dimensional (scale breaking) gravitational coupling constant need be introduced[9]. We thus conjecture that such "minimal Lagrangians" are physically relevant and shall try to trace the origin of masses back to dynamical spontaneous breakdown of chirality, scale invariance and possibly other symmetry groups. In what follows, we shall reserve the name "dynamical symmetry breakdown" for such a dynamical mass generation.

The discussion will be organized as follows : in section 2 we shall consider an abelian model of massless fermions interacting with massless vector and axial vector fields [3], [10],[11] . This prototype of minimal Lagrangian will serve to illustrate the characteristic features of models of dynamical symmetry breakdown which have been used in the last decade. We shall see that these models are faced with impressive difficulties which makes it very hard to draw a definite conclusion regarding their mathematical consistency.

In section 3 we shall see that many of the difficulties disappear if one chooses equal bare coupling constants for the vector and the axial gauge field. The physical mechanism leading to mass generation is now entirely different and results in an eigenvalue equation for the physical coupling constant. This we shall call a B-type mechanism to distinguish it from the previous case which will be referred to as an A-type mechanism.

In section 4 we shall see that B-type symmetry breakdown occurs when the underlying symmetry group belongs to the special class of "chiral groups" (not to be confused with the abelian group of chiral transformation). These are the groups for which each separate gauge fields couples to fermions of definite chirality. We shall also briefly mention the possibility of mass generation for the case of pure Yang-Mills fields. Finally in section 5 we shall outline how these results may eventually be relevant to physics.

This presentation of dynamical symmetry breaking was elaborated in collaboration with J. M. Frère and P. Nicoletopoulos.

II. A SIMPLE MODEL FOR DYNAMICAL SYMMETRY BREAKDOWN

a. The dynamical symmetry breakdown mechanism

Let us consider the minimal Lagrangian constructed from a massless fermion field interacting with its vector and axial-vector abelian gauge fields. This is

$$\mathcal{L} = i\,\bar{\psi}\,\gamma^{\mu}\partial_{\mu}\psi - g_V\,\bar{\psi}\,\gamma^{\mu}\psi V_{\mu} - g_A\,\bar{\psi}\,\gamma^{\mu}\gamma_5\psi\,A_{\mu}$$

$$-\frac{1}{4}\,F^{(V)}_{\mu\nu}\,F^{\mu\nu(V)} - \frac{1}{4}\,F^{(A)}_{\mu\nu}F^{\mu\nu(A)}$$

with

$$F^{(V)}_{\mu\nu} = \partial_{\mu}\,V_{\nu} - \partial_{\nu}\,V_{\mu};\quad F^{(A)}_{\mu\nu} = \partial_{\mu}A_{\nu} - \partial_{\nu}A_{\mu}\,. \tag{2.1}$$

We write the fermion propagator S(p) as

$$S(p) = \frac{i}{(2\pi)^4}\,\frac{1}{\not p - \Sigma(\not p)} = \frac{i}{(2\pi)^4}\,\frac{1}{\not p\,[1-\Sigma_1(p^2)] - m\Sigma_2(p^2)} \tag{2.2}$$

Here m is a scale parameter with the dimension of a mass. We shall choose for m the physical mass, that is the solution of $m = \Sigma\,(m)$. Of course m remains arbitrary because (2.1) does not contain any scale parameter; the important point is that as we shall see m can be different from zero if a "phase transition" does occur.

S(p) obeys the Schwinger-Dyson equation

$$\Sigma\,(\not p) = \frac{-i}{(2\pi)^4}\,g_V^2 \int d^4k\,D^{(V)}_{\mu\nu}\,\gamma^{\mu}\,S\Gamma^{\nu}$$

$$+\frac{-i}{(2\pi)^4}\,g_A^2 \int d^4k\,D^{(A)}_{\mu\nu}\,\gamma^{\mu}\,\gamma_5\,S\Gamma^{\nu}_5 \tag{2.3}$$

where $D^{(V)}_{\mu\nu}$ and $D^{(A)}_{\mu\nu}$ represent the vector and axial vector propagators; Γ^{ν} and Γ^{ν}_5 the corresponding proper vertices. (2.6) may be viewed as coupled integral equations for Σ_1 and Σ_2 and we shall write

$$m\,\Sigma_2(p^2) = \frac{-i}{(2\pi)^4} \int d^4k\; K(p;k)\; m\Sigma_2(k^2) \qquad (2.7)$$

where K(p;k) is a fonctional depending on Σ_1 and Σ_2 . Its explicit form can be obtained from (2.6) by expanding the right hand side in Feynman graphs involving bare vertices, bare meson propagators, and dressed fermion propagators (fig.1).
If in fig.1, the dressed fermion propagators S(p) would be expanded in terms of the bare ones $S^\circ(p) = \frac{i}{(2\pi)^4}\,\frac{1}{\not p}$ one would recover standard perturbation theory. In each order one would then have $m\,\Sigma_2(p^2) = 0$. The possibility of having $m\,\Sigma_2(p^2) \neq 0$ and hence (2.7) in a non perturbative treatment of (2.6) hinges on a spontaneous breakdown of the chiral invariance $\psi(x) \to (\exp i\,\alpha\gamma_5)\,\psi(x)$ due to the attractive vector interaction as will be manifest in subsection II.b. Such a chiral "phase transition" leading to a non vanishing fermion mass is the field theoretic analog of superconductivity [2].

We now show how the axial vector acquires mass as a consequence of $m \neq 0$. The $\Gamma^\mu{}_5$ satisfies the Ward identity[2] characteristic of chiracl invariance

$$q_\mu\;\Gamma^\mu{}_5(p-q/2,p+q/2) = \frac{i}{(2\pi)^4}\,[S^{-1}(p+q/2)\gamma_5+\gamma_5 S^{-1}(p-q/2)\,] \qquad (2.8)$$

Thus, when $q \to 0$ and $m\,\Sigma_2(p^2) \neq 0$ a pole develops in the induced pseudoscalar part of Γ^μ_5. This massless excitation describes a dynamical Nambu-Goldstone bound state. As in the case where spontaneous breakdown occurs through non vanishing expectation value of scalar fields this singularity produces a massive vector particle in a gauge invariant way [3) 10) 11)]. To see this we write

$$D^{(A)}_{\mu\nu} = \frac{-i}{(2\pi)^4}\;\frac{g_{\mu\nu} - q_\mu q_\nu/q^2}{q^2 - q^2\,\pi^A(q^2)} + (1-\xi)\,\frac{q_\mu q_\nu}{q^4} \qquad (2.9)$$

where ξ is an arbitrary gauge parameter and the gauge invariant polarization tensor has the form

$$\pi^{(A)}_{\mu\nu}(q) = [g_{\mu\nu}\,q^2 - q_\mu q_\nu]\;\pi^A(q^2) \qquad (2.10)$$

and obeys the Schwinger-Dyson equation

$$\pi^{(A)}_{\mu\nu}(q) = \frac{-i}{(2\pi)^4}\,g_A^2\;\mathrm{tr}\int d^4k\;\Gamma^{(A)}_{\mu 5}\;S\gamma_\nu\gamma_5\,S \qquad (2.11)$$

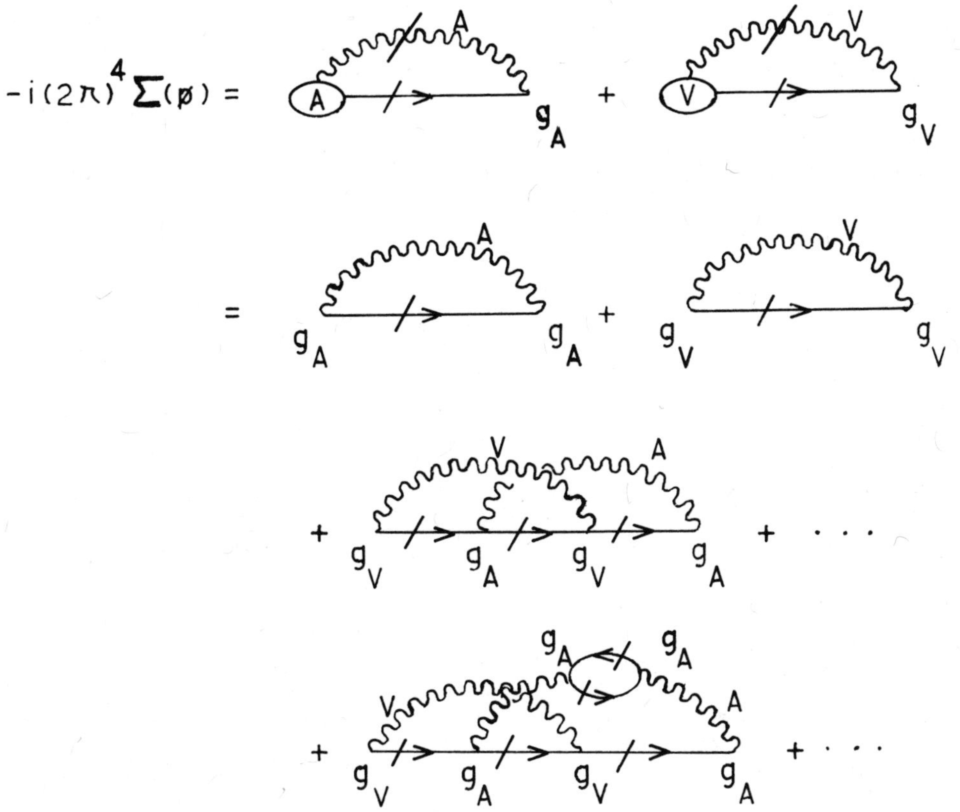

Fig. 1. Schwinger-Dyson equation for the self-energy.

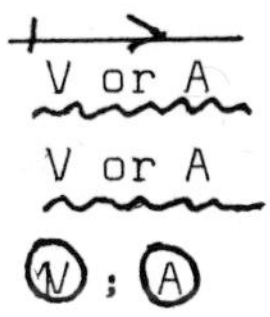

dressed fermion propagator S(p)

dressed V- or A-propagator

bare V- or A-propagator

proper vertices

As $q \to 0$ the contribution to (2.11) of the Nambu-Goldstone singularity when $m\Sigma_2(p^2) \neq 0$ leads to a singularity in $\pi^A(q^2)$ in (2.10) Thus

$$\pi^A(q^2) \underset{q^2 \to 0}{\to} S/q^2 + \text{regular} \tag{2.12}$$

and

$$D^{(A)}_{\mu\nu} \underset{q\to 0}{=} \frac{-i}{(2\pi)^4}[g_{\mu\nu} - q_\mu q_\nu/q^2]\frac{-1}{S} + \text{gauge terms} \tag{2.13}$$

and the axial vector may acquires mass M; approximatively

$$M^2 \simeq ZS \tag{2.14}$$

where Z is the residue of the pole in (2.9).

Let us exhibit an explicit relation between S and the symmetry breaking term $m\Sigma_2(p^2)$. For simplicity we shall approximate $\Sigma_1(p^2)$ by zero. In subsection II.b this will be seen to be exact when $g_A^2 \to 0$, $g_V^2 \to 0$, $\infty > g_V^2/g_A^2 > 1$ if the evaluation is carried out in the Landau gauge $\xi = 1$. In general no new qualitative effect results from $\Sigma_1(p^2)$ as long as there is a gauge in which it is finite. Such a specific gauge may exist to any order in g_V^2 (and g_A^2) as in the Baker-Johnson-Wiley electrodynamics[12]; otherwise the result (2.16) below is directly meaningful only in an approximation for which there exists a gauge rendering $\Sigma_1(p^2)$ finite. The following result for S will be expressed in terms of $m\,\Sigma_{2L}(p^2)$ where the subscript L refers to such a particular gauge. Note that while $m\Sigma_2(p^2)$ and $\Sigma_1(p^2)$ are gauge dependent, m as defined above is gauge independent.

Putting $\Sigma_1(p^2)=0$ in a specific gauge, the solution of (2.8) is as $q \to 0$

$$\Gamma^\mu{}_5 \to \gamma^\mu \gamma_5 - 2m\Sigma_{2L}(p^2)q^\mu/q^2 \tag{2.15}$$

Inserting (2.15) into (2.11) one obtains[3)10)]

$$S = \frac{-i}{(2\pi)^4}\, 8\, g_A^2 m^2 \int d^4k \, \frac{\Sigma_{2L}^2(k^2)}{[k^2 - m^2\Sigma_{2L}^2(k^2)]^2} \tag{2.16}$$

The transversality of $\pi^{(A)}_{\mu\nu}$ as $q \to 0$, which is required by gauge invariance, follows automatically from (2.15) when the integrals have been regularised in a gauge invariant way[13)].

The above example illustrates quite generally the dynamical breakdown mechanism : the breakdown of a global symmetry group (e.g. the chiral group) causes some fermions to acquire mass. Because of the local symmetry (gauge invariance of the second kind) the dynamical Nambu-Goldstone bosons so generated do not in fact appear as physical particles[11]. Instead they ensure that mass is generated in a gauge invariant way for those Yang-Mills fields which are coupled to them : these are the Yang-Mills fields coupled to apparently non conserved currents. By "apparently" we mean that the current would not be conserved if the fermion mass would arise from a bare mass term in the Lagrangian instead of a self consistent mechanism. In the present case only the axial current is "apparently" not conserved.

To discuss the validity of the above mechanism we shall first solve in the present model, the relevant equations in a simple "ladder approximation". There no U-V divergences appear. We shall then see how these U-V divergences may qualitatively modify the results.

b. Ladder approximation

We now consider only the "lowest order" graphs of fig. 1 together with all self energy insertions in the meson propagators. These graphs are depicted in fig. 2, together with their formal expansion in terms of graphs involving bare fermion propagators: these are all "ladder graphs". This ladder approximation will enable us to carry calculations explicitly.

It is easily checked[12] that in the Landau gauge [ξ = 1 in (2.9)] $\Sigma_1(p^2)$ is finite for all ladder graphs. Therefore as stated above we shall put $\Sigma_1(p^2) = 0$ in this gauge. It follows then from (2.6) that (2.7) takes now the form

$$m\Sigma_{2L}(p^2) = \frac{-3i}{(2\pi)^4} \int d^4k \; \frac{m\Sigma_{2L}(k^2)}{k^2 - m^2\Sigma^2_{2L}(k^2)}$$

$$\left[\frac{g_V^2}{(p-k)^2(1-\pi^V)} - \frac{g_A^2}{(p-k)^2(1-\pi^A)} \right] \tag{2.17}$$

where $m\Sigma_{2L}(p^2)$ is thus evaluated in the Landau gauge. The approximation for $q^2\pi^V(q^2)$ and $q^2\pi^A(q^2)$ must be chosen consistently with the transversality condition (2.10) ensuring gauge invariance. This is inconsistent with the ladder approximation except if we approximate $q^2\pi^V(q^2)$ and $q^2\pi^A(q^2)$ by their value at $q^2=0$ .This means that Z=1 in (2.13) and more generally

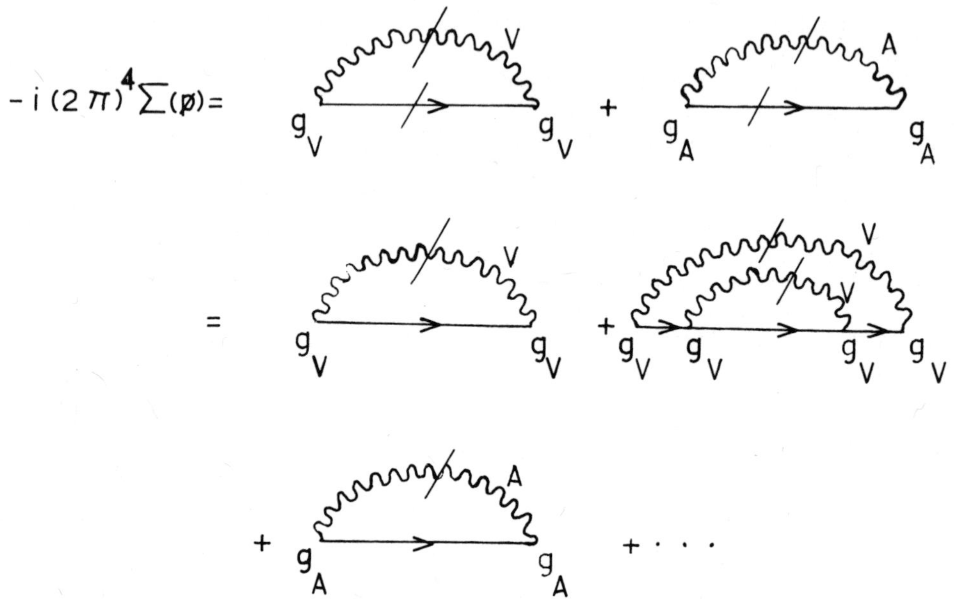

Fig.2. Ladder approximation

that $S = M^2$; we then obtain the coupled equations

$$m\Sigma_{2L}(p^2) = \frac{-3i}{(2\pi)^4}\int d^4k \, \frac{m\Sigma_{2L}(k^2)}{k^2-m^2\Sigma_{2L}(k^2)} \left[\frac{g_V^2}{(p-k)^2} - \frac{g_A^2}{(p-k)^2-M^2} \right] \tag{2.18}$$

$$M^2 = \frac{-i}{(2\pi)^4} \; 8g_A^2m^2\int d^4k \; \frac{\Sigma_{2L}^2(k^2)}{[k^2-m^2\,\Sigma_{2L}^2(k^2)]^2} \tag{2.19}$$

The non-linearity of (2.18) precludes a straightforward analytic solution. It is quite easy however to obtain the asymptotic behaviour of the solution for $p^2 \to \infty$ [11]; indeed the main contribution to the integral originates from the region of asymptotic momenta, so one may neglect $m\,\Sigma_{2L}(k^2)$ and M in the denominators, to obtain,

$$m\Sigma_{2L}(p^2) \underset{p^2\to\infty}{=} -\frac{i}{(2\pi)^4}\, 3(g_V^2 - g_A^2) \int \frac{d^4k}{k^2}\, \frac{m\Sigma_{2L}(k^2)}{(p-k)^2} \tag{2.20}$$

As is easily verified[12] this linear homogeneous integral equation has two solutions $\tilde{\Sigma}_{2L}(p^2)$

$$m\tilde{\Sigma}_{2L}(p^2) = Am\left(\frac{-p^2}{m^2}\right)^{\varepsilon/2} \tag{2.21}$$

$$\varepsilon = -1 \pm \sqrt{1-4G}\ ; \qquad G = \frac{3}{16\pi^2}\,(g_V^2 - g_A^2) \tag{2.22}$$

As m in (2.21) has been identified to the (arbitrary) fermion mass introduced before, $A(g_A^2, g_V^2, m/M)$ is in principle calculable through the complete (unlinearized) integral equation. Its precise evaluation is discussed in section III. It is seen there that $A\to 1$ when $g_V^2 \to 0$, $g_A^2 \to 0$ and $\infty > g_V^2\, g_A^2 > 1$. When G is small, the solution with negative square root is inconsistent with the hypothesis which led to the linearization of eq.(2.18), namely that the integral is dominated by the ultraviolet region. Thus, one is left for G small enough ($0 < G < 1/4$), with

$$m\,{}_{2L}(p^2) \underset{p^2}{=} Am\left(\frac{-p^2}{m^2}\right)^{1/2(-1+\sqrt{1-4G})}$$

$$A = 1 + O(g_A^2;\ g_V^2) \tag{2.23}$$

Notice that eq.(2.20) provides no constraint on the coupling constants (apart from the convergence conditions $0 < G < 1/4$) . This is due to the non-Fredholm character of the kernel. One might wonder whether this lack of an eigenvalue condition persists in the full non-linear eq.(2.18). Could the non-linearity add a sufficiently powerful constraint to single out specific values of G ? The answer appears to be no : in fact, by solving the non linear eq. (2.18) numerically on a computer[15], one checks that its solution is qualitatively the same as the solution of the following linear integral equation

$$m\Sigma_{2L}(p^2) = \frac{-i}{(2\pi)^4}\, 3(g_V^2-g_A^2) \int \frac{d^4k}{k^2-m^2}\, \frac{m\Sigma_{2L}(k^2)}{(p-k)^2} \tag{2.24}$$

This equation was presented as a good approximation to (2.18) in both the low and high p^2 regions, and solved exactly, in ref.16. Thus the asymptotic behaviour of $m\Sigma_{2L}(p^2)$ as obtained from (2.20), is given by (2.23) and connects smoothly to the low energy regime $m\Sigma_{2L}(p^2) \sim m$ for values of G such that $-1 < \varepsilon < 0$. This is the characteristic property of the dynamical symmetry breakdown mechanism described here. The fermion acquires a mass independent of the value of the coupling constants at least as long as U-V divergences are not considered.

We now evaluate M^2 from (2.19) in the limit $G \to 0$, $\infty > g_V^2/g_A^2 > 1$. In this case (see section III) $A \to 1$ and (2.23) reads

$$m\Sigma_{2L}(p^2) \underset{p^2\to\infty}{=} m\left(\frac{-p^2}{m^2}\right)^{-G} \tag{2.25}$$

Note that in this limit $\Sigma_1(p^2)$ is of order G and therefore (2.19) becomes exact. Inserting (2.25) in (2.19) and using an infrared cut-off $\lambda^2 = -k^2_{min}$, one obtains the contribution to M^2 due to asymptotic momenta

$$M^2 \simeq \frac{-i}{(2\pi)^4}\, 8\, g_A^2 m^2 \int_{-\lambda^2}^{\infty} \frac{d^4k}{k^4}\left(\frac{-k^2}{m^2}\right)^{-2G} \tag{2.26}$$

Performing a Wick rotation one obtains

$$\frac{M^2}{m^2} \simeq \frac{1}{(2\pi)^2}\,\frac{g_A^2}{2G}\left(\frac{\lambda^2}{m^2}\right)^{-2G} \tag{2.27}$$

When $G \to 0$ the cut-off dependence disappears and the integral stems thus entirely from the asymptotic region; one gets the exact result [11] for the ladder approximation

$$\frac{M^2}{m^2} \underset{G\to 0}{=} \frac{4}{3}\,\frac{1}{g_V^2/g_A^2 - 1} \tag{2.28}$$

Does this remarkable weak coupling result survives U-V divergences and are g_A and g_V still free parameters in the full theory ? These problems and the consistency of the whole approach when U-V divergences are taken into account will now be discussed.

c) U-V divergences.

In handling higher order graphs in the expansion of fig.1, one is faces as usually with U.V. divergences. As we do not know a priori if there are genuine or if they cancel in the complete summation it is usefull to device a renormalization program which gives meaning to the individual graphs. Such a renormalization program should not destroy the basic chiral symmetry so that the fundamental Ward-Identities such as (2.8) may be carried out in terms of renormalized quantities. Such a procedure is usually realized by introducing only the counter terms of the symmetric theory[17]. However as our method rests on non-perturbative solutions of the Schwinger-Dyson equations one has to modify the conventional procedure. This can be done, at least in principle, because the unrenormalized Green's functions have been expanded in terms of bare vertices and dressed propagators. To obtain the renormalized Green's functions, it then suffices to perform all vertex subtractions in the symmetric limit at a finite $p^2 = -\lambda^2$ in order to avoid introducing infrared divergences. The dressed unrenormalized propagators are then replaced by their renormalized counterparts and the bare coupling constant by a renormalized coupling $g(\lambda)$. This procedure clearly avoids any overcounting. In the present case one has for instance

$$g_V(\lambda) = Z_1^{-1}\{g_V(\lambda);\ g_A(\lambda);\ \frac{\lambda}{\Lambda}\}\ Z_2\{g_V(\lambda);g_A(\lambda);\lambda/\Lambda\}\ \cdot$$
$$Z_3^{1/2}\ \{g_V(\lambda);g_A(\lambda);\lambda/\Lambda\}\ .$$

While $g_V(\lambda)$ is not to be identified with the physical coupling constant, it differs from it only by a finite quantity if the theory is renormalizable. This will be the case if the symmetry breaking terms such as $m\Sigma_2(p^2)$ are made finite by symmetric counterterms only, or equivalency if they may be obtained as finite solution of integral equation involving renormalized quantites only. For this study it will be convient to substitute for the Schwinger-Dyson equation (2.7) the homogeneous Bethe-Salpeter equation obtained by projecting the Ward Identity (2.8) for $\Gamma^\mu{}_5$ on the pseudoscalar channel at $q \to 0$. This can be done as follows : the Bethe-Salpeter equation for $\Gamma^\mu{}_5$ may be expressed in terms of irreducible kernels and proper vertex functions as depicted in fig.3. When contracted with q^μ, the above equation greatly simplifies because one has

$$\frac{\partial}{\partial x^\mu} < \Omega|T\ S_i^\mu(x)S_{i_1}^{\mu_1}\ (z_1)\ldots\ S_{i_n}^{\mu_n}(z_n)|\Omega> \qquad (2.29)$$

where the indices i label the vector or the axial-vector character of the current considered. Equation (2.29) is an immediate consequence of the conservation laws

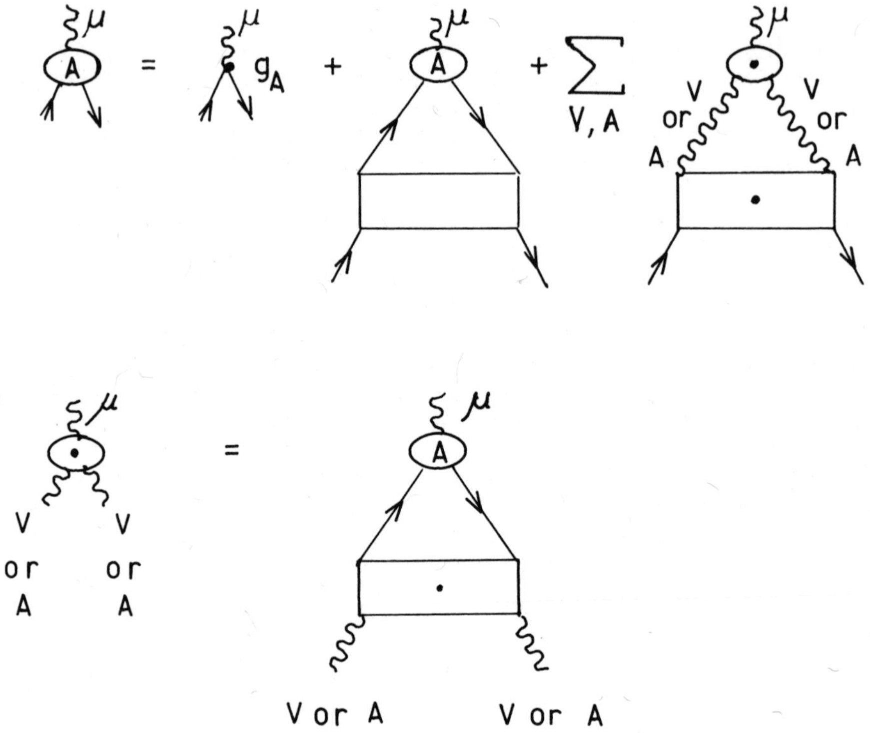

Fig. 3

Bethe-Salpeter equation for Γ_5^μ

3-point proper vertices for vectors and axial-vectors

irreducible Bethe-Salpeter kernels

$$\frac{\partial}{\partial x^\mu} S_5^\mu(x) = 0 \qquad\qquad \frac{\partial}{\partial x^\mu} S^\mu(x) = 0 \tag{2.30}$$

and of the equal time commutation relations

$$[S_5^\circ(\vec{x},x_o),\ S_{i_j}^{\mu_j}(\vec{z}_j,x_o)] = 0 \qquad ; \tag{2.31}$$

from (2.31) it indeed follows that the time d rivative of the T-symbol in (2.29) vanishes. Now (2.29) implies that the n-point vector (and axial-vector) proper vertex function is transverse with respect to any external momentum. One may check this directly from our Feynman graph expansion using the Ward Identity (2.8) (and the corresponding Ward Identity for the vector current).

Thus we see that when Γ_5^μ is contracted with q_μ the 3-point vector (axial-vector) proper vertex functions vanish. Moreover the remaining irreducible Bethe-Salpeter kernel can be simplified by omitting from it all graphs that fall into two pieces by cutting vector (or axial vector) meson propagato s only. Taking the limit $q \to 0$, using (2.8), we thus obtain after multiplication with γ_5

$$m\Sigma_2(p^2) = \frac{-i}{(2\pi)^4} \int d^4k \; \mathcal{K}'(p,k)\, \mathcal{J}(k^2) m\Sigma_2(k^2) \tag{2.32}$$

with

$$g(k^2) = S(k)\; S(-k)$$

$$\mathcal{K}'(p,k) = \frac{i(2\pi)^4}{4} \quad + \quad K'(p,k)\; \gamma^5$$

Here K'(p,k) is the simplified irreducible Bethe-Salpeter kernel at zero momentum. It is examplified in fig. 4. In particular the ladder approximation of subsection II.b is recovered from the graphs 4 a (in the Landau gauge) : one easily verifies that in this approximation equation (2.32) reduces to (2.18).

The advantage of using (2.32) instead of (2.7) is that it is an homogeneous vertex equation and therefore is valid also for renormalized quantities without having to introduce explicitly renormalization constants : we shall denote these by the subscript R . Thus the dynamical symmetry breakdown mechanism is consistent if the symmetry breaking terms are finite solutions of the renormalized integral equations obtained by substituting in (2.32) the renormalized kernels. In particular, the ladder approximation in the Landau gauge of subsection II,b corresponds to $Z_1 = Z_2 = Z_3 = 1$; $m\,\Sigma_{2L}(p^2)$ is finite and so is the resulting M^2.

In higher orders however the Z_i,s differ from 1 and U.V. divergences may spoil the results in two different ways.

1. There is no finite solutions to the renormalized version of (2.32).
2. The Ward identity (2.8) acquires anomalous terms in the right hand side; in other words the axial current is not conserved. The equation (2.32) looses its validity.

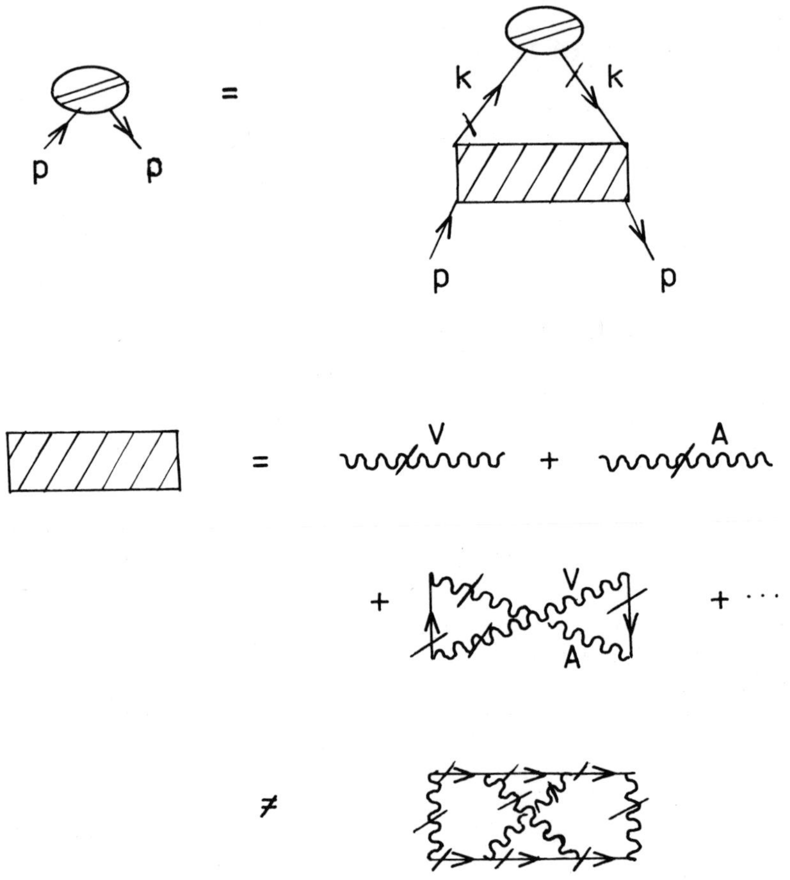

(4a)

(4b)

(4c)

Fig. 4.

Bethe-Salpeter equation form $\Sigma_2(p^2)$

= simplified irreducible Bethe-Salpeter kernel K' (p,k)

4 a = ladder approximation
4 b = corrections to the ladder approximation
4 c = graphs not contained in K'pr,k).

It is hard to discuss the point 1 above without assuming a particular form for $m\Sigma_2(p^2)$. If the scaling form (2.23) has to survive radiative correction then all logarithmic dependence in the kernel introduced by the renormalization procedure must go away to leave at the end a scale invariant integral. This is possible if, as in the BJW electrodynamics[12], $g_r(\infty)$ (and $g_A(\infty)$) obey a Gell-Mann Low condition[18] in which case the theory is not only renormalizable but even finite in a well chosen gauge[12]. There is not as yet any indication that this is indeed the case and mathematical progress in that direction is hard especially because the asymptotic coupling constant should be an essential zero of the Gell-Mann Low function [12] [19]. Of course we may try to give up the scaling form for $m\Sigma_2(p^2)$ but then one can hardly say anything and the weak coupling solution (2.28) looses its significance. For non abelian gauge group the situation is slightly different because one may have asymptotic freedom[20] and thus a vanishing of the asymptotic coupling constants without having to satisfy a Gell-Mann Low equation. However in this case the scaled form (2.23) should not persist because the slow decrease of the renormalized coupling constants will inevitably introduce logarithmic terms in the Green's functions[20]. Thus we see that despite its attractive features the ladder approximation of section II.b is entirely unreliable except if some Gell-Mann Low condition is fulfilled and this remains, for the time being, only a remote possibility.

As far as anomalies in Ward Identities one are concerned (point 2), these may be of two types

a) the well known triangle anomaly[21]. This effect is present in our model; it not only invalidates 2.32) but in fact also the renormalisability already at the level of the symmetric theory; however one can always compensate this effect by introducing more fermions as in reference 11. Such a procedure would merely introduce notational complications and we shall therefore simply ignore the triangle anomaly in the formal discussion;

b) the Johnson-Pagels mass anomaly[22]. This comes about because it is not clear that the above theory has zero bare mass or simply $\lim_{\Lambda\to\infty} m_0(\Lambda) \to \infty$ where Λ is an ultra-violet cut-off.

This point can be understood as follows, already in the context of the ladder approximation of section II.

We know that the expansion of fig. 2 in terms of graphs involving $S^\circ(p)$ is not valid in any finite order because $m\Sigma_2(p^2)$ is then zero. This is of course expected in the optic of a phase transition; however an expansion in terms of a bare "renormalized" propagator $\bar{S}(p) = \frac{i}{(2\pi)^4} \frac{1}{\not{p} - m}$ should be valid. Such an expansion is given by the usual mass renormalized perturbation series for

$$-i(2\pi)^4 \Sigma_2(p^2) = -i(2\pi)^4 m \; + \; \ldots$$

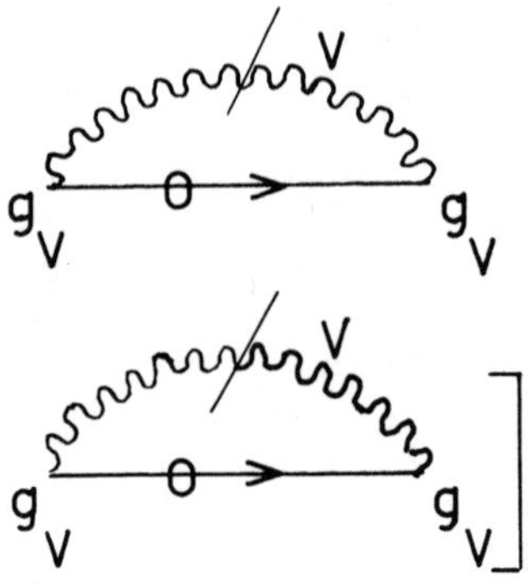

$+ \cdots$

Fig. 5.

Mass-renormalized perturbation series for $m\Sigma_2(p^2)$.

—θ—>— bare renormalized fermion propagator $\bar{S}(p)$

[ ] subtracted diagram at $p^2 = m^2$.

the Schwinger -Dyson equation (2.7) and is depicted in fig. 5 (remember $\Sigma_1(p^2) = 0$).

However, the individual terms are now divergent and one has to introduce a U-V cut-off Λ to carry out the expansion. Thus we write

$$m\Sigma_2(p^2) = \lim_{\Lambda\to\infty} [m\Sigma^\Lambda(p^2) - m\Sigma^\Lambda(m^2)] + m \tag{2.33}$$

$$m_o(\Lambda) = m\Sigma^\Lambda(m^2) - m \tag{2.34}$$

Thus if one wishes to evaluate $m\Sigma_2(p^2)$ using conventional mass - renormalized perturbation theory, one is lead to introduce a non vanishing cut-off dependent $m_o(\Lambda)$. In fact one may calculate $m_o(\Lambda)$ to find[23)]

$$m_o(\Lambda) = A'm \left(\frac{\Lambda^2}{m^2}\right)^{-\epsilon/2} \tag{2.35}$$

where A' is constant differing from A in (2.23) by terms of order G. Thus as $\Lambda \to \infty$, $m_o \to \infty$ and the result (2.21) remains valid. However as long as Λ is finite one has to modify the Ward Identity (2.8) as the axial vector current is not conserved :

$$q_\mu \Gamma_5^\mu \,(p-q/2,p+q/2) = \frac{i}{(2\pi)^4}\,[S^{-1}(p+q/2)\gamma_5 + \gamma_5 S^{-1}(p-q/2)$$
$$+ 2m_o\Gamma_5(p-q/2,p+q/2)\,] \qquad (2.36)$$

Here Γ_5 is the pseudoscalar proper vertex function.

When $\Lambda \to \infty$, $m_o \to 0$ but $\Gamma_5 \to \infty$ in such a way as the product remains finite and will stay finite in renormalized theory beyond the ladder approximation[22]. This is because the unrenormalized (cut-off dependent) Lagrangian has no chiral symmetry : $m_o\Gamma_5$ cancels the scalar part of the fermion self energy thereby ensuring the vanishing of $q_\mu\Gamma^\mu{}_5$ as $q \to 0$ without a Goldstone singularity[22]. Thus the fermion acquires mass but the axial vector remains massless and the dynamical mechanism described in II.a breaks down; such a conclusion however is based on the unnecessary assumption that the expansion described in fig.5 is valid for Λ finite; nevertheless there is an ambiguity between two apparently possible solutions : massive fermions and massive or massless axial vector meson. One feels that the ambiguity would only be resolved by a detailed stability analysis but this has not been performed so far.

In conclusion, we see that only in exceptional cases (Gell-Mann Low type conditions) can the weak coupling limit of the ladder approximation survive further radiative corrections. However even if this turns out to be a consistent theory one is not guaranteed that the axial meson acquires mass. In more general cases nearly nothing can be said about what happens in the full theory.

These mathematical difficulties have their physical counterparts. The mechanism for mass generation described here depends critically on the behaviour of the field theory at short distances. Thus the physics is strongly dependent on the model Lagrangian and this weakens the hope of solving approximately a physical problem by an approximate model. Therefore one would be tempted to abandon the whole program were it not for the fact that we have mist in this section a particular type of dynamical symmetry breakdown which does not share these deficiencies. This will now be discussed.

III. THE CASE $g_V^2 = g_A^2 = g^2$ [24]

a. Ladder approximation

In the preceeding section the solutions of (2.18) and (2.19) were qualitatively governed by the large loop momenta, for which one may disregard the masses in the denominator.

The situation is drastically different if $g_A^2 = = g_V^2 = g^2$. In this case the neglect of M^2 would imply $m\Sigma_{2L}^A(p^2) = 0$, namely no symmetry breaking is possible. In other words, the fermion acquires a mass if and only if the axial vector meson does. Eq. (2.18) now reduces to

$$m\Sigma_{2L}(p^2) = \frac{-i}{(2\pi)^4}\, 3g^2M^2 \int d^4k \;\frac{m\Sigma_{2L}(k^2)}{[k^2-m^2\Sigma_{2L}^2(k^2)]\;(p-k)^2\;[(p-k)^2-M^2]} \tag{3.1}$$

Because of the extra factor of $1/k^2$ in the integrand this equation is no longer dominated by the asymptotic domain of integration. Were $m\Sigma_{2L}(k^2)$ be equated to a constant in the denominator, Eq.(3.1) would now be a Fredholm linear integral equation, leading to an eigenvalue condition of the type

$$F(g^2,\ M/m) = 0 \tag{3.2}$$

This property should persist despite the non linearity of (3.1)[24]. We may obtain an approximate solution to (3.2) in the following way. Since the integral in (3.1) is entirely dominated by the infrared domain of integration we shall approximate $m\Sigma_{2L}(p^2)$ by its value at $p^2 = 0$, $m\Sigma_{2L}(p^2) \to m\Sigma_{2L}(0) \simeq m$.

Upon performing a Wick rotation, we easily obtain the relation

$$1 = \frac{\tilde{g}^2}{1-(m/M)^2}\ \log(M/m)^2 \tag{3.3}$$

with

$$\tilde{g}^2 = 3g^2/16\,\pi^2 \tag{3.4}$$

For weak coupling, equation (3.3)takes the form

$$(M/m)^2 = \exp(1/\tilde{g}^2) \tag{3.5}$$

The approximation (3.3) to the exact eigenvalue condition (3.2) is in fact a very good one. Indeed, solving iteratively (3.1) on a computer [15] yields for $\tilde{g}^2 \gtrsim 1$ the result (3.3) up to a few percent; for $\tilde{g}^2 < 1$ (3.5) is rapidly qualitatively correct (for $\tilde{g}^2 \simeq 0,3$ the error is a factor of 2). The essential singularity in the weak-coupling limit may be of interest for weak and electromagnetic interaction model building where large meson masses are desired.

However, we shall soon find that a contrast to the case $g_V^2 > g_A^2$ there is now no small coupling solution in ladder approximation. To see this we must consider equation (2.19) for M^2. In the absence of the exact form of $m\Sigma_{2L}(p^2)$ we may obtain a reasonable estimate for M^2 by the following considerations. We have seen that most of the contribution to the integral (3.1) comes from the region of small momenta. Hence when p is large we may replace the factors $(p-k)^2$ by p^2 in the integrand, to obtain

$$m\Sigma_{2L}(p^2) \underset{p^2\to\infty}{\longrightarrow} -\frac{i}{(2\pi)^4}\, 3g^2M^2\, \frac{1}{p^4}\int \frac{m\Sigma_{2L}(k^2)}{k^2}\, d^4k \qquad (3.6)$$

so that

$$m\Sigma_{2L}(p^2) \underset{p^2\to\infty}{\longrightarrow} \frac{\text{Constant}}{p^4} \qquad (3.7)$$

provided the integral in (3.6) converges; but such is indeed the case if (3.7) holds. (This $1/p^4$ behaviour was confirmed in the course of computer calculations [15]. If we now introduce eq.(3.7) for $m\Sigma_{2L}(k^2)$ in eq.(2.19) we observe that the integral converges rapidly. The high momentum contributions are thus negligible and we obtain a reasonable estimate of M^2 if we replace in (2.19) $m\Sigma_{2L}(k^2)$ by a constant while we limit the integration range by a suitable cut-off. We obtain in this way :

$$(M/m)^2 = B\, \tilde{g}^2 \qquad (3.8)$$

where B is a constant of order 1. Therefore, from eq.(3.3) and (3.8) we have

$$M/m \simeq \tilde{g}^2 = O(1) \qquad (3.9)$$

In fact from a computer calculation [15] one has

$$\tilde{g}^2 = 1,3 \pm 0,1 \qquad (3.10)$$

So only the strong coupling solution seems possible in ladder approximation. How such a result could eventually be reconciled with a description of weak and electromagnetic interaction will be discussed in section V.

To understand how a new condition (3.2) has arisen when $g_V^2 = g_A^2$ it is interesting to follow the solution (2.23) when g_V^2 approaches g_A^2 from above. In order to do this it is necessary to evaluate the constant A. For this purpose we make use of the form (2.30) and (2.31) in the $\Lambda \to \infty$ limit. From the first two graphs of fig.4, one gets (in the Landau gauge)

$$m\Sigma_L^\Lambda(p^2) = m\,[G \log \Lambda^2/p^2 + C + f(p^2/m^2; g_A^2; g_V^2) + O(G^2)] \tag{3.11}$$

The function $f(p^2/m^2;\ g_A^2;\ g_V^2)$ represents the term which vanishes when $p^2/m^2 \to \infty$. In particular

$$m\Sigma_L^\Lambda(m^2) = m\,[G \log \frac{\Lambda^2}{m^2} + C + f(1; g_A^2; g_V^2) + O(G^2)] \tag{3.12}$$

and thus

$$\begin{aligned} m_o &= m - m\,\Sigma_L^\Lambda(m^2) \\ &= m\,[1 - G \log \Lambda^2/m^2 - C - f(1;\ g_A^2; g_V^2) + O(G^2)] \end{aligned} \tag{3.13}$$

The quantity we are in is $m\Sigma_{2L}(p^2)$

$$\begin{aligned} m\Sigma_{2L}(p^2) &= \lim_{\Lambda\to\infty} [m\Sigma_L^\Lambda(p^2) + m_o] \\ &= m\,[1 - G \log m^2/p^2 + f(p^2/m^2; g_A^2; g_V^2) \\ &\qquad - f(1;\ g_A^2;\ g_V^2) + O(G^2)] \end{aligned} \tag{3.14}$$

We compare this last expression with the expansion of (2.23) in terms of G

$$m\Sigma_{2L}(p^2) \underset{p^2\to\infty}{=} A(1 + \frac{\varepsilon}{2} \log p^2/m^2 + O(G^2)) \tag{3.15}$$

The term $f(p^2/m^2; g_A^2; g_V^2)$ in (3.14) vanishes when $p^2/m^2 \to \infty$. Thus we get

$$A = 1 - f(1;\ g_A^2;\ g_V^2)$$

The condition $A = 0$ is obviously satisfied in type B symmetry breakdown. In this case indeed, neither the logarithmic nor the constant term appear in (3.11), so that (3.13) together with the condition $m_o = 0$ results in the vanishing of (3.16). Thus the vanishing of A in (2.21) may be seen as a rephrasing of the eigenvalue condition and this obviously goes through to all orders in G.

Thus the new equation obtained when $g_A^2 = g_V^2$ corresponds to the vanishing of the asymptotic part of $m\Sigma_{2L}(p^2)$. The interest of the above derivation is that it shows that now $m_o = 0$ is true independent of the U-V cut-off. Therefore there is no bare mass anomaly and this confirms the fact stated in the beginning of this section that now both fermion and axial-vector meson necessarily acquire mass simultaneously.

b) Full theory

The qualitative features of the ladder approximation are

1° there is an eigenvalue condition on the coupling constant originating from fast decrease of the electron self energy for asymptotic momentum (A=0);

2° there is no bare mass anomaly; the fermion and the axial vector meson mass ratio is determined;

3° the "bootstrap" for coupling constant and masses is a property of the low momentum region, independent of asymptotic consideration.

We shall see that under reasonable assumptions these properties survive in the full theory and are in fact consequences of a new group theoretical property arising when $g_A^2 = g_V^2$.

Let us consider (2.31) (in its renormalized version). Recall that K_R' is expended in terms of bare vertices, bare meson propagators and dressed fermion propagators; we thus write

$$K_R' = K_R'(p,k;\ g(\lambda);\ m\{\Sigma_{2R}\}) \qquad (3.17)$$

where we have made explicit the functional dependence of K' on the symmetry breaking terms $m\Sigma_{2R}$ arising from dressed (renormalized) fermion propagators. For $m\Sigma_{2R} = 0$, the evaluation of K_R' reduces to the summation of the renormalized perturbation series. Now when $g_A^2 = g_V^2$ we may use ad indep dent fields

$$X_\mu = (V_\mu + A_\mu)/\sqrt{2}$$

$$Y_\mu = (V_\mu - A_\mu)/\sqrt{2} \qquad (3.18)$$

X_μ is thus coupled only to left-handed two components spinors through $\sqrt{2}\, g\, \bar\psi\, \gamma_\mu(1+\gamma_5)\psi\, X^\mu$ and Y_μ only to right handed ones. $K'(p;k;g(\lambda);0)$ contains no diagram involving both X_μ and Y_μ

propagators. This is because K' is constructed out of meson exchanges between two fermions lines and no left-right flip in the fermion propagator can occur when $m\Sigma_2(p^2) = 0$; therefore all mesons originating from one fermion line are all $X_{\mu,s}$ (or $Y_{\mu,s}$). Now these $X_{\mu,s}$ (or $Y_{\mu,s}$) cannot be connected to $Y_{\mu,s}$ (or $X_{\mu,s}$) by fermions loops because such conversion again necessitates a chirality flip; thus all meson lines arriving to the second fermion propagator are also $X_{\mu,s}$ (or $Y_{\mu,s}$).

As the γ_5 vertex induces a chirality flip, $\mathcal{K}_R(p,k;g(\lambda);0)$ is zero to all orders in perturbation theory and at least to $m\Sigma_{2R}(p^2)$ insertions in fermion lines (e.g. in vacuum polarization loops as in III.a) are needed for a non vanishing result. A consistent assumption is that $m\Sigma_{2R}(p^2)$ cannot diverge in a suitable gauge $p^2 \to \infty$; therefore it follows from power counting that $\mathcal{K}_R(p,k;\ g(\lambda);\ m\{\Sigma_{2R}\}) \to 0$ as $k \to \infty$ at least as k^{-4} (up to log terms), thus $\mathcal{J}_R(k^2)$ $\mathcal{K}_R(p,k;g(\lambda),m\ \{\Sigma_{2R}\}\ )$ vanishes at least as k^{-6} (up to log terms). This is in contradistinction to the case $g_A^2 < g_V^2$, where one get a k^{-4} behaviour (up to log terms) for this kernel. The unbounded nature of d^4k/k^4 as $k \to \infty$ was responsible for the continuous spectrum of coupling constants and the related dependence of the fermion self energy on asymptotic momentum. the boundedness of the kernel when $g_A^2 = g_V^2$ leads on the contrary to a fast decrease of $m\Sigma_{2R}(p^2)$ when $g_A^2 = g_V^2$ and in general an eigenvalue condition for $g(\lambda)$ emerges [24]. This is the generalisation of (3.2) and then determines a renormalized low energy coupling constant of the theory; in terms of $g(\lambda)$ one may then evaluate M^2/m^2 and the whole symmetry breakdown mechanism is confined to the low momentum regime. The theory is thus expected to be renormalisable independently of specific assumptions about the short distance behaviour of the symmetric theory.

The reason behind these fast convergence properties is that because X_μ and Y_μ are gauge fields associated to two-component spinors, it is impossible to construct a non vanishing linear response to an external bare mass since this bare mass induced a chirality flip. This implies that mass generation stems only from the left-right communication induced in the exchanged kernel in (2.31) by the generated mass itself. Such a mass factor in the exchanged kernel provides a natural ultra-violet cut-off, resulting in the boundedness of the kernel.

We shall call such a low energy dominated symmetry breaking, through bounded kernels, a B-type dynamical symmetry breaking. To contrast with the asymptotically governed mechanism studied in section II we shall refer to the former as to A-type dynamical symmetry breaking.

IV. DYNAMICAL SYMMETRY BREAKDOWN FOR CHIRAL GROUPS [24) 28)]

From the discussion at the end of section III, one sees how the B-type mechanism may arise for other minimal Lagrangian. If each gauge field is coupled to pure left-handed or pure right-handed massless fermions, the linear response characteristic of asymptotic mass generation (type A) will vanish. Thus one expects type B behaviour (and the related constraint on low energy coupling constants) for all minimal lagrangians which couple each separate gauge field to massless fermions of definite chirality. We shall call chiral groups those which give rise to such couplings and these may or may not be parity conserving. As an example of chiral minimal lagrangians one may take the one obtainable from the weak $SU_3 \times SU_3$ group[26)] built upon the Konopinski-Mahmoud triplet [27)].

We cannot proof in general the above conjecture but we shall give a plausibility argument. The origin of the difficulty is that the analog of (2.29) does not hold for non abelian groups. Therefore if one wishes to construct from Ward identities the analog of (2.32) one cannot straightforwardly as in the abelian case eliminate all proper vertices not connected to external fermion lines. However let us assume that the breakdown is driven by the fermion mass instability; this means that a non perturbative solution already exists in some approximation where all proper vertex function not involving external fermion lines have been neglected (such an explicit approximation has been carried out before[10)]). Then one can solve iteratively for these vertices and one recovers an equation of the type (2.32) in fermion space. Now the kernel K' is no more irreducible except with respect to pairs of fermion propagators; however one easily verifies that for chiral groups the previous argument leading to boundedness and thus to B-type behaviour still goes through, provided no new singularity is generated by the reducible K'.

In fact, this result seems to be much more general and not to rely so heavily on the assumption that it is the fermion mass that drives the instability. Indeed there are indications that even pure Yang-Mills fields associated to a compact simple Lie group may acquire mass by this self-interaction through a B-type mechanism [28) 29)]. A "ladder-approximation" calculation with the octet of pure Yang-Mills vector fields associated with SU_3 give a solution where all "strange" vector meson acquire mass with the other rmeain massless; this happens for a specific value of the coupling constant, namely $g^2 = 64\,\pi^2/9$. (Of course this number is not to be taken seriously).

V. POSSIBLE RELEVANCE TO PHYSICS.

Thus it appears plausible that a mass generation for a well defined class of minimal Lagrangians is mathematically possible in the context of type B behaviour. What about physics "

The popularity of spontaneous symmetry breakdown rests largerly on the possibility of constructing unified gauge theories of weak and electromagnetic interactions. The motivations are CVC and the renormalisability of the theory. In the present context where we have precluded the inclusion of scalar fields as fundamental fields, the renormalisability of the theory is not straightforward and rests in fact on the very existence of solutions to the renormalized Bethe-Salpeter equations. A more fundamental difficulty is perhaps the following : how could one reconcile the smallness of the experimental weak coupling constant $G_W \sim e^2/M_W^2 \simeq 10^{-5}\ m^{-2}_{proton}$ in absence of a large mass parameter in the theory ? This difficulty is overcomed in the Weinberg model by the use of the vacuum expectation value of scalar field which is an arbitrary phenomenological parameter. In certain models of models of type A high ratios can be obtained, but this introduces an arbitrary small coupling constant ξ of order 10^{-3} for which there is not yet any rational[30]; moreover these models are subjected to all the ambiguities discussed in section II b) which might well ruin the whole approach. What happens if we stick to type B models such as that the $SU_3 \times SU_3$ model mentioned in section IV.

At first sight the situation is even worse because we are led to $\tilde{g}$ ($\simeq$ e?) of order 1 from the low energy eigenvalue equation. Of course more refined treatments are necessary to confirm this ladder approximation result but one feels that there is no great hope that an improved eigenvalue equation would modify qualitatively such a result : the convergent dimensionless integrals determining $\tilde{g}^2$ are normally expected to be of order 1. One might wish to overcome this difficulty by using complicated models involving many coupling constants so that the constraints implied by the boundedness of the kernels leaves some coupling free. However there is perhaps a more interesting approach which leads to an amusing suggestion on the relation between hadron and lepton physics.

In the simplified model of section III the value of $\tilde{g}^2$ results from two competing effects:

a) the self-energy equation for the fermion which tends to give large M/m ratios for $\tilde{g}^2 \ll 1$;

b) the vacuum polarization loop of the vector mesons which tends to give small M/m ratios for $\tilde{g}^2 \ll 1$.

Now for a given fermion the effect a) is quite insensitive to the addition of other fields (fermions and vector-mesons) into the theory. On the contrary the effect b) depends on the cumulative action of all "elementary" particles which couple to the vector mesons.

One is thus tempted to look for models containing both hadrons and leptons : if the high W-mesons mass originates within the hadron section, through the b-effect, then the a-effect would result in a weak coupling between leptons of small mass. Of course the question is : how could hadrons generate such high vector meson masses ? Simply counting the number of "elementary quarks" would hardly do the job and we must search for a new qualitative property of the quarks (confinement?). This is of course at the present stage only a speculation but it might be usefull to construct a model where the vector mesons mass is treated phenomenologically by including in the vacuum polarization loop of the vector mesons the effect of the hadron currents.

If of course gauge field theories will turn out to be relevant for strong interactions also via some confinement mechanism then the B-type symmetry breakdown described here may be also usefull to avoid introducing phenomenological scalar fields.

REFERENCES AND FOOTNOTES

1. See for instance R. Brout "phase transitions", W.A. Benjamin Inc., New York NY, (1963).
2. Y. Nambu, Phys.Rev.Lett.4, 380 (1960).
   Y. Nambu and G. Jona-Lasinio, Phys.Rev. 122, 345 (1961).
3. F. Englert and R. Brout, Phys.Rev. 13, 321 (1964).
4. P.W. Higgs, Phys.Lett. 12, 132 (1964); Phys.Rev.Lett. 13, 508 (1964); Phys.Rev. 145, 1156 (1966).
5. S. Weinberg, Phys.Rev.Lett. 19, 1264 (1967).
   For a recent review, see S. Weinberg, Rev.of Mod.Physics 46, 255 (1974).
6. H. Nielsen and P. Olesen, Nucl.Phys. B61, 45 (1973).
7. G. 't Hooft, Nucl.Phys. B 79, 276 (1974).
8. An exception is the model discussed by S. Coleman and E. Weinberg, Phys.Rev. D7, 1888 (1973).
   However the absence of scale breaking counterterms in this model must be imposed by an ad-hoc procedure; this comes about because scale invariance is usually broken by the renormalization program. However the scale dependence of the renormalized Lagrangian does not introduce a measurable scale if dimensional counterterms are prohibited by other symmetry considerations. This is the case for our minimal Lagrangians where the absence of mass counterterms is required by chiral invariance and by gauge invariance of the second kind.
9. F. Englert, E. Gunzig, C. Truffin and P. Windey, Phys.Lett. 57B, 73 (1975).

10. F. Englert, R. Brout and M.F.THiry, Il Nuovo Cimento 43, 244 (1966).
11. R. Jackiw and K. Johnson, Phys.Rev. D8, 2386 (1973); see also J.M. Cornwall and R.E. Norton, Phys.Rev. D8, 3338 (1973).
12. M. Baker and K. Johnson, Phys.Rev. D3, 2541 (1971) and references therein.
13. The regularization method of references 3 and 10 makes use of gauge invariant non local current operators. One may verify in case $\Sigma_{2L}(p^2)$ is a constant that this yields the same result as the dimensional regularization[14].
14. G.'t Hooft and M. Veltman, Nucl.Phys. 44B, 189 (1972).
15. J-M. Frère and P. Nicoletopoulos, private communication and J-M. Frère, Thesis - Brussels University (1975).
16. Th.A. Maris, V. Herscowitz and G. Jacob, Phys.Rev.Lett. 12, 313 (1964).
17. See for instance B.W. Lee and J. Zinn-Justin, Phys.Rev. D7, 1049 (1973).
18. M. Gell-Mann and F.E. Low, Phys.Rev. 95, 1300 (1954).
19. For an alternative but related possibility, see S.L. Adler, Phys.Rev. D5, 3021 (1972).
20. D.J. Gross and F. Wilczek, Phys.Rev.Lett. 30, 1343 (1973); and H.D. Politzer, Phys.Rev.Lett. 1346 (1973).
21. S. Adler, Phys.Rev. 177, 2426 (1969).
22. K. Johnson, 9th Latin American School of Physics, Santiago Chili (1967), edited by I. Saavedra (Benjamin, New York 1968). H. Pagels, Phys.Rev.Lett. 28, 1482 (1972).
23. S.L. Adler and W.A. Bardeen, Phys.Rev. D4, 3045 (1971).
24. F. Englert, J-M. Frère and P. Nicoletopoulos, Phys.Lett. 52B, 443 (1974).
25. It is easy to check that, after Wick rotation and angular integration, the integrand (3.1) is mapped on the interval [0,1] by the change of variable $y = 1/(-k^2/M^2) + 1$ and that the kernel remains bounded.
26. R. Gatto, Il Nuovo Cimento, 28, 567(1963).
27. E.J. Konopinski and H.M. Mahmoud, Phys.Rev. 92, 1045 (1953).
28. F. Englert, J-M. Frère and P. Nicoletopoulos, "Dynamical Symmetry breakdown in pure Yang-Mills field theory", to be published in Nucl.Phys. B.
29. For an alternate possibility see
30. F. Englert and R. Brout, Phys.Lett. 49B, 77(1974).

SUPERSYMMETRY

B. ZUMINO

CERN-Geneva

1. INTRODUCTION

Fermi-Bose supersymmetry was introduced by Wess and the author[1]. It is a symmetry which connects particles of integral spin with particles having half-integral spin, or bosons with fermions. The possibility of defining such a symmetry was suggested to us by the existence of supergauge transformations in dual models[2].

The supersymmetry algebra is very simple. Let Q_i be a constant Majorana spinor (we may use the Majorana representation where the γ matrices are real; then Q_i is a Hermitian spinor). Then the algebra is

$$\{Q_i, \bar{Q}_j\} = -2(\gamma^\mu)_{ij} P_\mu \ , \qquad \bar{Q} = \tilde{Q}\gamma^\circ$$
$$[P_\mu, Q_i] = [P_\mu, P_\lambda] = 0 \ , \tag{1}$$

where $P^\mu = (H, \vec{P})$ is the energy momentum operator, which generates four-dimensional translations. In a supersymmetry invariant field theory, the spinor charges are given as integrals

$$Q_i = \int J^\circ_i \, d^3x , \tag{2}$$

where the vector-spinor current is conserved

$$\partial_\mu J^\mu_i = 0 \ . \tag{3}$$

Lorentz transformations and parity operate as isomorphisms of the algebra (1), transforming Q_i as a spinor and P_μ as a vector. The algebra (1) is not a Lie algebra, since it contains both commutators and anticommutators (it is what mathematicians call a "graded algebra"). If one introduces parameters α_1, α_2, etc., which are totally anticommuting Majorana spinors (they commute with tensors and anticommute with spinors and among themselves), the anticommutation relation (1) can be written as a commutation relation

$$[\bar{\alpha}_1 Q, \bar{\alpha}_2 Q] = -2\, \bar{\alpha}_1 \gamma^\mu \alpha_2 P_\mu \,. \qquad (4)$$

The supersymmetry algebra can therefore also be described as an "extended Lie algebra", with parameters belonging to a Grassmann algebra. This kind of object has been studied in the mathematical literature[3].

Observe that, if one multiplies the anticommutation relation (1) by γ° and takes the trace over the spinor indices, one finds an expression for the total Hamiltonian in terms of the spinor charges

$$H = \frac{1}{4} \sum_{i=1}^{4} Q_i^2 \qquad (5)$$

valid, in presence of interaction, for any supersymmetric theory. Similar expressions can be obtained for the components of the momentum.

The fact that supersymmetry is not an ordinary Lie algebra allows it to avoid the difficulties and no-go theorems[4] which have plagued the various forms of relativistic SU(6). Actually, as we shall see, there exist non-trivial (and renormalizable) Lagrangian theories which are exactly invariant under the supersymmetry algebra.

Our motivation in introducing Fermi-Bose supersymmetry was to show the feasibility of constructing supermultiplets containing interacting particles with both integral and half integral spin. From a rather different point of view the same algebra (1) was considered independently by Volkov and Akulov[5]. They gave a non-linear realization of it in terms of a single massless spinor field and suggested that it may be relevant as a description of the properties of the neutrino. Their non-linear Lagrangian is non-renormalizable and describes a theory in which supersymmetry is spontaneously broken. Spontaneous breaking of supersymmetry can also occur in renormalizable field theory models[6] in which the supersymmetry is realized linearly. In any case, corresponding to the conserved spinor current (3), spontaneous breaking gives rise to a massless Goldstone spinor[7]. It is therefore very tempting to identify this massless spinor with the (electron) neutrino. Unfortunately, as pointed out by Bardeen[8], the Goldstone spinor satisfies low energy theorems analogous to the low energy theorems for the pion, with the important difference that here those theorems are exact because the mass is really zero. These low energy theorems contradict experimental facts on the spectrum in beta decay.

This situation with the Goldstone spinor could prove a serious difficulty for supersymmetry. The point is that, in a theory which is rigorously supersymmetric, the bosons and the fermions in each supermultiplet must have exactly the same mass, contrary to what happens in the real world. This follows directly from (1), because application of the spinor charge Q_i to a one-particle state will change it into another one-particle state having the same mass and spin differing by one-half. So, supersymmetry cannot be exact, it must be broken. However, if the breaking is spontaneous, it gives rise to a massless Goldstone spinor which, since it cannot be identified with the neutrino (by Bardeen's argument), does not exist in nature.

There are three possible ways out of this difficulty. The first is that some anomaly invalidates the low energy theorems, as in the case of $\pi^\circ \to 2\gamma$. This seems unlikely at present. The second is that supersymmetry is broken not spontaneously but explicitly and softly, so that the renormalized quantities keep satisfying interesting relations. That this is possible was shown by Iliopoulos and the author[9]. The third way out is connected with the possibility of making a supersymmetric theory of gravitation[10][11]. In such a theory the graviton must be put in a supermultiplet, together with a massless particle of spin three-halfs. The interest in such a theory is that it is hoped to be more convergent than the usual theory of gravitation, due to compensations between the divergences due to the spin two field and those due to the spin three-halfs field. The spin three-halfs field will have a universal coupling of gravitational strength, to the spinor current, which, as we shall see, is a transition current between a spinor and a boson. If it is really massless, we have here another difficulty, because particles would decay into it, contrary to experimental evidence. However, if the gauge supersymmetry of this theory is spontaneously broken, there is the possibility of a Higgs-like effect, by which, instead of the appearence of a massless spin one-half-Goldstone spinor, this field is absorbed by the spin three-halfs field which becomes massive. This is in perfect analogy with the usual Higgs effect, in which a Goldstone scalar disappears to provide the additional degrees of freedom needed to give mass to a massless vector. The spin one-half spin three-halfs Higgs-like effect would then cure both problems, that mentioned in the previous paragraph as well as that of the massless spin three-halfs particle.

In the following we describe some examples of supersymmetric field theories. Observe that, in ref. 1), an algebra larger than (1) was described, which contains also Lorentz transformations, dilatations, conformal and chiral transformations. In that algebra, a second set of spinor charges occurs, which, together with the Q_i, forms an eight-component conformal Majorana spinor. That larger algebra was later abandoned[12], in order to avoid the problems arising from scale and conformal anomalies. Anyway, if one is interested in the formal construction of Lagrangians invariant under the larger algebra, we need only observe that a Lagrangian invariant under (1) and under the conformal algebra is automatically invariant under the larger algebra.

2. SUPERMULTIPLETS AND LAGRANGIANS

The simplest supermultiplet consists of a scalar field A, a pseudoscalar field B, a Majorana spinor ψ, and two auxiliary fields F and G. Writing

$$\delta A = [\bar{\alpha} Q, A] \qquad \text{etc.} \tag{6}$$

for an infinitesimal supertransformation, one has[1]

$$\begin{cases} \delta A = i\bar{\alpha}\psi \\ \delta B = i\bar{\alpha}\gamma_5 \psi \\ \delta\psi = \partial_\mu (A - \gamma_5 B)\gamma^\mu \alpha + (F + \gamma_5 G)\alpha \\ \delta F = i\bar{\alpha}\gamma^\mu \partial_\mu \psi \\ \delta G = i\bar{\alpha}\gamma_5 \gamma^\mu \partial_\mu \psi \quad . \end{cases} \tag{7}$$

The commutator of two supertransformations is a translation. For instance

$$\delta_2 \delta_1 A = i\bar{\alpha}_1 \delta_2 \psi = i\bar{\alpha}_1 \gamma^\mu \alpha_2 \partial_\mu A + \ldots \tag{8}$$

where the dots denote terms symmetric in 1 and 2. Therefore

$$[\delta_2, \delta_1] A = 2i\, \bar{\alpha}_1 \gamma^\mu \alpha_2 \partial_\mu A \; . \tag{9}$$

With this supermultiplet, Wess and the author[12] constructed the first non-trivial supersymmetric model, with Lagrangian

$$L = -\frac{1}{2}\ [(\partial_\mu A)^2 + (\partial_\mu B)^2 + i\bar{\psi}\gamma^\mu\partial_\mu\psi - F^2 - G^2]$$

$$+ m(FA + GB - \frac{i}{2}\ \bar{\psi}\psi)$$

$$+ g(FA^2 - FB^2 + 2GAB - i\bar{\psi}\psi A + i\bar{\psi}\gamma_5\psi B).$$

The various terms of the Lagrangian, that is the kinetic term, the term proportional to m and that proportional to g, each change by a divergence under (7). Therefore the action integral is invariant. The auxiliary fields F and G can be eliminated by using their own equations of motion

$$F + mA + g(A^2-B^2) = 0$$

$$G + mB + 2g\ AB = 0 \qquad (11)$$

and the Lagrangian takes the more familiar form

$$L = -\frac{1}{2}[(\partial_\mu A)^2 + (\partial_\mu B)^2 + i\bar{\psi}\gamma^\mu\partial_\mu\psi + m^2A^2 + m^2B^2 + im\bar{\psi}\psi]$$
$$- gmA(A^2+B^2) - \frac{g^2}{2}(A^2+B^2)^2 - ig\bar{\psi}(A-\gamma_5B)\psi . \qquad (12)$$

Observe that, as a consequence of supersymmetry invariance, the scalar, the pseudoscalar and the spinor have the same mass, and all the couplings are expressed in terms of the single coupling constant g. One can easily verify that the supercurrent

$$J^\mu = \gamma^\lambda\partial_\lambda(A-\gamma_5B)\gamma^\mu\psi - (F+\gamma_5G)\gamma^\mu\psi \qquad (13)$$

is conserved as a consequence of the equations of motion.

This model has been studied in great detail by Ferrara, Iliopoulos and the author [9)13)]. It can be regularized in a supersymmetric way by introducing higher order derivatives in the kinetic term of the Lagrangian. The Ward identities corresponding to the conservation of the supercurrent (13) can be written down and used to prove that renormalization does not spoil the relations among masses and coupling constants due to supersymmetry. The model is found to be less divergent than the generic theory with the same kind of couplings. In particular the sum of all vacuum diagrams vanishes identically in each order in perturbation theory[14)]. This fact can be interpreted as saying that the vacuum polarization in a gravitational field due to supersymmetric matter does not generate a cosmological term in Einstein's equations.

Furthermore only one renormalization constant is required, the single wave function renormalization constant Z common to all fields. The renormalized mass and coupling constant are given by

$$m_r = Zm_o , \qquad g_r = Z^{3/2} g_o . \tag{14}$$

The Callan-Symanzik equations take a particularly simple form for this model, the functions β and γ being proportional to each other. As a consequence one can argue that the function $\beta(g_r)$ cannot vanish except at the origin and that the effective coupling constant increases indefinitely with energy.

Supersymmetry can be broken softly, by adding to the Lagrangian a term proportional to the field A. Just as in the analogous case of the σ model, the renormalization program can still be carried out. The masses of the various fields of the multiplet are now no longer equal. Instead one finds, in the tree approximation, the mass relation

$$m_A^2 + m_B^2 = 2m_\psi^2 . \tag{15}$$

In higher orders this relation can be shown to be corrected by finite terms only.

The fact that in the above supersymmetric model cancellations of divergences occur, which make it less divergent than the generic theory of its kind, leads one to ask whether a supersymmetric theory might not be renormalizable even if it does not appear to be so by simple power counting. This is the reason for the hope, mentioned in the Introduction, that a supersymmetric theory of gravitation might be less divergent than the usual theory. One must mention, however, that Lang and Wess[15] have investigated a model which is the simplet generalization of (10) with an interaction non renormalizable by power counting. They have indeed found cancellations of divergences, but not enough to make the model renormalizable. Perhaps this is not so surprising. Supersymmetry, with its single additional conservation law (3), cannot be expected to determine the infinitely many arbitrary constants of a non-renormalizable theory. This does not mean, however, that the gauge supersymmetry of a generalized theory of gravitation could not achieve that much.

3. GAUGE INVARIANCE AND SUPERSYMMETRY

The existence of Lagrangian theories which are both gauge invariant and supersymmetric was first shown by Wess and the author[16]. Their model makes use of a supermultiplet consisting of a vector field v_μ, a Majorana spinor λ and an auxiliary field D, transforming as

$$\begin{cases} \delta v_\mu = i\bar{\alpha}\gamma_\mu\lambda \\ \delta\lambda = -\frac{1}{4} v_{\mu\nu}\gamma^\mu\gamma^\nu\alpha + D\,\gamma_5\alpha \\ \delta D = i\bar{\alpha}\gamma_5\gamma^\mu\partial_\mu\lambda \quad , \qquad v_{\mu\nu} = \partial_\mu v_\nu - \partial_\nu v_\mu , \end{cases} \tag{16}$$

under a supersymmetry transformation. The commutator of two transformations (16) is a translation accompanied by a gauge transformation, but, by enlarging the multiplet, one could arrange that it be exactly a translation [1],[16]. This supermultiplet is put in interaction with a complex multiplet, or a pair of real multiplets, of the kind discussed in the previous section. Using real fields the Lagrangian can be written as

$$\begin{aligned} L = & -\frac{1}{4} v_{\mu\nu}^2 - \frac{i}{2}\bar{\lambda}\gamma^\mu\partial_\mu\lambda + \frac{1}{2} D^2 \\ & -\frac{1}{2}\sum_{i=1}^{2} [(\partial_\mu A_i)^2 + (\partial_\mu B_i)^2 + i\bar{\psi}_i\gamma^\mu\partial_\mu\psi_i + m^2(A_i^2+B_i^2)+im\bar{\psi}_i\psi_i] \\ & + g[D(A_1B_2-A_2B_1)-v^\mu(A_1\partial_\mu A_2-A_2\partial_\mu A_1 + B_1\partial_\mu B_2 - B_2\partial_\mu B_1 - i\bar{\psi}_1\gamma_\mu\psi_2) \\ & -i\bar{\lambda}\,\{(A_1+\gamma_5 B_1)\psi_2 - (A_2+\gamma_5 B_2)\psi_1\}] \\ & -\frac{g^2}{2} v_\mu^2 (A_1^2 + A_2^2 + B_1^2 + B_2^2) \end{aligned} \tag{17}$$

where the fields F_i and G_i $(i=1,2)$ have already been eliminated. The gauge transformation rotates the fields with the subscripts 1 and 2 into each other and changes v_μ by a four-gradient. The fields λ and D are gauge invariant. The Lagrangian (17) can be considered as a sort of supersymmetric extension of the quantum electrodynamics of scalars, pseudoscalars, and spinors. Observe that all couplings are expressed in terms of the single coupling constant g, as a consequence of the supersymmetry of the Lagrangian. The Lagrangian (17) has been shown to be renormalizable to all orders[17] in a manner consistent with gauge invariance and supersymmetry. If one adds to it a (parity violating) gauge and supersymmetry invariant terms proportional to the field D, one finds[6] that it provides a model of spontaneous breaking of supersymmetry with the corresponding appearance of a Goldstone spinor.

It is also possible to construct Lagrangian theories which are both supersymmetric and invariant under non-Abelian gauge transformations[18]. We skip the technical details and give only the results. The simplest supersymmetric and gauge invariant theory is the ordinary Yang-Mills theory of vectors in interaction with a multiplet of Majorana spinors belonging to the adjoint representation of the internal symmetry group. For instance, for SU(N), using N x N traceless hermitian matrices,

$$L = \mathrm{Tr}\left(-\frac{1}{4} v_{\mu\nu}^2 - \frac{i}{2} \bar{\lambda}\gamma^\mu \mathcal{D}_\mu \lambda + \frac{1}{2} D^2\right),$$

$$v_{\mu\nu} = \partial_\mu v_\nu - \partial_\nu v_\mu + ig\,[v_\mu, v_\nu] \,, \tag{18}$$

$$\mathcal{D}_\mu \lambda = \partial_\mu \lambda + ig\,[v_\mu, \lambda] \,.$$

That this theory is supersymmetric can be seen most easily by checking that the supercurrent

$$J^\mu = -\frac{1}{4} \mathrm{Tr}(v_{\nu\rho}\,[\gamma^\nu, \gamma^\rho]\gamma^\mu \lambda) \tag{19}$$

is conserved as a consequence of the equations of motion.

The multiplet v_μ, λ, D can be coupled to "matter multiplets" A,B, ψ, F, G belonging to any representation of the internal group. For the case of a matter multiplet in the adjoint representation, the Lagrangian is given by (18) plus

$$\mathrm{Tr}\,\{-\frac{1}{2}[(\mathcal{D}_\mu A)^2 + (\mathcal{D}_\mu B)^2 + i\bar{\psi}\gamma^\mu \mathcal{D}_\mu \psi - F^2 - G^2]$$

$$+ m(FA + GB - \frac{i}{2}\bar{\psi}\psi) + ig\,D\,[A,B] \tag{20}$$

$$+ g\bar{\lambda}\,[A + \gamma_5 B, \psi\,]\} \,.$$

Theories of this kind contain scalar and pseudoscalar fields. Since all coupling constants, including the self-interaction of the scalars and the pseudoscalar, are expressed in terms of the single Yang-Mills coupling g, they will be asymptotically free provided the Callan-Symanzik function β is negative, and provided the theories are renormalizable in accordance with supersymmetry (see below). For (18) plus n matter multiplets like (20), it turns out that

$$\beta = -\frac{g^3}{16\,\pi^2}(3-n)N \tag{21}$$

where N refers to SU(N). Therefore, for n=1,2 the theory is asymptotically free (for n=3, β vanishes to this order, which means that the renormalization of the coupling constant g is finite,

This fact however does not persist in higher order).

If one adds (18) plus (20) for m = 0, eliminates the field D, and combines the two Majorana fields λ and ψ into a complex spinor

$$\varphi = \frac{1}{\sqrt{2}} (\lambda + i\psi), \tag{22}$$

one obtains the Lagrangian

$$\mathrm{Tr}\,\{-\frac{1}{4} v_{\mu\nu}^2 - \frac{1}{2} (\mathcal{D}_\mu A)^2 - \frac{1}{2} (\mathcal{D}_\mu B)^2 - i\bar{\varphi}\gamma^\mu \mathcal{D}_\mu \varphi$$

$$-ig\ \bar{\varphi}\,[A + \gamma_5 B, \varphi] - \frac{g^2}{2} (i\,[A,B]\,)^2\} \tag{23}$$

which is invariant under the fermion number transformation

$$\varphi \rightarrow e^{i\omega} \quad , \qquad \varphi^* \rightarrow e^{-i\omega}\,\varphi^*. \tag{24}$$

The conserved supercurrent for (23) is

$$J^\mu = \mathrm{Tr}\,\{-\frac{1}{4} v_{\nu\rho}\,[\gamma^\nu, \gamma^\rho]\gamma^\mu\varphi + ig[A,B]\ \gamma_5\gamma^\mu\varphi - i\gamma^\lambda \mathcal{D}_\lambda (A-\gamma_5 B)\gamma^\mu\varphi\} \tag{25}$$

One can also obtain a V $\pm$ A scheme with fermion number. One need only combine two multiplets $v_\mu^{(1)}, \lambda^{(1)}$ and $v_\mu^{(2)}, \lambda^{(2)}$ described by Lagrangians like (18). The total Lagrangian can be rewritten in terms of the complex spinor

$$\varphi = \frac{1}{2} (1 - i\gamma_5)\lambda^{(1)} + \frac{1}{2} (1 + i\gamma_5)\lambda^{(2)} \tag{26}$$

and of the vector and axial vector fields

$$v_\mu = \frac{1}{\sqrt{2}} (v_\mu^{(1)} + v_\mu^{(2)}), \qquad a_\mu = \frac{1}{\sqrt{2}} (v_\mu^{(1)} - v_\mu^{(2)}) \tag{27}$$

and is again invariant under a fermion number transformation.

The above examples indicate how one can attempt to construct realistic supersymmetric models. One of the difficulties is that one does not have a handy device for generating masses without spoiling the renormalizability of the theory. One also must keep in mind that fermions in the adjoint representation of the internal symmetry group are unavoidable, a fact which is perhaps at variance with the successfull quark picture of elementary particles.

4. SUPERSPACE AND SUPERFIELDS

The idea of superspace[5)18)] is very simple. Consider a space (superspace) whose points are labelled by co-ordinates x ,θ, $\bar{\theta}$, where x_μ are the usual space-time co-ordinates, θ is a (totally anticommuting) two-component spinor and $\bar{\theta}$ its complex conjugate. If one wishes, one can arrange θ and $\bar{\theta}$ into a single four-component Majorana spinor

$$\begin{pmatrix} \theta \\ \bar{\theta} \end{pmatrix}$$

Using a (complex) representation of the γ matrices in which γ_5 is diagonal (one could also use the Majorana representation as in (1)). The parameters α of a supersymmetry transformation now take the form

$$\alpha = \begin{pmatrix} \zeta \\ \bar{\zeta} \end{pmatrix} . \tag{28}$$

Supersymmetry transformations are geometrical transformations in superspace

$$\begin{cases} \delta x_\mu = i\theta\,\sigma_\mu \bar{\zeta} - i\zeta\,\sigma_\mu\,\bar{\theta} & \sigma_0 = 1 \\ \delta\theta = \zeta \\ \delta\bar{\theta} = \bar{\zeta} & . \end{cases} \tag{29}$$

It is easy to see that the commutator of two such transformations is a translation. Furthermore (29) and translations leave invariant the differential forms

$$\omega_\mu = dx_\mu + i\bar{\theta}\sigma_\mu\, d\theta - i\theta\,\sigma_\mu\, d\bar{\theta}\ ,\quad d\theta,\quad d\bar{\theta}\ . \tag{30}$$

On the other hand, Lorentz transformations leave invariant the "line element"

$$\omega_\mu \omega^\mu \tag{31}$$

Now, a superfield is a field in superspace, $V(x,\theta,\bar{\theta})$. If one expands it in θ and $\bar{\theta}$, the power series terminates after a finite number of terms, since the square of each component of θ and $\bar{\theta}$ vanishes

$$V(x,\theta,\bar{\theta}) = C(x) + i\theta\chi(x) - i\bar{\theta}\bar{\chi}(x) + \ldots + \theta\theta\bar{\theta}\bar{\theta}\,\frac{1}{2}\,D(x) \tag{32}$$

Therefore a superfield corresponds to a finite supermultiplet of ordinary fields. A superfield is taken to transform as a scalar in superspace under (29)

$$V'(x,\theta,\bar{\theta}) = V(x',\theta',\bar{\theta}') \tag{33}$$

and can have spinor or vector indices which determine how it transforms under Lorentz transformations. From (33) and the expansion (32), one can derive the transformation property of the fields C, X, etc. of the supermultiplet. Observe that, if one introduces the new coordinates

$$z_\mu = = x_\mu + i\theta\sigma_\mu \bar{\theta} , \tag{34}$$

(29) gives

$$\delta z_\mu = 2 i\theta\sigma_\mu \bar{\zeta} , \tag{35}$$

which does not contain $\bar{\theta}$. Therefore, it is consistent to require that a superfield be a function only of z_μ and θ, $S(z,\theta)$, or that it satisfies

$$\bar{D} S = 0 \tag{36}$$

where

$$\bar{D} = - \left.\frac{\partial}{\partial\bar{\theta}}\right|_z = - \left.\frac{\partial}{\partial\bar{\theta}}\right|_x - i \theta\sigma^\mu\partial_\mu . \tag{37}$$

$\bar{D}$ is a covariant derivative (under (29)) and so is

$$D = \left.\frac{\partial}{\partial\theta}\right|_x + i\sigma^\mu \bar{\theta} \partial_\mu . \tag{38}$$

A superfield satisfying (36) is called left-handed, one satisfying the covariant constraint

$$DS = 0 \tag{39}$$

is called right-handed. A left-handed superfield (together with its right-handed complex conjugate) corresponds to the multiplet described in section 2.

One can develop a geometry of superspace. For instance, the larger supersymmetry group of ref.1) can be characterized as containing those transformations in superspace which multiply the line element (31) by a rescaling factor. As mentioned in the Introduction, Volkov and Soroka[10] have developed a description of curved superspace which combines gravitational theory with the interactions of particles of spin 3/2, 1, and 1/2. A different approach is that followed by Arnowitt, Nath and the author[11] who have succeeded in working out a complete Riemannian theory of superspace. This permits to formulate a unified theory of gravitation and matter. It appears now that the correct superspace should have a non-Riemannian geometry. Work along these lines appears very fruitful at this point.

In these lectures we have not touched upon a different and very interesting way of combining supersymmetry with an internal symmetry. It consists in endowing the spinor charges (2) themselves (not the supermultiplets as in the non-Abelian theories of section 3) with an internal symmetry index. The spinor charges then belong to a representation of the internal symmetry group and the algebra is no longer (1) but a non-trivial extension of it. So far this approach has difficulties. In particular, it has not been possible to construct local renormalizable Lagrangians which are invariant under the extended algebra.

No attempt has been made here to give a complete list of references. For that we refer to a forthcoming review article by S. Ferrara in the Rivista del Nuovo Cimento 1975.

REFERENCES

1. J. Wess and B. Zumino, Nuclear Phys. B70, 39 (1974).
2. A. Neveu and J.H. Schwarz, Nuclear Phys. B31, 86 (1971); P. Ramond, Phys.Rev. D3, 2415 (1971); Y. Aharonov, A. Casher and L. Susskind, Phys.Letters 35B, 512 (1971); J.L. Gervais and B. Sakita, Nuclear Phys. B34, 633 (1971).
3. F.A. Berezin and G.I. Katz, Mathemat.Sbornik (USSR) 82, 343(1970) English traduction Vol.11.
4. See e.g. L. O'Raifeartaigh, Phys.Rev.Letters 14, 575 (1965); Phys.Rev. 139, B1052 (1965). S. Coleman and J. Mandula, Phys.Rev. 159, 1251 (1967).
5. D.V. Volkov and V.P. Akulov, Phys.Letters 46B, 109 (1973).
6. P. Fayet and J. Iliopoulos, Phys.Letters 51B, 461 (1974).
7. A. Salam and J. Strathdee, Phys.Letters 49B, 465 (1974); J. Iliopoulos and B. Zumino, Nuclear Phys. B76, 310 (1974).
8. See B. de Wit and D.Z. Freedman, Stony Brook Preprint (1975).
9. J. Iliopoulos and B. Zumino, Nuclear Phys. B76, 310 (1974).
10. D.V. Volkov and V.A. Soroka, JETP Letters 18, 529 (1973).
11. R. Arnowitt, P. Nath and B. Zumino, Phys.Letters 56B, 81 (1975). P. Nath, R. Arnowitt, Phys.Letters 56B, 177 (1975).
12. J. Wess and B. Zumino, Phys.Letters 49B, 52 (1974).
13. S. Ferrara, J. Iliopoulos and B. Zumino, Nuclear Phys. B77, 413 (1974).
14. B. Zumino, Nuclear Phys. B89, 535 (1975).
15. W. Lang and J. Wess, Nuclear Phys. B81, 249 (1974).
16. J. Wess and B. Zumino, Nuclear Phys. B78, 1(1974).
17. A.A. Slavnov, DUBNA Preprint (1974); S. Ferrara and O. Piguet, CERN preprint (1975).

18. A. Salam and J. Strathdee, Phys.Letters 51B, 353 (1974);
S. Ferrara and B. Zumino, Nuclear Phys. B79, 413 (1974).
19. A. Salam and J. Strathdee, Nuclear Phys. B76, 477 (1974).
S. Ferrara, J. Wess and B. Zumino, Phys.Letters 51B, 239 (1974).

INFLUENCE OF HADRONIC DECAY CHANNELS IN CONFINEMENT MODELS OF THE NEW RESONANCES *

E. EICHTEN

Laboratory of Nuclear Studies, Cornell University

Ithaca, New York 14853

INTRODUCTION

The spectrum of hadrons is usually discussed in the framework of naive quark or dual models possessing purely discrete spectra. However by their very nature hadrons interact strongly with each other and have large decay amplitudes into other hadrons when allowed by kinematic and symmetry considerations. How can the effects of decay channels both real and virtual be taken into account in the context of these models? In particular what is the influence of hadronic decay channels on the naive spectrum of states and on the properties of the surviving states ?

These general questions may be studied in a concrete form within the charmonium model[1)] of the new resonances[2)]. In this model the new resonances are interpretated as bound states of a charmed quark (c) and antiquark ($\bar{c}$). The large mass of the charmed quark suggests that the bound states may be described by a simple non-relativistic potential interaction. Furthermore, the decay products for states above charm threshold are distinct from their parents. The parent states being much smaller than their daughters. This is in contrast to the lighter mesons where one can not avoid a fully symmetric treatment of decay effects. For these reasons the ($c\bar{c}$) system may be especially well suited to an investigation of the influence of coupling to decay channels on the naive results in a quark confinement model.

---

* Work supported in part by the National Science Foundation.

2. REVIEW OF NON-RELATIVISTIC POTENTIAL MODEL

Before describing a model which includes both quark confinement and hadronic decay channels in a unified treatment, it is useful to review the application of a non-relativistic potential model of charmonium to the spectrum of new resonances.

The charmonium model of T. Appelquist and H.D. Politzer is described in detail in the lectures of T. Appelquist in these proceedings. The fundamental field theory of strong interactions is taken to be an asymptotically free non-Abelian gauge interaction between quarks and gluons. The dynamical gauge group is SU(3) and the associated degrees of freedom denoted color. The usual quantum numbers, isospin and hypercharge, along with charm are associated with a global SU(4) symmetry of the Lagrangian broken only by the quark masses. The rise in $R \equiv \sigma_{e^+e^- \to \text{hadron}}/\sigma_{e^+e^-\to\mu^+\mu^-}$ in the region of center of mass 4-5 GeV is associated with onset of thresholds for hadronic final states carrying charm ; and the new narrow resonances ψ/J and ψ' are bound states of charmed quark-antiquark pairs.

In this picture the large mass associated with the charmed threshold and the proximity of the ψ/J and ψ' resonances to the assumed charm threshold suggest that bound states may be non-relativistic. It has been suggested by a number of authors that a non-relativistic potential model might be applicable[3,4,5]. The form of such a phenomenological potential can be chosen to reproduce the features of the underlying dynamics. The short distance behaviour of an asymptotically free gauge theory is calculable and is dominated by one gluon exchange. This suggests that at short distance the potential should vary as 1/r. At large distance the potential should be confining for colored states if quarks are not to be observed as asymptotic states. A specific form of this potential is suggested by lattice gauge model calculations, two dimensional models of confinement, and a cluster decomposition argument to vary as r. A simple potential having these properties is given by[6]

$$V(r) = -\frac{\alpha_s}{r} + r/q^2 \tag{2.1}$$

The free parameters of the model are α_s, q and the charmed quark mass m_c.

The energies, $\{E_{NL}\}$ and wavefunctions

$$\psi_{NLM}(\vec{r}) \equiv \frac{U_{NL}(r)}{r} Y_{LM}(\theta,\varphi) \tag{2.2}$$

for the $c\bar{c}$ bound states in this potential are given by the eigenvectors and eigenfunctions of the ordinary Schroedinger equation with appropriate boundary conditions.

Within this model the ψ/J resonance is assigned to the 1^3S_1 state and ψ' is assigned to the 2^3S state. States are designated as $n^{2S+1}L_J$ where n-1 is the number of radial nodes. Details of the fit of parameters can be found in reference 4. The values of the parameters obtained are

$$\alpha_S = .18, \quad a^{-1} = .19 \text{ GeV}, \quad \text{and } m_c = 1.6 \text{ GeV}. \qquad (2.3)$$

One can check that the solution obtained is consistent with the non relativistic assumption; $\langle v/c \rangle \sim 1/5$.

3. SHORTCOMINGS OF NAIVE MODEL

The simple model described above makes a number of predictions about the spectrum of $c\bar{c}$ bound states which have since been experimentally verified. The spectrum of low-lying states predicted by this model is shown in fig. 1. The next radial excitation above ψ' is the 3^3S_1 state at E_{CM} = 4.18 GeV. The data on R indicates a broad resonance (∼200 MeV) at E_{CM} = 4.2 GeV[7]. Also as reported at this conference there are preliminary indications[8] that at least two of the P-states have been found. One lying at E_{CM} = 3.43 GeV and the other at E_{CM} = 3.51 GeV. This is in good agreement with the predicted center of gravity for P-states at 3.465 GeV[9]. Note that no spin dependent have been included in this model. The model can be extended to include these forces[10] giving splitting of the order of 50-100 MeV but leaving the general structure of the spectrum unchanged.

This success in prediction of the spectrum must be tempered by the inadequacy of the model above charmed threshold. How does one understand the width of ψ'' (3^3S_1), which is presumably above charm threshold ? For example, the 4^3S_1 lies at E_{CM} = 4.6 GeV in the model but there is no resonance visible in R near that energy. How can $\psi, \psi', \psi'', \psi'''$, ... be described in a unified way where they bear very different relationships to decay channels ?

The success of the model with respect to the properties of states which depend on details of the wavefunctions is less spectacular. I will mention two important examples :
1) The electronic decay width of the ψ/J is used as an input to the model however the ψ' width is predicted to be 3.2 keV. Experimentally, the electronic width of the ψ' is 2.2 ± .03 keV[7]. Here the agreement is not very good.

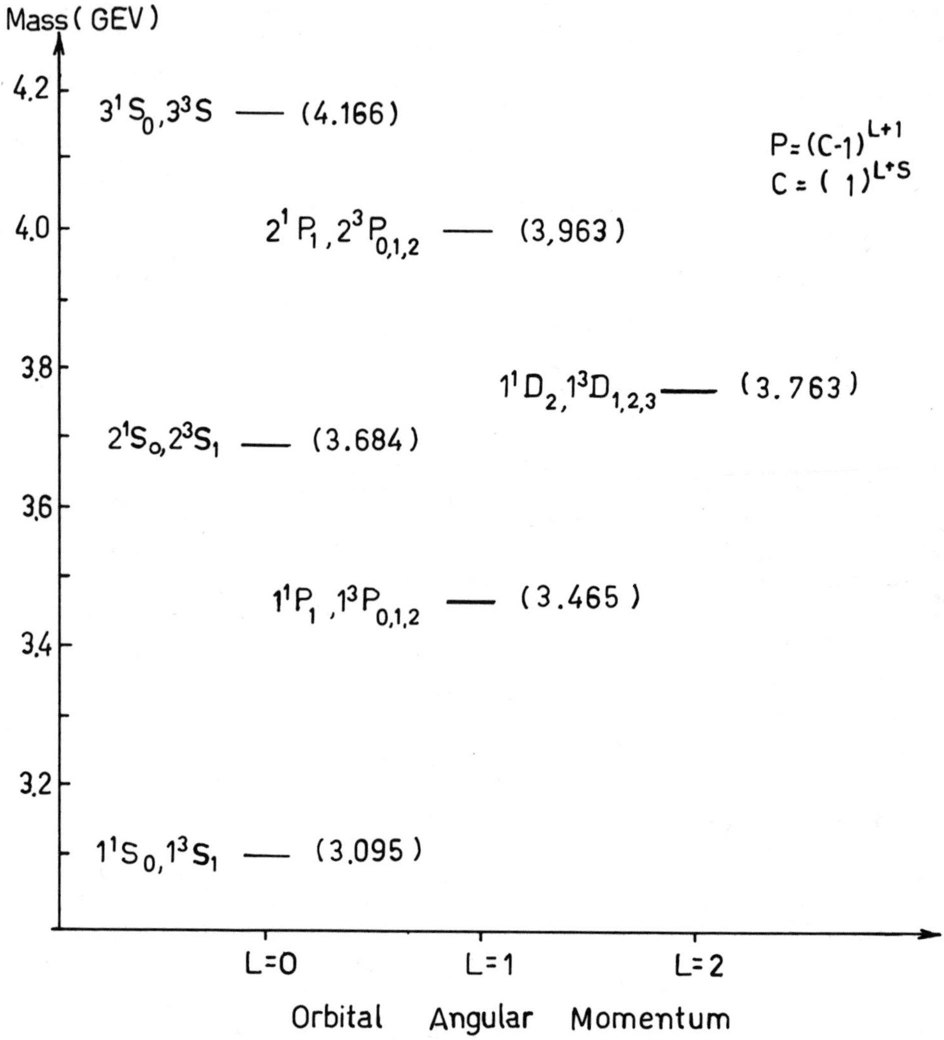

Fig. 1. Spectrum of low-lying states in a non-relativistic potential model of charmonium

2) The radiative transitions expected for the states below threshold are shown in fig. 2. The rates for the electric dipole transitions (E1) can be simply computed using the usual dipole transition formula

$$\Gamma_{E1:i\to f+\gamma} = 4/3 \ \frac{\alpha k^3}{(2j_i+1)} \ |\langle i \,\|\vec{D}\|\, f\rangle|^2 \qquad (3.1)$$

where $k = E_i - E_f$ and $\vec{D}$ is the dipole moment (charges measured in units of e).

The actual transition rates are very sensitive to the position of each state. For the transitions $2^3S_1 \to 1^3P_J+\gamma$ with J=0,1,or 2 we can estimate the transitions using the energy of the center of gravity of the P states :

$$\Gamma_{2^3S_1\to 1^3P_J+\gamma} = \frac{16}{243}\,\alpha k^3 (2J+1)\left|\int_0^\infty dr\; u_{20}(r)\; r\; u_{21}(r)\right|^2 \qquad (3.2)$$

where $u_{20}(r)$ and $u_{21}(r)$ are the radial wavefunctions of 2^3S_1 and 1^3P_J states respectively and k = 230 MeV. The transition rates are 120 keV for J = 2, 73 keV for J=1, and 24 keV for J=0. These rates are much larger than the present experimental upper limits[11]. One transition may have been observed at DESY; the rate seen suggests a rate a factor of 4 smaller than predicted[12].

The rate for the magnetic dipole (M1) transition $2^3S_1\to 1^1S_0+\gamma$ can also be calculated. The orthogonality of the 1S and 2S radial wave function greatly suppresses this transition. The rate is calculated to be about 1 keV; in agreement with the present experimental upper limits[11] on this transition.

Thus some of the properties of states depending on the details of their wavefunctions are in disagreement with experiment, particularly the E1 transition rates for ψ'. What is the effect of virtual decay channels on the spectrum of states below threshold and on the wave function of these states ?

4. INCLUDING HADRONIC DECAY CHANNELS[13,14]

In order to include coupling to decay channels consider the following basic Hamiltonian :

$$H \equiv \int d^3x\; \psi^+(x)(\vec{\alpha}\cdot\vec{\nabla}+\beta M)\;\psi(x)$$

$$+\frac{1}{2}\sum_a \int d^3x\; d^3x'\; :\rho_a(\vec{x}):U(\vec{x},\vec{x}'):\rho_a(\vec{x}'): \qquad (4.1)$$

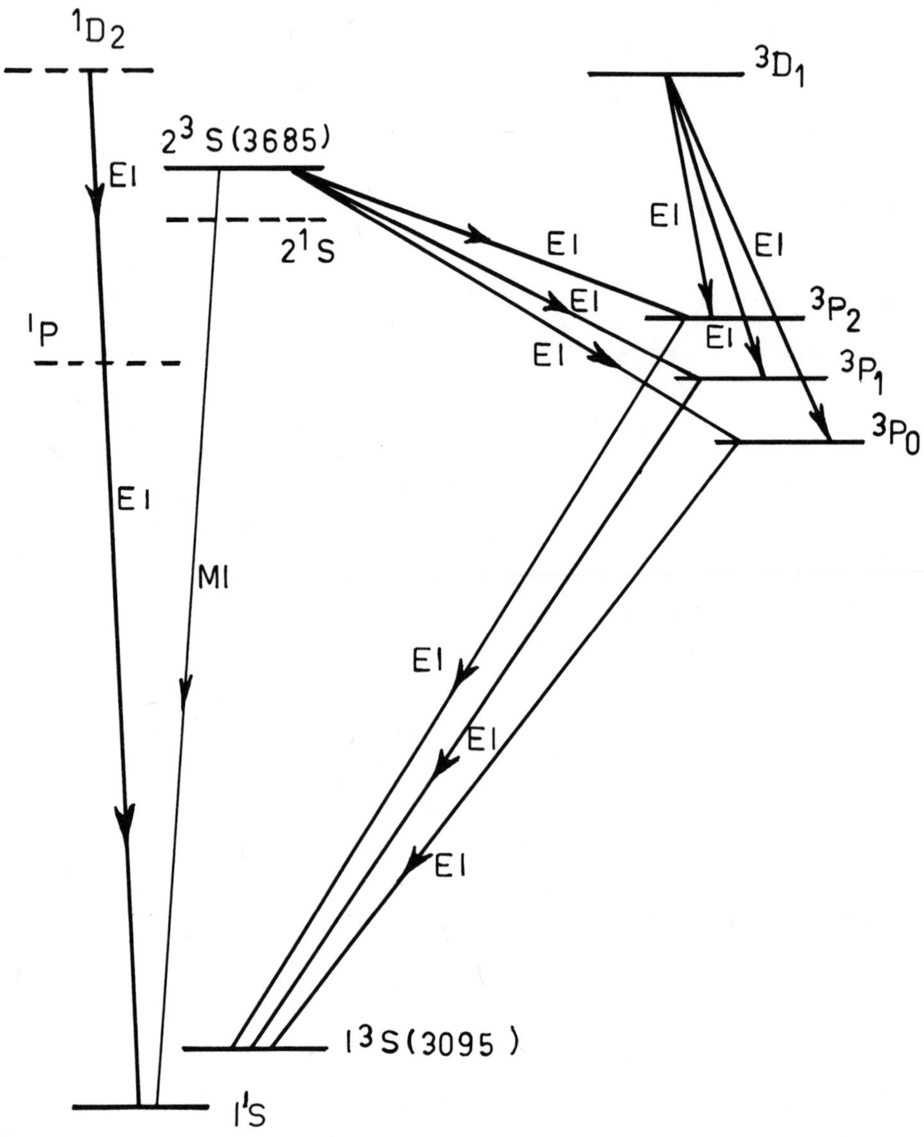

Fig.2. Radiative transitions. Heavy lines are allowed E1 γ-transitions; the $2^3S \rightarrow 1^1S$ decay is a highly suppressed M1 transitions. Dashed levels are unlikely to be produced or fed from above at an e^+e^- storage ring. Transitions among levels of LS multiplet are expected to be small and are not shown.

This Hamiltonian has the following properties designed to simulate the interactions due to colored gauge gluon exchange in an underlying field theory :

a) The Fermion field ψ belongs to the (4,3) representation of SU(4) (flavor) x SU(3) (color).

b) The mass matrix M is given by

$$M = \begin{pmatrix} m_p & & & 0 \\ & m_n & & \\ & & m_\lambda & \\ 0 & & & m_c \end{pmatrix}$$

and breaks the global SU(4) symmetry of the theory.

c) The quark color density $\rho_a(\vec{x})$ is given by

$$\rho_a(\vec{x}) = \psi^+(x)\ \lambda^a/2\ \psi(x)$$

where λ^a are the usual λ matrices for SU(3) color.

d) In order to obtain confinement of quarks, hopefully a property of an asymptotically free non-Abelian theory, and to recover the simple potential model in the appropriate limit the potential $U(\vec{x},\vec{x}')$ is taken to be a universal instantaneous interaction of the form :

$$U(\vec{x},t;\vec{x}',t') = \delta(t-t')\ |\vec{x}-\vec{x}'|\ /(4a^2/3) \tag{4.2}$$

Here we have not included any Coulomb term since the effects on the spectrum and wavefunctions were not dramatic and we are interested mainly in the qualitative effects of coupling to decay channels. The Hamiltonian given in eq.(4.1) is a straightforward generalization of the simple non-relativistic potential model considered in section 2.

The interaction term in the Hamiltonian may be rewritten using the following identities :

$$\sum_a \left(\frac{\lambda^a}{2}\right)_{ij} \left(\frac{\lambda^a}{2}\right)_{k\ell} = \frac{4}{9}\ (\delta_{ik}\delta_{kj}) - \frac{1}{6} \sum_a \left(\frac{\lambda^a}{2}\right)_{i\ell}\left(\frac{\lambda^a}{2}\right)_{kj} \tag{4.3a}$$

$$= \frac{1}{6}\ (\delta_{i\ell}\delta_{kj} + \delta_{ij}\delta_{k\ell}) + \frac{1}{3}\ (\delta_{i\ell}\delta_{kj} - \delta_{ij}\delta_{k\ell}) \tag{4.3b}$$

Using identity (4.3a) it is easy to see that the interaction leads to binding for $\bar{\psi}\psi$ color singlet states and a repulsive force for $\bar{\psi}\psi$ color octet states. The identity (4.3b) leads to an attractive force for $\bar{\psi}\bar{\psi}$ in a color triplet state and $\psi\psi$ in a color antitriplet state[15]. Since only color singlet states are expected to exist in confinement models we will consider only the part of the interaction leading to binding for color singlet states[16].

Decomposing quark fields into creation and destruction operators a variety of terms result. First there are terms which do not involve pair creation (destruction) terms. These terms bind a quark-antiquark ($q\bar{q}'$) color singlet states into mesons. These terms may be depicted graphically as shown below :

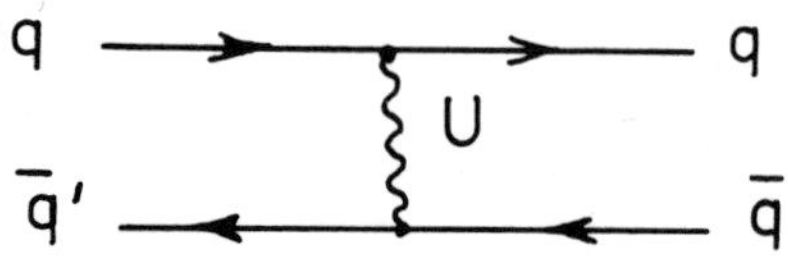

In the $c\bar{c}$ sector (i.e. q a charmed quark and $\bar{q}'$ a charmed antiquark) this interaction gives rise to the bindings of the naive charmonium model in the non-relativistic limit. The bound state wave functions can be calculated using the Bethe-Salpeter equation in the non-relativistic limit[17]. So the naive model is recovered. In the $c\bar{\ell}$ sector (ℓ denotes a lighter quark : p,n,λ) this interaction binds quarks into charmed mesons.

The second type of terms in H_I are pair creation (destruction) terms. Graphically these terms are

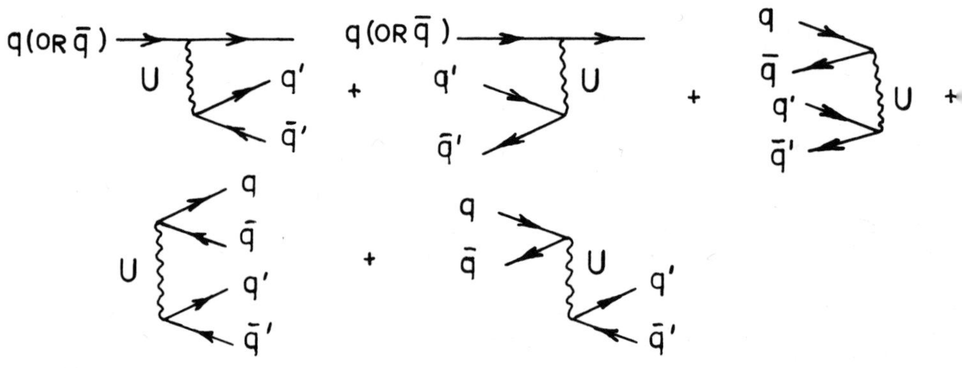

Since I am interested in the coupling of $c\bar{c}$ bound states to decay channels in the threshold region, i.e. when the charmed quarks are not relativistic, I will neglect terms involving creation or destruction of charmed quark pairs. Finally all pair terms involving only light quarks (p,n,λ) are included in the part of the Hamiltonian associated with the light mesons.

Thus for mesons the interaction Hamiltonian, H_I, may be rewritten as follows :

$$H_I \equiv H_\psi + H_D + W + H_L + H'_\psi \tag{4.4}$$

where

H_ψ = first type of terms in $c\bar{c}$ sector - binds charmonium states.
H_D = first type of terms in $c\bar{\ell}$ and $\ell\bar{c}$ sector - binds charmed mesons.
W = second type of terms in which a c (or $\bar{c}$) quark is propagated and $\ell\bar{\ell}$ quark pair is created (or destroyed) - couples charmonium states to charmed mesons.
H_L = first type and second type of terms involving only light quarks - binds ordinary mesons.
H'_ψ = second type of terms in which a $c\bar{c}$ pair is created or destroyed - small in energy range of interest.

One more assumption will be made in treating the decays of $c\bar{c}$ bound states, i.e. that final state interactions between charmed mesons due to coupling to lighter mesons can be ignored. These can be including in principle in this approach, however the cost is great as this introduces all the dynamics of the light meson part of the interaction Hamiltonian (H_L). Furthermore it is known that such interactions hardly influence R and only produces slowly varying modulations of scattering amplitudes. This restriction can be formally implimented by defining a projection operator on the $c\bar{c}$ sector, denoted P_ψ, and on the two bound state systems $(c\bar{\ell},\ell\bar{c})$, denoted P_c. Thus defining

$$W = P_c W P_\psi + \text{h.c.} \tag{4.5}$$

the "effective" Hamiltonian considered for calculating the influence of hadronic decay channels on the naive model of charmonium is

$$H_I = H_\psi + H_D + W \tag{4.6}$$

This model has two attractive features :

1) A universal interaction for calculation of bound states and decay amplitudes.
2) The interaction W is treated to all orders.

5. DECAY AMPLITUDES - LOWEST ORDER IN W

Before outlining the procedure for obtaining the general solution for the interaction Hamiltonian eq. 4.7, it is instructive to look at the lowest order decay amplitudes. As an example, consider the off energy shell decay amplitude for a $c\bar{c}$ bound state in the 3^3S_1 state, i.e. ψ'', decaying into a $c\bar{n}$ bound state in the 1^1S_0 state, denoted D^+, and a $\bar{c}n$ bound state in the 1^1S_0 state, denoted D^-. The lowest order decay amplitude

$$<\psi''(p,\vec{\lambda})|W|D^+(p_1)D^-(p_2)> \tag{5.1}$$

is shown graphically in figure 3. The analytical expression is given by writing the bound states in terms of the appropriate quark creation operators. For ψ'' you obtain

$$|\psi''(p,\vec{\lambda})> = \frac{1}{\sqrt{3}} \sum_{S_1 S_2} \int d^3p_1 d^3p_2 \, \delta(\vec{p}-\vec{p}_1-\vec{p}_2) \int d^3x \, e^{i(\vec{p}_1-\vec{p}_2)\cdot\vec{x}} \tag{5.2a}$$

$$[\psi^{c\bar{c}}_{3^3S_1}(\vec{p};\vec{x}) \quad \chi_\lambda^{J=1}(S_1,S_2) \sum_\alpha b^+_{c\alpha}(\vec{p}_1,S_1) d^+_{c\alpha}(p_2,S_2)] |0>$$

where $\psi^{c\bar{c}}_{3^3S_1}$ is the non-relativistic wave function for a 3S $c\bar{c}$ bound state of total momentum $\vec{p}$, and $\chi_\lambda^{J=1}(S_1,S_2)$ is a spin function coupling to spin 1/2 particles into a total spin 1 state. Note that $b^+_{c\alpha}(\vec{p}_1,S_1)$ is the creation operator for a charmed quark with color index α, momentum $\vec{p}_1$, and spin S_1 ; $d^+_{c\alpha}(\vec{p}_2,s_2)$ is the creation operator for a charmed antiquark with color index α, momentum $\vec{p}_2$ and spin s_2.

For $|D^+(p)>$ you can write

$$|D^+(p)> = \frac{1}{\sqrt{3}} \sum_{S_1,S_2} \int d^3p_1 d^3p_2 \, \delta(\vec{p}-\vec{p}_1-\vec{p}_2) \int d^3\vec{x} \, e^{i(\vec{p}_1-\vec{p}_2)\cdot\vec{x}} \tag{5.2b}$$

$$[\phi^{c\bar{n}}_{1^1S_0}(\vec{p},\vec{x})\chi^{J=0}(S_1,S_2) \sum_\alpha b^+_{c\alpha}(p_1,s_1)d^+_{n\alpha}(p_2,S_2)]|0> .$$

The expression for $|D^-(\vec{p})>$ is obtained by charge conjugation.

The decay amplitude, eq. 5.1, may be reexpressed using equations 5.2a, 5.2b, 4.1, and 4.7 as

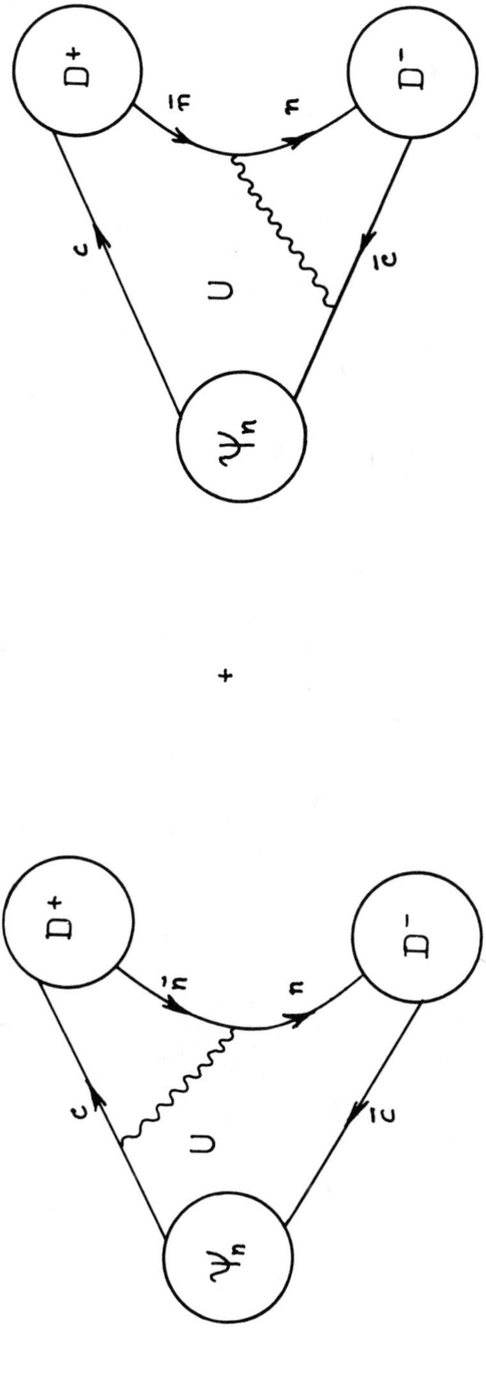

Fig.3. Lowest order decay amplitude into D^+D^- final state. H_I produces an $n\bar{n}$ pair with amplitude $\pm\,\vec{\sigma}\cdot\vec{\nabla}\,U$; the sign depending on whether the source is c or $\bar{c}$.

$$\langle\psi''(0,\lambda)|\; W|D^+(\vec{p})D^-(\vec{p}\,')\rangle = \frac{1}{(2\pi)^3}\,\delta^3(\vec{p}+\vec{p}\,')\,\frac{1}{\sqrt{3}}\,\frac{1}{\sqrt{2}}\,\frac{1}{m_n}$$

$$\int d^3x\, d^3y\, [\,(-i\vec{\varepsilon}_\lambda\cdot\vec{\nabla}_x)U(x)\,\phi^{c\bar{n}}_{1\,{}^1S_0}(\vec{p},\vec{x})\,\phi^{\bar{c}n}_{1\,{}^1S_0}(\vec{p}\,',\vec{x}-\vec{y})$$

$$\psi^{c\bar{c}}_{3\,{}^3S_1}(\vec{0},\vec{y})\; e^{i(m_c\vec{p}\cdot\vec{y}\,(m_c+m_n))}\,] \tag{5.3}$$

Figure 4 shows the shape of the decay amplitude, eq.5.3, for initial states n^3S_1 (n=1,2 or 3), i.e. the ψ,ψ', and ψ'' states. The decay amplitudes have a complicated momentum dependence.

In the numerical calculations I will describe in sections 6 and 7 the possible final states considered were limited to the lowest lying charmed meson states. Using the standard notation the $1^1S_0(c\bar{n})$, $1^3S_1(c\bar{n})$, $1^1S_0(c\bar{p})$, $1^3S_1(c\bar{p})$, $1^1S_0(c\bar{\lambda})$ and $1^3S_1(c\bar{\lambda})$ bound state mesons are denoted by D^+, D^{+*}, D°, $D^{\circ *}$, F^+, F^{+*} respectively [18]. The final states considered are D^+D^-, $D^\circ\bar{D}^\circ$, D^+D^{-*}, D^-D^{+*}, $D^\circ\bar{D}^{\circ *}$, $\bar{D}^\circ D^{\circ *}$, $D^{+*}D^{-*}$, $D^{\circ *}\bar{D}^{\circ *}$, F^+F^-, F^+F^{-*}, F^-F^{+*} and $F^{+*}F^{-*}$. Furthermore since the final state interactions of charmed mesons with ordinary mesons were not included in the model the masses of the charmed mesons m_D, m_{D^*}, m_F, m_{F^*} can not be consistently calculated and must be taken as arbitrary parameters.

6. OBTAINING THE GENERAL SOLUTION

A. General formalism

All quantities of interest can be extracted from the resolvent

$$G(Z) \equiv P_\psi(Z-H)^{-1}P_\psi\,. \tag{6.A.1}$$

$G(E+i\varepsilon)$ is the propagator for $c\bar{c}$ states including the effects of coupling to decay channels. Defining the bare propagator for $c\bar{c}$ bound states as

$$G_o(Z) \equiv P_\psi(Z-H_\psi)^{-1}P_\psi \tag{6.A.2}$$

the resolvent G may be reexpressed as

$$G(Z) = G_o(Z) + G_o(Z)\Omega(Z)\,G(Z) \tag{6.A.3}$$

where

$$\Omega(Z) \equiv W\,\frac{1}{Z-H_D}\,W^+ \tag{6.A.4}$$

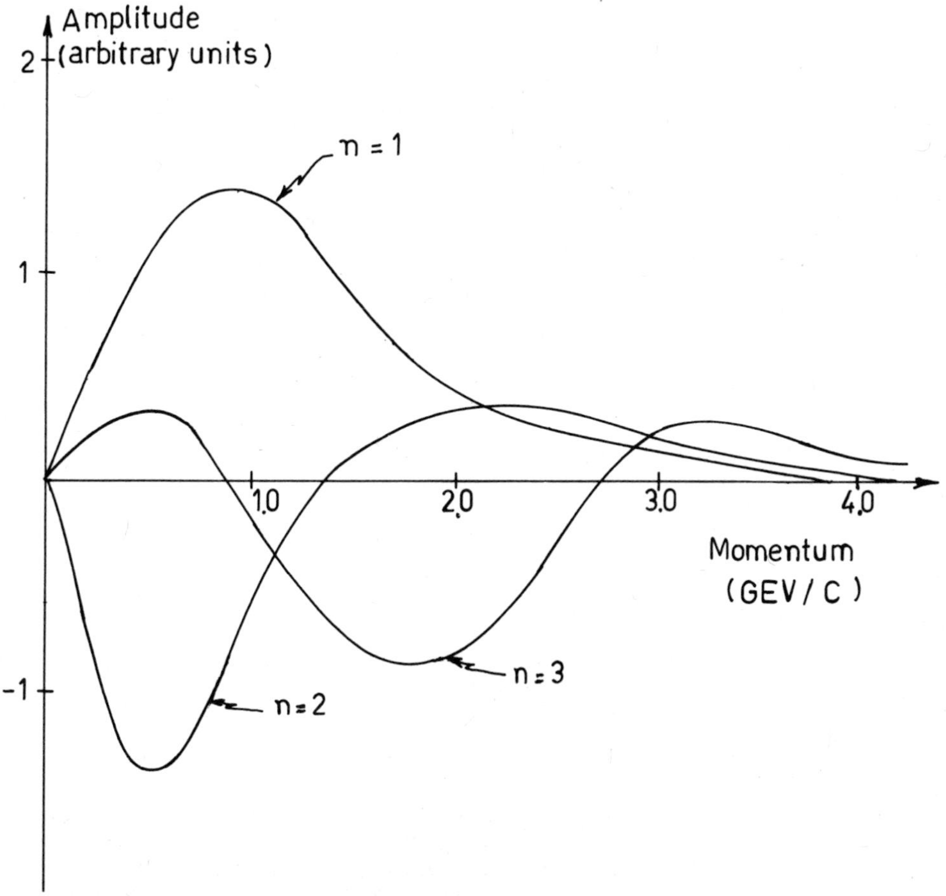

Fig. 4. General shape of the off energy shell decay amplitudes $n^3S_1 \to D^+D^-$ for n = 1,2, and 3.

and H, H_ψ and H_D are defined by equation 4.4 and W by equation 4.5

In practice it is more convenient to express 6.A.3 and 6.A.4 as matrix equations. Let $\{|n\rangle\}$ be a complete set of $c\bar{c}$ bound states with given quantum numbers J^{PC}. In particular for $J^{PC}=1^{--}$ we have the states ψ, ψ', ψ'', ... and also the 3^3D_1 state and its radial excitations. Converting the above equations into a set of infinite matrix equations gives

$$G_{nm}(E+i\epsilon) \equiv \langle n|G(E+i\epsilon)|m\rangle \tag{6.A.5a}$$

$$G_{onm}(E+i\epsilon) = \frac{\delta_{nm}}{E-\epsilon_n+i\epsilon}\ ,\ H_\psi|n\rangle = \epsilon_n|n\rangle \tag{6.A.5b}$$

$$\Omega_{nm}(E+i\epsilon) = \langle n|\ W\frac{1}{E-H_o+i\epsilon}\ W^+|m\rangle \tag{6.A.5c}$$

and

$$[G^{-1}(E+i\epsilon)]_{nm} = (E+i\epsilon-\epsilon_n)\delta_{nm} - \Omega_{nm}(E+i0). \tag{6.A.6}$$

The calculation of $\Omega_{nm}(E+i\epsilon)$ is done by means of a Hilbert transform.

$$\Omega_{nm}(E+i\epsilon) \equiv \Delta_{nm}(E) - i/2\ \Gamma_{nm}(E)$$

$$= \frac{1}{2\pi}\int_{E_{\text{threshold}}}^{\infty} \frac{dw}{E-w+i\epsilon} \sum_{\substack{\text{final states}, f, \\ \text{with energy } w}} \langle n|W|f(w)\rangle\langle f(w)|W^+|m\rangle \tag{6.A.7}$$

The real and imaginary parts of Ω denoted Δ and Γ in 6.A.7 are called the shift and width matrix respectively since in the diagonal approximation Δ_{nn} is the mass shift and Γ_{nn} is the total hadronic width of the state $|n\rangle$. The decay amplitudes appearing in 6.A.7 are simply the off energy shell decay amplitudes discussed in section 5.

B. Extraction of physically relevant quantities from the resolvent

Once G(Z) is known all quantities of physical interest can be easily determined both below and above threshold, E_{TH}, for the decay of charmonium states into charmed mesons.

Below threshold ($\text{Re}Z < E_{TH}$) :

1) The poles of G(Z) locate the levels that survive coupling to decay channels.
2) The residue of G(Z) at the pole determines the wave function as modified by coupling to decay channels, i.e.

$$G(Z;\vec{r},\vec{r}') \equiv \sum_{n,m} \langle\vec{r}|n\rangle\, G_{nm}(Z)\, \langle m|\vec{r}'\rangle$$

$$\sim \frac{\psi(r)\psi^*(\vec{r}')}{Z-M_\psi} \qquad \text{near } Z = M\psi\ . \tag{6.B.1}$$

Above threshold ($ReZ > E_{TH}$) :

1) In the $J^{PC} = 1^{--}$ sector, the discontinuity of $G(Z,\vec{0},\vec{0})$ across the cut gives the probability of finding a $c\bar{c}$ pair at zero separation and hence the contribution to R due to channel final states.

$$\Delta R = -\frac{24\pi}{(E)^2}\, 3Q^2 \text{Im}\, G(E+i\epsilon;\ \vec{0},\vec{0}) \tag{6.B.2}$$

2) The charmed meson-antimeson scattering amplitude and phase shifts can be explicitly calculated from G(Z). For example, in the case of $D\bar{D} \to D\bar{D}$ scattering the amplitude is simply given by

$$\langle D\bar{D}|W\ G(E+i\epsilon)\ W^+|D\bar{D}\rangle\ . \tag{6.B.3}$$

C. Calculational procedure

In order to explicitly preform the calculation of the resolvent there are two problems that must be faced. First the bare parameters must be adjusted so that the poles in G(Z) associated with the ψ/J and ψ' resonances have the experimentally observed masses and that the electronic width of ψ/J is correct. Second a suitable way to reduce the matrix equation to a finite matrix equation must be found.

Consider the $J^{PC} = 1^{--}$ channel and order the states in this channel by energy beginning with the lowest state. So that the states are ordered $|1\rangle = |\psi\rangle$, $|2\rangle = |\psi'\rangle$, $|3\rangle = |3^3D_1\rangle$, $|4\rangle = |\psi''\rangle$, etc.. The following proecedure is used to set the bare values of charmed quark mass, m_c, and linear potential slope parameter, a :

1) Evaluate bound state eigenfunctions and eigenvalues in the $c\bar{c}$ sector for input mass m_a and slope a.
2) Evaluate the elements of $\langle n|W|f\rangle$ for all two body charmed meson final states considered and for $n \leq N_{max}$, i.e. the lowest $N_{max}(c\bar{c})$ bound states in energy.
3) Evaluate $\Omega_{nm}(Z) \equiv \langle n|W(Z-P_cHP_c)^{-1}\, W^+|m\rangle$ by use of equation

6.A.6 for $n,m \leq N_{max}$.

5) Find the poles of $G(Z)$ below threshold corresponding to ψ/J and ψ' by finding zeros of $\det(G^{-1}(Z))$. Use equation 6.B.1 to find the function squared at the origin at ψ/J pole and from this obtain electronic width.
6) Adjust m_c and a to correct disagreement between results obtained and physical values. Repeat procedure until results agree with physical values.
7) Enlarge the number of states N_{max} until the results stabilize in the energy region of interest.

This procedure determines the number of states which contribute significantly in the energy region near charm threshold, given by N_{max} and thus determines how to truncate the infinite matrix equations. Once these parameters : m_c, A, and N_{max} are determined (for a given choice of initial parameters : the charmed meson masses and light quark masses) all the physical quantities in this and other J^{PC} channels can be calculated within the model.

7. RESULTS OF COUPLING TO DECAY CHANNELS

I would like to report some preliminary results on the influence of coupling hadronic decay channels to the naive charmonium model described in section 2. My emphasis is on the general character of these effects rather than details which are sensitive to exact values of charmed meson and light quark masses used.

Summary of general results :

1. For energies $\leq$ 4.7 GeV an essentially exact evaluation of G in the $J^{PC} = 1^{--}$ channel requires including the lowest fine $^3S_1$ states and the lowest three $^3D_1$ states. The bare masses are

$$m_c = 1.64 \text{ GeV} \quad \text{and} \quad a = 1.84 \text{ GeV}^{-1}$$

where the charmed meson masses and light quark masses are taken to be :

$$m(D^+) = m(D^\circ) = 1.90 \text{ GeV} \qquad m(F^+) = 2.08 \text{ GeV}$$

$$m(D^{+*}) = m(D^{\circ *}) = 2.08 \text{ GeV} \qquad m(F^{+*}) = 2.15 \text{ GeV}$$

$$m_n = m_p = .375 \text{ GeV} \qquad m_\lambda = .440 \text{ GeV}.$$

There are two reasons that energies are limited to $E \leq 5$ GeV in the model as presented. First, of course, the non relativistic approximations made in calculating the decay amplitudes will begin to fail at higher energies. Second only the lowest charmed meson two body final states have been included. At higher energies more thresholds must be included.

2. The decay amplitude for each two body final state has the expected form factor behaviour, dropping to zero for large momentum. The decay channels also provide substantial off-diagonal couplings probably invalidating the so-called diagonal approximation of generalized vector dominance for higher radial excitations than ψ'.

3. Although there are large energy shifts due to virtual decay :

$$\Delta_{1S} \equiv m^o_{1S} - m_{1S} = 90 \text{ MeV}$$

$$\Delta_{2S} \equiv m^o_{2S} - m_{2S} = 160 \text{ MeV}$$

the "renormalized" bound state spectrum does not differ markedly from the naive model. For the parameters given in (1.) the lowest P_{states} are

$$m(1^3P_J=0) = 3.42 \text{ GeV}$$

$$m(1^3P_J=1) = 3.43 \text{ GeV}$$

$$m(1^3P_J=2) = 3.44 \text{ GeV}$$

compared to the unperturbed degenerate value of 3.45 GeV. Lowering the charmed meson thresholds increase this splitting between different J values. No explicit spin-dependent forces were retained in the original interaction Hamiltonian, eq. (4.6); these mass splittings are induced as a result of chosing the pseudoscalar and vector charmed mesons to have different masses. The splittings would vanish if the pseudoscalar and vector masses were chosen degenerate. The mass difference between pseudoscalar and vector charmed mesons presumably arises both from explicit spin dependent forces in this $(c\bar{\ell})$ and $(\ell\bar{c})$ bound state sector and due to coupling through decay channels of charmed mesons to ordinary mesons.

4. The wave function of ψ' is strongly modified by coupling to decay channels. There are two types of modification. First there is a mixing within the $c\bar{c}$ sector to other states with the same J^{PC}. So that in terms of bare states ψ' may be written

$$\psi'_{c\bar{c}}(r) = \alpha\psi^o_{2S}(r) + \beta_1\,\psi^o_{1S}(r) + \beta_2\psi^o_{3S}(r) + \beta_3\psi^o_{4S}(r)$$

$$+ \ldots + \gamma_1\psi^o_{1D}(r) + \gamma_2\psi^o_{2D}(r) + \ldots \qquad (7.1)$$

In the no coupling limit $\alpha = 1$, $\beta_1 = \beta_2 = \ldots = 0$, $\gamma_1 = \gamma_2 = \ldots = 0$. In the presence of decay channels with the choice of masses given in (1.)

$$\alpha = .89, \ \beta_1 = -.11, \ \beta_2 = .02, \ \beta_3 = -.02, \ \gamma_1 = -.05$$

and all other β's and γ's essentially zero. The ratio of the function squared at the origin after coupling to that before coupling is .94.

The second effect is that the state is sometimes a virtual charmed meson-antimeson state, i.e. the wave function is no longer completely in the $c\bar{c}$ sector. A parameter which gives a measure of this effect is Z_2 defined by

$$Z_2 \equiv \int d^3\vec{r} \ \psi'^{*}_{c\bar{c}}(\vec{r}) \ \psi'_{c\bar{c}}(\vec{r}) \leqslant 1 \qquad (7.3)$$

where $\psi'_{c\bar{c}}$ is given by (7.1). It is easy to check using (7.2) that for the choice of parameters given in (1.) $Z_2 = .81$.

The choice of charmed meson thresholds closer to the ψ' mass result in a larger magnitude for these vitual decay channel effects; the mixing in the $c\bar{c}$ sector will increase and Z_2 will decrease. The more tightly bound 1S and 1P states are less influenced by coupling to decay channels.

5. The modification of the 2S(ψ') wave function and (to a lesser extent) the modification of the wave functions of the 1P and 1S states affect the gamma transition rates $\psi' \to 1^3P_{J=2,1,\mathrm{or}\ 0^+}\ \gamma$ and $\psi' \to |{}^1S_0 + \gamma$.

The rate for the E1 transitions : $\psi' \to 1^3P_J + \gamma$ are influenced by four factors :

a) The change in energy levels of the P states due to coupling to decay channels and due to explicit spin dependent forces not included.

b) Mixing of states within the $c\bar{c}$ sector and decrease of the norm of the wavefunction in $c\bar{c}$ sector.

c) Photon emission while ψ' is in a virtual charmed meson-antimeson state.

d) Photon emission taking ψ' (or a P state) from the $c\bar{c}$ sector to the virtual charmed meson sector.

The numerical calculations for the factors c) and d) have not yet been completed. The effect of b) has been calculated and leads to substantial decrease in the E1 transition rates; the decrease is approximately 30-40%, for each of the final P states. The influence of a) is difficult to determine since the size of explicit spin dependent force has not been determined. If the P states are assigned masses to agree with present experimental indications

$$m(1^3P_0) = 3.43 \ \text{GeV} \quad m(1^3P_1) = 3.51 \ \text{GeV} \quad m(1^3P_2) \gtrsim 3.53 \ \text{GeV}$$

then the effect of a) is to increase the $\psi' \to 1^3P_0 + \gamma$ transition rate by a factor of 2 and decrease the $\psi' \rightarrow 1^3P_1 + \gamma$ and $\psi' \rightarrow 1^3P_2+\gamma$ transition rates by factors of 3.6 and 5 respectively. The net effect of factors a) and b) is to substantially improve agreement between theory and experiment for these transitions.

The M1 transition $2^3S_1 \rightarrow 1^1S_o + \gamma$ is enhanced by coupling to decay channels as the orthogonality of the wave functions for 1S and 2S state is no longer true. The rate increases by a factor of 5; however this rate is still well within experimental limits.

6. If ψ' is very close ($\lesssim$ 20 MeV) to threshold R will show an enhancement in the threshold region, as is shown in figure 5.

7. There is induced S-D mixing between the 1^3D_1 and 2^3S_1 states on the order of 5 %. Therefore there should be a small spike in R at the mass of the 1^3D_1 state, $m(1^3D_1)$ = 3.780 GeV. However the area under the spike is small making it difficult to see at spear.

8. A typical curve of the charm contribution to R is shown in fig. 6. The details of the curve are sensitive to the position of the various charmed meson thresholds. The general features are :
a) The resonance at 4.2 GeV has a complex structure. Threshold effects distort the shape of the resonance greatly and their is mixing between the 3^3S_1 state and the 2^3D_1 state ($m(2^3D_1)$=4.375).
b) The model cannot account for the slow rise in R observed between 3.7 and 4.0 GeV.
c) The averaged value of ΔR_{charm} over the region 4.0-4.7 GeV remains relatively constant independent of the positions of various charmed meson thresholds. This value of $\langle \Delta R_{charm} \rangle$ is ≃ 1.3. If the background of 2.4 due to ordinary hadronic final states is subtracted from the data on R the average value ΔR_{exp} in the same energy region is ≃ 2.5. This discrepancy between theory and the data is unavoidable in our model. This suggests that there is a new threshold besides charm in this energy region.
d) If there is a heavy charged lepton with a mass about 1.8 GeV[19], one finds adding the contribution to R for this heavy lepton to the contribution from charmed final states and a 2.4 background value of R from ordinary hadronic final states gives a curve of R in good agreement with the experimental values. The difficulties mentioned in b) and c) disappear.

9. The $D\bar{D}$ elastic $J^{PC} = 1^{--}$ channel S-matrix element is shown in figure 7. Two inelastic resonances associated with the 3^3S_1 and 2^3D_1 states are clearly visible. The model appears to be capable of producing hadronic scattering amplitudes having the sort of struction found in phase shift analysis.

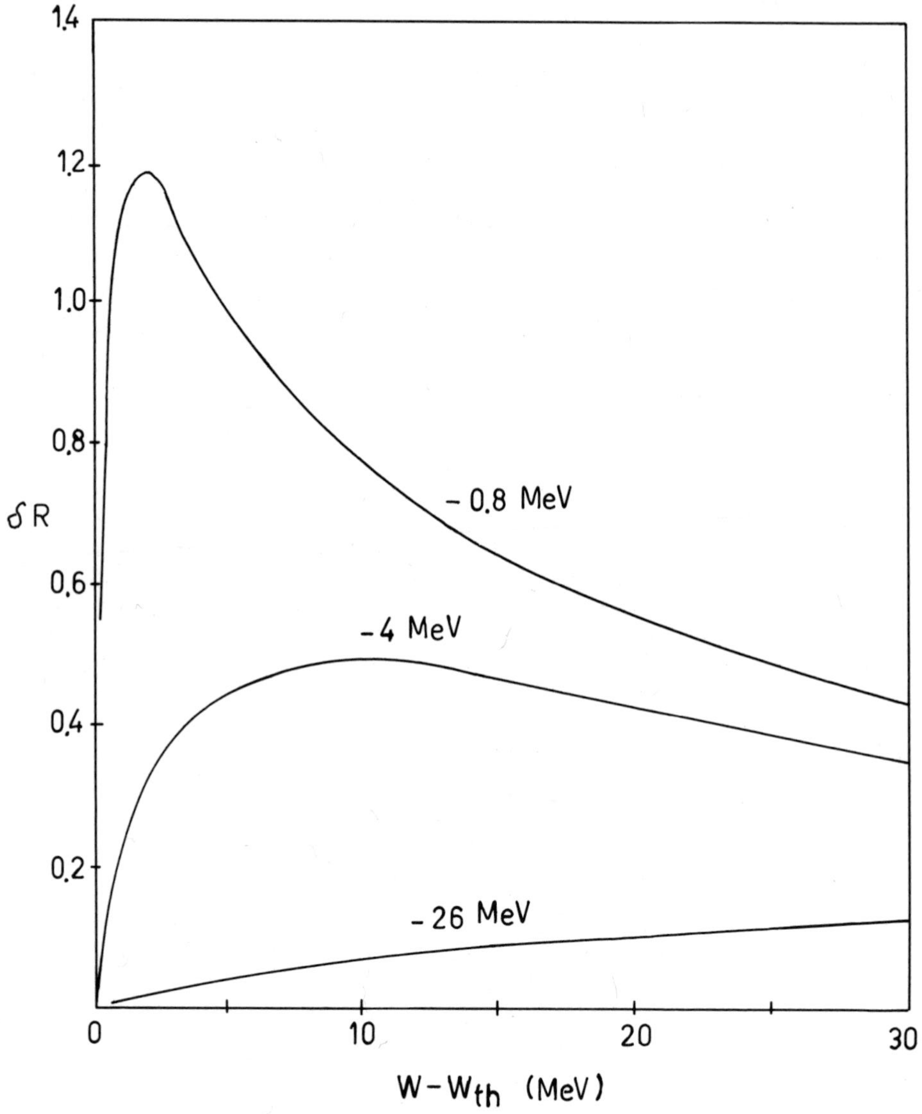

Fig.5. Behaviour of ΔR when ψ' lies just below the charm threshold W_c. The curves are labeled by $m_{\psi'} - W_c$.

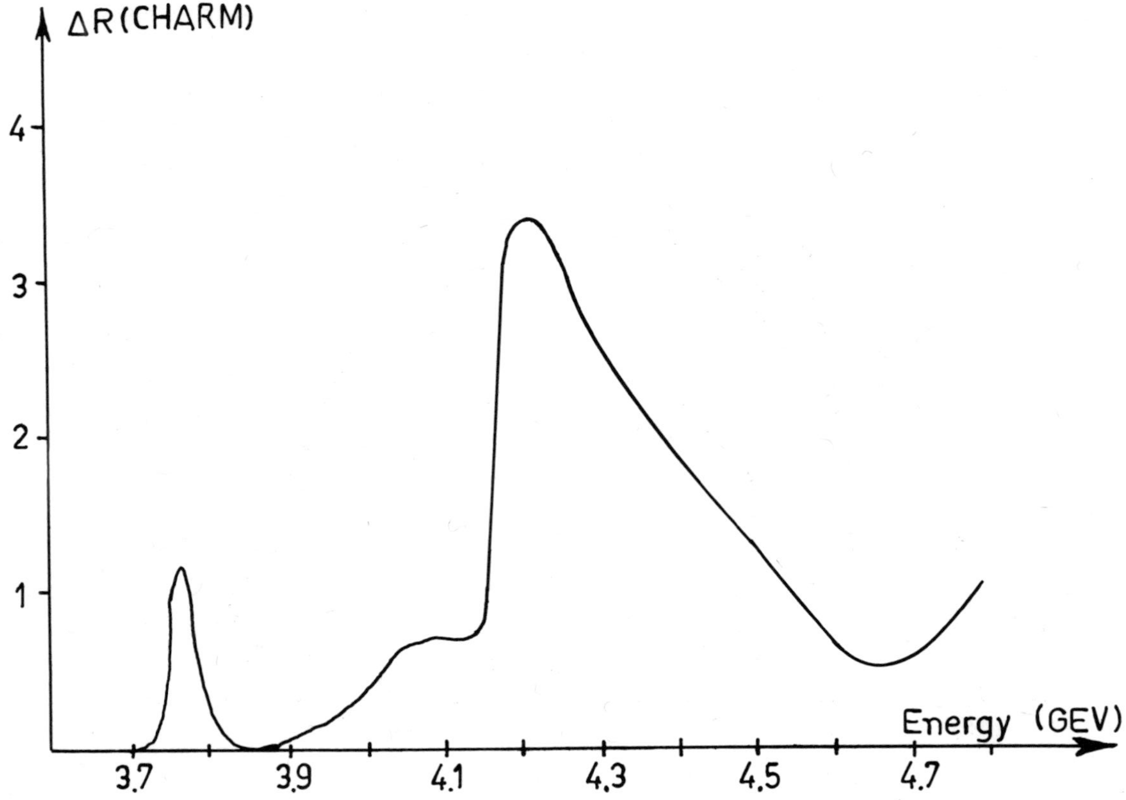

Fig. 6. The contribution ΔR of charm to R. In this case lowest charm threshold is at 3.7 GeV; other thresholds are given in the figure. The 1^3D_1 state is visible at 3.774 GeV.

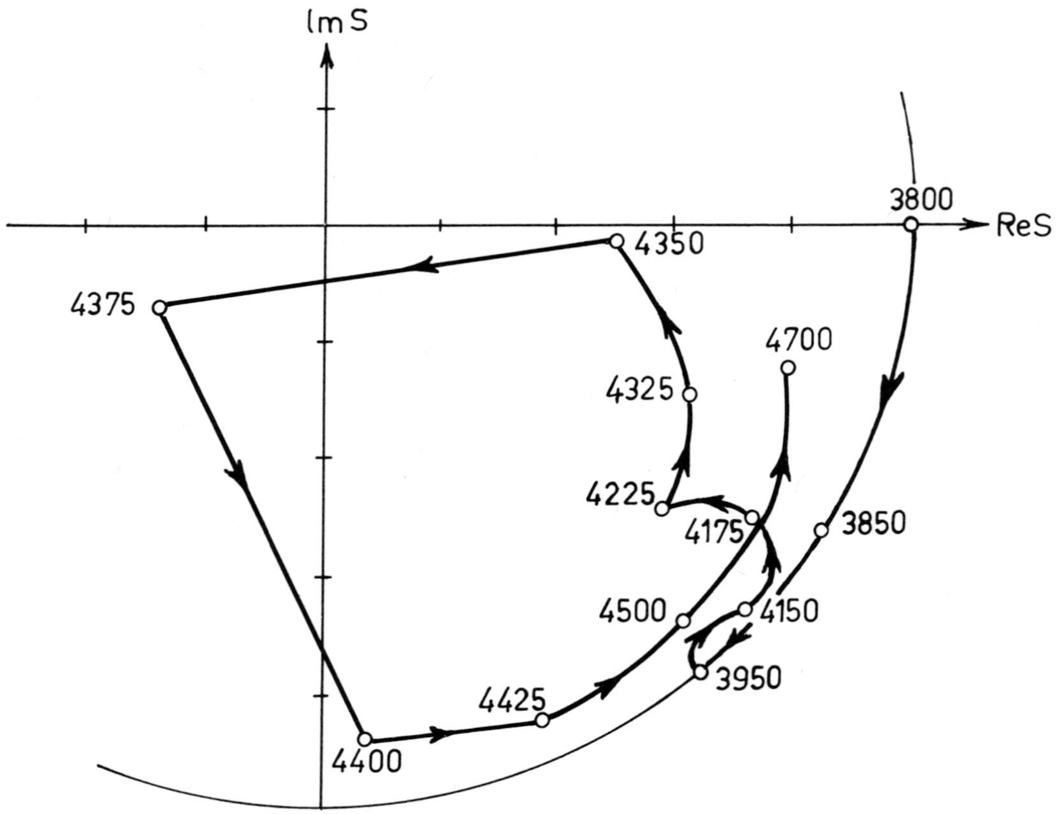

Fig.7. Argand plot of $e^{2i\delta}$ for $D\bar{D}$ elastic scattering in the $J^{PC} = 1^{--}$. The peak at 4.2 GeV in figure 6 appears as a somewhat inelastic resonance. There is a narrow ($\Gamma \sim 50$ MeV) highly inelastic resonance due to the 2^3D_1 state at 4.375 GeV.

8. CONCLUSION

In conclusion the preliminary results are very encouraging in two respects. First the model seems capable of producing a general structure for the charm contribution to R which compares favorably with the data. Second the modification of the wave function of states below threshold may be significant in reducing the radiative transitions and thus improving agreement with experiment. Hopefully an understanding of the effects of coupling to decay channels within the charmonium model will also shed some light on the general nature of these effects.

REFERENCES AND FOOTNOTES

1. T. Appelquist and H.D. Politzer, Phys.Rev.Lett. 33, 1404 (1974). See also the lectures of T. Appelquist in these proceedings.
2. J.J. Aubert et al., Phys.Rev.Lett. 33, 1404 (1974); J.-E. Augustin et al.,ibid. 33, 1406 (1974); G.S. Abrams et al., ibid. 33, 1453 (1974). See the lectures of H.L. Lynch in these proceedings for the present experimental situation concerning the new resonances.
3. T. Appelquist, A. De Rujula, S.L. Glashow, and H.D. Politzer, Phys.Rev.Lett. 34, 365 (1975).
4. E. Eichten, K. Gottfried, T. Kinoshita, J. Kogut, K.D. Lane, and T.M. Yan, Phys.Rev.Lett. 34, 369 (1975).
5. B. Harrington, S.Y. Park and A. Yildiz, Phys.Rev.Lett. 34, 168 (1975); H.J. Schnitzer, Harvard preprint, unpublished (1974).
6. This is the model proposed by K. Gottfried, T. Kinoshita, J. Kogut, K.D. Lane, T.M. Yan, and myself. See ref. 4.
7. See the data presented by H.L. Lynch in these proceedings.
8. W. Braunschweig et al., DESY preprint 75/20 (1975); G.J. Feldman et al., SLAC-PUB-LBL-4220 (1975).
9. The existence of P-states lying at about 3.400 GeV was also suggested by a loaded string model by C.G. Callen, R.L. Kingsley, S.B. Treiman, F. Wilczek and A. Zee, Phys.Rev.Lett. 34, 52 (1975).
10. See the talk of R. Barbieri in these proceedings. Also see R. Barbieri, R. Kögerler, Z. Kunszt and R. Gatto, CERN preprint ref. TH. 2036 (1975); A. De Rujula, H. Georgi, and S. Glashow, "Hadron masses in a gauge theory" Harvard University preprint (1975).
11. J.W. Simpson et al. HEPL Report No. 759.
12. See reference 8. The two photons resulting in the double transition $\psi' \to \gamma+X$, $X \to \gamma+\psi$ are observed at DESY so only the branching ratio $(\Gamma_{\psi'\to\gamma+X}/\Gamma_{\psi'})\,(\Gamma_{X\to\gamma+\psi}/\Gamma_X)$ can be directly measured; however the charmonium model this X should be identified as the $^3P_1$ state and furthermore the ratio

$\Gamma_{^3P_1 \to \gamma+\psi} / \Gamma_{^3P_1}$ should be near one.

13. The work I will be reporting on in the rest of my seminar was done in collaboration with K. Gottfried, T. Kinoshita, K.D. Lane and T.M. Yan.
14. Two other treatments of the effects of coupling to decay channels have appeared recently : J. Kogut, and L. Susskind, Phys.Rev.Lett. 34, 767(1975) and Cornell preprint CLNS-303 (1975); R. Barbieri, R. Kögerler, Z. Kunszt, and R. Gatto, CERN preprint, ref. TH. 2026 (1975). The former is based on the Born-Oppenheimer approximation which makes detailed comparison with the approach presented here somewhat difficult. In the latter the $q\bar{q}$ production amplitude does not depend on $c\bar{c}$ separation, the energy dependence of decay amplitudes is ignored and the influence of virtual decay on ψ' is not considered. The form factor effects (i.e. energy dependence) of decay amplitudes is found to be very important to results reported here.
15. The attractive force between ψ and ψ'in a color antitriplet state may be an important component in the force that binds baryons : see A. De Rujula, H. Georgi and S. Glashow, ref. 10, also see talk of R. Barbieri in these proceedings.
16. Actually it may well be that only states which are overall color singlets will be bound even if the full inte action term is retained.
17. E.E. Salpeter, Phys.Rev. 87, 328 (1962).
18. The notation for charmed mesons is that of M.K. Gaillard, B.W. Lee and J.L. Rosner, RMP 47, 277 (1975).
19. The possibility of a heavy lepton is suggested by the observation of μe events with $\gtrsim 2$ neutrals in e^+e^- annihilation : M.L. Perl *et al.*, SLAC-PUB-1626 (1975).

HADRONS (USUAL AND NEWLY DISCOVERED) IN GAUGE THEORIES A DREAM ?

R. BARBIERI

CERN - Geneva

CONTENT

1. INTRODUCTION

In this talk I will concentrate on the following two questions:

i) Given the relative success of the description within the framework of a non-Abelian gauge field theory of the newly discovered phenomena in e^+e^- physics (*), can we also learn something about usual hadrons in the same theoretical framework ?

ii) In particular, is it possible to incorporate in the phenomenological scheme proposed for the new particles, interpreted as "charmonia", also the usual hadrons ? This achievement might

(*) See the lectures by T. Applequist in these Proceedings.

produce restrictions that would be relevant to understanding the same starting point of the analysis (ψ particles and e^+e^- physics above 3 GeV).

It is worth while to start by giving a qualification of the ideas and the calculations that I will present in the following when trying to answer these questions. They have a high speculative content, and the resulting picture of hadron dynamics is so simple that many people will find it unbelievable; furthermore, the bridge between this phenomenological picture and the supposed underlying field theory is far from being built. Nevertheless, the importance of the questions that these ideas try to answer, as well as some of the results that will come out, make me confident about the usefulness of their presentation.

2. UNDERLYING FIELD THEORY FOR STRONG INTERACTIONS

The reference field theory for strong interactions will be the SU(4) ⊗ SU(3) non-Abelian gauge theory -- about which we have heard a lot in Appelquist's lectures -- involving four types of fractionally charged quarks, each in three colours and coupled to an SU(3) octet of massless gauge gluons. Here SU(4) is the minimal extension of the usual SU(3) group to incorporate the new quark, whose bound states with a corresponding antiquark we are eventually seeing in the e^+e^- channel. By this I mean that the theory is likely to be, but it is not necessarily, the original SU(4) model of Glashow, Iliopoulos and Maiani; what I am going to say can easily be modified to apply to more elaborate "charmed" quark models.

I would like to remind you that, in usual notations, the gluon-quark coupling is given in the theory as

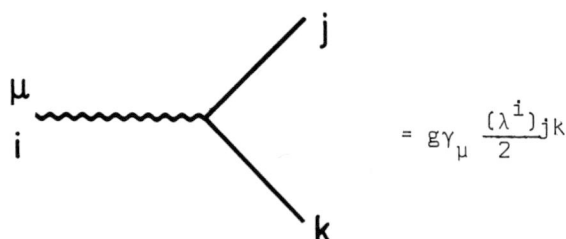

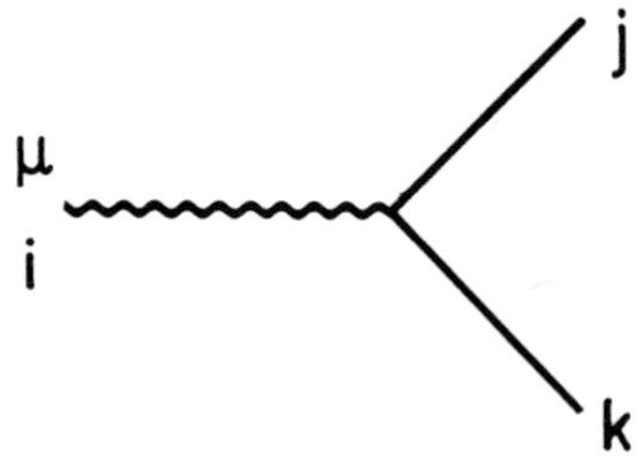

$$= g\gamma_\mu \frac{(\lambda^i)_{jk}}{2}$$

where μ stands for the Lorentz index of the vector gluon and i,j,k are colour indices.

Standard concepts associated with this kind of theory are : i) the asymptotic freedom; ii) the "infrared slavery", the latter being a conjecture. They have already been discussed here by several people. As to (i), I will assume that the effective "strong fine-structure constant" $g^2(M)/4\pi \equiv \alpha_S(M)$, renormalized at a running energy scale M, is small compared to 1 already at $M \sim 1$ GeV. From this point on, asymptotic freedom sets in and allows us to compute a smaller and smaller coupling $\alpha_S(M)$ in terms of $\alpha_S(1)$. The smallness of $\alpha_S(1)$, which seems to be required by deep inelastic experiments, is within the gauge theory an open question which has eventually become more mysterious with the need to introduce a mass parameter for the "charmed" quark of about 1.6 GeV (see below).

As to the speculation of "infrared slavery" and the consequent permanent confinement of quarks and gluons in colourless hadrons, let me only point out the possible relevance of the recent investigation by Cornwall and Tiktopoulos[1], who give arguments for the vanishing in the zero-mass limit of Yang-Mills theories both of the exclusive and inclusive cross-sections of colour non-neutral objects.

3. PHENOMENOLOGICAL SCHEME

The phenomenological scheme proposed from the theory by De Rujula, Georgi and Glashow[2] can be summarized in the following four points :

i) The hadrons (usual, "charmed" and "hidden charmed") are mainly described by non-relativistic dynamics. This hypothesis has clearly nothing to do with the non-Abelian gauge theory, but only with the actual values of the parameters appearing in the theory. Its self-consistency has to be checked.

ii) The "infrared slavery" produces an effective long-range binding force which is independent of quark spins and quark masses. The breaking of SU(4) enters only in the kinetic term of the effective Hamiltonian since different masses for quarks of different "flavours" are allowed ($m_p = m_n \equiv m_d, m_\lambda, m_c$).

iii) Since short distances are a regime where perturbations theory is meaningful (asymptotic freedom), the short -distance interaction is mediated by the one-gluon exchange with a small coupling α_S. The non-relativistic reduction of the one-gluon potential, $g^2\gamma^1_\mu\gamma^2_\mu/k^2$ gives then in coordinate space the usual Coulomb interaction α_S/r plus the Breit-Fermi potential S_{BF}, which contains spin-spin and spin-orbit interactions.

iv) The role played by the colour group in getting out of the theory the phenomenological scheme that we are discussing is presumably fundamental, but escapes our present knowledge fof non-Abelian gauge field theories. Part of this ignorance is expressed in making the assumption that hadrons are bound states of colour singlet wave functions. The graphical notation that I will use in the following for the meson and baryon wave functions is, respectively,

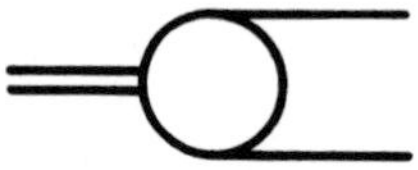

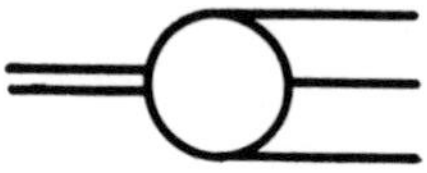

The properly normalized colour part of the meson wave function is $M_{ij} = (1/\sqrt{3})\delta_{ij}$, whereas for baryons $B_{ijk} = (1/\sqrt{6})\varepsilon_{ijk}$. The non-Abelian structure of the theory is reflected here only in the different coefficient that appear in front of the one-gluon exchange potential due to the colour coupling.
For mesons

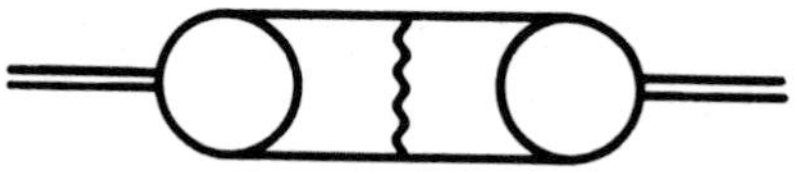

$$\mathrm{Tr}\,[M(\tfrac{g}{2}\lambda^a)M(-\tfrac{g}{2}\lambda^a)]\;\frac{\gamma_\mu^1\,\gamma_\mu^2}{k^2} = -\frac{4}{3}\,g^2\;\frac{\gamma_\mu^1\,\gamma_\mu^2}{k^2}$$

and for baryons

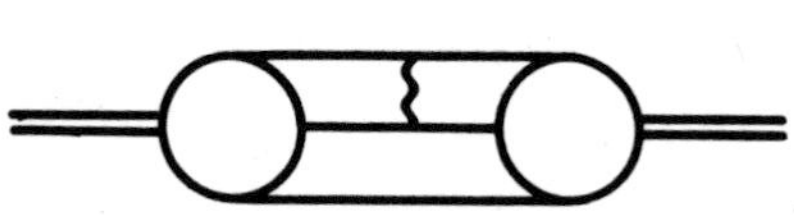

$$[B_{ijk}(\tfrac{g}{2}\;\lambda^a_{a\ell})B_{\ell mk}\;(\tfrac{g}{2}\,\lambda^a_{mj})]\;\frac{\gamma_\mu^1\,\gamma_\mu^2}{k^2} = -\frac{2}{3}\,g^2\;\frac{\gamma_\mu^1\,\gamma_\mu^2}{k^2}$$

4. APPLICATIONS TO USUAL S-WAVE BARYONS

Let us assume that $(m_\lambda - m_d)/(m_\lambda + m_d)$ is a small parameter. The SU(3) breaking term in the kinetic energy of the effective Hamiltonian is then treatable as a perturbation of an SU(6) symmetric theory. One has therefore a degenerate 56-plet of S-wave baryons, with a common O(3) wave function $\psi_0(\vec{r}_1,\vec{r}_2,\vec{r}_3)$.

We also assume that this wave function ψ_0 contains predominanly S-waves in any of the relative quark coordinates $(\vec{r}_i-\vec{r}_j)$.

It follows for states of equal quark content that the breaking in mass is given only by the Fermi spin-spin interaction which is contained in S_{BF} and is a contact interaction

$$\langle \delta M \rangle = -\frac{2}{3}\alpha_s \langle \psi_0 | \delta^3(\vec{r}_1 - \vec{r}_2) | \psi_0 \rangle \sum_{i>j} \frac{1}{m_i m_j} \langle \vec{S}_i . \vec{S}_j \rangle$$

The following mass relations are then easily derived[2] (particle names stand for particle masses) :

$$\frac{2(\Sigma^* - \Sigma)}{2\Sigma^* + \Sigma - 3\Lambda} = \frac{m_d}{m_\lambda} = 0.622 ,$$

$$\Sigma - \Lambda = \frac{2}{3} (1 - \frac{m_d}{m_\lambda})(\Delta - N) .$$

If the first equation is taken as fixing the ratio m_d/m_λ, the second relation is a successful prediction. It goes without saying that more common SU(3) or SU(6) predictions, such as the Gell-Mann/Okubo formula or the equal spacing rules for the decuplet, are also recovered here.

Coming to the electromagnetic properties of baryons, from our non-relativistic picture and the knowledge of the SU(6) wave functions we get for the magnetic moments[2]

$$\mu(N) = -\frac{2}{3}\mu(P); \quad \mu(\Lambda) = -\frac{m_d}{3m_\lambda} \quad \mu(P) \simeq -0.6 \ ; \ \mu(P) = \frac{P}{m_d}$$

The first relation is the usual successful SU(6) relation, which gets corrections in our scheme only by electromagnetism. The second relation is closer to the experimental values than the "naive" SU(6) relation $\mu(\Lambda) = -\mu(P)/3$, because of the previously determined m_d/m_λ mass ratio; it gets corrections in the theory of order $\alpha_S(m_\lambda - m_d)/(m_\lambda + m_d)$. Finally, the prediction for the proton magnetic moment $\mu(P) = P/m_d$ fixes m_d = 336 MeV (and consequently m_λ = 540 MeV); this is the least firm of the three relations because corrections to it of order α_S are expected.

5. SPECIFICATION OF THE FULL HAMILTONIAN

Given the relative success of this starting point, and leaving aside for the moment the mesons, we would like to extend the analysis to excited multiplets of different O(3) content and also to incorporate the "charmed" states. However, because of our ignorance of the binding potential, both these extensions require the introduction of new free parameters. In fact : i) the spatial wave functions for different multiplets are not correlated; and ii) the expansion in the quark mass breaking is untenable if, as expected $m_c >> m_d$. On the other hand, the actual dependence of the wave functions on the constituent quark masses is totally related to the potential itself.

Since we want to avoid the introduction of too many free parameters, we choose to work with a definite binding potential[3]. For a two-quark mesonic bound state -- a technically simpler problem to handle than the three-body problem for a baryonic system -- the fully specified total Hamiltonian that we are going to diagonalize is

$$H = m_1 + m_2 + \frac{1}{2\mu}\,\vec{p}^{\,2} + V_1 + V_2 \;; \qquad \mu = \frac{m_1 m_2}{m_1+m_2}$$

$$V_1 = \lambda r + V_o - \frac{4}{3}\,\frac{\alpha_s}{r}$$

$$V_2 = -\left(\frac{1}{m_1^3} + \frac{1}{m_2^3}\right)\frac{\vec{p}^{\,4}}{8} - \frac{4}{3}\,\alpha_s\, S_{BF}$$

A linear binding potential, together with a Coulomb component, has also been used successfully by Eichten et al. (*) to analyse the $c\bar{c}$ bound states. The constant V_0 represents the effects of intermediate range forces. The potential V_2, which will be treated in first order of perturbation, contains, together with the Breit-Fermi interaction, the $\vec{p}^{\,4}$ term coming from a consistent expansion of the relativistic kinetic energy. The coupling strengths λ and V_0 are taken as universal constants, i.e. independent of the quark masses.

As to the value of α_S, we should not forget that the presence in H of the term proportional to α_S relies on the use of perturbation theory at short distances. Since the mean distances that are relevant for bound states of different quark content are different, to optimize the convergence of the perturbation expansion I will take $\alpha_S(M)$ renormalized at the mass scale M, when describing a bound state of the same mass M. The reference value of α_S is given at the $\psi(3100)$ mass by the ratio $\Gamma(\psi \to e^+e^-)/\,\Gamma(\psi \to \text{hadrons})$ analysed as in Appelquist's lectures. Typical values are then : $4/3\ \alpha_S(c\bar{c}) = 0.27$, $4/3\ \alpha_S(\lambda\bar{\lambda}) = 0.36$, $4/3\ \alpha_S(d\bar{d}) = 0.42$.

6. MESON SPECTROSCOPY

The diagonalization of the Hamiltonian H is straightforwardly done numerically. The results are shown in table 1 for the S-wave spin-triplet (vector) and spin-singlet (pseudoscalar) mesons.

(*) See E. Eichten's talk in these Proceedings.

Table 1

	M_V	M_V^{exp}	M_P	M_P^{exp}	E_{RC}	ξ
cc	*3095	ψ(3095)	3041	η_c	-29	0.17
(cc)'	*3680	ψ'(3684)	3635	-	-94	0.19
(cc)"	4111	ψ"(4100)?	4061	-	-175	0.25
(cc)"'	4471	-	4435	-	-287	0.30
cλ	2105	F*	2034	F	-144	0.60
cd	1912	D*	1842	D	-301	0.90
λλ	1131	ϕ(1020)	959	η'(958)	-163	0.53
(λλ)'	1685	-	1541	E(1416)?	-517	0.70
λd	918	K*(892)	678	K(494)	-281	0.90
dd	716	ω(783) $\vec{\rho}$(770)	448	η(549) $\vec{\pi}$(140)	-370	0.90
(dd)'	1033	$\vec{\rho}$'(1250)?	755	-	-1100	1.30

The column denoted E_{RC} gives the relativistic corrections to the triplet states coming from the perturbative treatment of the potential V_2. The triplet-singlet splitting is produced by the spin-spin Fermi interaction contained in S_{BF}, which in the baryon case was responsible for the Σ-Λ mass splitting. In this table the ψ(3095) and ψ'(3684) masses are inputs. In fact, after the discussion of baryons and having taken the value of α_S from the analysis of the ψ-decay branching ratio, we are left in the Hamiltonian H with three free parameters : λ, V_0, m_c, which are fixed by the ψ,ψ' masees and by the leptonic decay width of the ψ (see below). The full set of parameters used in the calculation with their values is then : m_d : 336 MeV, m_λ = 540 MeV, m_c = 1.640 GeV, λ = 0.25 GeV^2, V_0 = -0.76 GeV, α_S(3100) = 0.20.

Let me make the following comments to table 1. The agreement with the experimental data for the vectors is quite good. The same cannot be said for the low-lying pseudoscalars, but this is not a un-understandable; relativistic corrections and mixings of quark content should be very important in this case (see below).

The recently reported[4)] μe events with ≥ 2 neutrals in e^+e^- annihilation can perhaps be interpreted in this scheme as the weak decay products of two charmed vector mesons (D*, F* in the usual notation (*). However, in order to have vector mesons stable against all but the weak interactions, we should not have their pseudoscalar partners (D,F) lying below them, as is the case in table 1. This may be a difficulty of our specific phenomenological analysis.

Without introducing any new input, the calculations can also be performed for the P- and D-wave bound states. The results are shown in table 2, where, for the D-waves, only the $c\bar{c}$ bound states are reported. Again, even if the experimental knowledge is quite uncertain in several cases, the agreement is acceptable. Note in particular the tentative insertion in this table of the very recently found[5] C-even objects, seen from the radiative decays of the ψ' particle, and interpreted here as P-wave charmonia.

Restricting to the states that are likely to be produced in an e^+e^--initiated reaction, the spectroscopic table of the $c\bar{c}$ bound states is almost filled up. We are waiting for the finding of a third P-wave state (2^{++}?) and for the pseudoscalar partners (η_c, η_c') of the ψ, ψ' particles.

As a last remark, note that the tensor term present in the Breit-Fermi potential introduces a mixing between the almost degenerate S- and D-wave 1^{--} charmonia of masses 3680 and 3778 MeV, respectively. From the computation[3], it turns out, however, that the mixing angle is very small $\theta = 2 \times 10^{-2}$, implying an essential decoupling of the D-wave in e^+e^- annihilation because of the vanishing at the origin of the L = 2 wave function.

7. POSSIBLE SOURCES OF CORRECTIONS TO THE SPECTROSCOPY

I would like to point out and briefly comment on three possible sources of corrections to the results that I have presented.

i) Importance of the relativistic corrections : if, in tables 1 and 2, one naively compares the size of E_{RC} with the over-all mass value, one concludes that the relativistic corrections, although non-negligible, are indeed relatively small except for the low-lying pseudoscalars. Let us look, however, in the last column of both tables, where the values are reported of the ratio ξ between the mean value of the $\vec{p}^4$ term present in V_2 and the mean non-relativistic kinetic energy ($\vec{p}^2/2\mu$). Since ξ^2 is a sort of relativistic expansion parameter, a better criterion for deciding whether the non-relativistic expansion is tenable or not is to see how small ξ is with respect to 1. And now the conclusion is that whenever a bound state contains a d-type quark, its description as a mainly non-relativistic system seems quite

(*) The heavy lepton pair production is also an appealing interpretation for these events.

Table 2

	M(0^{++})	M(1^{++})	M(1^{+-})	M(2^{++})	ξ
$(cc)_P$	3425 X(3440)?	3466 X(3530)?	3482 -	3503 -	0.11
$(c\lambda)_P$	2385 -	2456 -	2488 -	2528 -	0.49
$(\lambda\lambda)_P$	1326 -	1480 D(1285)	1540 -	1619 f'(1516)	0.45
$(\lambda d)_P$	1030 K(∿1300)?	1220 Q_A(1242)?	1297 Q_P(∿1300)?	1400 K**(1420)	0.70
$(dd)_P$	781 S^*(993)? $\vec{\delta}$(976)	1039 $\vec{A}_1$(1100)	1141 $\vec{B}$(1237)	1274 f(1270) $\vec{A}_2$(1310)	0.72
	M(1^{--})	M(2^{--})	M(2^{-+})	M(3^{--})	
$(cc)_D$	3778	3794	3798	3811	

inadequate. If we want to do better, given the underlying field theoretic framework, a Bethe-Salpeter equation treatment should be the only adequate one. And we would be immediately faced with at least serious technical, if not fundamental, difficulties : in brief "big relativistic corrections" is equivalent to "lack of dynamical information". The observation that gluonic states could be a non-negligible component for some of the known particles is perhaps relevant here.

(ii) The force due to strong th ee-meson couplings is completely neglected in our potential-model calculation. Considering, for example, the ψ particles, we expect in the real case for their wave functions an admixture of $c\bar{c}$ and $D\bar{D}$, $F\bar{F}$ components[6], with the pure quark-antiquark component presumably decreasing for the higher radial excitations. We could also say that the effective quark-antiquark potential should not go to infinity at large distances, but from some distance on it should show a Yukawa-like behaviour appropriate to meson-meson interaction[6]. To make these observations quantitative is difficult but not impossible (*).

(*) See E. Eichten's talk in these Proceedings.

iii) Finally, let us note that the admixture of quark content for the physical particles is neglected in the results of tables 1 and 2. And this admixture is known to be important at least for the low-lying pseudoscalars.

8. MESON DECAYS

The explicit knowledge of the meson wave functions enables us to compute some of the meson decays.

8.1 Leptonic decays of neutral vector mesons

From the diagram

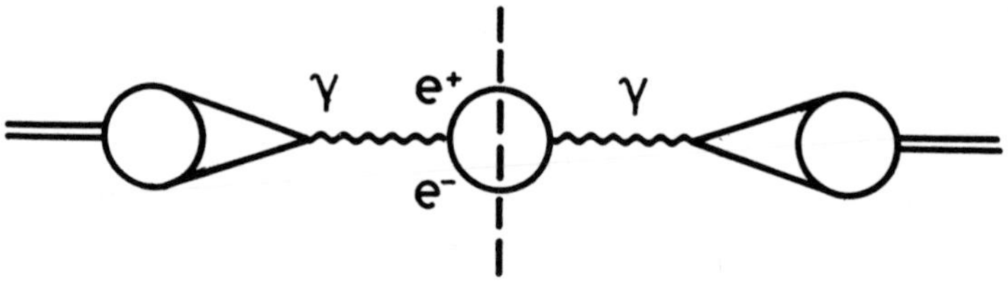

the well-known result is produced for the widths of these decays :

$$\Gamma(V \to e^+e^-) = e_V^2\, \alpha^2\, 16\pi \frac{|\psi_V(0)|^2}{m_V^2} \qquad (1)$$

where $e_V = (V|Q|V)$ is the mean value of the SU(4) charge operator over the state $|V>$. Including first-order corrections in the one-gluon exchange, one gets[7)]

$$|\psi_V(0)|^2 = |\psi_V^{NR}(0)|^2 \left(1 - \frac{16}{3\pi}\alpha_S(M_V)\right)$$

where ψ_V^{NR} are the non-relativistic wave functions obtained from the diagonalization of the potential V_1. Here again $\alpha_S(M_V)$ is changing with the mass of the vector meson. The results compared with the experimental values are

	ψ	ψ'	ψ"(4100)	φ	ρ	ω
Γ^{th}(keV)	5.2 *	3.1	2.7	2.1	10.6	1.2
Γ^{exp}	5.2	2.2	-	1.44	6.0	0.8

$\Gamma(\psi \to e^+e^-)$ is the input, as I have already stated. The relative value of the widths for the usual vector mesons (ρ,ω,ϕ) is a known successful prediction of SU(3). The introduction of the one-gluon correction in eq.(1) is then quite important for having a satisfactory description for the new mesons too. Note that despite the reduction of the widths owing to the α_s correction, the predicted values for the usual mesons are still higher than the experimental ones.

8.2 Weak decays of charmed mesons

In the specific GIM model for the weak interactions, we can go through the calculation of the leptonic weak decays quite similarly to the calculation of the electromagnetic widths. The relevant diagram here is the following one

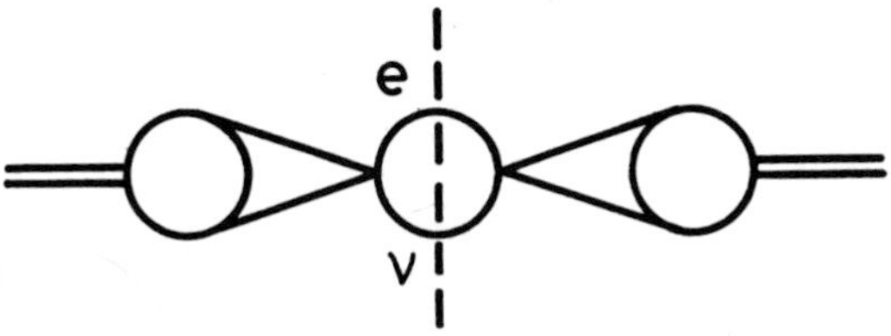

For a meson bound state of mass M, wave function (r), and total spin S = 0,1 we get (to leading order in m /M)

$$\Gamma(M \to \ell\nu) = \frac{3}{2\pi} G^2 \; |\psi_M^{NR}(0)|^2 \; [\frac{2M^2}{3} \; \frac{<\underline{S}^2>}{2} + m_\ell^2 \; \frac{<2-\underline{S}^2>}{2}]$$

where G should be read as $G_F \cos\theta_c$ for pn and cλ type states and $G_F \sin\theta_c$ for pλ and cn type bound states. The first term on the right)hand side, proportional to $<\underline{S}^2>$is contributing to the vector decays, whereas the second term, proportional to $<2-S^2>$, gives the pseudoscalar decays. Incidentally, this formula makes clear why the already mentioned μe events in e^+e^- annihilation could come, in the chain

$$e^+ + e^- \to M^+ + M^-$$
$$M^- \to \mu^- + \bar{\nu}_\mu \; (e^- + \bar{\nu}_e)$$
$$M^+ \to e^+ + \nu_e (\mu^+ + \nu_\mu)$$

from the decays of two charmed vector mesons, and not from pseudoscalar meson pair production : the electronic decays of the pseudoscalars are suppressed with respect to the muonic decays by the

$$\Gamma(D^{*+} \to \mu^+ + \nu_\mu) \simeq \Gamma(D^{*+} \to e^+ + \nu_e) \simeq 2 \times 10^{11}\ \mathrm{sec}^{-1}$$

$$\Gamma(F^{*+} \to \mu^+ + \nu_\mu) \simeq \Gamma(F^{*+} \to e^+ + \nu_e) \simeq 7 \times 10^{12}\ \mathrm{sec}^{-1}$$

$$\Gamma(D^{+} \to \mu^+ + \nu_\mu) \simeq 9 \times 10^{8}\ \mathrm{sec}^{-1}$$

$$\Gamma(F^{+} \to \mu^+ + \nu_\mu) \simeq 3 \times 10^{10}\ \mathrm{sec}^{-1}$$

The predictions for the pseudoscalars are roughly an order of magnitude higher than the analogous ones by Gaillard, Lee and Rosner[8], essentially based on extrapolation from K-leptonic decay. Indeed, in our calculation the decay width comes out an order of magnitude higher than the experimental value, as in the pion case. If, however, this well-known failure of the non-relativistic quark model for the low-lying pseudoscalars is simply interpreted as a reflection of the already observed inadequate description of these light particles in the model, the predictions for the heavier charmed particles should be considered on a sounder basis.

Coming to the semi-leptonic and non-leptonic decays -- by now a crucial problem for the charmed vector mesons -- our computational ability becomes lower and lower, since here the strong interactions effects can play a major role.

For the semi-leptonic decays of any charmed meson, Gaillard, Lee and Rosner give the following estimate :

$$\Gamma(M \to e\nu + \mathrm{had}) \simeq \Gamma(M \to \mu\nu + \mathrm{had}) \simeq \left(\frac{m_c}{m_\mu}\right)^5 \Gamma(\mu \to e\nu\bar{\nu}) \simeq 3\times10^{11}\mathrm{sec}$$

This is what we would expect if the charmed particle decays are viewed as occuring due to elementary quark processes, such as $c \to \lambda + \ell^+ + \nu$ or $c \to n + \ell^+ + \nu$, followed by quark decays with unit probability into stable hadrons.

A similar attitude can perhaps be taken when discussing non-leptonic decays. For processes due to charm quark decays, such as $c \to \lambda + p + \bar{n}$, one would have

$$\Gamma_1(M \to \mathrm{had}) \simeq 9 \times 10^{11}\ \mathrm{sec}^{-1}$$

In particular, I neglect here any asymptotic freedom enhancement factor, which for the heavy charmed mesons is suggested[9] not to be as important as for lighter strange particle decays. In addition to $c \to \lambda + p + \bar{n}$, a second kind of elementary process, such as $c + \bar{\lambda} \to p + \bar{n}$, can contribute to the non-leptonic decays of charmed mesons. This second kind of mechanism, which adds incoherently with the previous one to the total rate, is similar to the one leading to leptonic decays. Its contribution to the total rate is then particularly important for the vector decays (to the extent that the quark masses m_λ, m_d are small compared with the charm meson masses) and can be estimated as

$$\Gamma_2(F^{*+} \to \text{had}) \simeq 3 \cos^2\theta_c \; \Gamma(F^{*+} \to \mu\nu) \simeq 21 \times 10^{12}$$

$$\Gamma_2(D^{*o} \to \text{had}) \simeq 3 \cos^4\theta_c / \sin^2\theta_c \; \Gamma(D^{*+} \to \mu\nu) \simeq 9 \times 10^{12}$$

$$\Gamma_2(D^{*+} \to \text{had}) \simeq 3 \cos^2\theta_c \; \Gamma(D^{*+} \to \mu\nu) \simeq 6 \times 10^{11}$$

Summing up

$$\Gamma_{tot}(F^+) \simeq \Gamma_{tot}(D^+) \simeq \Gamma_{tot}(D^o) \simeq 10^{12}\text{sec}^{-1}$$

$$\Gamma_{tot}(F^{*+}) \simeq 3 \times 10^{13}\ \text{sec}^{-1};\ \Gamma_{tot}(D^{*+}) \simeq 2 \times 10^{12}\text{sec}^{-1};$$

$$\Gamma_{tot}(D^{*o}) \simeq 10^{13}\text{sec}^{-1}$$

Which of these two lines is going to be relevant for the experimental comparison depends obviously, as we discussed already, on the actual spectrum. As to the branching ratios, an obvious general conclusion is that, not including any enhancement factor of the charmed non-leptonic decays, makes the semi-leptonic decays (for pseudoscalars) and the leptonic decays (for vectors) an important factor of the total weak decays. And this may be quite consistent with recent experimental evidence[4,10].

8.3 Radiative decays of $c\bar{c}$ states

The importance of these decays for the newly found particles has been emphasized since the first suggestions of their interpretation as charmonia. And in fact the discovery of some of these decays has recently been reported[5]. As to their quantitative predictions, the use of our wave functions would not lead to any substantial difference with respect to the Cornell group results[11]. The problem then remains of understanding why the experimental values are definitely lower than these predictions.

The criticism to the calculation of the spectrum, especially in connection with the neglected meson-pair contamination of the $c\bar{c}$ wave function, applies here as well (*).

(*) See E. Eichten's talk in these Proceedings.

8.4 Hadronic decays and Zweig's rule

As we have heard in Appelquist's lectures, the asymptotic freedom mechanism is also likely to give an explanation of Zweig's rule for the hadronic decays. Indeed, the extrapolation of the lepton/hadron decay branching ratio from the ψ to the ϕ case according to the asymptotic freedom formulae is quite in agreement with the experimental values[7]. However, I think that I would feel better with the asymptotic freedom explanation of Zweig's rule if I would not be disturbed by the following problem, in order of importance : i) The smallness of the $\psi' \to \psi + 2\pi$ decay. (Although the quark diagram for this process is a "disconnected" one, it may be connected by relatively soft gluon lines.) ii) The apparent difficulty in seeing ψ' decays into normal hadrons. (The three-gluon decay process does not substantially distinguish the ψ from the ψ'). iii) The experimental limit on the η' width comparable to the $\phi \to 3\pi$ decay width. (One would argue that the two-gluon decay of the η', leading to pionic final states, should be bigger than the three-gluon ϕ-decay.) As to this last point, it is clear that a clean test of these ideas would be the finding of a definitely broader ($\gtrsim$ 1 MeV width) pseudoscalar partner of the ψ .

9. e^+e^- ANNIHILATION AT THE CHARM THRESHOLD

Let me very briefly address my attention finally to the e^+e^- annihilation total cross-section data in the region $\sqrt{s} \gtrsim 3.5$ GeV. A fit to the experimental curve can be made[3] for the total charmed hadron production cross-section by assuming that : i) this cross-section is mainly given by quasi-two-body charmed meson final states; ii) their form factors are dominated by the vector meson poles (ψ's excitations), and iii) the three meson vertices ($\psi \to DD$, $\psi' \to DD$, $\psi \to FF$, etc.) are given in the quark pair creation model by diagrams of the type

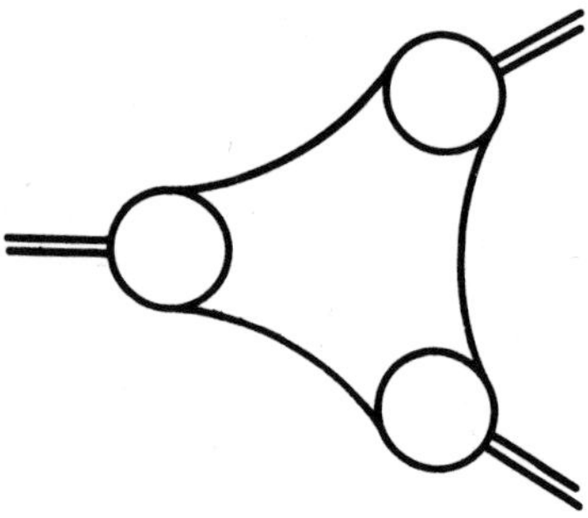

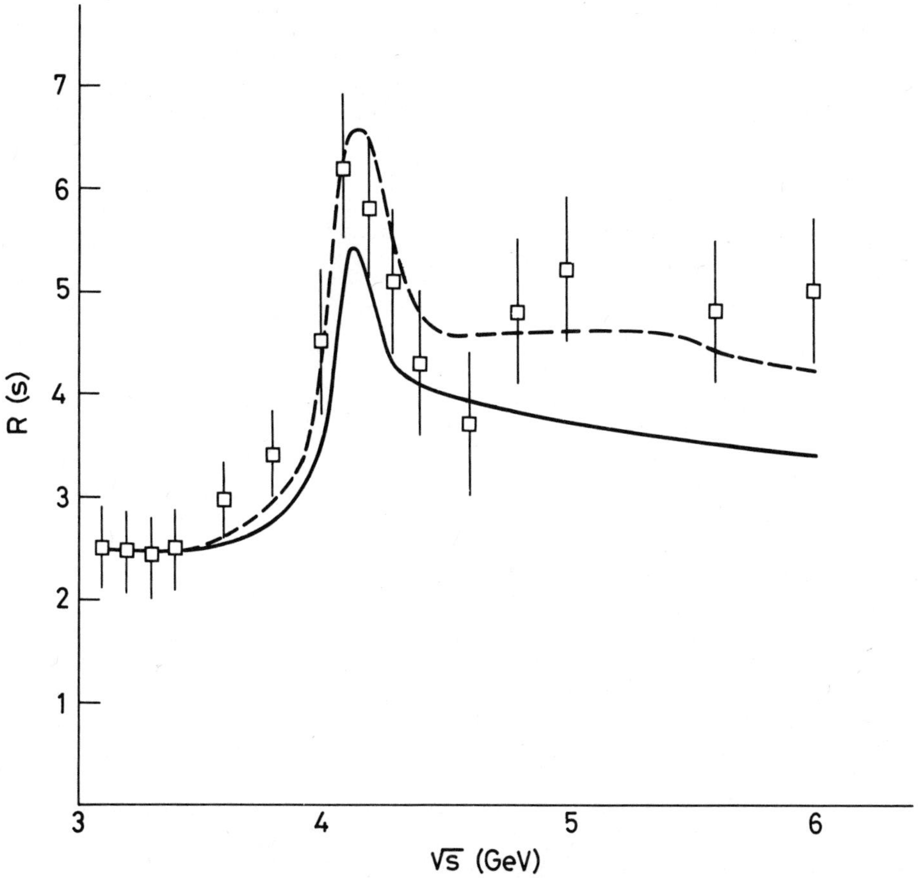

FIG. 1

The results are shown in fig.1 : the full line for no interference between the various ψ-poles, and the dotted line with the interference taken into account. In this last curve the bump at $\sqrt{s}$ = 4.1 GeV is accounted for despite the smallness of the predicted leptonic width (2.7 keV) of the ψ"(4100) resonance. Note also that the opening of the numerous thresholds keeps the value of R(s) substantially high even at $\sqrt{s}$ = 6 GeV. Despite this fact, the four-quark model is very probably getting into trouble with preliminary date from SLAC at higher values of $\sqrt{s}$ (up to 7.4 GeV), which show R essentially constant between 5 and 6[12]. As a matter of fact, a claim that even the data shown in fig. 1 call for new quarks (and/or for new heavy leptons) comes both from asymptotic freedom arguments (*) and from the application to e^+e^- annihilation of finite energy sum rules[13]. This is essentially because of the observation that, averaging R(s) over energy intervals of order of 1 GeV, we find remarkably constant values $\bar{R}$ around 5, whereas we would expect $3.3 \lesssim \bar{R} \lesssim 4$. This may support the existence of other quarks or of new heavy leptons.

REFERENCES

1. J.M. Cornwall and G. Tiktopoulos, Phys.Rev.Letters 35, 338(1975).
2. A. De Rujula, H. Georgi and S. Glashow, Hadron masses in a gauge theory, Harvard University preprint, 1975.
3. R. Barbieri, R. Kögerler, R. Kunszt and R. Gatto, CERN preprint ref. TH. 2036 (1975).
4. M.L. Perl et al., SLAC-PUB-1626 (1975).
5. W. Braunschweig et al., DESY preprint 75/20 (1975).
   G.J. Feldman et al., SLAC-PUB-LBL-4220 (1975).
6. J. Kogut and L. Susskind, the electron-positron annihilation cross-section near the new threshold (unpublished).
7. R. Barbieri, R. Kögerler, Z. Kunszt and R. Gatto, Meson hyperfine splittings and leptonic decays, to appear in Phys.Letters B
8. M.K. Gaillard, B.W. Lee and J.L. Rosner, Rev.Mod.Phys. 47, 277 (1975).
9. J. Ellis, M.K. Gaillard and P.V. Nanpoulos, CERN preprint, TH. 2030 (1975).
10. A.M. Boyarski et al., Phys.Rev.Letters 35, 195 (1975).
    J.J. Aubert et al., Phys.Rev.Letters 35, 416 (1975).
11. E. Eichten, K. Gottfried, T. Kinoshita, J. Kogut, K.D. Lane and T.-M. Yan, Phys.Rev.Letters 34, 369 (1975).
12. G.J. Feldman and M.L. Perl, Physics Reports C 19, 233 (1975).
13. V. Barger, W.F. Long and M.G. Olsson, University of Wisconsin-Madison preprint COO-452 (1975).

(*) See T. Appelquist's lectures in these Proceedings.

ASYMPTOTIC BEHAVIOUR IN QUANTUM FIELD THEORY

R.J. CREWTHER

CERN *

INTRODUCTION

The organizers have asked me to talk about the renormalization group and its applications. I shall concentrate on explaining the rules for short-distance behaviour in field theory and current

* The hospitality of the Aspen Center for Physics is gratefully acknowledged

algebra. Nothing will be said about applications to form factors, on-shell scattering amplitudes, charmonium, etc.[1]

The renormalization group originated in papers by Stueckelberg and Petermann[2] (SP) and Gell-Mann and Low[3] (GML) in 1953-54 and was revived, after a long quiescent period, by Wilson[4] (operator-product expansions and broken scale invariance) and Callan and Symanzik[5] (CS) (1968-70). Many people prefer the CS approach because there is no need to introduce unfamiliar renormalization prescriptions, but I think that it is unwise to completely ignore the original ideas. For example, a comparison of the GML and CS methods led to the use of mass-independent renormalization prescriptions (such as [6] dimensional renormalization) which generate "improved" CS equations[7]. Consequently, all of these methods will be reviewed. Applications will be restricted to the results of assuming asymptotic freedom [8] or broken scale invariance. I shall also report progress on an awkward technical problem, the properties of operator-product expansions for non-Abelian gauge theories.

I thank T. Applequist, J. Collins, O. Nachtmann, S. Rudaz, J. Zinn-Justin and B. Zumino for helpful comments.

I. RENORMALIZATION GROUP : CALLAN-SYMANZIK METHOD

Any renormalized amplitude A possesses a corresponding Callan-Symanzik (CS) equation[5]. The main ingredient in the derivation is knowing the mode of renormalization of A : is it multiplicative, subtractive, or something more complicated? So we begin with a mini review of the renormalization procedure itself [9].

Renormalization

Let Λ represent a cutoff mass which regulates ultraviolet divergences in Feynman diagrams. Typically, these divergences appear as powers and logarithms of Λ which blow up as Λ tends to infinity. Divergences are caused by loop integrations, so they can be isolated as divergences in one-particle-irreducible (1PI) (or "proper") subdiagrams*. Therefore, we consider the Feynman integral for an ℓ-loop 1PI amplitude with L external legs :

---

* A 1PI diagram is a connected graph wich cannot be separated into two disconnected pieces by cutting one of its internal lines. The corresponding 1PI amplitude is understood not to include propagators for external lines.

$$\Gamma(q_1,\ldots,q_{L-1}) = \int^{\Lambda} d^4p_1 \ldots d^4p_\ell \; I(p_1\ldots p_\ell;q_1\ldots q_{L-1}) \qquad (1.1)$$

The integrand I is a linear combination of products of vertices and internal propagators which depend on loop moments $p_1 \ldots p_\ell$ and external momenta $q_1 \ldots q_{L-1}$.

Convergence is tested by applying power-counting arguments[10,11]. Imagine that a subset $S = \{p'_1,\ldots,p'_m\}$ of the loop momenta $\{p_1,\ldots,p_\ell\}$ is scaled to infinity (all $p'_j = O(\eta)$ with $\eta \to \infty$) with all other momenta held fixed* . The integrand I develops a power-law behaviour in η for which the formula

$$|I(\eta)| \leqslant \eta^{c(S)} \cdot \{\text{constant}\} \qquad (\eta \to \infty)$$

provides an asymptotic bound. Here c(S) is the sum of individual powers contributed by each vertex and propagator depending on $p'_1,\ldots,p'_m$; (for example, there is a power -2 for each boson propagator involved). Thus each subset S can contribute cutoff-dependence

$$\int^{\Lambda} d^4p'_1 \ldots d^4p'_m \; I \leqslant \Lambda^{d(S)} \; \{\text{logs of } \Lambda\}$$

to Γ , where the power

$$d(S) = 4m(S) + c(S) \qquad (1.2)$$

is called the superficial degree of divergence of the subintegration over S. The integral (1.1) converges if d(S) is negative for <u>all</u> S.

Consider the special case in which the only divergence in Γ is that caused by all of the loop momenta $p_1,\ldots,p_\ell$ growing large together; in other words, Γ is superficially divergent ($d(\Gamma) \geqslant 0$) but internally convergent (e.g. $d(S) < 0$ for all $S \neq \{p_1,\ldots,p_\ell\}$). Suppose that Γ is differentiated with respect to one of the external momenta q_i. When $\partial/\partial q_i$ acts on the integrand I, the characteristic exponent $c(\Gamma)$ for its asymptotic behaviour as $p_1 \ldots p_\ell$ become large is reduced by at least 1, while no increase occurs in the other c(S). Therefore, if $\partial/\partial q_i$ is applied a sufficient number of times to Γ , the result is completely convergent. Since $\Gamma_{div.}$ (the divergent part of Γ) is annihilated by these differentiations, it must be a polynomial of degree $\leqslant d(\Gamma)$ in $q_1 \ldots q_{L-1}$. In diagram-

---

* More precisely, write $p'_j = \eta r_j$ and let η tend to ∞ , keeping all r_j, $p_i(\neq \text{any } p'_j)$ and q_i fixed and not allowing any partial sum $\sum' r$ of the r_j to vanish. The latter condition ensures that a p'-dependent propagator (which carries momentum $\eta\sum' r + \sum'_{p\neq p'} p + \sum' q$) has the expected asymptotic dependence on η . The special directions $\sum' r = 0$ are covered by looking at appropriate subsets S' of S.

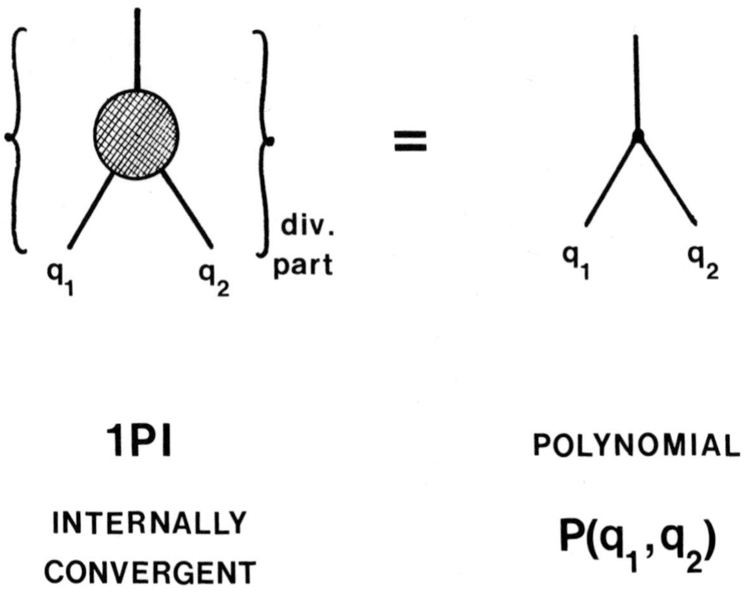

Fig. 1

matic language (fig. 1), this means that $\Gamma_{div.}$ is a local vertex produced by shrinking the blob Γ to a point.

An ℓ-loop divergence of this type is easily cancelled by including a counterterm $\Delta\mathcal{L}_\ell$ in the Lagrangian $\mathcal{L}$ with bare vertex equal to $-\Gamma_{div.}$. However, many 1PI graphs do not possess this property of being "primitively divergent"* . An arbitrary ℓ-loop 1PI graph may contain divergent ℓ'-loop 1PI subgraphs ($\ell' < \ell$) which must first be made convergent by including suitable counterterms

---

* A primitively divergent graph becomes convergent if any internal line is cut.

$\Delta\mathcal{L}_\ell$, in $\mathcal{L}$; i.e., it will be necessary to include internal counterterm vertices in the blob in fig.1. So we try the following prescription [10,12] :

(i) Start with $\mathcal{L}_0$, a Lagrangian from which propagators and vertices can be constructed.

(ii) Construct counterterms $\Delta\mathcal{L}_1$ which remove all divergences in 1-loop 1PI graphs generated by $\mathcal{L}_0$.

(iii) Use a new Lagrangian $\mathcal{L}_1 = \mathcal{L}_0 + \Delta\mathcal{L}_1$, to generate 2-loop graphs and construct $\Delta\mathcal{L}_2$ to get rid of the resulting 1PI divergences... and so on, to any finite number of loops and vertices [13] :

$$\mathcal{L}_\ell = \mathcal{L}_{\ell-1} + \Delta\mathcal{L}_\ell \quad .$$

Obviously, we obtain finite results for diagrams in which, for each pair of divergent 1PI subgraphs, one subgraph is entirely contained within the other ("nested" divergences) or the two subgraphs are disjoint. The difficult step in the proof [12] of convergence is to disentangle overlapping divergences (sets of divergent 1PI graphs which are neither disjoint nor nested). The result, valid for all polynomial interactions (renormalizable or otherwise), is that the above procedure renders all subintegrations convergent by powercounting* . In other words, the Lagrangian

$$\mathcal{L} = \mathcal{L}_\infty = \mathcal{L}_0 + \Delta\mathcal{L} \tag{1.3}$$

with counterterm Lagrangian $\Delta\mathcal{L}$ given by

$$\Delta\mathcal{L} = \Delta\mathcal{L}_1 + \Delta\mathcal{L}_2 + \ldots + \Delta\mathcal{L}_\ell + \ldots$$

generates cutoff-independent perturbative amplitudes.

The form of $\Delta\mathcal{L}$ is restricted by the fact that the degree of the polynomial $\Gamma_{div.}(q_1 \ldots q_{L-1})$ in fig. 1 cannot be greater than $d(\Gamma)$. The set of allowed vertices is easily determined by counting mass dimensionalities. Each term in the Lagrangian has dimensionality 4, in four space-time dimensions. Boson and fermion fields φ and ψ have dimensionalities 1 and 3/2 corresponding to $\overline{\varphi\varphi} \sim p^{-2}$ and $\overline{\psi\bar{\psi}} \sim p^{-1}$ for the propagators at large momentum p. (Higherspin fields obey this rule if special circumstances, such as gauge invariance, ensure that longitudinal terms like $iq_\mu q_\nu/q^2m^2$, in the spin-1 propagator $\overline{B_\mu B_\nu}$ (m = mass) are absent (m = 0) or cancelled off). These rules fix the effective dimensionality dim(g)

* For the general case, the involved combinatories of Hepp's proof[12] seem to be unavoidable. Authors of textbooks often choose to discuss a specific theory such as quantum electrodynamics, where skeleton expansions and Ward identities permit some simplification.

of a coupling constant g; e.g., for the vertex $g\varphi^6$, $\dim(\varphi^6) = 6$ implies $\dim(g) = -2$. A typical term in $\Delta\mathcal{L}$ is

$$\Delta\mathcal{L} \propto \Lambda^j (\ln\Lambda)^p m^r g^N O_d(\varphi,\psi,\partial) \tag{1.4}$$

for integer powers $j, p, r \geq 0$ and $N > 0$, where the dimensionality d of the operator O_d is given by

$$4 = j + r + N \dim(g) + d$$

$$\text{i.e. } d \leq 4 - N \dim(g) \quad . \tag{1.5}$$

If dim(g) is negative, d is unbounded as the order of perturbation N increases. The inevitable result is a non-renormalizable theory in which the number of distinct vertices (and hence new coupling constants) is unbounded. Apart from the lack of predictive power of these theories[14], the prescription (1.3) implies a quantization procedure for high-derivative counterterms which is not obviously unitary. So we shall ensure renormalizability ($d \leq 4$) by requiring

$$\dim(g) \geq 0 \tag{1.6}$$

but frequently permit external composite operators[13,15] Q to couple to $\mathcal{L}$ via sources χ (to a fixed order in χ) :

$$\mathcal{L} \rightarrow \mathcal{L}(\chi) = \mathcal{L} + \chi(Q+\Delta Q) + O(\chi^2) \tag{1.7}$$

Given the constraint (1.6), the rule for renormalization counterterms ΔQ generated by 1PI diagrams with a single Q-insertion is simply

$$\dim(\Delta Q) \leq \dim(Q) \tag{1.8}$$

For example, consider

$$\mathcal{L}_0 = \frac{1}{2}(\partial\varphi)^2 - \frac{1}{2}m^2\varphi^2 - g\varphi^4 \tag{1.9}$$

as a trial Lagrangian. Only the two- and four-legged 1PI amplitudes $\Gamma^{(2)}$, $\Gamma^{(4)}$ are superficially divergent. According to (1.4), we have $d \leq 4$, so the only possibilities are counterterms proportional to $(\partial\varphi)^2$ and φ^2 (from $\Gamma^{(2)}_{\text{div.}}$) and φ^4 (from $\Gamma^{(4)}_{\text{div.}}$). Therefore, the most general expression for the complete Lagrangian is

$$\mathcal{L} = \frac{1}{2}Z_3(\partial\varphi)^2 - \frac{1}{2}(m^2 - \delta m^2)\varphi^2 - gZ_1\varphi^4 \tag{1.10}$$

where Z_1, Z_3 and δm^2 are Λ-dependent. Note that there can be vertices in $\mathcal{L}$ which do not appear in $\mathcal{L}_0$. For example, the Yukawa trial Lagrangian

$$\mathcal{L}_o = \frac{1}{2}(\partial\varphi)^2 - \frac{1}{2}m^2\varphi^2 + \bar{\psi}(i\partial\!\!\!/ - M)\psi + g'\bar{\psi}\gamma_5\psi\varphi \qquad (1.11)$$

produces an additional φ^4 vertex in $\mathcal{L}$ because of divergences in four-point 1PI boson amplitudes $\Gamma^{(4)}$. The result is a theory with two coupling constants, g and g':

$$\mathcal{L} = \frac{1}{2}Z_3(\partial\varphi)^2 - \frac{1}{2}(m^2-\delta m^2)\varphi^2 - gZ_1\varphi^4$$

$$+ Z_2\,\bar{\psi}\,i\partial\!\!\!/\psi - (M-\delta M)\bar{\psi}\psi + g'Z_1'\,\bar{\psi}\gamma_5\psi\varphi \quad . \qquad (1.12)$$

The Lagrangian $\mathcal{L}$ (Eq. 1.10) or (1.12)) produces finite 1PI amplitudes $\Gamma^{(L)}$ and connected Green's functions $G^{(L)}$ (with L legs). We have no further use for $\mathcal{L}_o$; it is just a crutch which helps us to arrive at a suitable $\mathcal{L}$.

Sometimes it is convenient to introduce a new set of Feynman rules (labelled with a B for "bare" or "unrenormalized") for the same $\mathcal{L}$ in terms of new quantities φ_B, m_B^2, g_B,... . The substitution

$$\begin{aligned} \varphi_B &= Z_3^{1/2}\varphi \\ m_B^2 &= Z_3^{-1}(m^2-\delta m^2) \\ g_B &= Z_3^{-2}Z_1 g \end{aligned} \qquad (1.13)$$

in eq.(1.10) produces a simple set of B rules:

$$\mathcal{L} = \frac{1}{2}(\partial\varphi_B)^2 - \frac{1}{2}m_B^2\,\varphi_B^2 - g_B\,\varphi_B^4 \qquad (1.14)$$

Let $G_B^{(L)}$ be the L-point Green's function computed using the B rules. Numerically, the only difference between $G_B^{(L)}$ and $G^{(L)}$ is the normalization of each external line - internally, everything cancels when propagators and vertices are changed according to (1.13), because the Lagrangian $\mathcal{L}$ remains the same. So we get:

$$G^{(L)}(q_1\ldots q_{L-1};g,m) = Z_3^{-L/2}\,G_B^{(L)}(q_1\ldots q_{L-1};g_B,m_B,\Lambda) \quad . \qquad (1.15)$$

Therefore, if some cutoff dependence is suitably absorbed in coupling constant and mass renormalization, the remaining divergence of an unrenormalized Green's function can be segregated as a multiplicative constant[10] $Z_3^{L/2}$. There is a wave-function renormalization $Z_3^{-1/2}$ for each external boson line, and a similar factor $Z_2^{-1/2}$ for each fermion (if eq.(3.12) is being considered) with

$$g' = (Z_1')^{-1} Z_2 Z_3^{1/2} g_B' \qquad (1.16)$$

Renormalization group

Prescriptions like (1.2), (1.10) and (1.12) still contain a Λ-independent ambiguity, because all we have done so far is to require that divergent parts of 1PI graphs be cancelled. The corresponding finite parts have to be fixed by imposing a renormalization prescription (call it R), and quantities such as $\delta m^2, \varphi, \ldots$ must be regarded as R-dependent : $\delta m_R^2, \varphi_R, g_R, Z_1(R), \Gamma_R^{(L)}, G_R^{(L)}$, and so on. In order to completely specify each distinct vertex appearing in $\Delta\mathcal{L}$, we must impose a normalization condition on the corresponding 1PI amplitude. The label R refers to this set of conditions. The points in momentum space to which the conditions refer are called subtraction points.

A popular choice for R is to normalize on-shell. For example, boson mass and wave-function normalizations are fixed by choosing

$$\Gamma_R^{(2)}(q) = 0, \ \partial\Gamma_R^{(2)}/\partial q^2 = 0, \quad (q^2 = m_R^2)$$

for the self-energy $\Gamma_R^{(2)}$, so that the dressed boson propagator

$$G_R^{(2)}(q) = i/(q^2 - m_R^2 - i\,\Gamma_R^{(2)}(q)) \qquad (1.17)$$

has a pole at $q^2 = m_R^2$ with conventional residue. Another example is the charge of the electron $\sqrt{4\pi\alpha}$ (with $\alpha^{-1} = 137.036\ldots$) in quantum electrodynamics (QED), which refers to the amplitude for an on-shell electron to absorb a zero-frequency photon.

Alternatively, R can be specified by intermediate renormalization[13,16] in which the subtraction point is at the origin of momentum space; e.g. R for eq.(1.10) is fixed by

$$\Gamma_R^{(2)}(q) = -i\,m_R^2\,, \ \partial\Gamma_R^{(2)}/\partial q^2 = i\,, \quad (q = 0)$$

$$\Gamma_R^{(4)}(q_1 = q_2 = q_3 = 0) = -\,i g_R\, 4! \qquad (1.18)$$

It does not matter that g_R and m_R^2 are not directly measurable; they parametrize the theory as effectively as on-shell variables.

More generally, R can be a function of several fixed "reference momenta" $\lambda_1, \lambda_2, \ldots$ which parametrize the subtraction points : consider the prescription $R = R(\lambda_1 \ldots \lambda_5)$ for eq.(1.10) :

$$\Gamma_R^{(2)}(\lambda_1) = -i\, m_R^2\ , \quad \partial\Gamma_R^{(2)}/\partial q^2\Big|_{q=\lambda_2} = i\ ,$$

$$\Gamma_R^{(4)}(\lambda_3,\lambda_4,\lambda_5) = -i\, g_R\, 4! \qquad (1.19)$$

Note :

(i) To avoid trouble with unitarity, we want g_R and m_R^2 to be real. However, for arbitrary λ in eq. (1.19), $\Gamma^{(2)}$ and $\Gamma^{(4)}$ may have absorptive parts. In that case, the normalization conditions (1.19) apply only to the real parts of the 1PI amplitudes at non-singular points λ. Additional conditions for absorptive parts are not needed : a 1PI absorptive part converges superficially because it is obtained by cutting internal lines of the full 1PI amplitude. The λ-momenta should be zero or spacelike if a continuation of Feynman amplitudes to Euclidean space is being attempted.

(ii) Suppose we use (1.19) to specify $G^{(L)}$ and Z_3 in eq.(1.15); then they depend on λ as well as the variables shown explicitly. Of course, the B rules make no reference to λ, so $G_B^{(L)}$, when written as a function of m_B and g_B, does not depend on λ.

(iii) Do not confuse the momenta $\lambda_3,\lambda_4,\lambda_5$ at which g_R is fixed with the momenta entering and leaving a vertex inside a Feynman diagram !

Now suppose that two theorists are thinking about the same physical situation (i.e. they use the same $\mathcal{L}$) but use different renormalization procedures, R and R' :

$$\mathcal{L} = \mathcal{L}_R(\text{R-quantities}) = \mathcal{L}_{R'}(\text{R'-quantities}).$$

Transformation equations relating R and R' are easily found by referring back to the B rules, which do not depend on R or R'. The equations

$$\varphi_R = Z_3(R)^{-1/2}\varphi_B\ , \quad \varphi_{R'} = Z_3(R')^{-1/2}\varphi_B\ , \text{ etc.,}$$

imply (e.g. for eq. (1.12))

$$\varphi_{R'} = \zeta_3(R',R)^{-1/2}\varphi_R$$

$$\psi_{R'} = \zeta_2(R',R)^{-1/2}\psi_R$$

$$g_{R'} = \zeta_1(R',R)^{-1}\,\zeta_3(R',R)^2 g_R$$

$$g'_{R'} = \zeta'_1(R',R)^{-1}\,\zeta_2(R',R)\,\zeta_3(R',R)^{1/2} g'_R$$

$$m^2_{R'} = \zeta_3(R',R)\, m^2_R + \delta(R',R)$$

$$M_{R'} = \zeta_2(R',R) M_R + \Delta(R',R) \qquad (1.20)$$

and

$$G_{R'}^{(b,f)} = \zeta_3(R',R)^{-b/2}\zeta_2(R',R)^{-f/2} G_R^{(b,f)} \quad ,$$

$$\text{(b bosons, f fermions)} \qquad (1.21)$$

where the transformation coefficients ζ_i, δ, Δ, given by

$$\zeta_i(R',R) = Z_i(R')/Z_i(R) \qquad \text{(ditto for } \zeta'_1\text{)}$$

$$\delta(R',R) = \delta m_{R'}^2 - \zeta_3(R',R)\delta m_R^2$$

$$\Delta(R',R) = \delta M_{R'} - \zeta_2(R',R)\delta M_R \qquad (1.22)$$

are all cutoff-independent. The ambiguity mentioned at the beginning of this subsection is now completely characterized in terms of these coefficients.

The set $\mathcal{G}$ of all transformations $R \to R'$ is called[2] the "Renormalization Group". Measurable quantities $\mathcal{M}$ like decay rates and cross-sections are renormalization-group invariant : $\mathcal{M}(R') = \mathcal{M}(R)$ In other words, $\mathcal{M}$ should not be theorist-dependent.

The group property is partially realised in the existence of an identity transformation $R \to R$, an inverse $R' \to R$ to $R \to R'$, and a product $R \to R''$ equivalent to successive transformations $R \to R'$ and $R' \to R''$, with

$$\zeta_i(R,R) = 1$$

$$\zeta_i(R,R') = \zeta_i(R',R)^{-1}$$

$$\zeta_i(R'',R) = \zeta_i(R'',R')\zeta_i(R',R), \text{ etc.}$$

However, we do not have a group at this stage[17], because there is no rule for multiplying two arbitrary transformations $R \to R'$ and $R'' \to R'''$. A satisfactory rule can be obtained for special subsets of prescriptions which can be explicitly parametrized : $R = R(\lambda)$ (e.g. $R = R(\lambda_1 \ldots \lambda_5)$ discussed above). Then it is sufficient to consider transformations in λ-space which possess the group property : inversions, translations, rotations, scale transformations, etc. So the group property is certainly a feature of many subsets of $\mathcal{G}$, but it is not clear whether $\mathcal{G}$ itself can be characterized as a group or not. (Henceforth, I shall forget about this fine distinction).

The idea of SP[2] was to look at sets of prescriptions R depending on continuous parameters λ from which a Lie group with associated Lie algebras can be obtained, and find equations giving the result of performing an infinitesimal λ-transformation. The only subgroup of any practical importance seems to be the trivial case U(1) for scale transformations of the subtraction points λ (with generator $\lambda \cdot \partial/\partial\lambda$). Non-Abelian cases are easy to construct, but the results appear to be uninteresting.

Eq. (1.21) means that $G^{(b,f)}$ transforms multiplicatively under the group. The 1PI amplitudes transform similarly (with $\zeta^{1/2}$ for each line instead of $\zeta^{-1/2}$)

$$\Gamma_{R'}^{(b,f)} = \zeta_3(R',R)^{b/2}\ \zeta_2(R',R)^{f/2}\ \Gamma_R^{(b,f)}$$

$$((b,f) \neq (2,0),(0,2)) \qquad (1.23)$$

except for self-energy amplitudes, where there is an additional subtractive renormalization: eqs. (1.17) (valid irrespective of R) and (1.21) imply

$$\Gamma_{R'}^{(2,0)} = \zeta_3(R',R)\ \Gamma_R^{(2,0)}(q) + i(\zeta_3(R',R) - 1)q^2 + i\delta(R',R) \qquad (1.24)$$

for the boson self-energy ($\Gamma^{(2)} = \Gamma^{(2,0)}$), and there is an analogous equation for fermions :

$$\Gamma_{R'}^{(0,2)} = \zeta_2(R',R)\Gamma_R^{(0,2)}(q) + i(\zeta_2(R',R) - 1)\not{q} + i\Delta(R',R) \qquad (1.25)$$

Callan Symanzik Equation

The CS equation[5] was the product of a search for the Ward identity of scale transformations in perturbation theory[18]. The generator of scale transformations[19], $D(x_o) = \int d^3x\ x^\mu \theta_{\mu o}(x)$, has a time-variation given by the trace of the energy momentum tensor $\theta_{\mu\nu}$. In free-field theory, θ^μ_μ is the sum of mass terms

$$\theta^\mu_\mu = m^2\varphi^2 + M\,\bar{\psi}\,\psi\,. \qquad (1.26)$$

To obtain something which looks like a scaling Ward identity for an amplitude A, we need the corresponding amplitude A(Δ) with an extra operator insertion Δ where, for free-field theory, Δ reduces to the zero-momentum mass operator $\int d^4x\ \theta^{\mu}_{\mu}(x)$.

Consider the L-leg boson Green's function $G^{(L)}$ (connected), which is multiplicatively renormalized (eqs. (1.15), (1.21)), and assume for convenience that there is only one mass m and dimensionless coupling constant g. If we use the B rules (1.14) to compute[20] $G_B^{(L)}$, the only place in which m_B appears is in undressed propagators $i/(p^2-m_B^2)$. Hence the identity

$$m_B \frac{\partial}{\partial m_B} \frac{i}{p^2 - m_B^2} = \frac{i}{p^2 - m_B^2} (-2\ i\ m_B^2) \frac{i}{p^2 - m_B^2}$$

implies that the operation $m_B\partial/\partial m_B$ (with external momenta and g_B, Λ fixed) is equivalent to the insertion of a new vertex $-i\ m_B^2\ \varphi_B^2$ at zero-momentum transfer (i.e., $-i\Delta$). So we define

$$G_B^{(L)}(\Delta) = i\ m_B \frac{\partial}{\partial m_B}\ G_B^{(L)}(q;\ g_B,\ m_B,\Lambda)\ , \qquad (1.27)$$

where q is shorthand for $(q_1 .. q_{L-1})$. Of course, $G_B^{(L)}(\Delta)$ is cutoff-dependent, but its renormalization is very simple. If the operator Q in (1.7) is chosen to be φ^2, the only counterterm operator ΔQ which can satisfy (1.8) is again φ^2 (φ does not appear because of $\mathcal{L}(\varphi) = \mathcal{L}(-\varphi)$) :

$$\mathcal{L}(\chi) = \mathcal{L}(0) + \chi Z_{\Delta}(\Lambda;R)\varphi_R^2 + \chi^2 Z_{\Delta\Delta} \quad ; \qquad (1.7')$$

i.e. the composite operator $\varphi^2(x)$ is multiplicatively renormalized. Hence there is a wave-function renormalization Z_{Δ}^{-1} for the external line represented by the source χ in addition to factors $Z_3^{-1/2}$ for each boson line :

$$G_R^{(L)}(\Delta) = Z_{\Delta}^{-1}\ Z_3^{-L/2}\ G_B^{(L)}(\Delta)\ . \qquad (1.28)$$

It is convenient (but not essential) to use the intermediate renormalization prescription (1.18), because then we do not have to worry about additional dependence on reference momenta λ. In order to fix the finite ambiguity in Z_{Δ}, we have to choose a normalization condition for $\Gamma^{(2)}(\Delta)$, the 1PI amplitude with two boson legs and a Δ insertion :

$$\Gamma_R^{(2)}(\Delta; q = 0) = 2\ m_R^2\ . \qquad (1.18')$$

Note the relation

$$G_R^{(2)}(\Delta;q) = (G_R^{(2)}(q))^2\ \Gamma_R^{(2)}(\Delta;q)\ . \qquad (1.29)$$

Now we construct two identities, each the result of applying $m_R d/dm_R$ (fixed q, g_B, Λ) to $G_R^{(L)}$ and changing variables :

$$m_R \left.\frac{d}{dm_R}\right|_{q,g_B,\Lambda} G_R^{(L)} = \left[m_R \frac{\partial}{\partial m_R} + m_R \left.\frac{dg_R}{dm_R}\right|_{g_B,\Lambda} \frac{\partial}{\partial g_R} \right] G_R^{(L)}(q;g_R m_R)$$

$$= -\frac{1}{2} L\, Z_3^{-1} \; G_R^{(L)} \; m_R(\partial/\partial m_R) Z_3\, (\Lambda/m_R, g_B)$$

$$+ Z_3^{-L/2}\, (m_R/m_B)(dm_B/dm_R)_{g_B,\Lambda} m_B\, \partial/\partial m_B\; G_B^{(L)}\, (q;g_B,m_B,\Lambda)\ .$$

The second identity depends on eq.(1.15). The result of combining these identities with (1.28) is the CS equation[5,21]

$$\left[m_R \frac{\partial}{\partial m_R} + \beta_R \frac{\partial}{\partial g_R} + L\, \gamma_R \right] G_R^{(L)}\, (q;\ g_R, m_R)$$

$$= -i\delta_R\, G_R^{(L)}(\Delta;q;g_R,m_R) \tag{1.30}$$

with

$$\beta_R = m_R\, (dg_R/dm_R)_{g_B,\,\Lambda}\ ,$$

$$\gamma_R = -m_R\, \partial/\partial m_R \;\ln\{ Z_3\, (\Lambda/m_R, g_B)^{-1/2} \}$$

$$\delta_R = Z_\Delta (m_R/m_B)(dm_B/dm_R)_{g_B,\,\Lambda} \tag{1.31}$$

The functions γ, δ are cutoff-independent, because we can apply the conditions

$$G_R^{(2)}\, (q=0) = -i/m_R^2$$

$$(\partial/\partial q^2) G_R^{(2)}\, (q=0) = -i/m_R^4$$

$$G_R^{(2)}\, (\Delta;q=0) = -2/m_R^2 \tag{1.32}$$

(implied by eqs. (1.17), (1.18), (1.18'), (1.29)) to eq. (1.30) with L = 2 :

$$\delta_R = 1 - \gamma_R ,$$

$$\gamma_R = 2 + \frac{1}{2} m_R^{\ 4} \ \delta_R (\partial/\partial q^2) G_R^{(L)} (\Delta ; q=0) . \qquad (1.33)$$

It follows from (1.30) that β is also cutoff-independent. Hence the dimensionless functions β,γ,δ can only depend on g_R, in this renormalization prescription. The formulas

$$\beta_R(g_R) = -g_R \Lambda \partial/\partial\Lambda \ln \ Z_g(\Lambda/m_R, g_B) , \qquad (g_R = Z_g \ g_B)$$

$$\gamma_R(g_R) = \Lambda\partial/\partial\Lambda \ln \ \{Z_3 \ (\Lambda/m_R, g_B)\}^{-1/2} \qquad (\Lambda \to \infty) \qquad (1.34)$$

(from (1.31)) show the connection between β and γ and infinities in coupling-constant and wave-function renormalization.

Let us introduce a parameter η which scales all the momenta together : $q \to \eta q$. Since $G_R^{(L)}$ has mass dimensionality 4 - 3L, purely dimensional arguments imply the identity

$$[\eta\partial/\partial\eta + m_R \ \partial/\partial m_R + 3L-4] \ G_R^{(L)} \ (\eta q; \ g_R, m_R) = 0 \qquad (1.35)$$

so the CS equation can be cast in a form which shows the connection with scale transformations :

$$[\eta\partial/\partial\eta - \beta_R(g_R) \ \partial/\partial g_R + 3L - 4 - L \ \gamma_R(g_R)] \quad G_R^{(L)}(\eta q; g_R, m_R)$$

$$= -i \ \delta_R(g_R) \ G_R^{(L)} \ (\Delta; \ \eta q; \ g_R, m_R) \qquad (1.36)$$

The term $\beta_R \ \partial/\partial g_R$, caused by coupling-constant renormalization, means that $\int d^4x\theta^\mu_\mu(x)$ effectively picks up a dimension 4 vertex[5,22] $-i \ \beta_R \int d^4x\varphi^4(x)$.

If there are several coupling constants $g_R(i)$ (i= 1, ...,j), the $\beta_R \ \partial/\partial g_R$ term becomes $\sum_{i=1}^{j} \beta_R^{(i)} \ \partial/\partial g_R(i)$; i.e. there is a β-function $\beta_R^{(i)}$ for each $g_R(i)$, and it depends on all of the $g_R(i)$'s :

$$\vec{g}_R = (g_R(1), \ldots, g_R(j)) , \quad \vec{\beta}_R = (\beta_R^{(1)} \ \ldots \ \beta_R^{(j)} \ );$$

$$\sum_{i=1}^{j} \beta_R^{(i)} \ \partial/\partial g_R(i) \equiv \vec{\beta}_R \cdot \partial/\partial \vec{g}_R \quad ;$$

../..

$$\vec{\beta}_R = \vec{\beta}_R(\vec{g}_R) \quad ; \qquad \gamma_R = \gamma_R(\vec{g}_R) \quad . \tag{1.37}$$

If there are also several masses $m_R(k)$, each mass can generate its own CS equation, β-functions $\beta_R^{(i,k)}$, and γ-function $\gamma_R^{(k)}$. Occasionally, these CS equations are separately useful [23], but in most cases, it is better to sum them because then $\vec{m}_R \cdot \partial/\partial\vec{m}_R$ can be converted to $\eta\partial/\partial\eta$ using the generalization of (1.35).

Subscripts R have been attached to the symbols β, γ, δ in eq. (1.36) to indicate that their functional forms depend on the renormalization prescription. To see this, consider a prescription R' (such as on-shell normalization) in which $g_{R'}, \zeta_3(R',R)$ and $m_{R'}/m_R$ depend on g_R alone; i.e., we avoid introducing dependence on a separate dimensionful quantity λ. Then a change of variables (g_R, m_R) to $(g_{R'}, m_{R'})$ in eq. (1.30) produces a CS equation of the same form, with R' replacing R and

$$\beta_{R'}(g_{R'}) = \beta_R \, dg_{R'}/dg_R \; / \; [1 + \beta_R \, d/dg_R \, \ln(m_{R'}/m_R)] \; ,$$

$$\gamma_{R'}(g_{R'}) = [\gamma_R + \zeta_3^{-1} \beta_R \, dg_{R'}/dg_R] / [1 + \beta_R \, d/dg_R \, \ln(m_{R'}/m_R)] \; . \tag{1.38}$$

Observe that β, γ are R-independent in the 1-loop approximation.

The main complication for $R = R(\lambda)$ is that $m_R\partial/\partial m_R$ in eq.(1.30) has to be replaced by $\{m_R \, \partial/\partial m_R + \lambda \cdot \partial/\partial\lambda\}$ so that we still get the infinitesimal scale transformation $\eta\partial/\partial\eta$ in (1.36). In general, the β and γ functions depend on dimensionless ratios such as m_R^2/λ^2, λ_1^2/λ_2^2, ..., as well as on g_R.

The generalization to amplitudes $G^{(b,f)}_{ABC\ldots}$ involving several composite operators A,B,C... can be readily carried through. The analogue of eq. (1.7) is

$$\begin{aligned} \mathcal{L}(\chi) = \mathcal{L}(0) &+ \chi_A(A+\Delta A) + \chi_B(B+\Delta B) + \ldots \\ &+ \chi_A^2 \, \Delta(AA) + \chi_A\chi_B\Delta(AB) + \chi_A\chi_C\Delta(AC) + \ldots + \chi_B^2\Delta(BB) + \ldots \\ &+ \chi_A^3 \, \Delta(AAA) + \ldots \\ &+ \ldots\ldots \end{aligned} \tag{1.39}$$

where $\Delta(ABC\ldots)$ is a local vertex with cutoff-dependent normalization.

Consider the terms linear in χ . For general A, there may be many counterterms ΔA with dimensionality less than or equal to that of A (see eq. (1.8)). In other words, the Z-factors have to be treated as matrices. However, by adding suitable linear combinations of the counterterm vertices $\Delta A, \Delta B$, ..., we can always redefine A, B, C... such that they are multiplicatively renormalized :

$$A + \Delta A = Z_A(\Lambda)A, \quad B + \Delta B = Z_B(\Lambda)B, \ \ldots \tag{1.40}$$

Eq.(1.39) also contains counterterms $O(\chi^p)$ $(p > 1)$ produced by superficially divergent 1PI subdiagrams with p composite-operator insertions. However, these terms are very easy to isolate. For example, if A(x) and B(y) (x,y = coordinates) couple to the same 1PI diagram, the induced counterterm must be proportional to $P(\partial_x)\delta^4(x-y)$, where P is a polynomial; this is so because a 1PI diagram (made convergent internally) shrinks to a point when the divergent part is taken (fig. 1). In momentum space, we have $P(-iq_{AB})$, where q_{AB} is the momentum transferred from A to B. So the mode of renormalization is given by

$$\{G^{(b,f)}_{ABC\ldots}\}_R = Z_A^{-1}\, Z_B^{-1}\, Z_C^{-1} \ldots Z_3^{-b/2}\, Z_2^{-f/2}\, \{G^{(b,f)}_{ABC\ldots}\}_{Bare}$$

$$+ \sum_i P_i\,(q_{subset\ i})\mathcal{A}_i \tag{1.41}$$

where P_i is a polynomial in momentum-transfer variables $q_{subset\ i}$ for the ith subset of {A,B,C,...} and $\mathcal{A}_i$ is a nontrivial amplitude involving the remaining operators plus the counterterm operator Δ(i^{th} subset). There is a factor Z_Q^{-1} for each insertion Q, in addition to the usual factors $Z_3^{-1/2}$, $Z_2^{-1/2}$ for each external boson and fermion line in a complete or connected Green's function. The subtractive renormalizations $\sum_i P_i \mathcal{A}_i$ lack absorptive parts in channels corresponding to the polynomial dependence $P_i(q_{subset\ i})$.

By repeating the previous analysis, we end up with a CS equation for $G^{(b,f)}_{ABC\ldots}$,

$$[\, m_R\, \partial/\partial m_R + \beta_R(g_R)\, \partial/\partial g_R + b\gamma_R(g_R) + f\, \bar{\gamma}_R(g_R) + \gamma^{(R)}_{ABC\ldots}(g_R)]$$

$$\cdot\{G^{b,f}_{ABC\ldots}(q;g_R,m_R)\}_R \quad + \quad \sum_i \mathcal{P}_i(q_{subset\ i})\ \mathcal{G}_i(\text{other } q;g_R,m_R)$$

$$= -i\delta_R(g_R)\ \{G^{(b,f)}_{ABC\ldots}(\Delta;q;g_R,m_R)\}_R \tag{1.42}$$

where all factors are cutoff-independent, $\mathcal{P}_i$ is a polynomial, and

$$\bar{\gamma}_R(g_R) = \Lambda\partial/\partial\Lambda \; \ell n \; \{Z_2(\Lambda/m_R,g_B\}^{-1/2} \tag{1.43}$$

is the fermionic analogue of $\gamma_R(g_R)$. The important thing to notice is that there is an addition rule satisfied by

$$\gamma^{(R)}_{ABC\ldots}(g_R) = -\Lambda\partial/\partial\Lambda \; \ell n \; [Z_A Z_B Z_C \ldots] \quad , \tag{1.44}$$

the γ-function describing the combined effects of the product ABC... on wave-function renormalization. For each operator Q, there is a characteristic γ-function

$$\gamma^{(R)}_Q(g_R) = -\Lambda\partial/\partial\Lambda \; \ell n \; Z_Q \; (\Lambda/m_R,g_B) \tag{1.45}$$

which satisfies the formula

$$\gamma^{(R)}_{ABC\ldots}(g_R) = \gamma^{(R)}_A(g_R) + \gamma^{(R)}_B(g_R) + \gamma^{(R)}_C(g_R) + \ldots \tag{1.46}$$

Conserved currents (e.g., $\theta_{\mu\nu}$ and the electromagnetic current J_μ) obey identities such as

$$i(p'-p)_\mu G^{(0,2)}_{J_\mu}(p',p) = G^{(0,2)}(p') - G^{(0,2)}(p) \tag{1.47}$$

irrespective of which rules (B or any R) are chosen. Hence there is no wave-function renormalization factor for a conserved current

$$\zeta_{J_\mu}(R',R) = Z_{J_\mu} = 1 \tag{1.48}$$

and the corresponding γ-function (1.45) vanishes identically :

$$\gamma_{J_\mu} = 0 \quad . \tag{1.49}$$

Of course, there can still be subtractive renormalizations induced by time-ordering products of conserved currents. For example, the renormalization of $T\langle 0|J_\mu(x)\, J_\nu(0)|0\rangle$ involves a subtraction proportional to $(\partial_\mu\partial_\nu - g_{\mu\nu}\partial^2)\delta^4(x)$. Partially conserved currents such as the chiral SU(3)xSU(3) currents $(\mathcal{F}^a_\mu, \mathcal{F}^a_{\mu5})$ of current algebra become conserved at short distances, so they also satisfy (1.48) and (1.49).

II. ASYMPTOTIC SOLUTIONS

This chapter concerns situations in which the mass-inserted amplitude $G(\Delta)$ can be neglected in the CS equation. In perturbation theory, an appropriate asymptotic limit is specified by Weinberg's theorem[11].

Logarithms in perturbation theory.

Weinberg's theorem extends the power-counting method to the problem of estimating the asymptotic behaviour of a Feynman amplitude as some of its external momenta become large. Actually, it is a theorem in real-variable analysis which describes the asymptotic behaviour of infinite multiple integrals, so its application to Feynman integrals is necessarily restricted to the Euclidean region. However, Pohlmeyer[24)] has proven a corresponding theorem for Minkowski space.

Let $\mathcal{A}$ be a Feynman amplitude which depends on external momenta $q_1 \ldots q_K$ and $r_1 \ldots r_N$, where the q-momenta are large :

$$\mathcal{A} = \mathcal{A}(q_1 \ldots q_K ; \ r_1 \ldots r_N) \quad ;$$

$$q_i = \eta \ell_i + c_i, \quad \eta \to \infty \qquad \text{with } \ell_i, c_i, r_j \text{ all fixed;}$$

$$i = 1 \ldots K; \quad j = 1 \ldots N. \tag{2.1}$$

All momenta are understood to be incoming, so momentum conservation requires

$$\sum_{j=1}^{N} r_j + \sum_{i=1}^{K} q_i = 0 \tag{2.2a}$$

$$\sum_{i=1}^{K} \ell_i = 0 \tag{2.2b}$$

Imagine the large $O(\eta)$ momenta (indicated by heavy lines in fig.2) percolating through all vertices and propagators of a subgraph $\mathcal{G}'$ of a typical graph $\mathcal{G}$. Naturally, we arrange that the external lines of $\mathcal{G}'$ include all external lines of $\mathcal{G}$ which carry the $O(\eta)$ momenta $q_1 \ldots q_K$, and that it is kinematically possible for all internal lines of $\mathcal{G}'$ to carry $O(\eta)$ momenta. Each subgraph can contribute to the asymptotic behaviour of $\mathcal{G}$. The idea is to count the powers produced by propagators, vertices and loop integrations in $\mathcal{G}'$, keeping the intermediate momenta $k_1 \ldots k_M$ fixed. The effect of loop integration is illustrated by the 1-loop boson self-energy amplitude (fig. 3) for a scalar Yukawa theory (interaction $g\bar{\psi}\psi\varphi$) :*

---

* Here, the relevant subgraph $\mathcal{G}'$ is the complete graph $\mathcal{G}$. All other subgraphs involve restricting $\int d^4p$ to a finite volume, so they contribute $O(q^{-1})$.

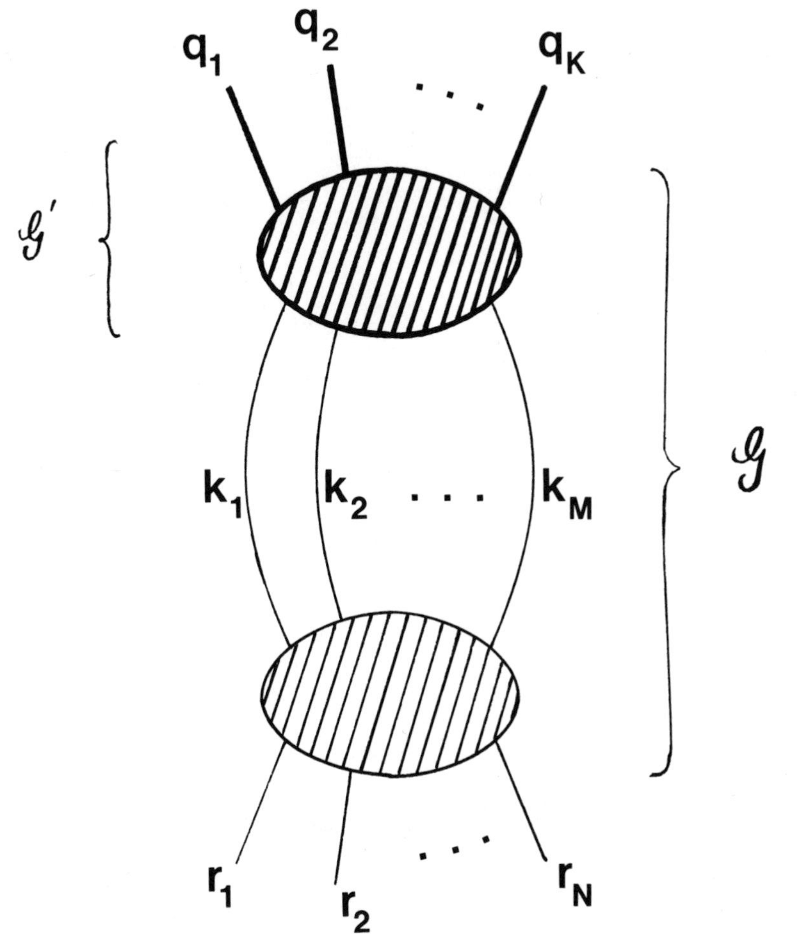

Fig. 2

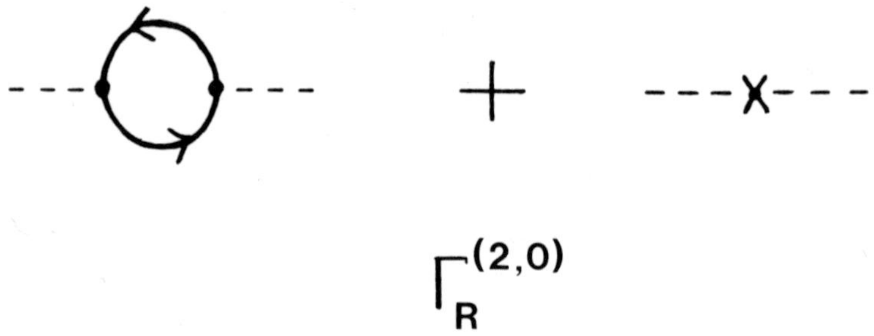

Fig. 3

$$\Gamma_R^{(2,0)}(q^2)_{1\text{-loop}} = -g_R^2 \int \frac{d^4p}{(2\pi)^4} \mathrm{Tr}\ \{(\not p - m)^{-1}(\not p + \not q - m)^{-1} - \text{subtraction}\}$$

$$\sim -(ig_R^2/8\pi^2)q^2 \ln q^2, \quad (q \to \infty) \qquad (2.3)$$

This example shows that it is necessary to count 4 powers of η for each loop integral $\int d^4p$ inside* $\mathcal{G}'$. Consequently, each $\mathcal{G}'$ contributes a characteristic power $O(\eta^{\mathcal{D}(\mathcal{G}')})$ (with logarithmic corrections), where the dimensionality $\mathcal{D}(\mathcal{G}')$ of $\mathcal{G}'$ is given by

$$\mathcal{D}(\mathcal{G}') = \sum \dim \{\text{propagators and vertices of } \mathcal{G}'\} + 4\ \{\text{number of independent loops in } \mathcal{G}'\} \qquad (2.4)$$

So far, the only restriction placed on the vectors ℓ_i is the momentum-conservation equation (2.2b). Another obvious requirement is that none of the ℓ_i should vanish - otherwise, some of the q-momenta would not be $O(\eta)$. However, the power-counting rules

---

* The intermediate lines $k_1 \ldots k_M$ may be parts of closed loops in $\mathcal{G}$ but this is not relevant for the subgraph $\mathcal{G}'$. It becomes relevant only when a larger subgraph is considered.

introduced above do not necessarily work unless additional conditions are imposed. For example, the 4 boson and 3 fermion propagators of the subgraph $\mathcal{G}'$ of fig. 4 (Yukawa theory) normally contribute $O(\eta^{-11})$. However, at the "exceptional" momentum point[25] $\ell_1 = -\ell_2$, the fermion propagator in the middle of $\mathcal{G}'$ no longer carries momentum $O(\eta)$:

$$\mathcal{A}(\mathcal{G}';\ \ell_i^2 \neq 0;\ \ell_1 = -\ell_2) = O(\eta^{-10}) \ .$$

Additional exceptional points exist in Minkowski space :

$$\mathcal{A}(\mathcal{G}';\ \ell_i^2 \neq 0;\ \ell_1 \neq -\ell_2;\ (\ell_1+\ell_2)^2 = 0) = O(\eta^{-9}) \text{ or } O(\eta^{-10})$$

(depending on whether $(\ell_1+\ell_2).(c_1+c_2+k_1)$ vanishes or not). The general conclusion is that tree graphs can be more singular than $O(\eta^{\mathcal{D}(\mathcal{G}')})$ if some of the partial sums $\Sigma'\ell$ happen to be lightlike.

Not surprisingly, loop integrals also misbehave at these exceptional points. Consider the 1-loop amplitude (fig. 5)

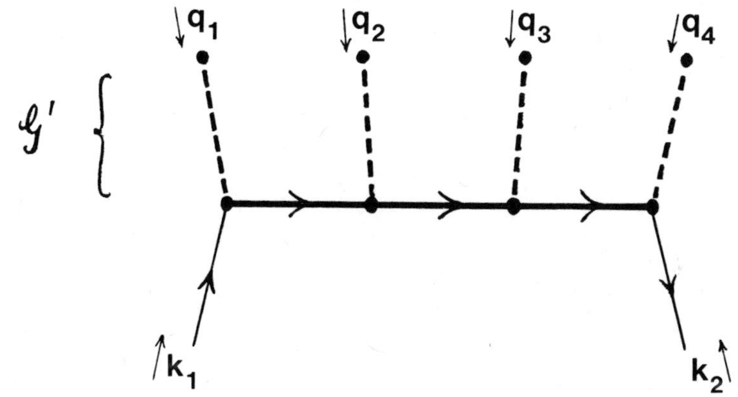

Fig. 4

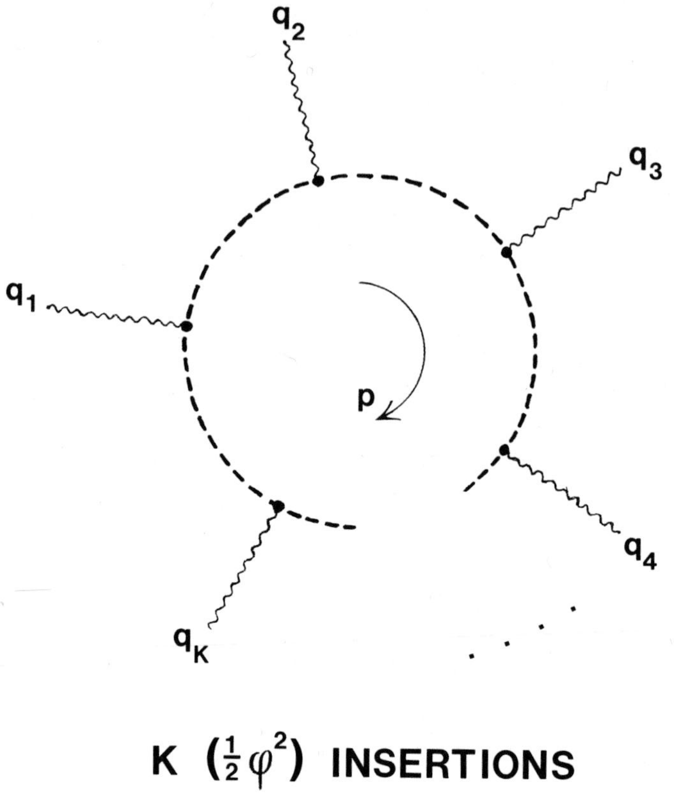

Fig. 5

$$I(q_1 \ldots q_K) = i^K \int (d^4p/(2\pi)^4 \prod_{i=1}^{K} [(p+Q_i)^2 - m^2]^{-1} \quad , \quad (K \geq 3),$$

$$(Q_i = \sum_{j=1}^{i} q_j, \quad q_i = \eta \ell_i + c_i, \quad \sum_{i=1}^{K} \ell_i = \sum_{i=1}^{K} c_i = 0), \qquad (2.5a)$$

which can be simply represented as a Feynman-parameter integral[26] :

$$I = (i^{K+1}(K-3)!/16\pi^2) \int_0^1 \prod_i d\alpha_i \;\; \delta(1-\Sigma\alpha)/\left[\sum_{i<j} \alpha_i\alpha_j(Q_i-Q_j)^2-m^2\right]^{K-2} \tag{2.5b}$$

The limit $\eta\to\infty$ is equivalent to setting m and c_i to zero in (2.5a) or (2.5b). If the massless amplitude $I(m=0=c_i)$ exists, the result is

$$I(\eta) \sim \eta^{4-2K}\, I(m=0=c_i;\ \eta=1)\ , \qquad (\eta\to\infty), \tag{2.6}$$

which agrees with naive power-counting :

$$\mathcal{D}(\mathcal{G}') = 4 - 2K \qquad (\text{with } \mathcal{G}' = \text{complete graph } \mathcal{G}).$$

However, $I(m=0=c_i)$ may not exist because of infrared singularities; in that case, $\eta^{2K-4}\, I(\eta)$ blows up as η tends to infinity. For example, if all partial sums $\Sigma'\ell$ are lightlike but non-zero, there is a singularity at $p = 0$:

$$I(\eta) \sim \eta^{4-2K} \int (d^4p/(2\pi)^4) \prod_{i=1}^{K} \left[p^2 + 2p\cdot\sum_{j=1}^{i} \ell_j + O(\eta)\right]^{-1} i^K$$

The true asymptotic behaviour can be found by substituting

$$(Q_i-Q_j)^2 = \eta r_{ij} + O(1), \qquad (\eta\to\infty)$$

into eq. (2.5b) with

$$r_{ij} = 2\,(\sum_{k=i}^{j} \ell_k)\cdot(\sum_{k=i}^{j} c_k) \qquad ;$$

thus, if r_{ij} all have the same sign, the answer is

$$I(\eta) \sim \eta^{2-K}\,(i^{K+1}(K-3)!/16\pi^2) \int_0^1 \prod_{i=1}^{K} d\alpha_i\, \delta(1-\textstyle\sum \alpha)/\left[\sum_{i<j}\alpha_i\alpha_j r_{ij}\right]^{K-2}$$

$$((\textstyle\sum{}'\ell)^2 = 0, \quad \sum{}'\ell \neq 0, \quad \eta\to\infty \quad ;\ K \geqslant 3), \tag{2.6'}$$

because the resulting α-integral is manifestly convergent.

In general, loop integrals become infrared singular at thresholds for the production of intermediate zero-mass particles. These thresholds occur when at least one of the momenta $\sum'\ell$ becomes lightlike. This is precisely the condition for exceptional asymptotic behaviour of tree-graph amplitudes, so we conclude that the general requirement for momenta to be non-exceptional[25] is

$$(\textstyle\sum{}'\ell)^2 \neq 0 \tag{2.7}$$

for all partial sums $\sum{}'\ell$ associated with connected amplitudes.

Note : (i) In coordinate space, the non-exceptional $\eta\to\infty$ limit corresponds to the short-distance limit

$$x_1 \ldots x_K \to y \qquad \text{(non-exceptional)}$$

where $x_1 \ldots x_K$ are coordinates conjugate to the momenta $q_1 \ldots q_K$. If K' partial sums $\sum{}'\ell$ become lightlike (0<K'<K), the x-coordinates are free to tend to K'+1 distinct y-coordinates :

$$x_1 \ldots x_K \to y_1 \ \ldots \ y_{K'+1} \qquad \text{(exceptional)} .$$

(ii) I have glossed over a complication caused by renormalization of ultraviolet divergences. Counterterms which remove logarithmic divergences develop logarithmic infrared singularities

$$\int^{\Lambda} d^{4\ell}p(\ln p)^{\ell'}/p^{4\ell} \qquad (\ell' < \ell)$$

as m, c_i, and all subtraction points λ are effectively scaled to zero by the $\eta \to \infty$ limit. Consequently, there are logarithmic corrections to naive power-counting at non-exceptional momenta (as in eq. (2.3)) :

$$\mathcal{G}' \ \text{contribution} = O\,[\eta^{\mathcal{D}(\mathcal{G}')}\{\text{Polynomial in } \ell n\ \eta\}] \tag{2.8}$$

Additional powers of η are not generated : the candidate

$$\text{counterterm } (m=\lambda=0) = \int^{\Lambda} d^{4\ell}\ p/p^{4\ell-\,d(\Gamma)} , \qquad (d(\Gamma) < 0)$$

is not permitted because it is ultraviolet convergent. *

The main features and limitations of the theorem have now been exposed. It states that the amplitude $\mathcal{A}(\mathcal{G})$ for a graph $\mathcal{G}$ satisfies the asymptotic bound

$$\mathcal{A}(\mathcal{G}) = O\,[\eta^{\text{Max}\,\mathcal{D}(\mathcal{G}')}\ \ln^{\beta(\mathcal{G})}\eta\,] \tag{2.9}$$

for the non-exceptional limit $\eta\to\infty$ specified by eqs. (2.1), (2.2), and (2.7). The power Max $\mathcal{D}(\mathcal{G}')$ denotes the maximum dimensionality attained in the set of all subgraphs $\mathcal{G}'$ of $\mathcal{G}$; (see fig. 2). The logarithmic power $\beta(\mathcal{G})$ (integer ≥ 0) is determined by counting powers of $\ln\Lambda$ in counterterms which remove the ultraviolet divergences of subgraphs with maximal dimensionality[27].

---

* In other words, we forbid "oversubtraction" :once a 1PI amplitude Γ becomes superficially convergent ($d(\Gamma) < 0$), counterterms which further reduce $d(\Gamma)$ are not introduced in eq. (1.3).

If propagators for external lines of $\mathcal{G}'$ are excluded, theories with dimensionless coupling constants obey the rule

$$\mathcal{D}(\mathcal{G}') = 4 - \{\text{number of bosons external to } \mathcal{G}'\} - (3/2)\{\text{number of fermions external to } \mathcal{G}'\} . \tag{2.10}$$

Hence the maximum dimensionality is found by minimizing the number M of intermediate lines in fig. 2. If composite operators (with sources χ) are present, the generalization of (2.10) is

$$\mathcal{D}(\mathcal{G}') = 4 - \sum \dim \{\text{external lines of } \mathcal{G}'\}, \tag{2.10'}$$

where each source χ is represented by a line (external or intermediate in fig.2) and given an appropriate dimensionality. For example, the source χ_{φ^2} for the composite operator $\varphi^2(x)$ carries dimensionality 2, so the complete graph shown in fig. 5 has dimensionality $4 - 2K$; (compare eq. (2.6)).

According to eqs. (2.10) and (2.10'), the value of $\mathcal{D}(\mathcal{G}')$ in renormalizable theories is completely specified by the number and nature of the lines carrying momenta $k_1 \ldots k_M$ and $q_1 \ldots q_K$ in fig.2. This means that fig.2 can be applied to a collection of Feynman diagrams for the amplitude $\mathcal{A}(q_1 \ldots q_K; r_1 \ldots r_N)$ if all sets of intermediate lines $\{k_1 \ldots k_M\}$ permitted by selection rules are considered.

Now we examine the limit $\eta \to \infty$ for the CS equation (eqs.(1.36), (1.42))

$$\partial_G G(\eta q) + \{\text{subtraction terms}\} = -i\delta\, G(\Delta;\eta q) ,$$
$$\partial_G = \eta\partial/\partial\eta - \beta\partial/\partial g - \gamma_G \tag{2.11}$$

for connected amplitudes $G(\eta q)$, $G(\Delta;\eta q)$ at non-exceptional momenta q. Graphs contributing to $G(\eta q)$ do not have external legs with fixed momenta (i.e., N=0 in fig.2), so for each graph $\mathcal{G}$, the dominant subgraph is $\mathcal{G}$ itself :

$$G(\eta q) = O[\eta^{\mathcal{D}(\mathcal{G})} \ell n^{\beta}\eta] , \qquad (\mathcal{D}(G) = \mathcal{D}(\mathcal{G})) \tag{2.12}$$

The dominant subgraphs * for $G(\Delta)$ are shown in fig. 6 :

---

*Graphs with one intermediate boson line and $\mathcal{D}(\mathcal{G}') = \mathcal{D}(G) - 2$ are possible if a φ^3 interaction is added to the Lagrangian (1.10) or (1.12).

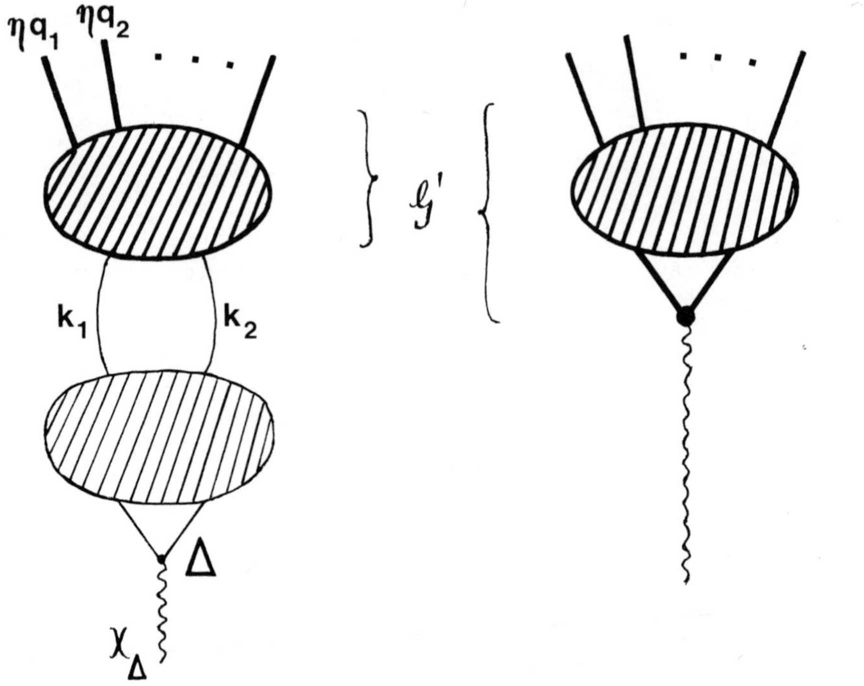

(a) (b)

Fig. 6

$$\mathcal{D}(\mathcal{G}'; \text{ fig. 6a}) = \begin{cases} \mathcal{D}(G) - 2 & (k_1, k_2 = \text{bosons}) \\ \mathcal{D}(G) - 3 & (k_1, k_2 = \text{fermions}); \end{cases}$$

$$\mathcal{D}(\mathcal{G}'; \text{ fig. 6b}) = \mathcal{D}(G) - \dim \chi_\Delta \tag{2.13}$$

Here χ_Δ is the source for $\bar{\psi}\psi$ or φ^2 at zero momentum :

$$\dim \chi_\Delta = 1 \text{ or } 2 . \tag{2.14}$$

Eqs. (2.13) and (2.14) imply

$$G(\Delta;\eta q) = O[\eta^{\mathcal{D}(G)-1} \ell n^{\beta'} \eta] \qquad \text{(fermions present)} \quad (2.15a)$$

$$G(\Delta;\eta q) = O[\eta^{\mathcal{D}(G)-2} \ell n^{\beta''} \eta] \qquad \text{(no fermions)} \quad (2.15b)$$

so the leading power $G^{as.}$ of G, defined by

$$G(\eta q) = G^{as.}(\eta q) + O[\eta^{\mathcal{D}(G)-1} \ell n^{\beta'''} \eta] \quad (2.16)$$

satisfies the equation*

$$\partial_G G^{as.}(\eta q) + \{\text{sub. terms}\}^{as.} = 0 , \quad (2.17)$$

with

$$\{\text{subtraction terms}\} = \sum_i \mathcal{P}_i \mathcal{G}_i \quad \text{(eq. (1.42))}$$

$$= \{\text{sub. terms}\}^{as.} + O[\eta^{\mathcal{D}(G)-1} \ell n^{\beta^{iv}} \eta] . \quad (2.18)$$

In other words, G(Δ) is asymptotically negligible[5].

The restriction to non-exceptional momenta[25] is essential. For example, apart from a few special cases[1], G(Δ) is not asymptotically negligible if G is an on-shell amplitude. Also, additional mass insertions usually do not reduce the asymptotic behaviour further : in most cases, the amplitudes G(n Δ's) and G(n+1 Δ's) in the CS equation**

$$\partial_{G(n\ \Delta's)} G(n\ \Delta's) + \{\text{sub.terms}\} = -i\delta\ G(n+1\ \Delta's),$$

$$\partial_{G(n\ \Delta's)} = \partial_G - n\ \gamma_\Delta ,$$

$$\gamma_\Delta = -\Lambda\partial/\partial\Lambda \ \ln Z_\Delta(\Lambda/m_R, g_B) \quad (2.19)$$

have the same asymptotic power $\mathcal{D}(G) - 2$. The situation for G(ΔΔ)

---

* Unless a selection rule forces it to vanish, $G^{as.}$ almost invariably receives contributions from the simplest Feynman graphs for G. If desired, the subtraction terms can be removed by considering unordered products instead of the T-product $G^{(b,f)}_{ABC}$ in eq. (1.42).

** See eqs. (1.28) and (1.45), and the addition rule (1.46).

is illustrated in fig. 7 :

$$\mathcal{D}(\mathcal{G}';\ \text{fig.7a}) = \begin{cases} \mathcal{D}(G) - 2 & (k_1,k_2 = \text{bosons}), \\ \mathcal{D}(G) - 3 & (k_1,k_2 = \text{fermions}); \end{cases}$$

$$\mathcal{D}(\mathcal{G}';\ \text{fig.7b}) = \mathcal{D}(G) - 2\ \dim \chi_\Delta \tag{2.20}$$

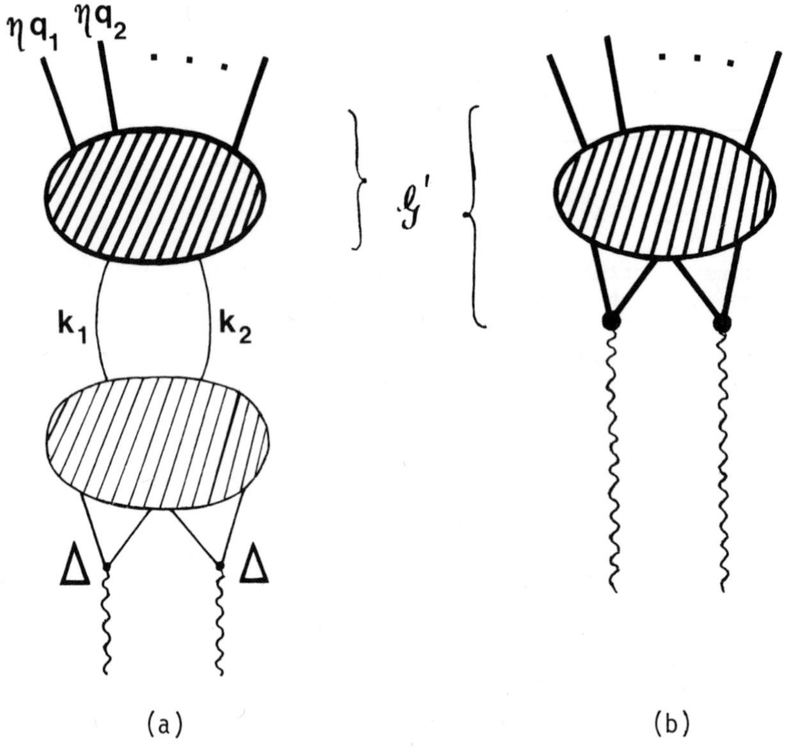

Fig.7

If Δ is proportional to $\int d^4x\ \bar{\psi}\psi(x)$, the asymptotic power is reduced from $\mathcal{D}(G) - 1$ for $G(\Delta)$ to $\mathcal{D}(G) - 2$ for $G(\Delta\Delta)$. Otherwise, there is no reduction; clearly, any number of Δ insertions can be attached to the lower blob in fig. 7a without changing the asymptotic power.

In perturbation theory, the general form of $G^{as.}$ consistent with eq.(2.17) is

$$G^{as.}(\eta) = \eta^{\mathcal{D}(G)} \sum_N g^N \sum_{p=0}^{\ell(N)} \ln^p\eta\ G_{N,p}\ , \tag{2.21}$$

where $\ell(N)$ is the number of loops in N^{th} order diagrams for the amplitude G. The presence of $\partial/\partial g$ in the CS operator ∂_G means that (2.17) relates the Nth order coefficient $G_{N,p}$ to lower-order coefficients. The consequences of this are best illustrated with a simple example, the photon propagator in QED[28] :

$$D'_{F\mu\nu}(q) = -(i/q^2)(g_{\mu\nu} - q_\mu q_\nu/q^2)d(q^2/m^2,\alpha) + \{q_\mu q_\nu/q^4 \text{ terms}\},$$
$$(\alpha = (\text{charge})^2/4\pi). \tag{2.22}$$

Charge renormalization is simpler than in eq.(1.16) because gauge invariance implies[9] $Z_1 = Z_2$:

$$\alpha_R = Z_3\ \alpha_B\ . \tag{2.23}$$

It follows that the combination $\alpha d(q^2/m^2,\alpha)$ has no wave function renormalization, so its CS equation has no γ-function or subtraction terms and the leading asymptotic power $d^{as.}(\eta^2q^2/m^2,\alpha)$ satisfies the equation *

$$[\eta\partial/\partial\eta - \alpha\beta(\alpha)\ \partial/\partial\alpha]\ (\alpha\ d^{as.})^{-1} = 0\ , \tag{2.24}$$

with

$$\beta(\alpha) = -\Lambda\partial/\partial\Lambda\ \ln Z_3(\Lambda/m_R,\alpha_B)\ . \tag{2.25}$$

In the perturbative expansions

$$[d^{as.}(\eta,\alpha)]^{-1} = \sum_{n=0}^{\cdots} \alpha^n\ a_n(\eta)\ ,\quad (a_0(\eta) \equiv 1)\ ,$$
$$\beta(\alpha) = \sum_{n=1}^{\cdots} \alpha^n\ b_n \tag{2.26}$$

---

*It is convenient to consider the inverse $(d^{as.})^{-1}$ because $(1-d^{-1})$ is the photon self-energy amplitude (1PI).

are substituted in eq. (2.24), the result is a set of equations

$$\eta\partial/\partial\eta\ a_{n+1}(\eta) = \sum_{r=0}^{n} a_r(\eta)\ (r-1)b_{n-r+1}, \quad (n=0,1,2,\ldots) \quad (2.27)$$

which imply that a two-loop calculation[29] of $\beta(\alpha)$ ($b_1 = 2/3\pi$, $b_2 = 1/2\ \pi^2$) is sufficient to determine the leading logarithm in <u>any</u> order :

$$a_1(\eta) = -b_1\ \ell n\ \eta + a_1(1)\ ,$$

$$a_n(\eta) = -b_2 b_1^{\ n-2}\ (\ell n\ \eta)^{n-1}/(n-1) + O\left[(\ell n\eta)^{n-2}\right]\ , \quad (n \geq 2)\ . \quad (2.28)$$

Examination of the cutoff dependence of photon self-energy diagrams shows that the n-loop graphs of fig. 8 (plus suitable counterterm graphs) are responsible for the $(\ell n\ \eta)^{n-1}$ dependence of $a_n(\eta)$ for $n \geq 2$.*

The next problem is to decide how to sum the logarithms in eq. (2.21). The leading-logarithm approximation is very easy to analyze, but it usually fails. For example, eq. (2.28) yields

$$(\alpha\ d^{as.})^{-1} = \{\text{leading logs}\} + \{1^{st}\ \text{non-leading}\} + \ldots \quad (2.29)$$

with

$$\{\text{leading logs}\} = \alpha^{-1}\ \{1 - (2\alpha/3\pi)\ \ell n\ \eta\ \}, \quad (2.30)$$

$$\{1^{st}\text{non-leading}\} = a_1(1) + (3/4\pi)\ \ell n\ \{1 - (2\alpha/3\pi)\ \ell n\eta\}. \quad (2.31)$$

Eq. (2.30) is unsatisfactory as an asymptotic approximation for $(\alpha d^{as.})^{-1}$ because it implies that $\alpha d^{as.}$ has a "ghost" pole[30] (i.e. a pole at a spacelike momentum point). This is not a genuine difficulty, because the "non-leading" term (2.31) dominates (2.30) in the relevant region ($\ell n\ \eta \sim 3\pi/2\alpha$). In fact, the leading-logarithm approximation works only if η is much smaller :

$$1 - (2\alpha/3\pi)\ \ell n\ \eta >> \alpha \quad . \quad (2.32)$$

---

* The sum of the two-loop graphs goes as $-b_2\alpha_B^2\ \ell n\Lambda$, and there is an additional power of $\ell n\ \Lambda$ for each of the n-2 fermion-loop insertions in the internal photon line : Z_3^{-1} (fig.8) $= O\left[(\ell n\ \Lambda)^{n-1}\right]$. All other subsets of n-loop graphs (with Z_1=Z_2 for each subset) are $O[(\ell n\Lambda)^{n-2}]$ so, according to the discussion of eqs.(2.8) and (2.9), they cannot influence the leading logarithmic behaviour of $a_n(\eta)$.

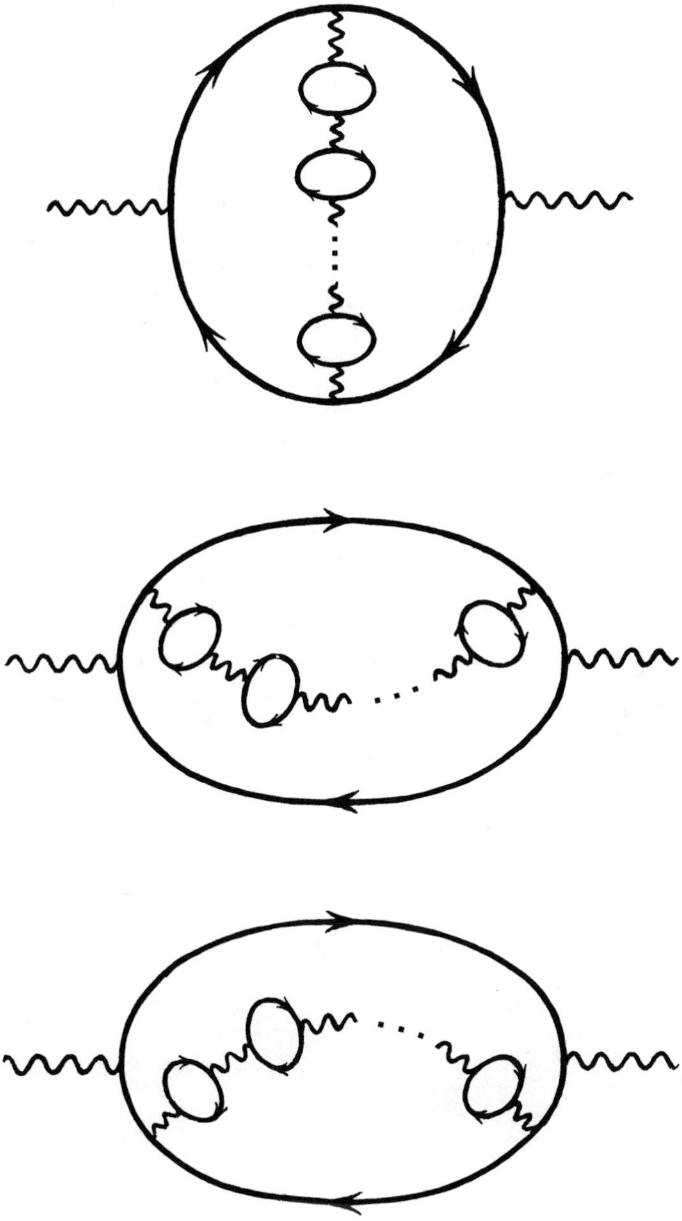

Fig. 8

As η increases, more and more non-leading logarithms (i.e. higher orders in the expansion of $\beta(\alpha)$) become important. When η becomes sufficiently large, the perturbation expansion for $\beta(\alpha)$ cannot be truncated at finite order, and all of the logarithms in (2.21) must be summed.

Further progress depends on the following assumptions :

(I) The theory exists at finite values of the coupling constant g, where $\beta(g)$, $\gamma_G(g)$, $G(g)$, $G^{as.}(g)$,... are differentiable functions of g. Perturbation theory is generated by asymptotically expanding these functions about $g = 0$.

(II) At finite g, the leading power of G is $G^{as.}$, as in perturbation theory.

Very little is known about the validity of (I) in non-trivial theories; for example,

(i) Does the radius of convergence of the perturbation series vanish[31]?

(ii) If so, does unitarity (or something equally respectable) uniquely specify the continuation from $g = 0$ to finite g ?

(iii) Are inequivalent theories generated by different orders of summation[32] (e.g. of diagrams, or of terms in (2.21))?

In general, the status of assumption (II) is equally uncertain. However, for asymptotically free theories[8], it is not difficult to show[7] that (II) works, provided that the validity of (I) is assumed.

Characteristics of solutions

Consider eq. (2.17) when subtraction terms are absent and there is only one coupling constant g :

$$[\eta\partial/\partial\eta - \beta(g)\ \partial/\partial g - \gamma_G(g)]\ G^{as.}(\eta q;g,m) = 0\ . \qquad (2.33)$$

The general solution of (2.33) is most conveniently written[18]

$$G^{as.}(\eta q;g,m) = G^{as.}(q;\bar{g},m)\ \exp \int_g^{\bar{g}} dx\ \gamma_G(x)\ /\beta(x) \qquad (2.34)$$

when the auxiliary function $\bar{g} = \bar{g}(g,\eta)$ (known as the "effective coupling constant") is defined by

$$\int_g^{\bar{g}} dx/\beta(x) = \ell n\ \eta \quad ; \qquad \bar{g}(g,1) = g\ . \qquad (2.35)$$

The η-dependence of $G^{as.}(\eta q)$ is entirely contained in the $\bar{g}$-dependence of the right-hand side of eq.(2.34). For consistency, we must suppose that the integral $\int dx/\beta(x)$ diverges at least once to $+\infty$ and once to $-\infty$. Otherwise, eq.(2.35) does not permit the variable η to run freely from 0 (the infrared limit) to ∞ (the ultraviolet limit).

If the divergence in $\int dx/\beta(x)$ is caused by the range of integration becoming infinite,

$$\left| \int_{g}^{\pm\infty} dx/\beta(x) \right| = \infty \quad , \tag{2.36}$$

then eq. (2.35) implies that $\bar{g}$ diverges as $\ell n\ \eta$ becomes infinite. There is no good argument for supposing that this does not happen in real life. Nevertheless, people usually do not bother with this case because of its lack of predictive power : additional assumptions for the $g \to \infty$ behaviour of the functions $G^{as.}(q;\ g,m)$ must be introduced in order to determine their asymptotic behaviour from eq.(2.34).

Alternatively, $\beta(x)$ may possess "eigenvalues" or "fixed points" g_∞,

$$\beta(g_\infty) = 0 \tag{2.37}$$

which make $\int dx/\beta(x)$ diverge; (e.g., the origin x = 0 is a fixed point). According to eq. (2.35), $\bar{g}$ approaches the nearest fixed point g_∞ as $\ell n\ \eta$ tends to $+\infty$ (ultraviolet-stable fixed point) or to $-\infty$ (infrared-stable fixed point)[33]; see fig. 9. A series of fixed points g'_∞, g''_∞, ...(as in fig. 10) produces independent regions I, II, III, ... in coupling-constant space. The effective coupling constant is restricted to the region in which the "physical" coupling constant g lies *. For example, if g lies within region III in fig. 10, the relevant IR- and UV-stable fixed points are g''_∞ and g'''_∞ :

$$g''_\infty = \lim_{\eta\to 0} \bar{g} < \bar{g}(g,\eta) < g''' = \lim_{\eta\to\infty} \bar{g} \quad . \tag{2.38}$$

Generally, IR-stable fixed points are less interesting because the correction $(G - G^{as.})$ from mass insertions is expected to dominate $G^{as.}$ when η becomes too small.

It is convenient to discuss the asymptotic properties of eq. (2.34) for the case $\bar{g} \to g_\infty$ in terms of the function[34,35]

$$\varepsilon(\eta) = (\ell n\ \eta)^{-1} \int_{g}^{\bar{g}} dx\{\gamma_G(x) - \gamma_G(g_\infty)\} / \beta(x) \tag{2.39}$$

which is closely related to the exponential factor in (2.34) :

---

* If g is exactly equal to g_∞, eq.(2.33) can be solved directly : $G^{as.}(\eta q;\ g_\infty,m) = \eta^{\gamma_G(g_\infty)}\, G^{as.}(q; g_\infty,m)$.

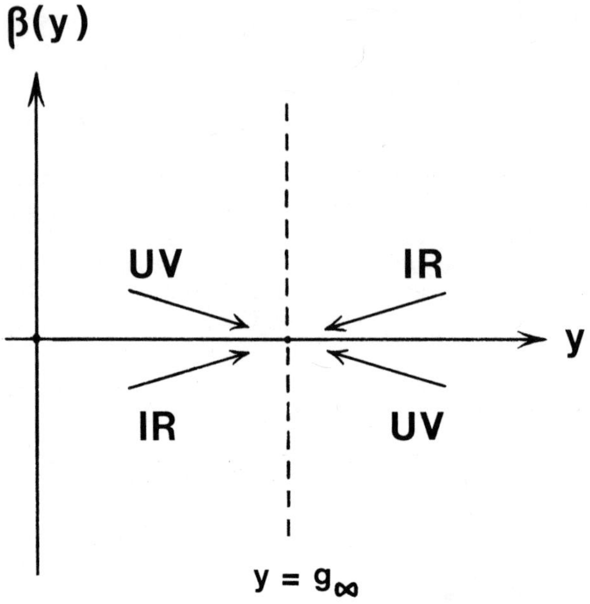

Fig. 9

$$\exp \int_{g}^{\bar{g}} dx \; \gamma_G(x)/\beta(x) = \eta^{\gamma_G(g_\infty) + \varepsilon(\eta)} \quad . \tag{2.40}$$

A change of integration variable $x \to v$ in eq.(2.39), with

$$x = \bar{g}(g,\eta^v) \; , \quad dx/\beta(x) = \ell n \; \eta dv \; ,$$

yields the formula

$$\varepsilon(\eta) = \int_0^1 dv \; \{\gamma_G[\bar{g}(g,\eta^v)] - \gamma_G(g_\infty)\} \tag{2.41}$$

From eq.(2.41), we conclude[34)]

$$\varepsilon(\eta) \to 0 \quad , \quad (\ell n \; \eta \to \pm\infty), \tag{2.42}$$

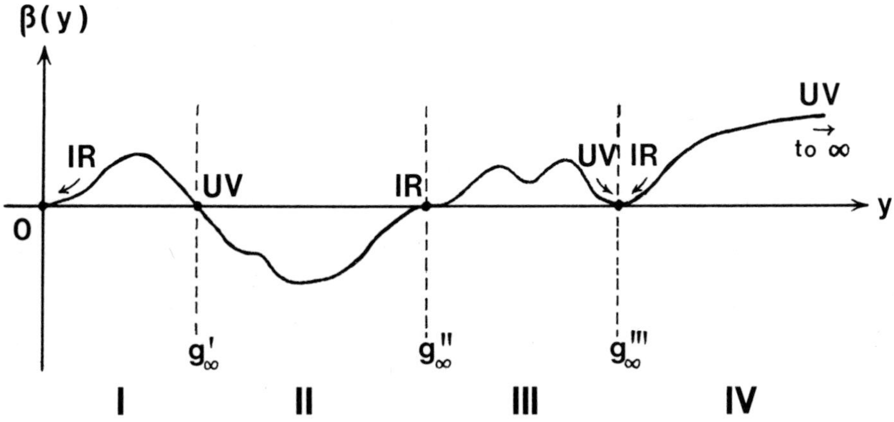

Fig. 10

because $\bar{g}(g,\eta^{v})$ tends to g_∞ for $v \neq 0$, and $\gamma_G(x)$ is bounded and continuous (according to assumption (I)).

The asymptotic behaviour of $G^{as.}(\eta q)$ is obtained by expanding $\bar{g}$ about g_∞ in eq. (2.34). The leading term

$$G^{as.}(\eta q;g,m) \sim \eta^{\gamma_G(g_\infty)+\varepsilon(\eta)} G^{as.}(q;g_\infty,m) \tag{2.43}$$

is almost entirely determined by properties of the theory at the fixed point[3-5,33]. All dependence on the region $g \leqslant x < g_\infty$ is contained in the factor $\eta^{\varepsilon(\eta)}$, which controls the overall normalization of the leading term. If the region $x \simeq g_\infty$ generates a singularity of the integral in eq.(2.39), this factor also modifies the leading power $\eta^{\gamma_G(g_\infty)}$ in eq.(2.43) : e.g.,

$$\eta^{\varepsilon(\eta)} \sim \eta^{c/(\ell n \eta)^p} \qquad \text{or} \qquad \eta^{c/(\ell n|\ell n\ \eta|)^p} \tag{2.44}$$

The precise form of the modification depends on[34] :

(i) the strength of the singularity in $\{\gamma_G(x) - \gamma_G(g_\infty)\}/\beta(x)$ as x tends to g_∞ ;

(ii) the rate at which $\bar{g}$ approaches g_∞. This is controlled by the strength of the zero in $\beta(x)$ at $x = g_\infty$; (see eq.(2.35)). For example[32], if g_∞ is an infinite-order zero of $\beta(x)$, $\bar{g}$ approaches g_∞ very slowly

$$\text{e.g.} \quad \bar{g} - g_\infty = O[(\ln|\ln \eta|)^{-p}] , \qquad (p > 0)$$

compared with the rate at which $\bar{g}$ tends to a simple zero :

$$\bar{g} - g_\infty = O[\eta^{\beta'(g_\infty)}] , \quad (\beta'(g_\infty) \neq 0). \tag{2.45}$$

Note : (a) If there are several coupling constants $\vec{g}$ as in eq. (1.37), the effective coupling constant $\vec{\bar{g}}$ is a vector in coupling-constant space defined by the equations

$$\eta\partial/\partial\eta \ \vec{\bar{g}}(\vec{g},\eta) = \vec{\beta}(\vec{\bar{g}}) ; \qquad \vec{\bar{g}}(\vec{g},1) = \vec{g} . \tag{2.35'}$$

The exponent in eq.(2.34) becomes

$$(\ln \eta)^{-1} \int_0^1 dv \ \gamma_G[\vec{\bar{g}}(\vec{g},\eta^v)] ,$$

and the condition for a fixed point $\vec{g}_\infty$ is

$$\vec{\beta}(\vec{g}_\infty) = 0 . \tag{2.37'}$$

A fixed point is UV-stable if it attracts $\vec{\bar{g}}$ from <u>all</u> directions in $\vec{g}$-space as η tends to infinity.

(b) The simplest example of a subtractively renormalized amplitude is the two-point function

$$\int d^4x \ e^{iq.x} \ T \langle 0|J_\mu(x)J_\nu(0)|0\rangle^{1PI} = (q_\mu q_\nu - g_{\mu\nu}q^2)\pi(q^2/m^2,g) , \tag{2.46}$$

whose absorptive part is measured in the inclusive process $e^+e^- \to$ hadrons. According to eq.(1.49), there is no g-dependent γ-function, so the analogue of (2.33) is

$$[\eta\partial/\partial\eta - \beta(g)\ \partial/\partial g] \ \pi^{as.}(\eta^2q^2/m^2,g) + K(g) = 0 , \tag{2.47}$$

with solution

$$\pi^{as.}(\eta^2q^2/m^2,g) = \pi^{as.}(q^2/m^2,\bar{g}) - \int_g^{\bar{g}} dx \ K(x)/\beta(x) . \tag{2.48}$$

Broken scale invariance.

Now we consider the possibility that Green's functions are asymptotically scale-invariant [4]. This means that the leading term (2.43) should be proportional to a pure power $\eta^{\gamma_G(g_\infty)}$ in the

UV limit $\eta \to \infty$.

To ensure that corrections of the type (2.44) are absent, we have to assume that the function $\varepsilon(\eta)$ tends to zero sufficiently rapidly :

$$\varepsilon(\eta) = O\,[(\ell n\ \eta)^{-1}]\ . \qquad (2.49)$$

Referring to eq.(2.39), we see that the condition for asymptotic scale invariance is[34]

$$I_G = \int_g^{g_\infty} dx\{\gamma_G(x) - \gamma_G(g_\infty)\}\ /\beta(x) = \text{convergent.} \qquad (2.50)$$

Eq.(2.50) is supposed to be valid for all Green's functions G, so this looks like a very strong assumption. However, if g_∞ is a simple zero of $\beta(x)$,

$$\beta(g_\infty) = 0\ , \qquad \beta'(g_\infty) < 0\ , \qquad (g_\infty = \text{UV-stable}) \qquad (2.51)$$

(2.50) is automatically satisfied because assumption (I) says that $\gamma_G(x)$ is differentiable.*

The asymptotic expansion of eq.(2.34) can now be specified more precisely than in eqs.(2.42) and (2.43) :[5,18]

$$G^{as.}(\eta q;\ g,m) = \eta^{\gamma_G(g_\infty)}\ \exp I_G\ G^{as.}(q;g_\infty,m)$$
$$+\ O[\eta^{\{\gamma_G(g_\infty)\ -|\beta'(g_\infty)|\}}]\ ,\ (\eta\to\infty) \qquad (2.52)$$

Eq.(2.52) indicates that non-leading terms decrease as a power $O[\eta^{-|\beta'(g_\infty)|}]$ relative to the leading term. This result is a consequence of eq.(2.45) and the fact that the most important non-leading contributions arise from $O(\bar{g}-g_\infty)$ terms in the expansions of $G^{as.}(q;\bar{g},m)$ and the exponential factor in eq.(2.34).

According to assumption (II), contributions to the asymptotic behaviour of $G(\eta q)$ due to mass insertions also decrease as a power $O\,[\eta^{-p(\Delta)}]$ $(p(\Delta) > 0)$ relative to the leading term in the expansion of $G^{as.}(\eta q)$. Eq.(2.52) and assumption (II) imply the result

---

* Conversely, if g_∞ is not a simple zero of $\beta(x)$, it is unlikely that asymptotic scale invariance can be valid; for every Green's function G ,$\{\gamma_G(x) - \gamma_G(g_\infty)\}$ would have to display a sufficiently strong zero at x = 0 to compensate for the zero of $\beta(x)$ in eq.(2.50).

$$G(\eta q;g,m) = \eta^{\gamma_G(g_\infty)} \{1 + O(\eta^{-P})\} \; \exp I_G \; G^{as.}(q;g_\infty,m),$$

$$P = \mathrm{Min}\ \{p(\Delta), |\beta'(g_\infty)|\} \ . \tag{2.53}$$

Now suppose that G is the complete Green's function $G_{ABC...}$, where A, B, C, ... are renormalized field operators (simple or composite). Eq.(2.53) is not directly applicable because of subtractive renormalizations (1.41) induced by the time-ordering operation in

$$G_{ABC...}\ \delta^4(q_1+q_2+...) = \int d^4x_1 d^4x_2 ... \exp i(q_1.x_1+q_2.x_2+ ...)$$
$$T <0|A\ (x_1)B(x_2)C(x_3)...\ |0> \ . \tag{2.54}$$

However, this problem can be trivially circumvented by considering the unordered product instead. According to the addition rule (1.46) and assumption (I), the characteristic power $\gamma_G(g_\infty)$ in Eq.(2.53) is given by the rule

$$\gamma_G(g_\infty) - \gamma_G(0) = \gamma_A(g_\infty) + \gamma_B(g_\infty) + ... \ . \tag{2.55}$$

Thus each operator Q contributes $\eta^{\gamma_Q(g_\infty)}$ to the asymptotic behaviour of G. In coordinate space, the result is

$$<0|A(\rho x_1)B(\rho x_2)C(\rho x_3)\ ...\ |0>$$
$$\sim \rho^{-(d_A+d_B+d_C+...)} f(x_1,x_2,x_3,...),\ (\rho = \eta^{-1} \rightarrow 0), \tag{2.56}$$

where

$$d_Q = \{\text{canonical dimension of } Q\} + \gamma_Q(g_\infty) \tag{2.57}$$

is the <u>dynamical dimension</u>[4,33] of the operator Q. Eq. (2.56) says that dynamical dimension is an additively conserved quantity at short distances. This rule, abstracted from the preceding field-theoretic machinery, is the basis for Wilson's theory of broken scale invariance[4].

If Q is a conserved operator, the "anomalous" term $\gamma_Q(g_\infty)$ vanishes because of eq. (1.49) : e.g.,

$$d_{J_\mu} = 3, \quad d_{\theta_{\mu\nu}} = 4 \ . \tag{2.58}$$

In particular, if J_μ is the electromagnetic current for hadrons, the two-point function $\langle 0|J_\mu J_\nu|0\rangle$ obeys the familiar condition

$$\langle 0|J_\mu(x)J_\nu(0)|0\rangle \sim (R/12\pi^4)(g_{\mu\nu}\partial^2 - \partial_\mu\partial_\nu)(x^2 - i\varepsilon x_0)^{-2}$$

$$(x_\alpha \to 0) \qquad (2.59)$$

so the total cross section for $e^+e^- \to$ hadrons is predicted to be asymptotically scale-invariant :[36)]

$$\sigma(e^+e^- \to \text{hadrons})/\,\sigma^{as.}(e^+e^- \to \mu^+\mu^-) \to R\ , \quad (q^2 \to \infty) \qquad (2.60)$$

Here q is the sum of the e^+ and e^- momenta, and

$$\sigma^{as.}(e^+e^- \to \mu^+\mu^-) = 4\pi\alpha^2/3q^2$$

is a convenient normalizing factor. Asymptotic corrections to eq. (2.60) are $O\,[\,(q^2)^{-P/2}\,]$, where P is the power defined in eq.(2.53). (In fact, eqs.(2.59) and (2.60) are valid for any UV-stable fixed point g_∞ ; if g_∞ is not a simple zero of $\beta(x)$, the corrections to (2.60) decrease logarithmically).

A practical difficulty of this theory is that we do not know how to compute quantities like R, because they are characteristic of the non-trivial interacting theory at $g = g_\infty$.

Asymptotic freedom

A theory is asymptotically free[8,37)] if

(i) the fixed point at the origin is UV-stable : for each coupling constant g_i, the condition

$$g_i\beta_i(g_1,g_2,\ldots) < 0\ , \qquad (i = 1,\ 2,\ \ldots) \qquad (2.61)$$

is satisfied in the neighbourhood of the origin $\vec{g} = 0$.

(ii) the value of $\vec{g}$ is chosen such that, as η increases from 1 to ∞, $\vec{\bar{g}}$ describes a path linking $\vec{g}$ with the origin :

$$\vec{\bar{g}} \to 0\ , \quad (\eta \to \infty)\ . \qquad (2.62)$$

Eq. (2.62) is not an automatic consequence of (2.61) because it assumes the absence of barriers such as non-trivial fixed points which can prevent $\vec{g}$ from reaching points close to the origin. The important feature of these theories is that most of the terms in eq.(2.43) can be computed explicitly. In particular,

$G^{as}(q;g_\infty,m)$ and $\gamma_G(g_\infty)$ are trivially given by free-field theory because the relevant fixed point g_∞ vanishes.

Checking eq.(2.61) is just a matter of computing one-loop (*) contributions to (g) in perturbation theory. Most theories do not obey eq.(2.61) : in QED, the coefficient b_1 of the one-loop term in (2.26) is positive (b_1 = +2/3π), and more generally[38], a four-dimensional renormalizable field theory cannot be asymptotically free unless non-Abelian gauge mesons are present. Furthermore, eq.(2.61) is satisfied only by special classes of gauge theories. The simplest and most interesting case[8,39] involves massless gauge fields A^a_μ interacting with themselves and with fermions ^{i} belonging to a representation R of the gauge group G. For applications to strong interactions, [1,8,39] the fermions are supposed to be current quarks[40] distinguished by properties of "colour" and "flavour" :

$$\{\psi^i\} = \begin{pmatrix} u_1 & d_1 & s_1 & c_1 & \cdots \\ u_2 & d_2 & \cdots & & \\ \vdots & & & & \end{pmatrix} \qquad (2.63)$$

The flavours u,d,s,c,... (= up, down, strange, charmed, ... quarks) are gauge-invariant and transform under observable symmetry groups $G_{obs.}$ such as chiral SU(3) x SU(3). Each flavour carries a colour index K which is transformed by G but not by $G_{obs.}$ (e.g. $u \to u_K$, K = 1,2,...). For example[40], if κ = 1,2,3 refers to the fundamental representation $\underline{3}$ of G = SU(3), the complete fermionic representation R is a direct sum :

$$R = \underline{3}_u \oplus \underline{3}_d \oplus \underline{3}_s \oplus \cdots\cdots \qquad (2.64)$$

The Feynman rules of the theory involve a coupling constant g, the structure constants c^{abc} of the gauge group G, and the matrix generators τ^a of G for the representation R. If necessary, the fermions can be provided with a gauge-invariant mass matrix $\mathcal{M}$ in flavour space to ensure that symmetries like chirals SU(3) x SU(3) are softly broken. Propagators and vertices are generated by the Lagrangian (**)

$$\mathcal{L} = -\frac{1}{4}F^2 + \bar{\psi}(i\not{D}_f - \mathcal{M})\psi + \mathcal{L}_{g.f.} + \mathcal{L}_{ghost} , \qquad (2.65)$$

(*) Two-loop, if the one-loop terms accidentally vanish.

(**) For background, consult review articles[41-43] on the quantization and renormalization of gauge theories.

where

$$F^a_{\mu\nu} = \partial_\mu A^a_\nu - \partial_\nu A^a_\mu + gc^{abc} A^b_\mu A^c_\nu \tag{2.66}$$

is the gauge-covariant field-strength tensor, and

$$(D_f)_\mu = \partial_\mu - igA^a_\mu \tau^a \tag{2.67}$$

is the gauge- covariant derivative for the fermionic representation R. The terms F^2, $\bar{\psi}\not{D}_f\psi$ and $\bar{\psi}m\psi$ are invariant under infinitesimal gauge transformations

$$\delta A^a_\mu = -D^{ab}_\mu \, \delta\omega^b + O[(\delta\omega)^2] \quad , \tag{2.68}$$

$$\delta\psi = -ig\delta\omega^a \tau^a \psi + O[(\delta\omega)^2] \quad ,$$

where

$$D^{ab}_\mu = \partial_\mu \delta^{ab} - gc^{abc} A^c_\mu \tag{2.69}$$

is the covariant derivative for the adjoint representation of G, and $\delta\omega^a(x)$ are arbitrary non-singular functions of the coordinate x. The term $\mathcal{L}_{g.f.}$ in eq.(2.65) specifies a gauge for which the propagator of A^a_μ is well defined; e.g., a Fermi-type gauge-fixing term

$$\mathcal{L}_{g.f.} = -\frac{1}{2\xi} (\partial^\mu A^a_\mu)^2 \tag{2.70}$$

generates the propagator

$$\overbrace{A^A_\mu A^b_\nu} = -i \, \delta^{ab} k^{-2} \{ g_{\mu\nu} - (1-\xi) k_\mu k_\nu / k^2 \} \tag{2.71}$$

Unphysical contributions of $\mathcal{L}_{g.f.}$ to loop integrals must be cancelled by including a ghost Lagrangian[44,45] $\mathcal{L}_{ghost}$ in eq. (2.65). The choice of $\mathcal{L}_{g.f.}$ in eq.(2.70) produces a ghost Lagrangian

$$\mathcal{L}_{ghost} = (\partial^\mu \varphi^*)^a \, D^{ab}_\mu(A) \varphi^b \tag{2.72}$$

where the ghost field φ^a is a Lorentz scalar with Fermi-Dirac statistics. Eqs.(2.65-72) refer to the unrenormalized (B) representation.

One way of computing coupling-constant renormalization is to compare $O(A^2)$ and $O(A^3)$ terms in $\mathcal{L}$. Wave-function renormalization

$$(A^a_\mu)_R = Z_3^{-1/2} (A^a_\mu)_B \tag{2.73}$$

and a rescaling of the gauge parameter(*)

$$\xi_R = Z_3^{-1}\xi_B \tag{2.74}$$

account for the $O(A^2)$ terms. Three-meson 1PI divergences produce a Z_1-factor (similar to that in eq.(1.10))

$$- \frac{1}{2} g_B c^{abc}(\partial^\mu A^\nu - \partial^\nu A^\mu)^a_B \; (A^b_\mu A^c_\nu)_B \; =$$

$$- \frac{1}{2} g_R Z_1 c^{abc}(\partial^\mu A^\nu - \partial^\nu A^\mu)^a_R \; (A^b_\mu A^c_\nu)_R \tag{2.75}$$

so coupling-constant renormalization is given by

$$g_R = Z_3^{3/2} \; Z_1^{-1} \; g_B \tag{2.76}$$

Note : Instead of (2.75), we could have considered one of the other g-dependent vertices in $\mathcal{L}$: $O(A^4)$, $O(\bar\psi \; A \; \psi)$ or $O(\varphi^* A \; \varphi)$. For example, if $\tilde{Z}_1, \tilde{Z}_3$ are the renormalization factors for ghost amplitudes,

$$\varphi_R = \tilde{Z}_3^{-1/2} \varphi_B$$

$$g_B c^{abc}(\partial^\mu \varphi^*)^a_B \; (A^b_\mu \varphi^c)_B = g_R \; \tilde{Z}_1 \; c^{abc}(\partial^\mu\varphi^*)^a_R (a^b_\mu \varphi)_R \; , \tag{2.77}$$

the result is

$$g_R = \tilde{Z}_1^{-1} \; Z_3^{1/2} \; \tilde{Z}_3 . g_B \quad ; \tag{2.78}$$

As long as the regularization procedure[46,47] is chosen such that gauge Ward identities[43,48-50] are satisfied, the consistency of equations such as (2.76) and (2.78) is assured[48] :

$$Z_1/\tilde{Z}_1 = Z_3/\tilde{Z}_3 \tag{2.79}$$

A computation of the one-loop contributions to g_R/g_B yields (after a bit of algebra) the basic result[8,39]

$$\beta(g) = -bg^3 + O(g^5) \quad ,$$

$$b = \frac{1}{16\pi^2} \left[\frac{11}{3} C_2(G) - \frac{4}{3} T(R) \right] \quad , \tag{2.80}$$

(*) The scaling factor in (2.74) is Z_3^{-1} because Ward identities forbid $(\; .A)^2$ counterterms[48] : i.e., $\xi_B^{-1}(\partial . A)^2_B = \xi_B^{-1}(\partial . A)^2_R$. Because of (2.74), the CS differential operator ∂_G for ξ-dependent Green's functions G contains an additional term $-2\nu(g,\xi)\xi\partial/\partial\xi$.

with

$$C_2(G)\delta^{ab} = c^{acd}c^{bcd} \quad , \quad (C_2(G) > 0), \qquad (2.81)$$

$$T(R)\delta^{ab} = \mathrm{Tr}\{\tau^a\tau^b\} \quad , \quad (T(R) \geq 0) \ . \qquad (2.82)$$

Hence eq.(2.61) ($b > 0$) is satisfied if there are not too many fermions in the theory. The rate at which $\bar{g}$ approaches the origin is a direct consequence of eqs.(2.35) and (2.80) :

$$\bar{g}^2 = (2b\ \ell n\ \eta)^{-1} + O\,[\ell n\ell n\eta/(\ell n\eta)^2] \ , \quad (\eta \to \infty) \ . \qquad (2.83)$$

Most Green's functions $G(\eta q)$ are not asymptotically scale-invariant because the one-loop term $c_G g^2$ in the perturbative expansion

$$\gamma_G(g) - \gamma_G(0) = c_G g^2 + O(g^4) \qquad (2.84)$$

causes the integral

$$\int dx\ \{\gamma_G(x) - \gamma_G(0)\}\ /\ \beta(x)$$

in eq.(2.50) to diverge logarithmically at $x = 0$. The rate at which the additional power $\varepsilon(\eta)$ (eq.(2.39)) approaches zero is governed by the equation

$$\vartheta_G \equiv \int_g^0 dx[\,\{\gamma_G(x) - \gamma_G(0)\}/\beta(x) + C_G/bx]$$

$$= \lim_{\eta\to\infty}\ [\ \varepsilon(\eta)\ \ell n\eta\ + \frac{C_G}{b}\ \ell n\ (\bar{g}/g)]\ \ ; \qquad (2.85)$$

i.e. $\varepsilon(\eta)$ decreases more slowly than $(\ell n\ \eta)^{-1}$:

$$\varepsilon(\eta) \sim (C_G/2b)\ \ell n\ell n\ \eta/\ell n\ \eta \quad , \quad (\eta \to \infty) \ .$$

This means that the leading singularity of $G^{as.}(\eta q)$ differs from the singularity $\eta^{\gamma_G(0)}$ observed in free-field theory; there exist logarithmic modifications of the form[8,39] $(\ell n\ \eta)^{C_G/2b}$:

$$G^{as.}(\eta q;\ g,m) = \eta^{\gamma_G(0)}\ [\,2bg^2\ell n\ \eta]^{C_G/2b}\ G^{as.}(q;0,m)\exp\vartheta_G$$

$$\{1 + O\,[\,\ell n\ell n\eta/\ell n\eta\,]\}. \qquad (2.86)$$

This is a non-perturbative result. It depends on the validity of assumption (I), because the factor $\exp\vartheta_G$ depends on the values of $\beta(x)$ and $\gamma_G(x)$ in the region $0 < x < g$. Also, a $(\ell n\ \eta)^{c_G/2b}$ singularity cannot be generated by finite-order perturbation theory for $G(\eta q)$, because in general, the power $c_G/2b$ is not an integer.

Of course, if there is no wave-function renormalization for G (i.e. $\gamma_G(x) \equiv \gamma_G(0)$), the leading singularity is the same as in free-field theory (parton model). In particular, the ratio

$$R(q^2/m^2;g) = \sigma(e^+e^- \to \text{hadrons})/\sigma^{as.}(e^+e^- \to \mu^+\mu^-)$$
$$= R^{as.}(q^2/m^2;g) + \{\text{non-leading powers}\} \qquad (2.87)$$

tends to the asymptotic value[8,39]

$$R = \sum_{\substack{\text{flavours}\\ \text{colours}}} Q^2 , \qquad (Q = \text{quark charge}). \qquad (2.88)$$

It is not difficult to find the leading asymptotic correction to this result. The two-loop contribution $R^{as.}(q^2/m^2;g)$ is $3g^2T(R)/16\pi^2$ relative to the one-loop contribution $\sum Q^2$; (the factor $3g^2/16\pi^2$ can be extracted from the Jost-Luttinger calculation[29] for QED). So, if the right-hand side of the equation

$$R^{as.}(\eta^2q^2/m^2;g) = R^{as.}(q^2/m^2;\bar{g}) \qquad (2.89)$$

is expanded in $\bar{g}$ about $\bar{g} = 0$, the result is[51]

$$R^{as.}(\eta^2q^2/m^2;g) = \sum Q^2 \{1 + 3T(R)\bar{g}^2/16\pi^2 + O(\bar{g}^4)\} , \ (\eta \to \infty) . \qquad (2.90)$$

Thus the first non-leading term is positive and decays logarithmically. This result illustrates the difference between asymptotic freedom and asymptotic free-field behaviour; in free-field theory the leading correction is negative and approaches zero very rapidly: $O(\eta^{-4})$.

Asymptotic freedom allows us to make very precise statements about amplitudes which involve a short-distance limit. Unfortunately the reverse is true elsewhere, especially for the problem of deciding whether the theory possesses a respectable S-matrix or not. Most Yang-Mills theories (e.g. eq.(2.65)) do not possess a perturbative S-matrix because of infrared singularities at thresholds for the production of massless gauge mesons, and in any case, we do not want massless states to appear in the hadronic spectrum. The perturbative solution to this problem is to break the gauge symmetry spontaneously, i.e. to introduce enough Higgs-Kibble[52] scalar fields Φ to make all of the gauge mesons massive. The presence of Φ^4 couplings means that eq.(2.61) is much more difficult to satisfy[39,53]. Nevertheless, asymptotically free theories with scalar mesons do exist[54] and in particular, there are models[55] in which all perturbative states are massive.

An immediate reason for not pursuing this line further is that it involves an unrealistic perturbative constraint. The spectrum of a summed-up theory is unlikely to bear much resemblance to the perturbative spectrum(*) and hadronic states are definitely not perturbative.

Another approach is to assume that there is a dynamical mechanism which breaks gauge invariance spontaneously[58]. There are no Φ fields; instead, the infrared singularities of perturbation theory are supposed to sum to scalar meson poles which simulate the Higgs-Kibble mechanism in a non-perturbative manner. Dynamical generation of mass from Lagrangians which contain no dimensionful parameters has been exhibited[59] for four-dimensional scalar QED and some two-dimensional models.

However, the most likely possibility is that the strong interaction gauge symmetry (colour) is exact[39,60]. One would break gauge invariance spontaneously only if it were desirable to produce a hadronic spectrum with loss of non-degenerate coloured states, including states with quark-like quantum numbers. Note that isospin and ordinary SU(3) have nothing to do with breaking a strong gauge group, because they are approximate degeneracy symmetries. Similarly, observable hadronic currents j_μ (electromagnetic, chiral, etc.) are colour-invariant; otherwise, instead of current conservation or partial conservation, we would have

$$D^{ab} \cdot j^{b} = \text{zero or soft operator} \qquad (2.91)$$

where D^{ab} is the appropriate covariant derivative (2.67) with strong coupling constant g.

So we return to the colour-flavour model mentioned at the beginning of this section. All flavours, observable operators and observable states are colour singlets. Perturbative amplitudes become infrared singular either as some of the external momenta are taken on-shell, or as a partial sum of the external momenta becomes lightlike. We are better off with the singularities because then we can at least contemplate the possibility[60] that these infrared effects, summed to all orders in perturbation theory, confine quarks, gluons and coloured "bound states" (constituent quarks, etc.). The precise criterion for this is poorly understood, but there are some hopeful signs :

(*) A variety of classes of graphs can be summed to produce non-perturbative states. The traditional bound-state picture[56] typically involves the ladder approximation for a scattering amplitude. The semiclassical approaches[57] which are currently fashionable treat observable states as coherent superpositions of infinitely many perturbative states.

(i) The Bloch-Nordsieck solution[61)] of the infrared problem in QED, which results in the existence of observable photon and fermion states, fails when applied to Yang-Mills theories[62,63)].
(ii) The lattice approximation[64)] for the colour-flavour model indicates quark confinement.

It is surprising that perturbative infrared singularities in gauge-invariant channels receive so little attention, because there we do have a precise criterion : observable amplitudes must not be infrared singular. In other words, when summed to all orders, the singularities should conspire to shift all gauge-invariant thresholds away from zero mass. Gauge-invariant singularities occur even if the fermions in eq. (2.65) are all massive ($\mathcal{M} \neq 0$). For example, consider the stress-energy tensor

$$\theta_{\mu\nu} = -F_{\mu\alpha} F^{\alpha}_{\nu} + \frac{1}{4} g_{\mu\nu} F^2 + \frac{i}{4} \bar{\psi}(\gamma_\mu \overleftrightarrow{D}^f_\nu + \gamma_\nu \overleftrightarrow{D}^f_\mu) - g_{\mu\nu}\bar{\psi}(i\not{D}_f - \mathcal{M})\psi$$
$$+ \{\text{renormalization corrections}\} . \qquad (2.92)$$

The diagram shown in fig. 11(a) is responsible for the singularity

$$i \int d^4x\, e^{iq.x} T\langle 0|\theta_{\alpha\beta}(x)\theta_{\gamma\delta}(0)|0\rangle \sim -\{d(G)/60\pi^2\}\, q_\alpha q_\beta q_\gamma q_\delta \ln q^2$$
$$[1 + O(g^2)], \quad (q^2 \to 0) \qquad (2.93)$$

where d(G) is the number of generations of the gauge group G. More generally, amplitudes $T\langle 0|O_1 O_2 O_3 \ldots|0\rangle$ involving gauge-invariant operators O_i can be infrared singular whenever the diagrams of fig. 11(b) (intermediate gluon or ghost-antighost pair) are permitted by selection rules. Of course, additional singularities appear if some of the fermions are massless.

Note that observable amplitudes must be regular (not merely non-singular) at lightlike momentum transfers. For example, if Q is the gauge-invariant operator

$$Q(x) = \varepsilon_{\alpha\beta\gamma\delta} F^{\alpha\beta}F^{\gamma\delta} + \{\text{renormalization corrections}\} , \qquad (2.94)$$

there is a non-singular branch cut at $q^2=0$ in lowest-order perturbation theory for the two-point function

$$i \int d^4x\, e^{iq.x}\, T\langle 0|Q(x)\, Q(0)|0\rangle \sim -\{2d(G)/\pi^2\}\, q^4 \ln q^2 [1+O(g^2)],$$
$$(q^2 \to 0) \qquad (2.95)$$

The theory makes sense only if branch cuts like $q^4 \ln q^2$ disappear when the sum to all orders in perturbation theory is performed. This means that a complete analysis of the infrared problem involves non-singular graphs with more than two intermediate gluons or ghosts (fig. 11(c)), as well as the diagrams of fig. 11(b).

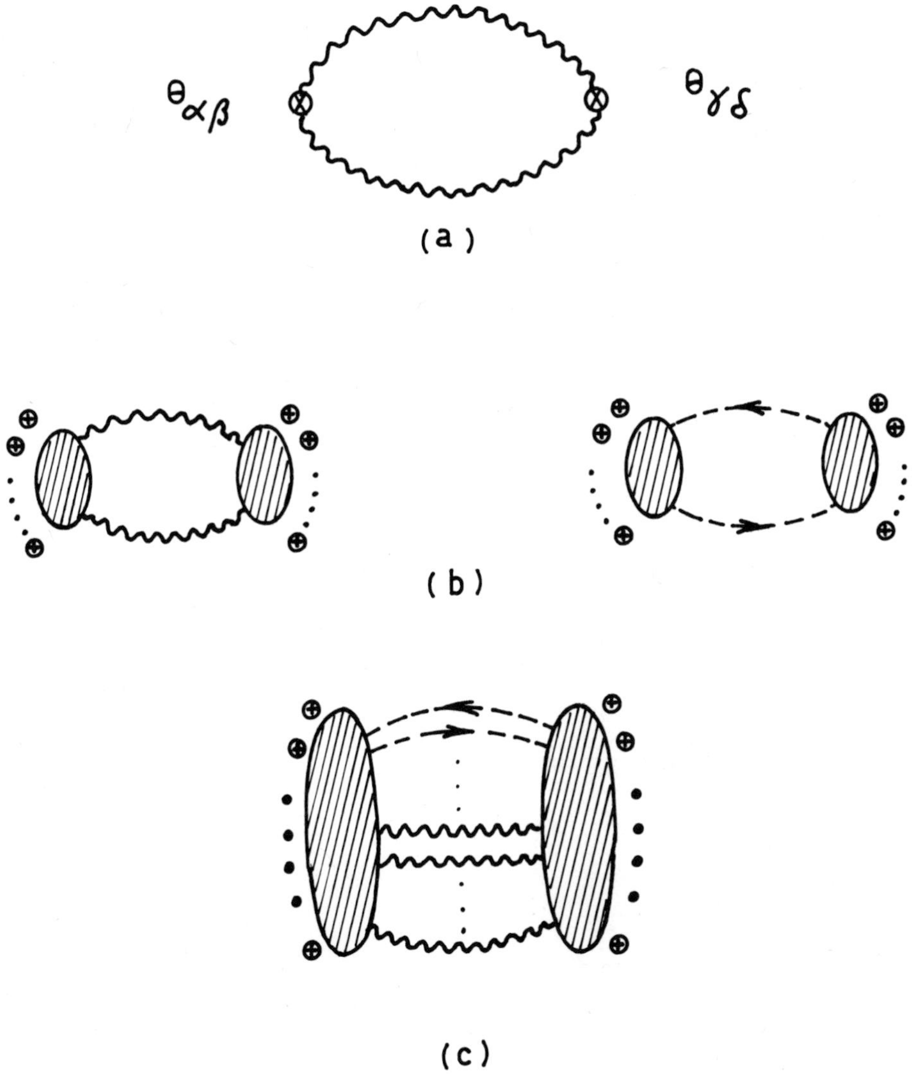

FIG. 11

The non-perturbative nature of the problem can be readily appreciated by considering the extreme case $\mathcal{M} = 0$ in (2.65), as advocated by Gross and Neveu[59] (but without breaking the colour symmetry(*). The only dimensional parameter in the theory is the renormalization subtraction point λ . Amplitudes obey equations such as (2.33) and (2.48) exactly (with λ replacing $\mathcal{M}$), so the infrared behaviour for the limit in which all external momenta approach zero can be discussed in terms of a non-trivial infrared-stable fixed point g_∞ (possibly at infinity). For example, consider eq.(2.48), (which is also valid for form factors of $T\langle\theta_{\alpha\beta}\theta_{\gamma\delta}\rangle$) . The criterion is that

$$\lim_{\eta\to\infty} \pi(\eta^2 q^2/\lambda^2; g)\Big|_{\mathcal{M}=0} = \pi(q^2/\lambda^2; g_\infty)\Big|_{\mathcal{M}=0} - \int_g^{g_\infty} dx\, K(x)/\beta(x) \tag{2.96}$$

should be finite with zero absorptive part, so $K(x)$ must approach zero sufficiently strongly as $x \to g_\infty$ to make $\int^{g_\infty} dx\, K(x)/\beta(x)$ converge.

The case $\mathcal{M} \neq 0$ is more complicated. Apart from the lattice approximation, the main lines of analysis will probably involve :
(i) theorems[63,65] that infrared singularities of amplitudes are given by the theory with massive fields omitted (except as external lines), up to finite renormalizations;
(ii) the improved CS equation[7], to be considered in the next chapter.

(*) In order to generate ordinary SU(3) breaking for the $\mathcal{M} = 0$ case, it must be supposed that there exist inequivalent methods of summing the theory, and that one of these methods is SU(3)-asymmetric.

III. OTHER METHODS

Gell-Mann-Low Analysis[3)]

Let us return to the subject of coupling-constant renormalization in quantum electrodynamics.

The coupling constant α used in practical calculations (α^{-1} = 137.036...) refers to the amplitude for an electron to emit a zero-frequency photon. According to eq.(2.23), charge renormalization is controlled by the wave-function renormalization of the photon propagator

$$D'_{F\mu\nu}(q) = -(i/q^2)(g_{\mu\nu}-q_\mu q_\nu/q^2)d(q^2/m^2,\alpha) - i\xi q_\mu q_\nu/q^4 \tag{3.1}$$

so the choice of α as coupling constant corresponds to the normalization condition

$$d(0,\alpha) = 1 \tag{3.2}$$

for $D'_{F\mu\nu}(q)$ at $q^2 = 0$.

Consider another renormalization procedure in which the coupling constant is given by the amplitude for the emission of a virtual photon with spacelike momentum λ . Now the photon propagator $(D'_{F\mu\nu})$ depends on q,λ,m and α_λ,

$$\alpha_\lambda(D'_{F\mu\nu})_\lambda = -(i/q^2)(g_{\mu\nu}-q_\mu q_\nu/q^2)D(q^2/\lambda^2,\ m^2/\lambda^2,\alpha_\lambda)-i\alpha_\lambda\xi_\lambda q_\mu q_\nu, \tag{3.3}$$

and is normalized at $q^2 = \lambda^2$:

$$D(1,m^2/\lambda^2,\alpha_\lambda) = \alpha_\lambda \tag{3.4}$$

When the renormalization prescription is altered by changing λ , the normalization of $D'_{F\mu\nu})$ changes by a factor ζ_3^{-1} (as in eq. (1.21)), but the combination $\alpha_\lambda(D'_{F\mu\nu})_\lambda$ is invariant :

$$D(q^2/\lambda_1^2,\ m^2/\lambda_1^2,\alpha_{\lambda_1}) = D(q^2/\lambda_2^2,\ m^2/\lambda_2^2,\alpha_{\lambda_2}) \tag{3.5}$$

$$= \alpha d(q^2/m^2,\alpha) \tag{3.6}$$

For $q^2 = \lambda_2^2$, eqs.(3.5) and (3.6) become

$$\alpha_{\lambda\sqrt{\rho}} = D(\rho,m^2/\lambda^2,\alpha_\lambda) \tag{3.7}$$

$$= \alpha d(\rho\lambda^2/m^2,\alpha) \tag{3.8}$$

where ρ is a convenient scale parameter :

$$\lambda_2 = \lambda_1\sqrt{\rho} \quad , \quad \lambda_1 = \lambda \quad . \tag{3.9}$$

The result of substituting (3.7) into (3.5) is a functional equation for the renormalization-group transformation[(*)] $\lambda \to \lambda\sqrt{\rho}$:

$$D(q^2/\lambda^2,\ m^2/\lambda^2, \alpha_\lambda) = D(q^2/\rho\lambda^2,\ m^2/\rho\lambda^2,\ D(\rho, m^2/\lambda^2, \alpha_\lambda)) \tag{3.10}$$

Application of $\partial/\partial\rho$ to eq.(3.10) at $\rho=1$ produces the renormalization-group equation

$$[X\partial/\partial X + M\partial/\partial M - \psi(\alpha_\lambda, M)\partial/\partial\alpha_\lambda]\ D(X, M, \alpha_\lambda) = 0 \tag{3.11}$$

with $X = q^2/\lambda^2$, $M = m^2/\lambda^2$, and

$$\psi(\alpha_\lambda, M) = [\partial D(\rho, M, \alpha_\lambda)/\partial\rho]_{\rho=1} \tag{3.12}$$

In general, eq.(3.11) cannot be integrated because ψ depends on two variables. However, GML[3] observed that all renormalized Feynman integrals remain convergent when M is s t equal to zero :

$$m = 0, \quad q^2 \neq 0, \quad \lambda^2 \neq 0 \quad . \tag{3.13}$$

For example, the subtracted integral

$$I(q, \lambda, m) = \int d^4p (p^2 - m^2)^{-1} \{[(p+q)^2 - m^2]^{-1} - [(p+\lambda)^2 - m^2]^{-1}\} \tag{3.14}$$

remains both infrared and ultraviolet convergent when the condition (3.13) is imposed. For the general case, ultraviolet power counting is not affected by setting m to zero, but infrared convergence has to be checked, either by inspection [3] or by appealing to Kinoshita theorem[1,67,68]. The result is an integrable equation of the same form as eq.(2.33),

$$[X\partial/\partial X - \psi(\alpha_\lambda)\partial/\partial\alpha_\lambda] D(X, 0, \alpha_\lambda) = 0 \tag{3.15}$$

where

$$\psi(Z) = \psi(Z, 0) \tag{3.16}$$

is known as the Gell-Mann-Low function.

Instead of analyzing $D(X, 0, \alpha_\lambda)$ further, let us make use of the connection (3.6-8) between $D(X, M, \alpha_\lambda)$ and the conventionally

---

(*) The connection between the GML analysis and the work of Stueckelberg and Petermann[2] was noted by Bogoliubov and Shirkov[28,66].

renormalized theory. Thus eq.(3.12) can be rewritten

$$\psi(\alpha d(\lambda^2/m^2,\alpha),m^2/\lambda^2) = [(\partial/\partial\rho)\alpha d(\rho\lambda^2/m^2,\alpha)]_{\rho=1}$$

$$= \partial/\partial\, \ell n(-\lambda^2)\, [\alpha d(\lambda^2/m^2,\alpha)] \quad (3.17)$$

Order-by-order in perturbation theory, the leading power in m^2 at $m=0$ can be isolated by substituting the expansions

$$\psi(Z,m^2/\lambda^2) = \psi(Z) + O\,[m^2\,\{\text{logs of } m\}] \quad (3.18)$$

$$\alpha d(\lambda^2/m^2,\alpha) = \alpha d^{as.}(\lambda^2/m^2,\alpha) + O\,[m^2\{\text{logs of } m\}] \quad . \quad (3.19)$$

Eq.(3.18) depends on the existence of the $M\to 0$ limit (3.13), while (3.19) is just a special case of (2.16); (i.e. $\alpha d^{as.}$ is the amplitude which appears in eqs.(2.24) and (2.26)). The result is an integrable equation

$$\psi(\alpha d^{as.}(\lambda^2/m^2,\alpha)) = \partial/\partial\,\ell n(-\lambda^2)[\alpha d^{as.}(\lambda^2/m^2,\alpha)] \quad (3.20)$$

with solution

$$\ell n(-\lambda^2/m^2) = \int_{\alpha d^{as.}(-1,\alpha)}^{\alpha d^{as.}(\lambda^2/m^2,\alpha)} dZ/\psi(Z) \quad (3.21)$$

The analysis of the GML equation (3.21) is very similar to that of eq.(2.35). The sign of the first term in the perturbation expansion [3,69]

$$\psi(Z) = Z^2/3\pi + Z^3/4\pi^2 + Z^4\{\xi(3) - 101/96\}/3\pi^3 + O(Z^5) \quad (3.22)$$

implies that $Z=0$ is an infrared-stable fixed point. If there is an ultraviolet-stable fixed point α_0,

$$\psi(\alpha_0) = 0 \qquad , \qquad (\alpha < \alpha_0) \ , \quad (3.23)$$

it follows from (3.21) that α_0 is the asymptotic charge of the photon propagator at short distances :

$$\alpha d^{as.}(\lambda^2/m^2,\alpha) \to \alpha_0 \qquad , \qquad (-\lambda^2 \to \infty). \quad (3.24)$$

If $\psi(Z)$ does not vanish for $Z > 0$, we must suppose that $\int dZ/\psi(Z)$ diverges at $Z = \infty$:

$$\alpha d^{as.}(\lambda^2/m^2,\alpha) \to \infty \quad , \qquad (-\lambda^2 \to \infty\ ). \quad (3.25)$$

Otherwise the theory is inconsistent because of the presence of a tachyonic singularity in $\alpha d^{as.}(\lambda^2/m^2,\alpha)$ at

$$(\lambda^2)_{\text{tach.}} = -\mathcal{m}^2 \exp \int_{q(\alpha)}^{\infty} dZ/\psi(Z) \qquad (3.26)$$

(where $q(\alpha)$ is shorthand for $\alpha d^{as.}(-1,\alpha)$). It is now obvious why the "ghost" (tachyon) problem[30] discussed previously (eq.(2.30)) is not a genuine difficulty : it is not reasonable to assume that the small-z behaviour $Z^2/3\pi$ is a good approximation for $\psi(Z)$ at large Z.

The connection between $\psi(Z)$ and the β-function (2.25) is found by applying the CS differential operator

$$[2\partial/\partial\ell n(-\lambda^2) - \alpha\beta(\alpha)\partial/\partial\alpha]$$

to eq.(3.21) :[32]

$$\psi(q(\alpha)) = \frac{1}{2}\,\alpha\beta(\alpha)dq(\alpha)\;/d\alpha,$$

$$q(\alpha) = \alpha d^{as.}(-1,\alpha) \quad . \qquad (3.27)$$

For conventional on-shell renormalization, the perturbative expansions of $q(\alpha)$ and $\beta(\alpha)$ are [3,29]

$$q(\alpha) = \alpha - 5\alpha^2/9\pi + (\zeta(3) + 65/648)\alpha^3/\pi^2 + O(\alpha^4) ,$$

$$\beta(\alpha) = 2\alpha/3\pi + \alpha^2/2\pi^2 - 121\,\alpha^3/144\pi^2 + O(\alpha^4) \quad . \qquad (3.28)$$

A feature of the GML formulation is that $\psi(Z)$ is a universal function. Unlike $\beta(Z)$, it does not depend on the renormalization prescription λ . Adler's proposal[32] for computing the fine-structure constant

$$\beta(\alpha_\infty) = 0 \qquad (\alpha_\infty^{-1} = 137.036\ldots\ ??) \qquad (3.29)$$

illustrates this point. Eq.(3.29) involves the β-function for on-shell renormalization. This makes sense because the fine-structure constant is the on-shell coupling constant. A different renormalization procedure would produce a different β-function ($O(\alpha^3)$ terms and higher) with an eigenvalue numerically different from α_∞. On the other hand, the corresponding asymptotic charge

$$\alpha_0 = q(\alpha_\infty) \qquad (3.30)$$

is renormalization-group invariant. It does not refer to a particular point in λ-space, so apart from the constraint $\alpha < \alpha_0$, its value has nothing to do with the value of the fine-structure constant.

To obtain renormalization-group equations for multiplicatively renormalized amplitudes G, one can use the fact that the ratio

$$G(p^2/\lambda^2,\ m^2/\lambda^2,\alpha_\lambda)/G(q^2/\lambda^2,\ m^2/\lambda^2,\alpha_\lambda)$$

is λ-independent for arbitrary sets of external momenta p and q. For example, the analogue of eq.(3.11) is

$$[X\partial/\partial X + M\partial/\partial M - \psi(\alpha_\lambda,M)\partial/\partial\alpha_\lambda - \chi(\alpha_\lambda,M)]\ G(X,M,\alpha_\lambda) = 0, \quad (3.31)$$

$$\chi(\alpha_\lambda,M) = -\ [(\partial/\partial\rho)\ \ln G(\rho^{-1},M/\rho,D(\rho,M,\alpha_\lambda))]_{\rho=1}\ . \quad (3.32)$$

The analysis of the M=0 limit is the same as that of eq.(2.33), so $\chi(\alpha_0,0)$ is the power of χ which characterizes the large-X behaviour of G :

$$\text{anomalous dimension} = \gamma_G(g_\infty) - \gamma_G(0) = [\chi(\alpha_0,0) - \chi(0,0)]/2 \quad (3.33)$$

Of course, we expect anomalous dimension to be an invariant of the renormalization group. This can be checked from eq.(1.38) and by considering transformations on eq.(3.32).

More complicated cases of coupling-constant renormalization can be handled by finding the appropriate generalization of the invariant function D in eq.(3.3). For example, consider the invariant

$$\bar{D} = [G^{(2)}(p_1)G^{(2)}(p_2)G^{(2)}(p_3)G^{(2)}(p_4)]^{1/2}\ \Gamma^{(4)}(p_1,p_2,p_3,p_4) \quad (3.34)$$

with $G^{(2)}$ = bosons propagator, $\Gamma^{(4)}$ = 4-log 1PI vertex, p_i = external momenta, $\Sigma p = 0$. In analogy with (3.4) the coupling constant g_λ for the φ^4 interaction is given by $\bar{D}$ at the subtraction point $p_i = \lambda_i$. The M→0 limit can be taken if p_i and λ_i are not exceptional. The resulting $\psi(g)$ function is universal with respect to transformations which scale all of the λ_i together. In gauge theories, one vertex (e.g. 3-gluon) can be used to define $\bar{D}$ and g_λ, and then gauge Ward identities fix the normalization of the other vertices (4-gluon, ghost-gluon, etc.).

The GML approach involves the same assumptions ((I) and (II) in the discussion after (2.32)) as the CS method, so the choice of CS or GML method (or variation thereof) is mostly a matter of convenience; (e.g. the GML method is ideal for two-dimensional QED[70]).

Improved CS Equation

According to assumption (II), non-leading powers in the asymptotic expansion of a perturbative amplitude do not compete with the leading power when summed to all orders. These non-leading powers reflect the presence of masses (or other dimensionful coupling constants) in the Lagrangian $\mathcal{L}$. The rule[4,7,33,71] for estimating these terms involves the maximal dynamical dimension d_Δ of the mass operators of $\mathcal{L}$; e.g.

$$d_\Delta = \begin{cases} 3 + \gamma_\Delta(g_\infty) & , \quad (\Delta = \bar{\psi}\psi) , \\ 2 + \gamma_\Delta(g_\infty) & , \quad (\Delta = \varphi^2/2) , \end{cases} \tag{3.35}$$

where γ_Δ is the γ-function (1.46) for the composite operator Δ. It states that the Nth set of non-leading powers of a perturbative amplitude $G(\eta q)$ sums to an amplitude which is *

$$O[\eta^{-N(4-d_\Delta)} \{\text{logs of } \eta\}] \quad , \qquad (N = 1,2,\ldots \text{not too large}) \tag{3.36}$$

relative to the summed-up leading power $G^{as.}(\eta q)$. Thus we must assume[4]

$$d_\Delta < 4 \tag{3.37}$$

in order to ensure the validity of assumption (II). Eq.(3.37) is automatically satisfied[7] if the theory is asymptotically free because the relevant anomalous dimension $\gamma_\Delta(0)$ vanishes.

Wilson introduced eq.(3.36) as one of the rules of his theory of broken scale invariance[4] and later justified it** within the context of the renormalization group by applying a variant of the GML analysis due to Eriksson[72]. The conventional GML procedure is to renormalize the mass on-shell; the renormalized mass is taken to be the position of the pole in the propagator and is regarded as being independent of the subtraction point λ. (This accounts for the absence of a non-trivial coefficient for the operator $M\partial/\partial M$ in eqs.(3.11) and (3.31)). Eriksson's method involves a mass parameter m defined in terms of the propagator at the subtraction point λ. In bare outline, what happens is that a change of λ induces a mass renormalization

---

* So $p(\Delta)$ can be replaced by $4-d_\Delta$ in eq.(2.53).

** See section IV of ref.33. Similar conclusions were obtained by Symanzik[71] for $\lambda\varphi^4$ theory in the CS formalism.

$$m_{\lambda'} = \zeta_m(\lambda',\lambda)m_\lambda \qquad (3.38)$$

as well as coupling-constant and wave-function renormalization. Because of the factor ζ_m, there is a function $(m_\lambda^2/\lambda^2)\chi_m(\alpha_\lambda, m_\lambda^2/\lambda^2)$ associated with $\lambda\partial/\partial\lambda\,(m_\lambda^2/\lambda^2)$ in the same way that $\psi(\alpha_\lambda, m_\lambda^2/\lambda^2)$ is related to $\lambda\partial/\partial\lambda(\alpha_\lambda)$ (as in eqs. 3.7) and (3.12), except that the change in renormalization prescription produces a different two-variable ψ-function). The variable m_λ^2/λ^2 acts like a coupling constant in the renormalization-group equations. Thus the anomalous dimension of Δ, $\chi_m(\alpha_o,0)/2$ (or $\chi_m(\alpha_o,0)$ for scalars), appears when the theory is expanded about the fixed point $(\alpha_o,0)$ in $(\alpha_\lambda, m_\lambda^2/\lambda^2)$ space. Eq.(3.37) is the condition for this fixed point to be UV-stable in the m_λ^2/λ^2 direction.

The result (3.36) was rediscovered by Weinberg[7] as a direct consequence of his improved CS equation

$$[\mu\partial/\partial\mu + \beta(g_R)\partial/\partial g_R + \gamma_m(g_R)m_R\partial/\partial m_R + \gamma(g_R)]\; G_R(\eta q; g_R, m_R, \mu) = 0 \qquad (3.39)$$

Eq.(3.39) is characteristic of "mass-independent" renormalization prescriptions in which there is a renormalization parameter μ but the β and γ functions do not depend on m_R/μ. Instead of a mass-insertion term, there is an ordinary derivative $m_R\partial/\partial m_R$ multiplied by a γ-function $\gamma_m(g_R)$ associated with mass renormalization. This means that eq.(3.39) can be directly integrated without taking a zero-mass limit.

The main step in the derivation of (3.39) is to find a suitable renormalization prescription. One possibility[7] is to multiply g_B, m_B and G_B by Z-factors given by setting $m_B = 0$ in the corresponding factors $Z(g_B, \Lambda/\mu, m_B/\mu)$ which relate bare and renormalized quantities in the GML prescription. Here, μ is the GML reference momentum λ and $Z_m = m_R/m_B$ is the Z-factor for the composite operator $\bar\psi\psi$. Assuming that cutoff-independent renormalized amplitudes are produced[7,73], this prescription is obviously mass-independent :

$$\begin{aligned}\beta(g_R) &= \lim_{\Lambda\to\infty} (\mu\partial/\partial\mu) g_R(g_B, \Lambda/\mu)\\ \gamma_m(g_R) &= \lim_{\Lambda\to\infty} (\mu\partial/\partial\mu)\; \ln Z_m(g_B, \Lambda/\mu)\\ \gamma(g_R) &= \lim_{\Lambda\to\infty} (\mu\partial/\partial\mu)\; \ln Z_G(g_B, \Lambda/\mu) \quad .\end{aligned} \qquad (3.40)$$

Eq.(3.39) is an immediate consequence of combining the formulas

$$G_R = Z_G\, G_B$$

$$(\mu\partial/\partial\mu) G_B(\text{fixed } q, g_B, m_B) = 0 \qquad (3.41)$$

Weinberg also observed that all of the properties needed in the derivation of eq.(3.39) are contained in 't Hooft's analysis[74] of the renormalization group for dimensionally renormalized[46,75-77] amplitudes. This line was subsequently studied in detail by many authors[78]. Briefly, the analysis depends on the following theorem [77,79] : in the dimensional renormalization scheme, counterterm vertices are polynomials in the renormalized mass parameter m_R as well as in the external momenta q . * Thus Z-factors depend on the unity of mass[77] (which accompanies each loop integral $\mu^{4-n} \int d^n p$), the complex parameter , and g_B, which has dimensionality $(4-n)\kappa$ (with $\kappa = 1/2$ for gauge theories, $\kappa = 1$ for φ^4 coupling):

$$Z = Z(g_B \mu^{(n-4)\kappa}, n) \qquad (3.42)$$

Once again, eq.(3.39) follows directly from (3.41) with[78]

$$\beta(g_R) = \lim_{n\to 4} (\mu\partial/\partial\mu)\, g_R(g_B\, \mu^{(n-4)\kappa}, n)$$

$$\gamma_m(g_R) = \lim_{n\to 4} (\mu\partial/\partial\mu) \ln Z_m(g_B\, \mu^{(n-4)\kappa}, n)$$

$$\gamma(g_R) = \lim_{n\to 4} (\mu\partial/\partial\mu) \ln Z_G(g_B\, \mu^{(n-4)\kappa}, n) \quad . \qquad (3.43)$$

The analogue of eq.(1.35) is

$$[\eta\partial/\partial\eta + m_R\partial/\partial m_R + \mu\partial/\partial\mu - d_G]\; G_R(\eta q; g_R, m_R, \mu) = 0 \qquad (3.44)$$

where the integer d_G is the mass dimensionality of G, so if d_G is absorbed in to the definition of the γ-function

$$\gamma_G(g_R) = d_G + \gamma(g_R) \quad , \qquad (3.45)$$

Eq. (3.39) can be written

---

* One proves (e.g. by induction in ℓ) that the ℓ-loop amplitude $(\partial/\partial m_R)^p \Gamma$ converges for sufficiently large integers p; here, Γ is a 1PI amplitude (fig. 1) with ℓ-loop counterterms ($\ell' < \ell$) included in the Lagrangian. Observe that the result is not true for conventionally renormalized theories : $Z = Z(\Lambda/m_R, g_B) =$ polynomial in g_B and $\ln(\Lambda/m_R)$.

$$[\eta\partial/\partial\eta - \beta(g_R)\partial/\partial g_R + [1-\gamma_m](g_R)][m_R \frac{\partial}{\partial m_R} - \gamma_G(g_R)]$$

$$G_R(\eta q; g_R, m_R, \mu) = 0 \tag{3.46}$$

The general solution of eq.(3.46) is[7)]

$$G_R(\eta q; g_R, m_R, \mu) = G_R(q; \bar{g}, \bar{m}, \mu) \exp \int_g^{\bar{g}} dx \gamma_G(x)/\beta(x) \tag{3.47}$$

where, in addition to the effective coupling constant $\bar{g} = \bar{g}(g_R, \eta)$, there is an "effective mass"

$$\bar{m} = m_R \eta^{-1} \exp \int_g^{\bar{g}} dx \gamma_m(x)/\beta(x) \quad . \tag{3.48}$$

The $\eta\to\infty$ limit of eq.(3.47) is controlled by the asymptotic behaviour of both $\bar{g}$ and $\bar{m}$:*

$$\bar{m} = m_R \eta^{\gamma_m(g_\infty)-1} \exp \int_g^{\bar{g}} dx\, [\gamma_m(x) - \gamma_m(g_\infty)] \,/\beta(x)$$

$$\sim m_R^{\gamma_m(g_\infty)-1} \{\text{logs of } \eta\} \quad , \quad (\bar{g} \to g_\infty) \; . \tag{3.49}$$

Now, apart from finite renormalizations, the massless amplitude $G_R(q; g_R, 0, \mu)$ may be identified with the leading-power amplitudes $G^{as.}_R(q; g, m)$ and $G(X, 0, g_\lambda)$ of the CS and GML analysis. Therefore the condition[7)]

$$\bar{m} \to 0 \, , \qquad \gamma_m(g_\infty) < 1 \tag{3.50}$$

is essential if assumption (II) is to be satisfied. Making the identification

$$d_\Delta = \begin{cases} 3 + \gamma_m(g_\infty) & (\Delta = \bar{\psi}\psi) \\ 2 + 2\gamma_m(g_\infty) & (\Delta = \varphi^2/2) \; , \end{cases} \tag{3.51}$$

we see that eq. (3.50) is equivalent to eq. (3.37).

Evidently, the terms N = 1,2,... of eq.(3.36) correspond to increasing powers of $\bar{m}$ in the Taylor expansion of eq.(3.47) about $\bar{m} = 0$:

---

* Note the analogy between the η-dependence fof $(\bar{g}, \bar{m}/m_R)$ and the λ-dependence of $(\alpha_\lambda, m_\lambda^2/\lambda^2)$ in Wilson's analysis[33)]. The mass-independence property simplifies the analysis : $\eta\partial/\partial\eta(\bar{g}, \ell n\, \bar{m}/m_R)$ is given by the integrable expression $(\beta(\bar{g}), \gamma_m(\bar{g})-1)$, whereas the equations for $\lambda\partial/\partial\lambda(\alpha_\lambda, m_\lambda^2/\lambda^2)$ are coupled.

$$G_R(\eta q; g_R, m_R, \mu) \sim \{\exp \int_g^{\bar{g}} dx\ \gamma_G(x)/\beta(x)\} \sum_{r=0}^{\cdots} (\bar{m}^r/r!)$$
$$[(\partial/\partial m)^r G_R(q; g_R, m, \mu)]_{m=0} \qquad (3.52)$$

However, it is necessary to insert the warning "not too large" in (3.36), because in general, the limit

$$\lim_{m\to 0} (\partial/\partial m)^r\ G_R(q; g_R, m, \mu) \quad , \text{(non-exceptional } q) \qquad (3.53)$$

does not exist if r is too large[7]. If m_R is a fermionic mass parameter, trouble first appear when $(\partial/\partial \bar{m})^3$ acts on a single internal propagator, producing a logarithmic infrared singularity in the limit $m \to 0$:

$$\bar{m}^{-3}(\partial/\partial\bar{m})^3 \int^{\Lambda} d^4p/(\not p - \bar{m}) = O(\bar{m}^{-3}\ \ell n\ \bar{m}) \quad . \qquad (3.54)$$

Higher derivatives $(r > 3)$ do not reduce the power further :

$$\text{e.g. } \bar{m}^{-5}(\partial/\partial\bar{m})^5 \int^{\Lambda} d^4p/(\not p - \bar{m}) = O(\bar{m}^{-3}) \ . \qquad (3.55)$$

For scalar masses m_R, the r=2 term (3.53) is infrared singular.

The restriction of (3.36) to small values of N means that the non-leading terms are not completely under control. In particular, the logarithmic powers β in $O(\bar{m}^{-3}\ \ell n^{\beta}\ \bar{m})$ (for fermions) should be further investigated; for example, are there models in which β is bounded as the order of perturbation grows ? I think that β is larger than 1 in general, because the coefficient of the term (3.54) in the expansion of $\bar{m}^3(\partial/\partial\bar{m})^3 G_R$ is a zero-mass amplitude evaluated at exceptional momenta :

$$\text{coefficient} = \lim_{m\to 0} G_R(p=0;\ -p=0,\ q;\ \bar{g}, m, \mu) \qquad (3.56)$$

Here (p,-p) are external momenta for the fermion-antifermion pair obtained by cutting the internal propagator $(\not p - m)^{-1}$ in (3.54). In particular, the zero-mass limit can produce logarithms in the fermionic self-energy which modify the denominator

$$(\not p - \bar{m})^{-1} \to (\not p\ \ell n^{\alpha}\ (p^2/\mu^2) - \bar{m})^{-1} \qquad (3.57)$$

and generate extra logarithms of $\bar{m}$ in eq. (3.54).

Eq.(3.46) is valid for all momenta q, so there is no need to restrict its application to ultraviolet limits. Thus the discussion of infrared singularities at the end of Chapter II can be extended in an obvious way to include the case $m \neq 0$. For example the analogous of eq.(2.96) is

$$\lim_{\eta\to 0} \pi_R(\eta^2 q^2; g_R, m_R, \mu) = \pi_R(q^2; g_\infty, m_\infty, \mu) - \int_{\bar{g}}^{g_\infty} dx\ K(x)/\beta(x) \quad (3.58)$$

Unfortunately, the asymptotic value m_∞ of $\bar{m}$ $(\eta \simeq 0)$ is not known because the IR-stable fixed point g_∞ is not at the origin.

IV. OPERATOR-PRODUCT EXPANSIONS

In order to apply the renormalization group to topics such as current algebra and deep inelastic electroproduction, it is necessary to consider ultraviolet limits for subsets of the external lines of amplitudes. The appropriate tool is Wilson's expansion for operator products at short distances[4)].

Short-Distance Limit

The short-distance expansion[4,15,80-83)]

$$\prod_{i=1}^{K} A_i(x+\rho\xi_i) \sim \sum_n \mathcal{C}_n(\rho\xi_j-\rho\xi_k)O_n(x) \ , \ (\rho\to 0, \text{ fixed } x,\xi_i) \tag{4.1}$$

is an asymptotic expansion of the operator product ΠA_i into local operators $O_n(x)$ (including the identity operator I), with c-number coefficient function $_n$ depending on the various coordinate differences $\rho(\xi_j-\xi_k)$. The product ΠA_i may be unordered (e.g. multiple commutators), or ordered (T-product, anti-T-product,...), or some combination thereof. The terms $\mathcal{C}_n O_n$ in the sum Σ_n are ordered according to the strength of the singularity of $\mathcal{C}_n$ in the scaling variable ρ :*

$$\lim_{\rho\to 0} \ [\{\Pi A_i - \sum_{n=0}^{N} \mathcal{C}_n(\rho)O_n(x)\}/\mathcal{C}_N(\rho)] = 0 \ , \quad (N=0,1,2,\ldots) \tag{4.2}$$

The effect of expanding about $x' = x-\rho\xi$ instead of x is to change the contributions of derivative operators :

$$O_n(x'+\rho\xi) \quad \sum_{r=0}^{\cdots} (\rho^r/r!)(\xi\cdot\partial/\partial x)^r O_n(x) \tag{4.3}$$

In free -field theory, the expression (4.1) is trivially obtained by Taylor-expanding the Wick expansion of ΠA_i. For example, consider the Wick expansion

$$\begin{aligned} T\{\mathcal{J}^i_\mu(\xi)\mathcal{J}^j_\nu(0)\} = \ & \{\overline{\psi}(\xi)\gamma_\mu(\lambda^i/2)\psi(\xi)\overline{\psi}(0)\gamma_\nu(\lambda^j/2)\psi(0)\}\ I \\ & + \ :\overline{\psi}(\xi)\gamma_\mu(\lambda^i/2)\psi(\xi)\ \overline{\psi}(0)\gamma_\nu(\lambda^j/2)\psi(0): \\ & + \ :\overline{\psi}(0)\gamma_\nu(\lambda^j/2)\psi(0)\overline{\psi}(\xi)\gamma_\mu(\lambda^i/2)\psi(\xi)\ : \\ & + \ :\overline{\psi}(\xi)\gamma_\mu(\lambda^i/2)\psi(\xi)\overline{\psi}(0)\gamma_\nu(\lambda^j/2)\psi(0)\ : \end{aligned}$$

---

* Unlike its cousin, the light-cone expansion[84,85)], the short-distance expansion does not involve performing an infinite summation $\sum_{n=0}^{\infty}$

for SU(3) currents

$$\mathcal{J}^i_\mu(x) = :\bar{\psi}(x)\gamma_\mu(\lambda^i/2)\psi(x): \ , \quad (\lambda^i\text{=matrices for ordinary SU(3)}) \tag{4.5}$$

with $\overbrace{\psi(0)\bar{\psi}(\xi)} = i \int\{d^4q/(2\pi)^4\}\ e^{iq\cdot\xi}/(\not{q}-m+i\varepsilon)$

$$\sim -\{i/2\pi^2\}\ \ \xi\cdot\gamma/(\xi^2-i\varepsilon)^2 \quad , \quad (\xi \to 0). \tag{4.6}$$

When expanded about $\xi = 0$, each term on the right-hand side of (4.4) contributes to the Wilson expansion :

$O_n = I$, (first term of (4.4));

O_n = two-fermion operators

$= :\bar{\psi}(x)\{\lambda,\gamma \text{ matrices}\}\ \{\overleftrightarrow{\partial\partial\ldots\partial}\}\ \ \psi(x):$

(second and third terms of (4.4));

O_n = four-fermion operators

$= :\bar{\psi}(x)\{\lambda,\gamma\}\ \{\overleftrightarrow{\partial\partial}\ldots\partial\}\ \psi(x)\bar{\psi}(x\}\ \lambda,\gamma\}\ \{\partial\partial\ldots\ \overleftrightarrow{\partial}\}\psi(x)\{\lambda,\gamma\}\ldots$
(last term of (4.4)). (4.7)

The expansion of $\overbrace{\psi(0)\bar{\psi}(\xi)}$ produces terms of the form ξ^p or $\xi^p \ln(\xi^2m^2)$, so we have

$$\mathcal{C}_n(\xi) = \xi_\alpha\xi_\beta\ldots\xi_\omega(\xi^2)^p\ \{\text{constant, or (const.)}\ln(m^2\xi^2)\}\ , \quad (p\text{=integer}). \tag{4.8}$$

The logarithm is at least $O\ (m^3 \ln m)$ in eq.(4.6) (compare (3.54)), so it contributes to unimportant non-leading terms in the Wilson expansion.

The expansion (4.1) remains valid in renormalized perturbation theory[71,81,83,86]. The operators A_i, O_n are provided with counter-terms so that they yield finite matrix elements. In gneral, there are more operators O_n than in free-field theory. For example, the expansion of $T\{\mathcal{J}_\mu \mathcal{J}_\nu\}$ involves six-fermion,eight-fermion... operators as well as renormalized versions of the operators in (4.7). The operators O_n can also depend on other fields appearing in the Lagrangian (e.g. gluon and ghost fields in gauge theories). The properties[71,87] of the coefficient functions $\mathcal{C}_n$ are very similar to those of the asymptotic parts $G^{as.}$ of renormalized Green's functions; they become coupling-constant dependent and exhibit extra logarithmic singularities compared with eq.(4.8),

$$\mathscr{C}_n(\xi) \sim \xi_\alpha \xi_\beta \ldots \xi_\omega (\xi^2)^p \text{ Polynomial } (\ell n \xi^2, g_R) \tag{4.9}$$

or more generally (in analogy with eq.(2.21))

$$\mathscr{C}_n(\rho\xi_j - \rho\xi_k) \sim \rho^p \text{ Polynomial } (\ell n\rho, g_R) \ , \quad (p\text{=integer}) \tag{4.10}$$

Since the most interesting applications of eq.(4.1) are non-perturbative a proof based on the general principles of axiomatic field theory would be very desirable. The main difficulty to be overcome is that there is no general argument which forbids the appearance of infinite oscillations in matrix elements of ΠA_i as ρ tends to zero[82]. A rather complete analysis is possible[82] if oscillations are assumed to be absent.

Most practical applications[4,88] involve products ΠA_i of observable hadronic currents : the electromagnetic current J_μ; the energy-momentum tensor $\theta_{\mu\nu}$, the currents $\mathscr{F}^i_\mu, \mathscr{F}^i_{5\mu}$ associated with chiral SU(3 x SU(3) , and so on. It is a consequence of (4.1) that the corresponding operators O_n are also observable. Naturally[4], the expansion respects the conservation and covariance properties of ΠA_i for exact symmetries (Poincaré invariance, isospin, charge conjugation, etc.) so it is often convenient to isolate a sector (J,I,..) of the expansion containing operators O_n with spin J, isospin I, etc. its a rule[4], the leading power of ρ in a given sector also conserves symmetries which are only approximate (e.g. ordinary SU(3), chiral SU(2) x SU(2)). For example, the leading power of ρ in a given sector also conserves symmetries which are only approximate (e.g. ordinary SU(3), chiral SU(2) x SU(2)). For example, the leading contribution proportional to $\theta_{\mu\nu}$ in the expansion

$$\mathscr{F}^i_{5\mu}(x+\xi)\ j_{5\nu}(x-\xi) \sim \ldots + \mathscr{C}^{ij}_{\mu\nu\alpha\beta}(\xi)\theta^{\alpha\beta}(x) + \ldots \ , \quad (\xi\to 0) \tag{4.11}$$

satisfies the constraints

$$\mathscr{C}^{ij}_{\mu\nu\alpha\beta} = \delta^{ij}\mathscr{C}_{\mu\nu\alpha\beta}$$
$$\partial^\mu \mathscr{C}_{\mu\nu\alpha\beta}(\xi) = 0 = \partial^\nu \mathscr{C}_{\mu\nu\alpha\beta}(\xi) \quad . \tag{4.12}$$

As in the discussion of the asymptotic properties of Green's functions, it is desirable to classify contributions to a given sector by powers of the scaling variable ρ, not by logarithms : *

$$[\mathscr{C}_{n+1}(\rho)/\mathscr{C}_n(\rho)]_{(J,I,\ldots)} = O\,[\rho^{-p(n)}] \ , \quad (p(n)>0). \tag{4.13}$$

---

* The consistency conditions used to analyse $\pi^\circ \to 2\gamma$ decay[89] involve coefficient functions classified according to (4.13)

The most important point is that the expansion (4.1) should not depend on the matrix element to which it is applied. The general condition for this to be true is

$$<F| \prod_{i,j} \{A_i(x+\rho\xi_i)B_j(y_j)\}|I> \sim \sum_n \mathcal{C}_n(\rho)<F| \prod_j\{O_n(x)B_j(y_j)\}|I> , \tag{4.14}$$

where $|I>$, $|F>$ are vacuum or on-shell states, the matrix element is complete (i.e. includes disconnected pieces), $\Pi_{i,j}$ and Π_j refer to unordered or ordered products *, and the limit $\rho\to 0$ is taken with

(i) all momenta and other parameters of $|I>,|F>$ fixed,

(ii) the coordinates x,ξ,y_j fixed, $(y_j\neq x)$.

Note : (a) When ordered products (such as T-products) are involved, there can be ambiguities proportional to products of $\delta(\xi_i-\xi_j)$ and $\delta(y_i-y_k)$ and derivatives thereof. For example, the renormalization of the left-hand side of eq.(4.14) may involve counterterms of the form

$$\chi_{A_1}\chi_{A_2} \ldots \chi_{B_1}\chi_{B_2} \ldots \Delta(A_1A_2\ldots B_1B_2 \ldots)$$

in eq.(1.39). The $\rho\to 0$ limit of these δ-function terms need not match the δ-functions associated with counterterms

$$\chi_{O_n}\chi_{B_1}\chi_{B_2} \ldots \Delta(O_nB_1B_2 \ldots)$$

on the right-hand side.

(b) Even in free-field theory, the state vector

$$|\varphi(\rho)> = \int d^4x \prod_i d^4\xi_i \; f(x,\xi_1,\xi_2,\ldots)[\{\Pi A_i - \sum_{n=0}^{N} \mathcal{C}_n O_n\}/\mathcal{C}_N]|I>, \tag{4.15}$$

(f = suitable smearing function)

does not vanish :

$$<\varphi(\rho)|\varphi(\rho)> \longrightarrow 0 \quad ,$$

$$<F \;|\varphi(\rho)> \longrightarrow 0 \quad , \qquad (\rho\to 0) \; ; \tag{4.16}$$

i.e. the limit $\rho\to 0$ is "weak".

---

* More precisely, to products which can be constructed from discontinuities of the completely time-ordered product.

For most operators $B_j(y_j)$, it is absolutely mandatory that the coordinates y_j (and not the corresponding momenta) be held fixed as ρ tends to zero. This point will be illustrated by considering the connected amplitude of fig. 12(a) in the tree approximation, with

$$B(y) = :\bar{\psi}(y)\ (\partial^2)^{25}\psi(y): \tag{4.17}$$

The most singular q-number contribution generated by the Wick expansion (4.4) is

$$T\{\mathcal{J}^i_\mu(\xi)\mathcal{J}^j_\nu(0)\} \sim f^{ijk}\mathcal{C}^\lambda_{\mu\nu}(\xi)\,\mathcal{J}^k_\lambda(0) \quad , \qquad (\xi \to 0), \tag{4.18}$$

where $\mathcal{C}_{\mu\nu\lambda}(\xi)$ scales as ξ^{-3} and f^{ijk} are the structure constants for ordinary SU(3). (This singularity generates the equal-time commutator

$$[\mathcal{J}^i_0, \mathcal{J}^j_0\,]_{E.T.} = if^{ijk}\delta^3(\vec{\xi})\mathcal{J}^k_0$$

of current algebra). Thus the momentum-space singularity is

$$\int d^4\xi\ e^{iq.\xi}\mathcal{C}_{\mu\nu\lambda}(\xi) = O(q) \ , \qquad (q \to \infty) \ . \tag{4.19}$$

The graphs responsible for this contribution are shown in fig.12(b).

We must also consider the graphs of fig.12(c). If the coordinate y, z_1, z_2 are held fixed as $\xi \to 0$, the leading term contributed by these graphs is independent of ξ (i.e. $O(\xi^{-3})$ relative to (4.18)) and is given by the graph of fig. 12(d), with

$$Q_{\mu\nu}(0) = :\bar{\psi}(0)\gamma_\mu(\lambda^i/2)\psi(0)\bar{\psi}(0)\gamma_\nu(\lambda^j/2)\psi(0): \quad . \tag{4.20}$$

This agrees with the result of applying the limit $\xi \to 0$ to the last term of eq.(4.4). Thus the contribution $Q_{\mu\nu}$ is always present in the expansion of $T\{\mathcal{J}_\mu\mathcal{J}_\nu\}$ and does not depend on whether B(y) is present or not.

On the other hand, if we transform to momentum space and let the momenta of the current $\mathcal{J}_\mu, \mathcal{J}_\nu$ become large ($q = \eta\ell + c; \eta\to\infty\ , \ell^2\neq 0$) with the momenta of ψ, $\bar{\psi}$ and B held fixed, the result

$$\{\text{Fig.12(c)graphs}\} = O(q^{48}) \quad . \tag{4.21}$$

dominates the "expected" term (4.19). Thus naive substitution of the operator-product expansion for $T\{\mathcal{J}_\mu\mathcal{J}_\nu\}$ gives the wrong answer. However, it is very easy to explain the result (4.21) in the language of operator-product expansions. For simplicity, let us set the momentum of the B operator equal to zero and consider the limit $\xi \to 0$ of the amplitude

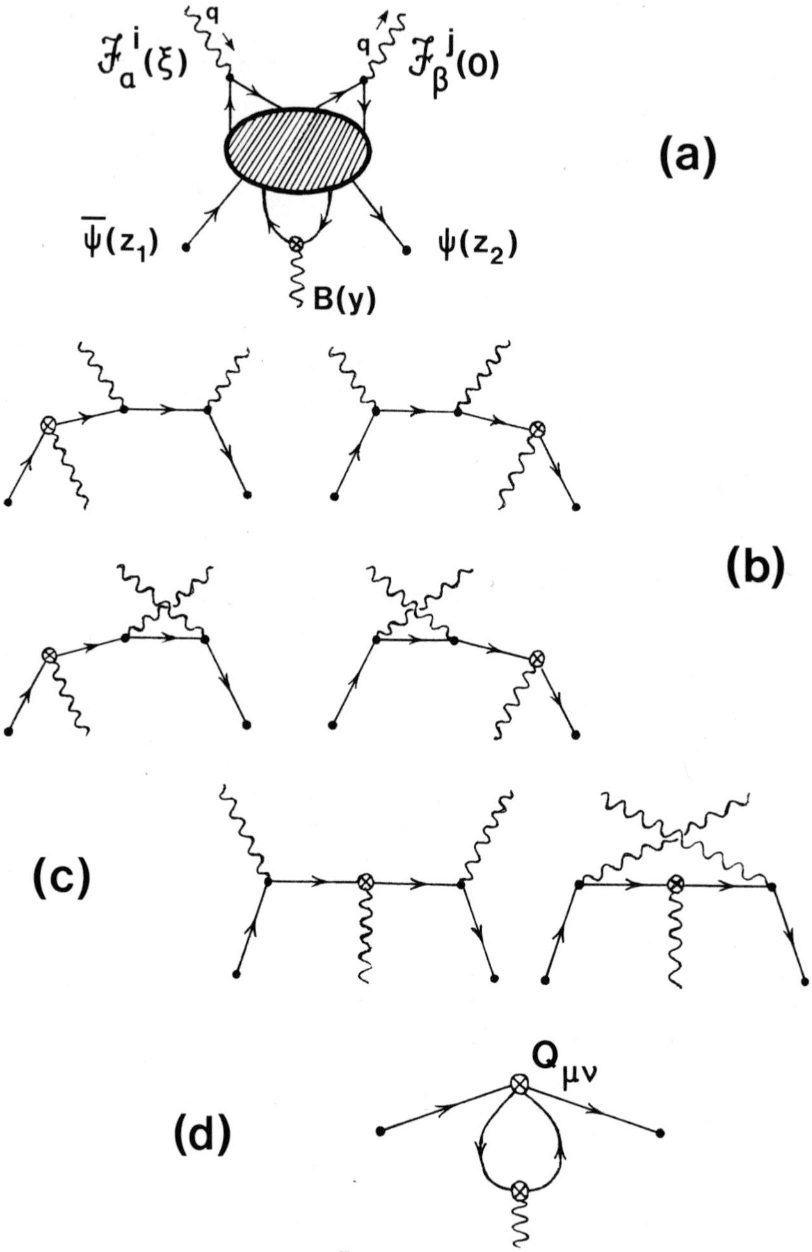
q
q
$\mathcal{J}_\alpha^i(\xi)$
$\mathcal{J}_\beta^j(0)$
(a)
$\overline{\psi}(z_1)$
$\psi(z_2)$
B(y)
(b)
(c)
$Q_{\mu\nu}$
(d)

Fig. 12

$$\mathcal{A}(Z_1,Z_2,\xi) = \int d^4y\ T\langle 0|\ \bar{\psi}(Z_1)\psi(Z_2)B(y)\mathcal{J}^i_\alpha(\xi)\mathcal{J}^j_\beta(0)|0\rangle\ . \tag{4.22}$$

The point is that, in addition to the expansion of $T\{\mathcal{J}_\mu\mathcal{J}_\nu\}$, it is necessary to consider the expansion

$$T\{B(y)\mathcal{J}^i_\alpha(\xi)\mathcal{J}^j_\beta(0)\} \sim \mathcal{C}^{ij}_{\alpha\beta}(\xi,y):\bar{\psi}(0)\psi(0): + \ldots,\ (\xi,y\to 0), \tag{4.23}$$

$$\mathcal{C}^{ij}_{\alpha\beta}(\rho\xi,\rho y) = O(\rho^{-56})\ . \tag{4.24}$$

Substituting into (4.22), we find

$$\mathcal{A}(Z_1,Z_2,\xi) \sim \int d^4\ \mathcal{C}^{ij}_{\alpha\beta}(\xi,y)\ T\langle 0|\bar{\psi}(Z_1)\psi(Z_2):\bar{\psi}(0)\psi(0):|0\rangle,\ (\xi\to 0). \tag{4.25}$$

Counting powers in coordinate space, we have 4 powers of ρ for the integral $\int d^4y$ to be added to the power -56 in (4.24) :

$$\int d^4y\ \mathcal{C}^{ij}_{\alpha\beta}(\xi,y) = O(\xi^{-52})\qquad,\quad (\xi\to 0) \tag{4.26}$$

Fourier transforming (4.26) yields the result

$$\int d^4\xi\ e^{iq.\xi}\int d^4y\ \mathcal{C}^{ij}_{\alpha\beta}(\xi,y) = O(q^{48})\quad,\quad (q\to\infty) \tag{4.27}$$

in agreement with eq.(4.21). In other words, it is not sufficient to substitute (4.18) into (4.22) because, for any non-zero value of ξ (no matter how small), the integral $\int d^4y$ always contains the region $y = O(\xi)$ in which the contribution (4.23) dominates.

It can be seen from the literature[15,71,81,86,90] that a complete derivation of the Wilson expansion in perturbation theory is necessarily non-trivial. Indeed, explicit derivations of (4.14) have been given only for the following special cases

$$\prod_{i=1}^{L} A_i(x+\rho\xi_i) = \prod_{i=1}^{L}\ \{\varphi(x+\rho\xi_i)\ \text{ or }\ \partial_\alpha\partial_\beta\ldots\partial_\omega\varphi(x+\rho\xi_i)\},$$

$$\prod_{j=1}^{N} B_j(y_j) = \prod_{j=1}^{N}\ \varphi(y_j) \tag{4.28}$$

in φ^4 theory; (for L arbitrary, see ref. 90). However, the main idea of the derivation is the same for all cases (i.e. simple or composite operators A_i, B_j in any renormalizable theory) and can be summarzied fairly simple.

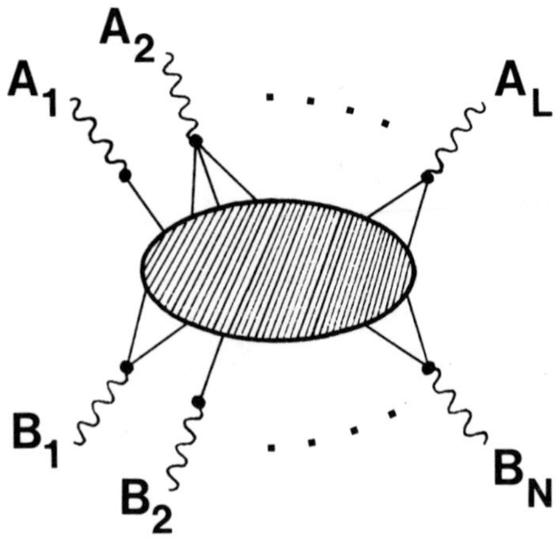

Fig. 13

Consider the complete time-ordered amplitude (fig. 13)

$$\mathcal{A} = T\langle 0 \left| \prod_{i=1}^{L} \prod_{j=1}^{N} A_i(\rho\xi_i) B_j(y_j) \right| 0\rangle \tag{4.29}$$

which corresponds to the choice x=0 , $|I\rangle = |F\rangle = 0$ in (4.14). In fig. 13, wavy lines correspond to sources for the operators A_i, B_j (as in eq.(1.7)) and non-wavy lines indicate propagators of fields (gluon, ghost, fermion, etc.) from which the Lagrangian and A_i, B_j are constructed. Some of the source functions X can be chosen to generate the field operator themselves; e.g., see the lines labelled A_1 and B_2 in fig. 13.

The amplitude $\mathcal{A}$ includes disconnected contributions, so the first step is to extract all factors of the form

$$G_S = T\langle 0 \left| \prod_{i\in S} A_i(\rho\xi_i) \right| 0\rangle \tag{4.30}$$

where S is any non-empty subset of {1,...,L} . We already know how to deal with the $\rho \to 0$ limit of G_S, because in momentum space, it becomes the $\eta \to \infty$ limit of $G_S(\eta q + c)$ at non-exceptional q . Summing over all possible sets S (including the complete set {1,..., }), we have [15,81]

$$T \prod_{i=1}^{L} A_i(\rho\xi_i) = :\prod_{i=1}^{L} A_i(\rho\xi_i): + \sum_S T\langle 0 | \prod_{i\in S} A_i(\rho\xi_i) | 0\rangle : \prod_{i\notin S} A_i(\rho\xi_i): \tag{4.31}$$

where the genralized Wick products*

$$:\prod_i{}' A_i(\rho\xi_i): , \qquad \left(\prod_i{}' = \prod_{i\notin S} \text{ or } \prod_{i=1}^{L}\right)$$

correspond to time-ordered amplitudes

$$\mathcal{A}' = T\langle 0 | \prod_{j=1}^{N} :\prod_i{}' A_i(\rho\xi_i): B_j(y_j) | 0\rangle \tag{4.32}$$

constructed from graphs in which every A_i operator in the product $\Pi' A$ is connected to a B_j operator. The problem is now reduced to that of obtaining expansions for the $\mathcal{A}'$ amplitudes. The expansion for the original amplitude $\mathcal{A}$ can then be reconstructed from eq. (4.31), with the amplitudes G_S appearing as factors in various coefficient functions.

---

* The name is appropriate because, in free-field theory, generalized Wick products are the same as ordinary Wick products.

(Observe that we chose to extract only those factors (4.30) which do not depend on the B_j operators).

Let us denote the Fourier transform of (4.32) by

$$F.T.[\mathcal{A}] = \tilde{\mathcal{A}}'(q_1 \ldots q_K;\ r_1 \ldots r_N)\ \delta(\sum_{i=1}^{K} q_i + \sum_{j=1}^{N} r_j)\ , \qquad (4.33)$$

where K is the number of A_i operators in (4.32) and the momenta $\{q_1 \ldots q_K\}$ and $\{r_1 \ldots r_N\}$ are conjugate to the coordinates $\{\rho\xi_i\}$ and $\{y_1 \ldots y_N\}$ respectively. The notation is chosen such that eqs.(2.1) and (2.2) and fig. 2 can be applied directly to the amplitude $\tilde{\mathcal{A}}'$. Fig.2 involves decomposing $\tilde{\mathcal{A}}'$ into an "upper blob" U ($\mathcal{G}'$ in fig.2) and a "lower blob" $\mathcal{L}$:

$$\tilde{\mathcal{A}}' = \int \prod_{i=1}^{M} d^4k_i\ U(q_1 \ldots q_K; k_1 \ldots k_M)\, \mathcal{L}(k_1 \ldots k_M; Y_1 \ldots Y_N;\ \Sigma r = -\Sigma q) . \qquad (4.34)$$

The U amplitude is defined in terms of graphs $\mathcal{G}'$ which become 1PI when the coordinates ξ_i are set equal : i.e. $\xi_i = \xi$, for all i in Π_i' (as shown in fig. 14). The rationale for this definition is that it isolates all graphs $\mathcal{G}'$ which are able to carry $O(\eta)$ momenta simultaneously along all internal lines; i.e. the decomposition (4.34) is designed to facilitate power-counting for the limit $q_1 \ldots q_K \to \infty$.

Note : (i) One of the consequences of the definition is that propagators for the intermediate lines $k_1 \ldots k_M$ (including self-energy corrections) are incorporated in the amplitude .

(ii) The expression for U includes a momentum-conservation U-function

$$\delta(\ \Sigma q + \Sigma k) \qquad (4.35a)$$

so U can be considered as a function of (K+M) independent variables $(q_1 \ldots q_K,\ k_1 \ldots k_M)$. Also, some of the graphs ' may be disconnected; each connected component gives rise to a factor

$$\delta(\Sigma' q + \Sigma'' k) \qquad (4.35b)$$

in the expression for U, where Σ' and Σ'' denote non-empty partial sums.

(iii) In fig. 2 (as applied to $\tilde{\mathcal{A}}'$), all external lines should be understood to be wavy sources lines. Previously (figs. 6(b), 7(b)), we allowed some of the source lines $r_1 \ldots r_N$ to be intermediate as well. However, for reasons which will shortly become evident, we now require that none of the intermediate lines $k_1 \ldots k_M$ be wavy, (as in fig. 14).

The amplitude U can be expanded in $k = (k_1 \ldots k_M)$ about the

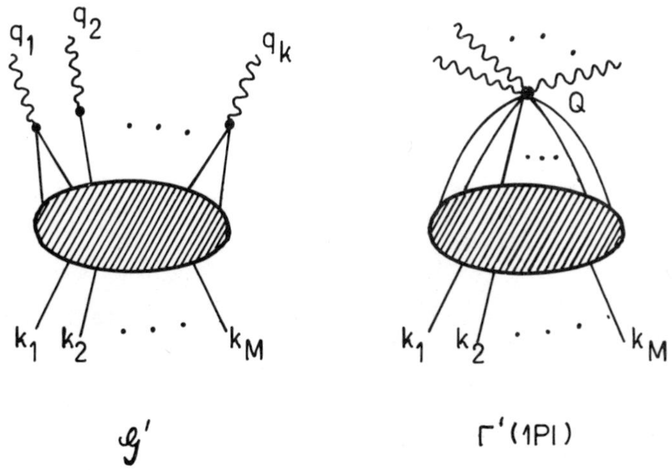

q_1
q_2
q_k
k_1 k_2 k_M
$\mathcal{G}'$
Q
k_1 k_2 k_M
Γ'(1PI)

FIG. 14

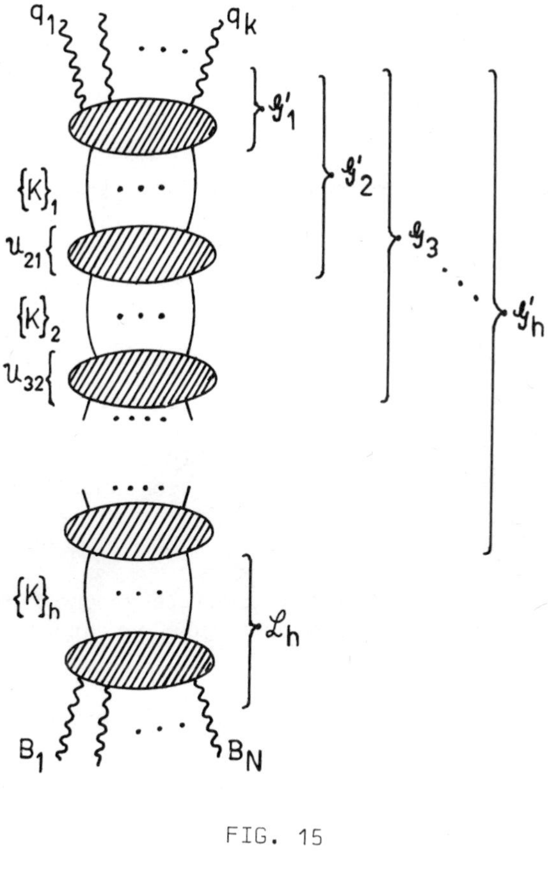

q_1
q_k
$\mathcal{G}'_1$
$\mathcal{G}'_2$
$\mathcal{G}_3$
$\mathcal{G}'_h$
$\{K\}_1$
$\mathcal{U}_{21}$
$\{K\}_2$
$\mathcal{U}_{32}$
$\{K\}_h$
$\mathcal{L}_h$
B_1
B_N

FIG. 15

point*

$$k = (0,0,\ldots,0) \tag{4.36}$$

with all momenta $q = (q_1 \ldots q_K)$ held fixed :

$$U(q,k) = \sum_{n=0}^{p-1} (n!)^{-1} (k\partial/\partial\lambda)^n U(q,\lambda)\Big|_{\lambda=0} + R_p(q,k), \quad (p\geq 1)$$

$$(\text{notation} : k\partial/\partial\lambda = \sum_{i=1}^{M} k_i^\mu \partial/\partial\lambda_i^\mu) . \tag{4.37}$$

Consider the dimensionality of the amplitudes appearing in (4.37). The definition of dimensionality given by eq.(2.4) does not take account of the extra δ-functions (4.35) which we have chosen to incorporate in the definition of U. Since each δ-function contributes a power -4, we have

$$\mathcal{D}(U) = \mathcal{D}(\mathcal{G}') - 4 \;\{\text{number of connected components of } \mathcal{G}'\} \tag{4.38}$$

Each derivative $\partial/\partial\lambda$ acting on U lowers its dimensionality by 1 :

$$\mathcal{D}\{(\partial/\partial\lambda)^n U(q,\lambda)\} = \mathcal{D}(U) - n \quad ,$$

$$\mathcal{D}\{R^{(p)}\} = \mathcal{D}(U) - p \quad . \tag{4.39}$$

We see that eq.(4.37), when substituted into (4.34), looks like a Wilson expansion with coefficient functions $(\partial/\partial\lambda)^n U(q,\lambda=0)$ multiplying operators O_n given by momentum-space vertices

$$(k)^n = (k_{i_1})_{\mu_1} (k_{i_2})_{\mu_2} \ldots (k_{i_n})_{\mu_n} \tag{4.40}$$

However, the prescription (4.37) is not genrally satisfactory : the remainder amplitude $\mathcal{R}_p$ need not be $O(\eta^{-p}\, \ell n^\beta\, \eta)$ relative to U if $\mathcal{G}'$ contains subgraphs with sufficiently large dimensionality. It is necessary to find a prescription for constructing a remainder amplitude, in which the dimensionalities of all subgraphs are simultaneously lowered.

The answer is obvious if the complete graph $\mathcal{G}$ for the amplitude $\mathcal{A}'$ can be uniquely decomposed into a nested set of subraphs (fig. 15)

---

* If zero-mass propagators are present, the expansion must be performed about a non-exceptional point $\lambda = (\lambda_1 \ldots \lambda_M)$ in k-space. The subsequent analysis is unchanged, apart from notational complications due to the λ-dependence of various amplitudes.

$$\mathcal{G}'_1 \subset \mathcal{G}'_2 \subset \; \ldots . \subset \mathcal{G}'_k \quad . \tag{4.41}$$

Let U_1, U_2,...,U_h be the corresponding "upper-blob" amplitudes, with

$$U_2(\{k\}_2,q) = \int\{dk\}_1 \, U_{21}(\{k\}_2,\{k\}_1)U_1(\{k\}_1,q) \; , \; \ldots$$
$$U_{(m+1)} = \int\{dk\}_m \, U_{(m+1)m} \, U_m \; , \text{ etc. } , \tag{4.42}$$

so that the complete amplitude can be written

$$\tilde{\mathcal{A}}' = \int\{dk\}_h \, \mathcal{L}_h \, (r,\{k\}_h) \; U_h(\{k\}_h,q)$$
$$= \int \prod_{m=1}^{h} \{dk\}_m \, \mathcal{L}_h \, U_{h(h-1)} \, U_{(h-1)(h-2)} \cdots U_{21}U_1 \tag{4.43}$$

We begin by subtracting the smallest blob U_1 as in (4.37),

$$U_1 \rightarrow (1-t_1)U \tag{4.44}$$

where t_m is shorthand for the Taylor operator

$$t_m = \sum_{r=0}^{p(m)-1} (r!)^{-1}(k.\overrightarrow{\partial/\partial\lambda})^r_m \Big|_{\lambda=0} \quad , \qquad (m=1,\ldots,h) \tag{4.45}$$

For the next graph $\mathcal{G}'_2$, we have

$$U_2 \rightarrow (1-t_2)\int\{dk\}_1 \, U_{21}(1-t_1)U_1 \quad , \tag{4.46}$$

and so on, until we arrive at a completely subtracted amplitude

$$\tilde{\mathcal{S}}'_p = \int \prod_{m=1}^{h} \{dk\}_m \, \mathcal{L}_h(1-t_h)U_{h(h-1)}(1-t_{h-1})\ldots(1-t_2)U_{21}(1-t_1)U_1 \tag{4.47}$$

in which the dimensionalities of the subgraphs $\mathcal{G}'_1 \ldots \mathcal{G}'_h$ are reduced by p(1)...p(h) respectively. The difference between $\tilde{\mathcal{A}}'$ and $\tilde{\mathcal{S}}'_p$ is a finite series in the form of an operator-product expansion :

$$\tilde{\mathcal{A}}' = \sum_{n=0}^{n(p)} \tilde{\mathcal{B}}'_n(q) \, \mathcal{A}_n(r) + \tilde{\mathcal{S}}'_p \tag{4.48}$$

The prescription for constructing $\tilde{\mathcal{S}}'_p$ must be generalized to include cases in which there are overlapping subgraphs. Recall (fig. 14) that the subgraphs $\mathcal{G}'$ are defined such that, when the coordinates ξ_i are identified, the result is a 1PI graph Γ' associated with the class of operators

$$Q(\xi) = :\prod_i{}' A_i(\xi) : ,\ldots\ldots, \quad :\prod_i{}'\{\partial\ldots\partial A_i(\xi)\} : ,\ldots . \qquad (4.49)$$

The superficial degree of divergence of Γ' is trivially related to the dimensionality of the amplitude U associated with $\mathcal{G}'$:

$$\mathcal{D}(U) = d(\Gamma') - 4K \qquad (4.50)$$

Furthermore, the subtractions $t_1, t_2, \ldots, t_h$ in (4.47) become renormalization counterterms ΔQ for the operators Q when the coordinates ξ_i are set equal to ξ. So for the general case, the prescription is that subtractions used to construct $\tilde{\mathcal{S}}'_p$ from $\tilde{\mathcal{A}}'(\xi_i \neq \xi)$ should correspond to renormalization counterterms ΔQ (possibly oversubtracted) for the amplitude $\mathcal{A}'(\xi_i = \xi)$. This ensures that all subdimensionalities of $\tilde{\mathcal{S}}'_p$ are reduced relative to those of $\mathcal{A}'(\xi_i \neq \xi)$, because the renormalization theorem[12] says that all of the $d(\Gamma')$ can be lowered to any desired value by including sufficient counterterms ΔQ in eq.(1.7). Thus eq.(4.48) remains valid for the general case. An example of overlapping subgraphs is given in fig. 16. There are two suited sets

$$\mathcal{G}'_1 \subset \mathcal{G}'_{2a} \subset \mathcal{G}'_3 \quad , \quad \mathcal{G}'_1 \subset \mathcal{G}'_{2b} \subset \mathcal{G}'_3$$

with $\mathcal{G}'_{2a}$ and $\mathcal{G}'_{2b}$ overlapping. The remainder amplitude is given by the formula

$$\tilde{\mathcal{S}}'_p = \iint \{dk\}_3 \{dk\}_1 \, \mathcal{L}_3 (1-t_3) \quad .$$

$$.\left[\int \{dk\}_{2a} U_{3(2a)}(1-t_{2a})U_{(2a)1} - \int \{dk\}_{2b} U_{3(2b)} t_{2b} U_{(2b)1}\right](1-t_1)U_1$$

$$(4.51)$$

No account has been taken of the need for renormalization counterterms for the operator O_n (for example), so this discussion applies only to unrenormalized (B) amplitudes * . In coordinate space, eqs. (4.48) and (4.31) imply

$$\mathcal{A}_B = \sum_{n=0}^{n(p)} (\mathcal{C}_n)_B (\mathcal{A}_n)_B + \mathcal{S}_p(\rho\xi_i, y_j)_B \quad ,$$

$$(\mathcal{A}_n)_B = T\langle 0| \prod_{j=1}^{N} O_n(0)\, B_j(y_j) |0\rangle_B \quad , \qquad (4.52)$$

---

* Unlike refs. 15, 71, 81, 86, 90, which deal directly with renormalized amplitudes. The present approach is less satisfactory but avoids some complicated algebra.

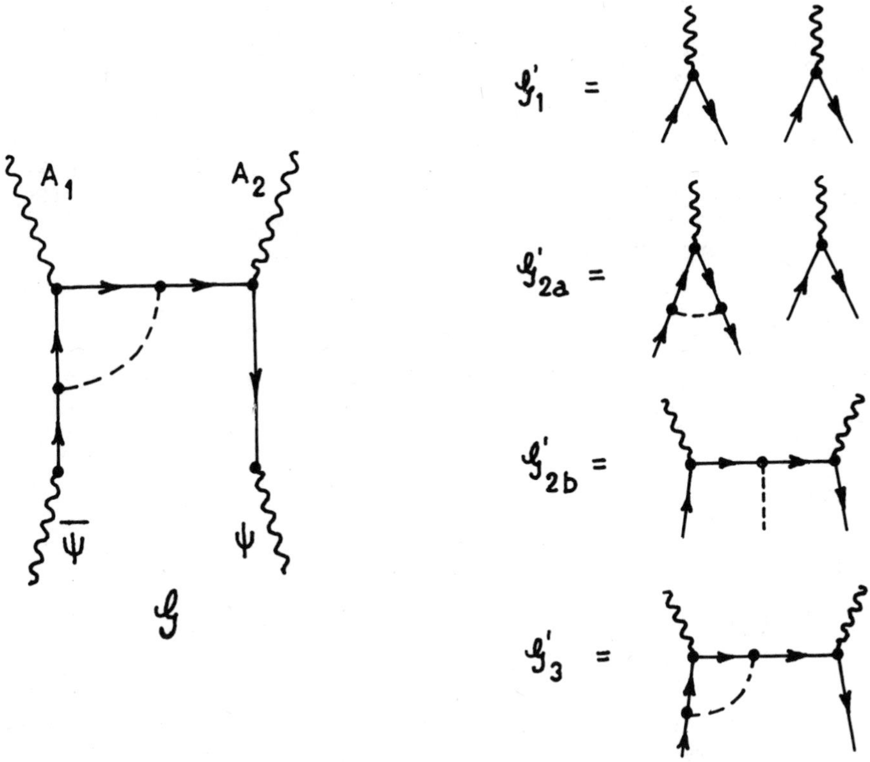

FIG. 16

where $\mathcal{S}_p$ is constructed from eq. (4.31) and the Fourier transforms of $\mathcal{S}_p'$, p amplitudes.

According to the rule (1.41), $\mathcal{A}$ and $\mathcal{A}_n$ are renormalized as follows :

$$\mathcal{A}_R = \prod_{ij} Z_{A_i}^{-1} Z_{B_j} \mathcal{A}_B + \delta[\xi_i - \xi_j, \rho\xi_i - y_j, y_i - y_j]$$

$$(\mathcal{A}_n)_R = Z_n^{-1} \prod_j Z_{B_j}^{-1} (\mathcal{A}_n)_B + \delta[y_i, y_i - y_j] \ . \tag{4.53}$$

Some of the Z-factors (e.g. Z_n) may be matrices. The terms $\delta[x_i, y_j, \ldots]$, which involve δ-functions of the coordinate differences $x_i, y_j, \ldots$, are subtractions generated by counterterms $\Delta(A_1 A_2 \ldots B_1 B_2 \ldots)$ and $\Delta(O_n, B_1 B_2 \ldots)$ in eq. (1.39); (see note (a) following eq. (4.14)). In general, the terms $\Sigma_n \mathcal{B}_n \mathcal{A}_n$ and $\mathcal{S}_p$ mix under renormalization. However, in renormalizable theories, renormalization does not increase the subdimensionalities of an amplitude. Therefore, the term $O(\mathcal{S}_p)$ in the equation

$$(\mathcal{B}_n)_R = Z_n \prod_i Z_{A_i}^{-1} (\mathcal{B}_n)_B + \delta[\xi_i - \xi_j] + O(\mathcal{S}_p) \tag{4.54}$$

can be absorbed into the definition of the renormalized amplitude $(\mathcal{S}_p)_R$ without changing its maximum subdimensionality. So the result is

$$\mathcal{A}_R = \sum_{n=0}^{n(p)} (\mathcal{B}_n)_R (\mathcal{A}_n)_R + \{\delta\text{-functions}\} + (\mathcal{S}_p)_R \ , \tag{4.55}$$

where the remainder amplitude obeys the bound

$$(\mathcal{S}_p/\mathcal{A})_R = O(\rho^p \ln^\beta \rho) \ , \quad (p = \text{positive integer}; \ \rho \to 0 \ \text{with} \ \xi_i, y_j \ \text{fixed}) \tag{4.56}$$

Note : (i) The terms $\mathcal{B}_n O_n$ do not depend on the operators B_j because the sets of intermediate lines $\{k_1 \ldots k_{M(m)}\}_m$ do not involve sources. Because of this restriction on intermediate lines, the coordinates ξ_i, y_j must be held fixed as ρ tends to zero in (4.56).

(ii) Strictly speaking, the term {δ-functions} (absent for the special case (4.28)) violates eq. (4.1) because it need not be negligible relative to $\mathcal{B}_n$ and may depend on the B_j operators. However, the violation is of the same form as the δ-function ambiguities of ordered products, so it is of little practical importance.

(iii) It is permissible to fix the momenta of simple (i.e. non-composite) operators B_j in eqs. (4.55) and (4.56). For example, consider the operator B_2 in fig. 13; even if its momentum is fixed, its source line cannot be intermediate because the corresponding subgraph would not be 1PI for $\xi_i = \xi$.

(iv) The on-shell states $|I>, |F>$ of eq(4.14) can be recovered by choosing some of the B_j in (4.55) to be suitable interpolating operators. If simple operators are used, remark (iii) is relevant. More generally, the on-shell limit isolates the set of graphs for which the interpolating operator (simple or composite) communicates with other external lines via a single internal line. For this subset, the source of the interpolating field cannot be an intermediate line, so the time-ordered case of (4.14) is obtained.

The set of allowed discontinuities of $T<F|\pi_{ij} A_i B_j|I>$ is restricted by the locality of the operator O_n on the right-hand side of eq.(4.14). If the ordering of the A_i operators with respect to each other (but not to the B_j operators) is partially or completely removed, $T<F|\pi_j O_n B_j|I>$ remains completely time-ordered, while the $i\varepsilon$-prescription for coefficient functions is adjusted accordingly; e.g.

$$T\{A_1(\xi)A_2(0)\} \sim (\xi^2-i\varepsilon)^{\alpha} O_1(0) \,,$$

$$A_1(\xi)A_2(0) \sim (\xi^2-i\varepsilon\xi_o)^{\alpha} O_1(0) \,, \quad (\xi \to 0) \,. \tag{4.57}$$

It is also possible to unorder part of $\pi_j B_j$ on each side of (4.14), retaining the time-ordering between $A_i(\rho\xi_i)$ (or O_n) and $B_j(y_j)$ for all i,j. Special precautions are necessary when unordering a $\overline{A_i B_j}$ pair. In general, the relevant B_j operator should not appear in the middle of the partially or completely unordered product $\pi_i A_i$: e.g.

$$<0|A_1(\xi)A_2(0)B(y)|0> \sim (\xi^2- i\varepsilon\xi_o)^{\alpha} <0|O_1(0)B(y)|0>,$$

$$<0|A_1(\xi)B(y)A_2(0)> \sim \;\; ?? \quad , (\xi \to 0, \; y \text{ fixed}). \tag{4.58}$$

These remarks are important[91] for the process

$$e^+ + e^- \to h + X \tag{4.59}$$

where h is an observed hadron (momentum p_h) and X denotes other final-state hadrons with combined momentum p_X. The one-particle inclusive cross-section is proportional to

$$\bar{W}_{\mu\nu} = \sum_X <0|J_\mu(0)|Xh> \;\; <Xh|J_\nu(0)|0> \; (2\pi)^4\delta^4(p_X+p_h-q), \tag{4.60}$$

where q is the momentum of the virtual photon. If Φ_h is an interpolating operator for h, the corresponding off-shell amplitude in coordinate space is

$$W_{\mu\nu} = <0|T\{J_\mu(\xi) \; \Phi_h(y_1)\} \;\; \bar{T} \;\; \{J_\nu(0)\Phi_h(y_2)\}|0> \quad , \tag{4.61}$$

where the coordinate ξ is conjugate to q and $\bar{T}$ denotes the anti-T-product. It is not possible to substitute an operator-product expansion for $J_\mu J_\nu$ because local operators O_n cannot be produced by the combination $T(A_1B_1)\bar{T}(A_2B_2)$ (in the notation of (4.14)). Mueller[91] has suggested an expansion of $W_{\mu\nu}$ in which there are coefficient functions (different from the Wilson coefficient functions) which do not multiply local operators O_n. Derivations have been given for some non-gauge theories [91,92].

Renormalization-Group Properties

A particular sector (J,I,...) of the operator-product expansion (4.1) is obtained by projecting out intermediate sets $\{k_1...k_M\}$ with total quantum numbers (J,I,...) or equivalently, by a judicious choice of operators B_j in eq.(4.29) :

$$\mathcal{A} \to \mathcal{A}_{(J,I,...)}$$

The corresponding operators $\{O_n\}$ have quantum numbers (J,I,...). For simplicity, let us assume that there is a single operator O_n (matrix element $\mathcal{A}_n$) which contributes the first p powers of the scaling parameter ρ to the asymptotic expansion of $\mathcal{A}$, order-by-order in perturbation theory :

$$\begin{aligned}\mathcal{A}_{(J,I,...)}/\mathcal{A}_n &= F_n(\rho\xi_i, y_j; g, m, \mu) \\ &= \mathcal{C}_n(\rho\xi_i; g, m, \mu)\ \{1 + O(\rho^p \ln^\beta \rho)\}\ , \\ &\quad (\rho \to 0\ ,\quad p = \text{positive integer})\ . \end{aligned} \tag{4.62}$$

As indicated by the notation, we choose to carry out the analysis using a mass-independent renormalization prescription *.

The improved CS equation for the ratio F is (neglecting δ-function subtractions)

$$[\mu\partial/\partial\mu + \beta(g)\partial/\partial g + \gamma_m(g)\ m\partial/\partial m + \gamma(g) - \gamma_n(g)]F(\rho\xi_i, y_j; g, m, \mu) = 0 \tag{4.63}$$

* See ref.7. The corresponding analysis for the ordinary CS equation [18,71,86,87] involves expanding $T\langle\Delta\Pi_{ij}A_iB_j\rangle$, where Δ is the zero-momentum mass-insertion operator (e.g. $\Delta = \int d^4x :\bar{\psi}(x)\psi(x):$). Straightforward substitution of (4.1) works for the leading power in a given sector, but for non-leading powers, the expansion of $:\bar{\psi}\psi:\Pi_i A_i$ also contributes. Compare eqs.(4.17-27).

where $\gamma(g)$ and $\gamma_n(g)$ are the γ-functions for the amplitudes $\mathcal{A}$ and $\mathcal{A}_n$ (with $\gamma(0) = 0 = \gamma_n(0)$). The addition rule (1.46) implies

$$\gamma(g) = \sum_i \gamma_{A_i}(g) + \sum_j \gamma_{B_j}(g)$$

$$\gamma_n(g) = \gamma_{O_n}(g) + \sum_j \gamma_{B_j}(g) \quad ,$$

so all dependence on γ-functions for the B_j operators drops out in the difference

$$\gamma(g) - \gamma_n(g) = \sum_i \gamma_{A_i}(g) - \gamma_{O_n}(g) \tag{4.64}$$

Eqs.(4.62-64) imply

$$[\mu\partial/\partial\mu + \beta\partial/\partial g + \gamma_m m\partial/\partial m + \sum_i \gamma_{A_i} - \gamma_{O_n}]\mathcal{C}_n = O[\rho^p \ln^\beta\rho\, \mathcal{C}_n(\rho)]. \tag{4.65}$$

Since $\mathcal{C}_n$ does not depend on y_j, $\mu\partial/\partial\mu$ can be eliminated in terms of $m\partial/\partial m$ and either $\rho\partial/\partial\rho$ or $\eta\partial/\partial\eta$ ($\eta = \rho^{-1}$) using a homogeneity equation of the form (3.44) : e.g.

$$[\mu\partial/\partial\mu + m\partial/\partial m + \eta\partial/\partial\eta - d_{\mathcal{C}}]\tilde{\mathcal{C}}_n(\eta q_i;\, g,m,\mu) = 0 \quad , \tag{4.66}$$

where $\tilde{\mathcal{C}}_n(q)$ is the Fourier transform of $\mathcal{C}_n(\xi)$ and $d_{\mathcal{C}}$ is the mass dimensionality of $\mathcal{C}_n$. Thus the analogue of eq.(3.46) is

$$[\eta\partial/\partial\eta - \beta\partial/\partial g + (1-\gamma_m)m\partial/\partial m - (d_{\mathcal{C}} + \sum_i \gamma_{A_i} - \gamma_{O_n})]\tilde{\mathcal{C}}_n =$$

$$= O[\eta^{-p} \ln\beta'_\eta\ \tilde{\mathcal{C}}_n] . \tag{4.67}$$

If m is a fermionic mass, the amplitudes

$$\tilde{\mathcal{C}}_n^{(r)}(\eta q; g,\mu) = (\partial/\partial m)^r\, \tilde{\mathcal{C}}_n(\eta q_i; g,m,\mu)\big|_{m=0} \tag{4.68}$$

exist in perturbation theory for $r = 0,1,2$ but not for $r = 3,4,\ldots$, so eq.(4.67) becomes

$$[\eta\partial/\partial\eta - \beta\partial/\partial g - \{d_{\mathcal{C}} + \sum_i \gamma_{A_i} - \gamma_{O_n} - r(1-\gamma_m)\}]\tilde{\mathcal{C}}_n^{(r)} = 0 \ ,$$

$$0 \leqslant r \leqslant \mathrm{Min}\,(2,p-1) \tag{4.69}$$

The consequences of (4.69) are analogous to those of eq.(2.33):

(i) Broken scale invariance :

$$\mathcal{C}_n^{(r)} \sim \text{const. } \rho^{-P(r,n)} \quad ,$$

$$P(r,n) = \sum_i d_{A_i} - d_{O_n} - r(4-d_\Delta) \qquad (4.70)$$

For r=0, eq.(4.70) extends to coefficient functions[4)] the rule (2.56) that dynamical dimensions are conserved at short distances. Non-leading terms r=1,2,... are damped by a power $r(4-d_\Delta)$ of ρ, as in eq.(3.36), if the condition (3.37) is satisfied. For the special case

$$A_i = \text{(partially) conserved currents, } O_n = \text{mass operator,} \qquad (4.71)$$

there is a cancellation [4,7)] of anomalous dimensions for the leading term (r = 1)

$$P(r=1,\ n=\Delta) = \sum_i d_{A_i} - 4 = \text{integer} \qquad (4.72)$$

which is important for the short-distance analysis of $\eta \to 3\pi$ decay and other weak and electromagnetic corrections to strong interactions.

(ii) Asymptotic freedom : [7,93)]

$$\mathcal{C}_n^{(r)} \sim \text{const. } \rho^{-p(r,n)} (\ell n\rho)^{\lambda(r,n)} \{1 + O(\ell n|\ell n\rho|/\ell n\rho)\}$$

$$p(r,n) = \sum_i \dim A_i - \dim O_n - r$$

$$\lambda(r,n) = \{\sum_i c_{A_i} - c_{O_n} + rc_m\}/2b \qquad (4.73)$$

Here dim Q denotes the canonical dimension of Q, b is defined in eq.(2.80), and c_Q is the one-loop coefficient in the expansion

$$\gamma_Q(g) = c_Q g^2 + O(g^4) \qquad (4.74)$$

The result (4.73) generalizes eq.(2.86). If A_i and O_n are conserved or partially conserved currents for r=0, or if eq.(4.71) is applicable[7)] with r=1, the logarithmic power $\lambda(r,n)$ vanishes and the leading contribution to $\mathcal{C}_n^{(r)}$ (including the constant of proportionality) is given by the free-field result. (The last statement assumes that the complete operator O_n is generated in the expansion of ΠA_i in free-field theory. This need not be the case for $O_n=\theta_{\mu\nu}$; e.g. if the operators A_i are chiral currents, only the fermionic part of $\theta_{\mu\nu}$ is generated. The result[93)] is that the constant of proportionality is computable and differs from the constant which normalizes the term $\theta_{\mu\nu}$ (fermionic) in free-field theory).

If we want to check these predictions for short-distance behaviour by measuring the amplitudes $\mathscr{C}_n(q_i)$, it is essential that the momenta $\{q_i\}$ be non-exceptional. For example, consider the connected matrix element

$$W_{\mu\nu} = (2\pi)^{-1} \int d^4x \; e^{iqx} \langle p| \; [J_\mu(x), J_\nu(0)] \, |p\rangle$$

$$= -(g_{\mu\nu}/M)F_1(\xi,q^2) + (p_\mu p_\nu/p.q)F_2(\xi,q^2)$$

$$+ (i \; \varepsilon_{\mu\nu\lambda\eta} p^\lambda q^\eta/2p.q)F_3(\xi,q^2)$$

$$+ \text{ other terms if } J_\mu \text{ not completely conserved;}$$

$$-1 \leqslant \xi = -q^2/2p.q = \omega^{-1} \leqslant 1 \; ;$$

$$|p\rangle = \text{ nucleon, mass } M, \text{ momentum } p \; ;$$

$$F_i(-\xi,q^2) = \pm F_i(\xi,q^2) \qquad (4.75)$$

For $0 \leqslant \xi \leqslant 1$, $W_{\mu\nu}$ is proportional to the total cross-section for deep inelastic lepton-nucleon scattering, where q is the momentum transferred by the current and F_i are the usual structure functions. As is well-known[94,95], the Bjorken limit $-q^2\to\infty$ at fixed ξ corresponds to $x^2 \to 0$, not $x_\mu \to 0$. In the language of eqs.(2.1) and (2.7), this is because the limit involves exceptional values of q :

$$q = \eta\ell + c \; ; \quad \eta \to \infty \quad , \; \ell^2 = 0 \; , \quad p\cdot\ell \neq 0, \quad c.\ell \neq 0 \; . \qquad (4.76)$$

Also well-known[96] is the fact that the leading Wilson coefficient function in the spin-J sector is asymptotically proportional to the moment

$$M_n^i(q^2) = \int_0^1 d\xi \; \xi^n \; F_i(\xi,q^2) \; , \quad (n = J-1, \; J-2, \; J-1, \ldots \text{ for } i = 1,2,3\ldots) ; \qquad (4.77)$$

i.e. the $x_\mu \to 0$ limit corresponds to the limit $-q^2\to\infty$ of M_n . It does not matter that the non-exceptional limit ($\ell^2 \neq 0$) is not kinematically consistent with the bound $|\xi| \leqslant 1$. This bound is relevant for the commutator amplitude (2.75), whereas the lack of exceptionality of the $\eta\to\infty$ limit should be checked for the time-ordered analogue of (2.75) :

$$T = T(\zeta,q^2)\ , \qquad \zeta = 2p.q\ \sqrt{-q^2}$$

$$-q^2 \to \infty \quad \text{at fixed } \zeta\ . \tag{4.78}$$

(For simplicity, the indices μ,ν have been dropped). Given a suitably normalized spin-J projection * of T,

$$T_n(q^2) \propto (q^2)^{\text{power}} \int_{-1}^{1} dv(1-v^2)^{1/2}\, C_n(v)T(-iv,q^2)\ , \tag{4.79}$$

the $-q^2 \to \infty$ limit of T_n is non-exceptional and is determined by $\mathcal{E}_n(q)$, and the commutator amplitude is recovered by taking the absorptive part :

$$\mu_n(q^2) = \text{Abs}\ T_n(q^2)$$

$$\mu_n/M_n \to 1\ , \qquad (-q^2 \to \infty). \tag{4.80}$$

The difference between μ_n and M_n is that only spin-J operators contribute to μ_n, whereas non-leading powers on the expansion of M_n involve a mixture of spins.

It would be desirable to directly measure the q-dependence of the momenta in order to distinguish the alternatives

$$M_n(q^2) \sim \begin{cases} K_n & \text{(Bj. scaling)} \\ K'_n(\ell n q^2)^{-\lambda(n)} & \text{(as.freedom, eq.(4.73))} \\ K''_n(q^2)^{-a(n)/2} & \text{(br.scale inv., eq.(4.70))}, \end{cases} \tag{4.81}$$

where K_n, K'_n, K''_n are q-independent and the powers $\lambda(n)$, $a(n)$ are given by

$$\lambda(n) = c_{on}/2b\ , \qquad a(n) = \gamma_{0_n}(g\ )\ . \tag{4.82}$$

However, it has been observed[98] that $M_n(q^2)$ and $\mu_n(q^2)$ (small n) differ significantly for $1 \leqslant -q^2 \leqslant 10\ (\text{GeV})^2$, so there is an ambiguity caused by non-leading terms not being negligible. Thus it is not surprising that all three possibilities** (4.81) continue to be discussed in the literature. For example, Parisi[100] and Nachtmann[48] present the case for large anomalous dimensions a(n)

---

* Details of eqs.(4.79) and (4.80), such as explicit formulas for μ_n, are given in ref.97. The functions $C_n(v)$ are Gegenbauer polynomials.

** It has been argued [38,99] that Bjorken scaling is not consistent with the renormalization group. Note that the possibility a(n)=0 (all n) for gauge theories with $g_\infty \neq 0$ has yet to be excluded.

($\simeq$ 1/2 or 1). Parisi's method (in which the moment equations (4.77) are inverted) has since been applied to asymptotically free theories[101,102].

The most important qualitative feature of (4.81) is that, if Bjorken scaling is not valid, $F_i(\xi,q^2)$ should rise at small ξ and fall near $\xi=1$ as $-q^2$ increases. This is because the smallest value of n (n=0) corresponds to a conserved operator such as $\theta_{\mu\nu}$ ($\lambda(0)=0=a(0)$) and positivity requirements[98] force a(n) and $\lambda(n)$ to increase with n ($\lambda(n)$, $a(n)>0$, $n>0$), so all momenta decrease except for the asymptotically constant n=0 moment. Eventually, F_i collapses to a δ-function :

$$F_i(\xi,q^2) \rightarrow 2(K_o' \text{ or } K_o'')\ \delta(\xi) \tag{4.83}$$

Recently, the NAL muon-scattering group[103] reported indications that the structure functions exhibit this behaviour.

Complications for Yang-Mills theories

In many cases, the operator O_m is not multiplicatively renormalized. Instead, there is a set of operators $\vec{O}_m$ which generate each other as renormalization counterterms

$$\vec{O}_m \rightarrow \vec{O}_m \overleftrightarrow{Z}_m^{-1} \quad , \tag{4.84}$$

where the Z-factor and its γ-function

$$\overleftrightarrow{\gamma}_m = \lim_{n\rightarrow 4} (\mu\partial/\partial\mu\ \overleftrightarrow{Z}_m)\ \overleftrightarrow{Z}_m^{-1} \tag{4.85}$$

are matrices. (I am using dimensional regularization, as in (3.43) so O_m has been substituted for O_n in order to avoid confusion with the complex parameter n).

According to eq. (1.8), the counterterms ΔQ of a vertex Q have canonical dimension dim(ΔQ) less than or equal to dim(Q). Counterterms with dim(ΔQ) < dim(Q) are not generated unless masses or other dimensionful coupling constants are present; e.g. $\bar{\psi}\not{\partial}\psi$ generates $m\bar{\psi}\psi$. So, if we restrict ourselves to the leading singularities $G^{as.}$ and $\mathscr{C}_n(r=0)$, it is sufficient to compute all counterterms of the zero-mass theory :

$$\dim(\Delta Q) = \dim(Q) \quad . \tag{4.86}$$

In particular, the asymptotic behaviour of the moments $M_m(q^2)$ in (4.81) is controlled by a term $\vec{\mathscr{C}}_m . \vec{O}_m$ in the operator-product expansion of $J_\mu J_\nu$, where all operators $(O_m)_i$ in the set $\vec{O}_m$ have the same canonical dimension*

$$\dim O_m = J + 2 \quad , \quad (J = \text{spin of } O_m) \ , \tag{4.87}$$

and where the coefficient functions $\vec{\mathscr{C}}_m(r=0)$ satisfy the renormalization group equation

$$[\eta\partial/\partial\eta - \beta(g)\partial/\partial g + \overleftrightarrow{\gamma}_m(g) - d(m)]\, \vec{\mathscr{C}}_m(\eta q; g, \mu) = 0 \ ,$$
$$(d(m) = \text{mass dim. of } \mathscr{C}_m) \ . \tag{4.88}$$

The new feature of the solution of eq.(4.88) (compared with the solution (2.34) of eq.(2.33)) is that matrices occuring in the exponential must be ordered :

$$\vec{\mathscr{C}}_m(\eta q; g, \mu) = \eta^{-d(m)} \{\exp - \int_g^{\bar{g}} dx\, \overleftrightarrow{\gamma}_m(x)/\beta(x)\}_{\text{ord.}}\, \vec{\mathscr{C}}_m(q; \bar{g}, \mu)$$

$$\{\exp\ldots\}_{\text{ord.}} = \sum_{r=0}^{\infty} (-1)^r \int_g^{\bar{g}} dx_1 \int_{x_1}^{\bar{g}} dx_2 \ldots \int_{x_{r-1}}^{\bar{g}} dx_r\, \overleftrightarrow{\gamma}_m(x_1)\ldots\overleftrightarrow{\gamma}_m(x_r) \Big/ \prod_{i=1}^{r} \beta(x_i) \tag{4.89}$$

The limit $\eta \to \infty$ applied to (4.89) produces the following results :
(i) Broken scale invariance : anomalous dimensions are given by the eigenvalues of $\overleftrightarrow{\gamma}_m(g_\infty)$.[93]
(ii) Asymptotic freedom : the constant c_{O_m} in eq.(4.73) should be replaced by an eigenvalue of the matrix $\overleftrightarrow{c}_m$ given by

$$\overleftrightarrow{\gamma}_m(g) = g^2\, \overleftrightarrow{c}_m + O(g^4) \tag{4.90}$$

In each case, the dominant contribution in eq.(4.81) is obtained by substituting the smallest eigenvalue in eq.(4.82).

The application of these rules to the Yang-Mills Lagrangian (2.65) is not entirely straightforward. The trouble is that, whatever the gauge, the mixing matrix $\overleftrightarrow{Z}$ for SU(3)-singlet twist-two operators is enormous. It would be an onerous task to compute all of the matrix elements of $\overleftrightarrow{c}_m$, extract the eigenvalues, and (by considering a sufficiently large class of gauges) determine which eigenvalues are physically relevant. Indeed, the original

---

* In perturbation theory, the "twist" $\tau(Q)$ of an operator[104] is defined as follows : $\tau(Q) = \dim(Q) - \text{spin}(Q)$. Thus eq.(4.87) refers to twist-2 operators.

calculations[93] are based on a simplified prescription which is assumed to be equivalent to the above rules. So the problem is to verify the correctness of the results for twist-two operators and more generally, to find labour-saving prescriptions for arbitrary-twist operators.

Let us begin with the tree approximation for the upper-blob graphs $\mathcal{G}'$ in figs. 2 and 14 (i.e., the zero-loop approximation for coefficient functions). Even for this case, the complete set of allowed operators is complicated. The intermediate sets $\{k_1 \ldots k_M\}$ may include Faddeev-Popov ghosts as well as fermions and gauge mesons, so some of the tree-approximation operators are ghost-dependent and hence not manifestly gauge-invariant. However, if we restrict ourselves to twist-two operators, it is easy to see that intermediate ghost lines cannot appear in the tree approximation for $\mathcal{G}'$. An analysis of Abelian gauge theories by Gross and Freiman[104] can be readily generalized to show that the allowed operators are given by symmetrizing the gauge-invariant combinations

$$O^{\mu_1 \ldots \mu_2} = \bar{\psi}\gamma^{\mu_1} D_f^{\mu_2} \ldots D_f^{\mu_J} \{1 \text{ or } \lambda^i/2\}\psi ,$$

$$(\lambda^i = \text{ordinary-SU(3) matrices}) \qquad (4.91)$$

in the Lorentz indices $\mu_1 \ldots \mu_J$ and removing traces. The symmetric-traceless part can be projected out by multiplying the operators in (4.91) by the source[93]

$$\chi_{\mu_1}\chi_{\mu_2} \ldots \chi_{\mu_J} \quad , \quad (\chi^2 = 0) \quad . \qquad (4.92)$$

The result is a pair of spin-J operators, one an SU(3) singlet, the other an octet :

$$Q_1 = \bar{\psi}\ \chi.\gamma(i\chi.D_f)^{J-1}\psi$$

$$Q(8) = \bar{\psi}\ \chi.\gamma(i\chi.D_f)^{J-1}(\lambda^i/2)\psi \qquad (4.93)$$

The next step is to consider the renormalization of these operators so that the one-loop contributions $\overleftrightarrow{c}$ to the mixing matrix can be isolated. In this context, the one-loop approximation means that we are considering one-loop counterterms $\Delta_1 Q$ of Q, plus one-loop counterterms $\Delta_1\Delta_1 Q$ of $\Delta_1 Q$, plus $\Delta_1\Delta_1\Delta_1 Q$ ... and so on, until no new vertices are produced. All of these counterterms necessarily appear as operators in the Wilson expansion* ;

*To simplify the discussion, I shall assume that operators not generated as counterterms of tree-approximation operators do not appear in the Wilson expansion. This problem is circumvented in ref. 105.

otherwise, the expansion would not transform consistently under renormalization-group transformations.

Renormalization produces complications because the Lagrangian (2.65) which generates the Feynman rules is not manifestly gauge-invariant. Instead, it is invariant under the following set of transformations, introduced by Becchi, Rouet and Stora[43,50] :

$$\delta A^a_\mu = -D^{ab}_\mu(A)\varphi^b\delta\lambda \ , \qquad \delta\psi = -iq\varphi^a\delta\lambda\tau^a\psi,$$

$$\delta\varphi^a = \frac{1}{2}\, c^{abc}\varphi^b\varphi^c\delta\lambda,$$

$$\delta\varphi^{a*} = C^a(A)\delta\lambda \tag{4.94}$$

Observe that A^a_μ and ψ undergo a gauge transformation (2.68) with gauge function $\delta\omega^a$ given by $\varphi^a\delta\lambda$. Since the ghost field obeys Fermi-Dirac statistics, the parameter $\delta\lambda$ should be treated as an anticommuting number[106]. The function $C^a(A)$ refers to the choice of gauge; according to the quantization rules[45,49], the gauge-fixing term can be written in the form

$$\mathcal{L}_{g.f} = -\frac{1}{2}\,(C^a(A))^2 \tag{4.95}$$

with corresponding ghost Lagrangian given by

$$\mathcal{L}_{ghost} = \varphi^{a*}\,\delta C^a(A)/\delta\lambda \quad . \tag{4.96}$$

As in eqs.(2.65-72), eq.(4.94) involves unrenormalized quantities.

The rules for counterterms ΔQ of a gauge-invariant operator Q are as follows :

(i) If a given order of perturbation theory produces no ghost-dependent terms in ΔQ, ΔQ is manifestly gauge-invariant. Roughly speaking, this is because $\varphi\delta\lambda$ acts as a c-number for the set of Feynman diagrams being considered.

(ii) The appearance of ghost-dependent counterterms in ΔQ implies a lack of explicit gauge invariance for the accompanying meson-dependent counterterms. A non-trivial analysis[107-109] based on the Ward identities associated with (4.94) shows that ΔQ is symmetric under a source-dependent generalization of (4.94).

Important special cases of rule (i) are :

(a) The operator $Q(8)$ is multiplicatively renormalized[93] in Lorentz-invariant gauges, because octet twist-two operators cannot depend on ghosts and $Q(8)$ is the only vertex which is both Lorentz- and gauge-invariant.

(b) All counterterms ΔQ in the axial gauge[110]

$$N^\mu A_\mu = 0 \qquad (N^\mu = \text{fixed 4-vector}) \tag{4.97}$$

are manifestly gauge-invariant because the relevant gauge-fixing term

$$\mathcal{L}_{g.f.} = -\lim_{\alpha\to 0} (2\alpha)^{-1} (N.A)^2 \tag{4.98}$$

produces a vanishing interaction term $g\varphi^*.(N.A \times \varphi)$ in (4.96).

Since the renormalization of Q(8) presents no difficulties, let us concentrate on the mixing matrix generated by the singlet operator Q_1 in eq.(4.93). The one-loop counterterms $\Delta_1 Q_1$ include Q_1 itself plus a mesonic operator generated by the set of 1PI diagrams displayed in fig.17. Only fermion propagators are involved, so the new counterterm is proportional to the gauge-invariant combination

$$Q_2 = \chi^\mu F_{\mu\alpha}(iX.D)^{J-2} F^{\alpha\nu} \chi_\nu \; . \tag{4.99}$$

The next set of counterterms $\Delta_1 Q_2$ includes Q_1 and vertices given by the divergent parts of the diagrams in fig. 18. As noted by Gross and Wilczek[93], the ghost contributions from fig.18(a) do not vanish. This means that the accompanying mesonic terms (fig. 18(b)) are necessarily not gauge-invariant[111].

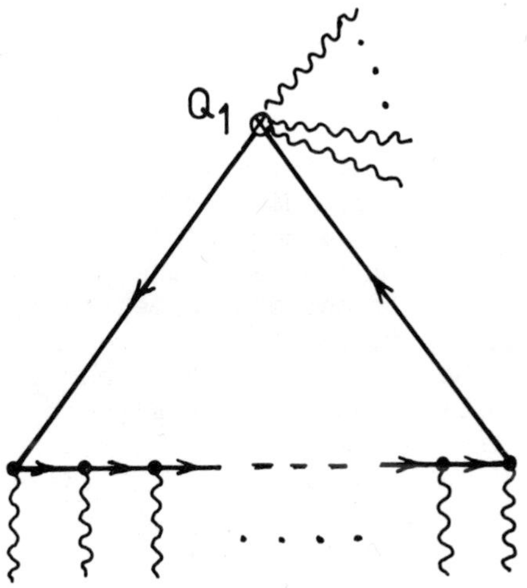

FIG. 17

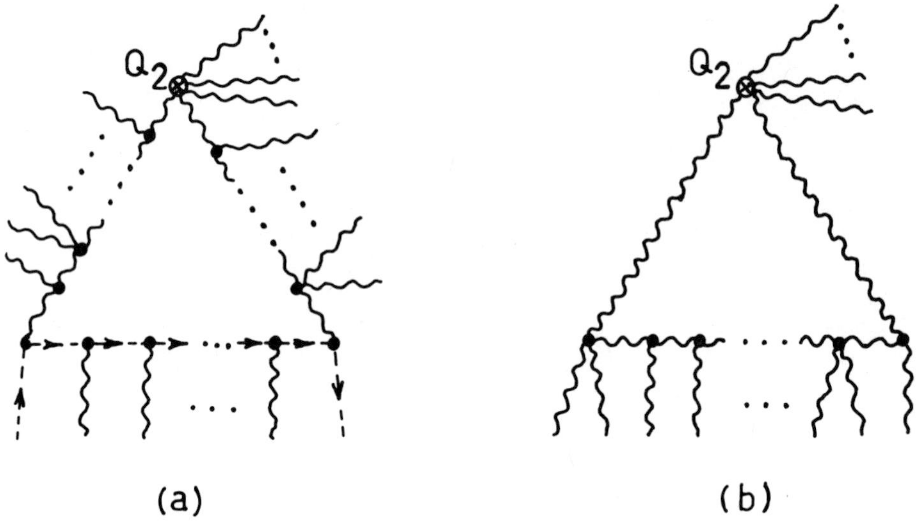

FIG. 18

The prescription adopted in ref. 93 is to construct a 2x2 submatrix $\overleftrightarrow{Z}(2)$ of $\overleftrightarrow{Z}$ corresponding to the gauge-invariant combinations Q_1, Q_2. Three of the elements of $\overleftrightarrow{Z}(2)$ ($Q_1 \to Q_1$, $Q_1 \to Q_2$, $Q_2 \to Q_1$) are obtained from $\Delta_1 Q_1$ and $\Delta_1 Q_2$ without ambiguity. The diagonal element $Q_2 \to Q_2$ is defined to be the contribution from 1PI diagrams in which Q_2 is coupled to only two external lines, both of which are put on-shell. Thus the rows and columns of $\overleftrightarrow{Z}$ are chosen to refer to a basis vector with elements $\{Q_1, Q_2$, $O(A^2)$ operators which vanish on-shell, other operators i.e. $O(A^3)$, $O(A^4), \ldots, O(\varphi^* \varphi)$, $O(\varphi^* A \varphi), \ldots\}$, and $\overleftrightarrow{Z}(2)$ is taken to be the 2x2 submatrix in the upper left-hand corner of $\overleftrightarrow{Z}$. The answer is supposed to be given by the eigenvalues of $\overleftrightarrow{Z}(2)$. Gross and Wilczek 93) checked their answer by repeating their on-shell prescription for the axial gauge (4.97) in the lightlike case $N^2=0$.

This prescription is not obviously correct because in general, the eigenvalues of a submatrix do not coincide with any of the eigenvalues of the complete matrix. The fact that Gross and Wilczek obtain the same answer in two sets of gauges is not conclusive; it merely suggests that the prescription is gauge-invariant for twist-two operators. No matter how rigorous the proof of gauge invariance may be, the answer is not correct unless it is an eigenvalue of the complete mixing matrix $\overleftrightarrow{Z}$.

Indeed, Kluberg-Stern and Zuber[112] have observed that the prescription fails for the twist-4, spin-0 operator $(F_{\mu\nu})^2$ in the usual Fermi gauges (2.70). Subsequent analysis[107,109] has shown this to be the rule rather than the exception. However, the exceptional case turns out to be that of twist-two operators. For reasons which can be understood only by consulting the detailed analysis [106], the eigenvalues of $\overleftrightarrow{Z}(2)$ are indeed eigenvalues of $\overleftrightarrow{Z}$. Instead of continuing in gauges with ghosts, I shall explain what happens for axial gauges[105].

All Green's functions are homogeneous in the fixed vector N_μ, because the only source of N_μ-dependence in eq.(4.98) which generates the free propagator for gauge mesons :

$$\Delta_{\mu\nu}(k,N) = -(i/k^2)\ \{g_{\mu\nu} - (k_\mu N_\nu + k_\nu N_\mu)(k.N + N^2 k_\mu k_\nu/(k.N)^2\}. \tag{4.100}$$

In order to carry out the renormalization program, it is essential that the lightlike case $N^2=0$ chosen by Gross and Wilczek be avoided. The chief characteristic of integrals involving denominators $(k.N)^{-\ell}$ for $N^2=0$ is that is is impossible to maintain power-counting [113]; there are insufficient scalar products available. For example, a momentum-dependent scalar counterterm involving N necessarily contains a factor

$$V(N,p,q) = \prod_{i=1}^{r} (N.p_i)/(N_i q_i) \qquad (p,q = \text{momenta},\ N^2=0) \tag{4.101}$$

because it must be homogeneous in N_μ. Since V is not a polynomial in p and q, power-counting does not work. In principle, renormalization is still permitted for the non-local vertex V, because the denominators $(N.q_i)^{-1}$, $(N.k)^{-\ell}$ are given principal-value singularities and hence have no absorptive part. However, explicit calculations[105] of meson and fermion self-energies demonstrate the existence of divergent parts proportional to $\ell n\ q^2$, where q is the momentum of the dressed propagator. The absorptive part does not vanish, so the $N^2=0$ gauge is not renormalizable. For $N^2 \neq 0$, these difficulties disappear[105,114]; the N_μ-dependence of counterterms always takes the form

$$\text{counterterm} = \{N_{\mu_1} \ldots N_{\mu_{2r}}/(N^2)^r\}\ \{N_\mu\text{-independent}\}, \quad (r = \text{integer}) \tag{4.102}$$

or some linear combination thereof, and all counterterms are polynomials in momentum space.

Consider the renormalization mixing matrix $\overleftrightarrow{Z}$ generated by the operator Q_1 in axial gauges $(N^2 \neq 0)$. In view of the trouble caused by ghosts in Lorentz-invariant gauges, it is tempting to assume that the absence of ghosts in axial gauges means that no spurious mixing accurs. The simplest way to disprove this idea is to compute the divergent part of the diagram shown in fig. 19 for the case spin J=2 [105],

$$\begin{aligned} P.P.\Gamma_{ij}(J=2) = \{g^2\delta_{ij}C_2(R)/4\pi^2(n-4)\}\{-4\not{\chi}\,\chi.q/3 + \not{N}\,\chi.N.q.\chi/N^2 \\ - 2\not{q}(N.\chi)^2 - \not{\chi}\,\chi.N\,q.N/N^2\}, \end{aligned} \quad (4.103)$$

where P.P. denotes the pole part of the 1PI amplitude Γ_{ij} and $C_2(R)$ is the value of the quadratic Casimir operator for the fermionic representation R of the gauge group :

$$(\tau^a\tau^a)_{ij} = \delta_{ij}C_2(R) \quad (4.104)$$

In addition to the matrix element $Q_2 \to Q_1$ given by the term proportional to $\chi\not{\chi}.q$, three N_μ-dependent vertices are generated. They appear in the following list of independent gauge-invariant operators P_r evaluated at zero momentum (i.e. $P_r \to$ zeroth-order 1PI amplitude $\langle\psi_j(-q)P_r\,\bar{\psi}_i(q)\rangle$) :

$$\begin{aligned} P_1 &= i\bar{\psi}\not{N}\,\chi.D_f\psi\,\chi.N/N^2 &&\longrightarrow \delta_{ij}\not{N}\,\chi.q\,\chi.N/N^2\,, \\ P_2 &= i\,\bar{\psi}\not{D}_f\,\psi(\chi.N)^2/N^2 &&\longrightarrow \delta_{ij}\not{q}(\chi.N)^2/N^2, \\ P_3 &= i\bar{\psi}\not{\chi}\,N.D_f\,\psi\chi.N/N^2 &&\longrightarrow \delta_{ij}\not{\chi}\,N.q\,\chi.N/N^2\,, \\ P_4 &= i\,\bar{\psi}\not{N}\,N.D_f\psi(\chi.N)^2/(N^2)^2 &&\longrightarrow \delta_{ij}\not{N}\,N.q(\chi.N)^2/(N^2)^2\,. \end{aligned} \quad (4.105)$$

(Here, and in the rules stated previously, the characterization "gauge-invariant" refers only to the operator-dependent parts of counterterms. The rules do not forbid dependence on gauge-dependent c-numbers such as N_μ.) Since we are considering the case J=2, the list (4.105) is restricted to operators which are quadratic in χ. Eqs.(4.103) and (4.105) become very complicated for arbitrary spins J. Similar results can be obtained for the diagrams in fig. 20; except for the case J=2, P.P. $\Gamma(J)$ is N_μ-dependent.

Evidently, the mixing matrix $\overleftrightarrow{Z}$ is very large. However, some simple features emerge if we define its rows and columns in terms of a well-chosen basis vector. Let us introduce operator classes C_p,

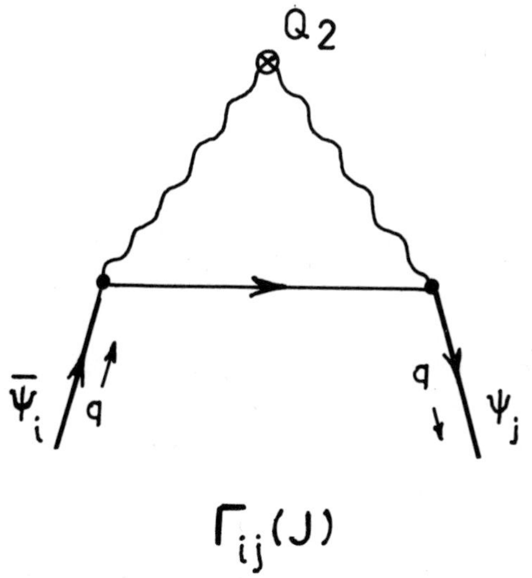

FIG. 19

where an operator Q belongs to C_p if it is of the form

$$Q = (\chi.N)^p \{\chi.N - \text{independent terms}\} , \quad (p=0,1,2,\ldots,J). \tag{4.106}$$

The basis vector is partitioned in the following way : $\{(C_0$ operators), (C_1 operators), (C_2 operators),..., (C_J operators)$\}$. The result is the partitioned matrix $\overleftrightarrow{Z}$ shown in fig. 21. There is a series of submatrices $M_o, M_1, M_2, \ldots, M_J$ which occupy the main diagonal, where M_p describes the mixing of operators within the class C_p.

All counterterms ΔQ of the operator Q in (4.106) contain the factor $(\chi.N)^p$, because $(\chi.N)^p$ commutes with loop integrals. Consequently, operators in the class C_p do not generate counterterms with smaller values of p :

$$C_p \to C_{p'} \ , \qquad (p' \geqslant p). \tag{4.107}$$

Eq.(4.107) implies the identity

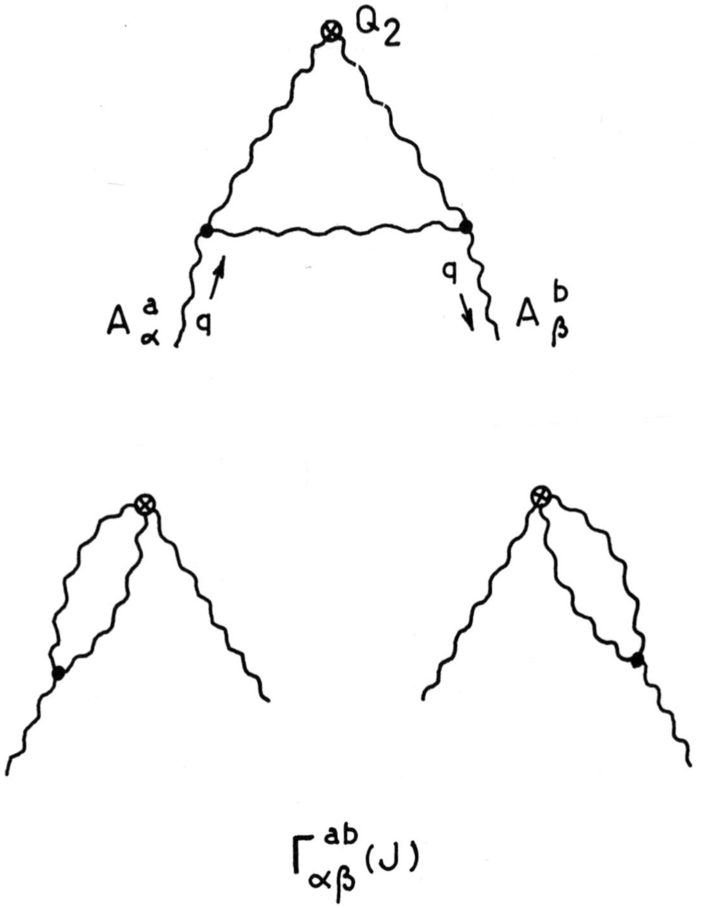

FIG. 20

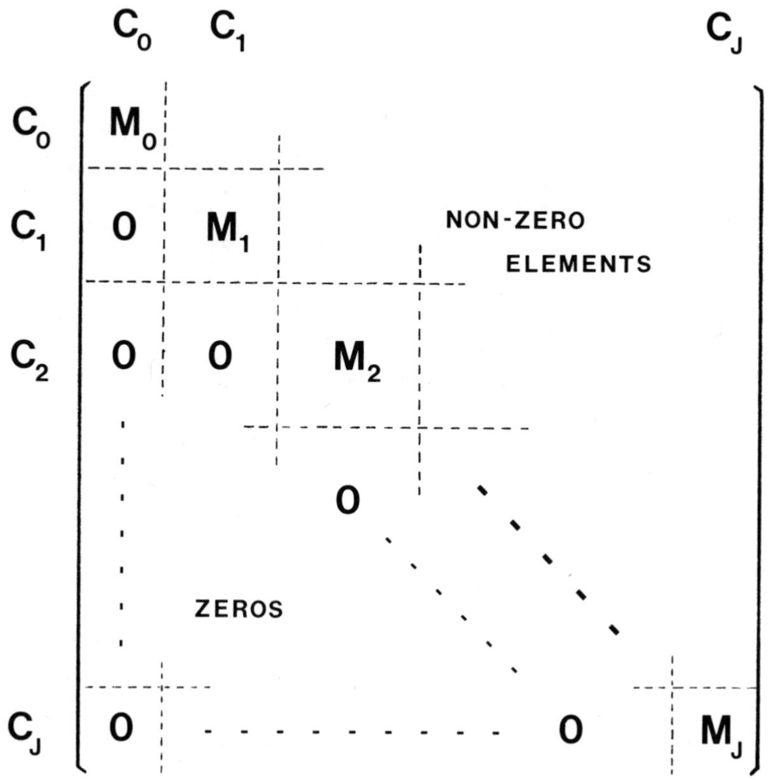

MIXING MATRIX $\overleftrightarrow{Z}$

Fig. 21

$$\det(\overleftrightarrow{Z} - \lambda\overleftrightarrow{I}) = \prod_{p=0}^{J} \det(M_p - \lambda I_p) \qquad (4.108)$$

because all submatrices below the diagonal set $(M_0, M_1, M_2, \ldots, M_J)$ vanish *. Hence the eigenvalues of $\overleftrightarrow{Z}$ are given by the eigenvalues of $M_0, M_1, M_2, \ldots, M_J$.

The only $\chi.N$-independent gauge-invariant combinations are Q_1, Q_2 and

$$Q_3 = (N^\alpha \chi^\beta F_{\alpha\beta})(i\chi.D)^{J-2}(N^\gamma \chi^\delta F_{\gamma\delta})/N^2 , \qquad (4.109)$$

so the matrix M_0 is 3x3. The diagrams in fig.17 are N_μ-independent so the matrix element $Q_1 \to Q_3$ vanishes. Also, an explicit calculation of the diagrams in fig. 20 yields the result[105]

$$P.P\Gamma^{ab}(J) = \{g^2 C_2(G)/2\pi^2(n-4)\}\langle A^a_\alpha(q)Q_2 A^b_\beta(-q)\rangle_{1PI,bare} \cdot \{1 - 1/J(J-1) - 1/(J+1)(J+2) + \sum_{j=2}^{J} 1/j\} + O(\chi/N) . \qquad (4.110)$$

There are non-zero terms proportional to the bare vertices of Q_2 and (for $J \geq 4$) $\chi.N$-dependent operators, but there is no term proportional to

$$\langle A^a_\alpha(q)Q_3\, A^b_\beta(-q)\rangle_{1PI,bare} = 2\delta^{ab}(N.q\chi_\alpha - \chi.qN_\alpha)(N.q\chi_\beta - \chi.qN_\beta) \cdot (\chi.q)^{J-2}/N^2 , \quad (J=2,4,6,\ldots). \qquad (4.111)$$

Hence the matrix element $Q_2 \to Q_3$ also vanishes. Furthermore, explicit calculations (with $N^2 \neq 0$) show that the 2x2 submatrix of M_0 generated by Q_1 and Q_2 is exactly the same as the submatrix $\overleftrightarrow{Z}(2)$, obtained in ref. 93; for example, the $Q_2 \to Q_2$ matrix element can be checked by combining Eq.(4.110) with the 1-loop result[116]

$$Z_3(\text{axial gauge}) = 1 - (11C_2(G) - 4T(R))g^2/24\pi^2(n-4) + O(g^4) \qquad (N^2 \neq 0) \quad (4.112)$$

for the Z-factor of the gauge-meson field A^a_μ .

---

* That is, $\overleftrightarrow{Z}$ is block-triangular; see fig. 21. The same property, involving just two operator classes, has been demonstrated for $\overleftrightarrow{Z}$ in Lorentz-covariant gauges[107-109,115].

The results for M_0 are summarized in fig. 22. Clearly, the eigenvalues of $\overleftrightarrow{Z}(2)$ are also eigenvalues of both M_0 and $\overleftrightarrow{Z}$. Therefore the results in ref.93 are correct, provided that the other eigenvalues of $\overleftrightarrow{Z}$ are unphysical. (A separate discussion is necessary in order to verify this last point[105].)

The same procedure works for operators with arbitrary twist. Further simplification can be achieved by applying the following theorems[105] :

(i) The presence of a factor $D^{\mu}F_{\mu\nu}$, $\not{D}_f\psi$ or $\bar{\psi}\not{D}_f$ in the gauge-invariant combination Q means that ΔQ also contains one of these factors.

(ii) All N_μ-dependent counterterms contain either $D^{\mu}F_{\mu\nu}$, $\not{D}_f\psi$ or $\bar{\psi}\not{D}_f$ as a factor.

Theorem (i) is the analogue of a theorem of Kluberg-Stern and Zuber[107] for gauges with ghosts. Some consequences of theorem (ii) are evident in the twist-two case :

(a) The results $Q_2 \not\leftrightarrow Q_3$, obtained in the 1-loop approximation in eq.(4.110), is true to all orders of perturbation theory.

(b) The operator P_4, defined in eq.(4.105), cannot contribute to eq. (4.103), and the operators P_1 and P_3 necessarily appear in the combination (P_1-P_3).

$\eta \to 3\pi$ Decay

Finally, let us consider a problem in current algebra[117], the soft-meson theorems for $\eta \to 3\pi$ decay. We shall see that it is essential that the analysis be carried out in terms of short-distance expansions. Only then does it become clear that there is an extra term[118] which must be added to the conventional predictions[119-121] for the decay amplitudes.

We begin by recalling the standard assumption[122] for chiral symmetry-breaking terms in the energy density for strong interactions :

$$\theta_{oo} = \bar{\theta}_{oo} + u_o + cu_8 , \qquad (c \simeq -1.25) \tag{4.113}$$

The operator $\bar{\theta}_{oo}$ is SU(3) x SU(3) invariant. The scalar-density operators u_o, u_8 belong to the $(3,\bar{3}) \oplus (\bar{3},3)$ representation formed by the set $\{u_i, v_j;\ i,j = 0,1,2,\ldots,8;\ u_i$ = scalar density, v_j = pseudoscalar density$\}$. In gauge theories with explicit chiral symmetry breaking ($m \neq 0$ in eq.(2.65)), $(u_o + cu_8)$ is given by the renormalized version of a mass term in $\bar{\psi} m \psi$.

It is natural to assume that the strong-interaction energy density induced by second-order electromagnetic effects is

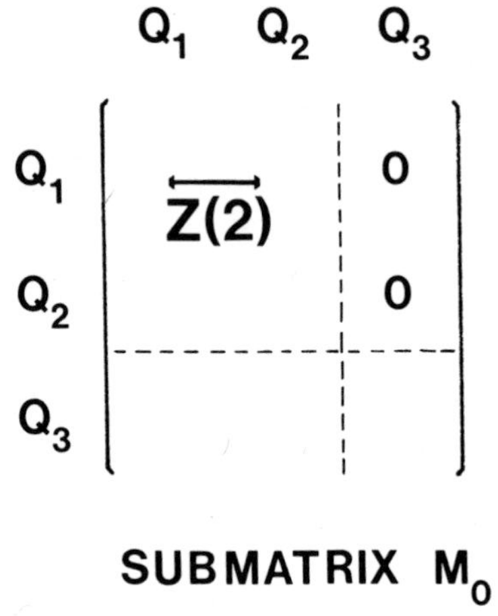

FIG. 22

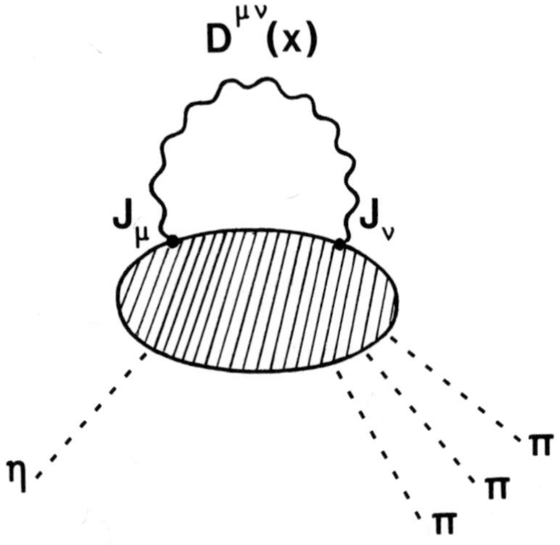

FIG. 23

$$\mathcal{H}_{em}(\text{pure}) = -e^2 \int d^4x\, D^{\mu\nu}(x) T\,\{J_\mu(x) J_\nu(0)\}, \tag{4.114}$$

where $D^{\mu\nu}(x)$ is the free photon propagator in coordinate space and J_μ is the electromagnetic current for hadrons, (see fig. 23). However, the consequences of (4.114) disagree with experiment :

(a) Sutherland[123] showed that the decay amplitudes $A(\eta \to \pi^+\pi^-\pi^\circ)$ and $A(\eta \to 3\pi^\circ)$ vanish in the SU(2) x SU(2) approximation in which one of the pions (charged or uncharged) is soft.

(b) Dashen[124] obtained the rule

$$\langle K^+|\, \mathcal{H}_{em}(\text{pure})\,|K^+\rangle - \langle K^\circ|\, \mathcal{H}_{em}(\text{pure})\,|K^\circ\rangle$$

$$\simeq \langle \pi^+|\, \mathcal{H}_{em}(\text{pure})\,|\pi^+\rangle - \langle \pi^\circ|\mathcal{H}_{em}(\text{pure})\,|\pi^\circ\rangle \quad , \tag{4.115}$$

which is not consistent with the observed K^+-K° and π^+-π° mass differences.

The standard remedy[119-121] is to include a u_3 tadpole[125] in the expression for the energy density :

$$\text{total energy density} = \theta_{oo} + \mathcal{H}_{em}(\text{tadpole}), \tag{4.116}$$

$$\mathcal{H}_{em}(\text{tadpole}) = \mathcal{H}_{em}(\text{pure}) + c_3 u_3 \; , \quad (c_3 = O(e^2)). \tag{4.117}$$

In this scheme, eq.(4.115) is retained and the tadpole is supposed to be responsible for the deviation

$$\delta m^2 = m^2(K^+) - m^2(K^\circ) - m^2(\pi^+) + m^2(\pi^\circ) \tag{4.118}$$

from Dashen's rule; e.g. the SU(3) approximation implies

$$\langle\pi|u_3|\pi\rangle \simeq 0$$

$$\langle K^+|u_3|K^+\rangle \simeq -\langle K^\circ|u_3|K^\circ\rangle \; ; \tag{4.119}$$

$$\delta m^2 = 2\, c_3 \langle K^+|u_3|K^+\rangle \simeq c_3 \sqrt{3}\langle\pi^\circ|u_3|\eta\rangle \; . \tag{4.120}$$

Similarly, Sutherland's null result for $\mathcal{H}_{em}(\text{pure})$ is retained and the decay amplitudes

$$T = -\langle 3\pi|\mathcal{H}_{em}(\text{tadpole})\,|\eta\rangle \tag{4.121}$$

are supposed to be given by

$$T \simeq -c_3 \langle 3\pi|u_3|\eta\rangle \tag{4.122}$$

in the SU(2) x SU(2) approximation. The amplitude $T(\eta \to \pi^+\pi^-\pi^\circ)$

still vanishes when a charged pion becomes soft, but the soft-π° limit produces a new equal-time commutator

$$i[F_5^3, u_3] = (\sqrt{2}\, v_0 + v_8)/\sqrt{3} = v\ , \qquad (4.123)$$

$$T(\eta \to \pi^\circ(\text{soft})2\pi) = (c_3/F_\pi)\langle 2\pi|v|\eta\rangle\ , \qquad (4.124)$$

where F^i, F_5^j $(i,j = 1,\ldots,8)$ are the charges which generate chiral $SU(3) \times SU(3)$[122],

$$F^i = \int d^3x\, \mathcal{J}_0^i(x)\ , \qquad F_5^i = \int d^3x\, \mathcal{J}_{50}^j(x)\ , \qquad (4.125)$$

the pion decay constant* $F_\pi \simeq 94$ MeV is given by

$$\langle \pi^\circ(q)|\mathcal{J}_{5\mu}^3(0)|0\rangle = -iq_\mu F_\pi \qquad (4.126)$$

and the matrix element $\langle 2\pi|v|\eta\rangle$ $(=\langle 2\pi^\circ|v|\eta\rangle = \langle \pi^+\pi^-|v|\eta\rangle)$ is evaluated at zero momentum transfer. If $\langle 2\pi|v|\eta\rangle$ is not negligibly small, the energy dependence of the decay spectra is given by the formulas[120,126]

$$T(\eta \to 3\pi^\circ) \simeq \text{constant} = (c_3/F_\pi)\ \langle 2\pi|v|\eta\rangle\ ,$$

$$T(\eta \to \pi^+\pi^-\pi^\circ) \simeq (1 - 2E_0/m_\eta)(c_3/F_\pi)\ \langle 2\pi|v|\eta\rangle,$$

$$(E_0 = \pi^\circ \text{ energy with } \eta \text{ at rest}), \qquad (4.127)$$

in agreement with experiment[127]. Contracting another pion yields the equal-time commutator

$$i[F_3, v] = -u_3\ , \qquad (4.128)$$

$$\langle 2\pi|v|\eta\rangle \simeq F^{-1}\langle \pi^\circ|u_3|\eta\rangle\ . \qquad (4.129)$$

Eqs.(4.120), (4.127) and (4.129) yield the result[119,120]

$$T(\eta \to 3\pi^\circ) \simeq \delta m^2/\sqrt{3}\, F_\pi^2\ ,$$

$$T(\eta \to \pi^+\pi^-\pi^\circ) \simeq (1-2E_0/m\ )\ \delta m^2/\sqrt{3}\, F_\pi^2\ . \qquad (4.130)$$

This corresponds to a partial decay width[119] $\simeq 65$ eV for $\eta \to \pi^+\pi^-\pi^\circ$ which is considerably smaller than the experimental value[128], 204 ± 29 eV. It can be argued that the prediction (4.130) is not expected to be as accurate as (4.127) because it involves approximate SU(3)xSU(3), not SU(2)xSU(2). Even so, the agreement with experiment is not impressive.

---

* Note the extra factor 2 in the definitions of chiral currents and F_π in ref. 119 : e.g., $F_\pi(\text{ref.}119) = 2\, F_\pi$ (4.126).

Now we consider the short-distance approach to $\eta \to 3\pi$ decay. Wilson[4] observed that the Sutherland analysis and its generalization to include tadpoles assume that the expression (4.114) for $\mathcal{H}_{em}$(pure) converges. For the theory of broken scale invariance or asymptotically free theories, this is not true[4,7]. When the operator-product expansion

$$T\{J_\mu(x)J_\nu(0)\} \sim \sum_n \mathcal{C}_{n\mu\nu}(x)O_n(0), \qquad (x_\alpha \to 0) \qquad (4.131)$$

is substituted in eq.(4.14) to test the ultraviolet convergence of the integral $\int d^4x$, Lorentz invariance implies

$$\int d^4x \; \mathcal{C}_{n\mu\nu}(x)D^{\mu\nu}(x) = 0 \qquad (4.132)$$

for all coefficient functions except those which multiply scalar operators. In the isospin-1 sector*, we have

$$T\{J_\mu(x)J_\nu(0)\} \sim \mathcal{C}_{\mu\nu}(x)u_3(0), \; (x_\alpha \to 0, \; (J,I)=(0^+,1)\text{sector}). \qquad (4.133)$$

This is an example of the special case (4.71), so the formula

$$\mathcal{C}_{\mu\nu}(x) \sim (\text{constant}) \; (g_{\mu\nu}x^2 + 2x_\mu x_\nu)/(x^2)^2 \; , \; (x_\alpha \to 0) \qquad (4.134)$$

is valid for both broken scale invariance[4] (irrespective of the anomalous dimension d of u_3, $1 \leqslant d < 4$) and asymptotic freedom[7]. Since $D^{\mu\nu}(x)$ goes as x^{-2}, we see that the integral (4.114) diverges logarithmically.

The integral must be properly subtracted so that the result is a convergent expression to which current-algebraic techniques can be applied. The divergence is logarithmic, so a single subtraction suffices[4] :

$$\mathcal{H}_{em} = -e^2 \int d^4x \; D^{\mu\nu}(x) \; [\, T\{J_\mu(x)J_\nu(0) - \mathcal{C}_{\mu\nu}(x)u_3(0)] \; + fu_3(0). \qquad (4.135)$$

The finite constant f is introduced to adjust the finite part of the u_3 contribution to fit ele tromagnetic mass differences.

Previously we saw that the tadpole term was introduced as an extra parameter in response to the difficulties associated with eq.(4.114). Here, the u_3-dependence of the electromagnetic interaction arises naturally. Indeed, the distinction between an electromagnetic term $\mathcal{H}_{em}$(pure) and a strong isospin-breaking

---

* Singularities in the SU(3)-singlet sector also have to be removed, but this is irrelevant for $\eta \to 3\pi$ decay and electromagnetic mass differences

term c_3u_3, which can be made for eqs.(4.116) and (4.117) no longer exists.

Wilson[4,129] concludes that the interaction (4.135) yields the same results for $\eta \to 3\pi$ decay as the tadpole interaction (4.117). He identifies the fu_3 term as the Coleman-Glashow tadpole[125] c_3u_3, and argues that the finiteness of the other term means that Sutherland's argument applies to it, in the same way that the Sutherland argument applies to $\mathcal{H}_{em}$(pure) if $\int d^4x$ in eq.(4.114) happens to converge. However, for reasons which will now be explained, I believe that the conclusion has to be modified.

First we note that the quantities $\mathcal{C}_{\mu\nu}(x)$, f in eq.(4.135) are not absolute; they can be specified only with respect to a particular subtraction condition. The choice of a subtraction condition from the infinite set of possibilities is purely a matter of convenience. A label i will be added to a symbol to indicate its dependence on a given subtraction condition i :

$$f, \mathcal{C}_{\mu\nu}(x) \to f_i, \mathcal{C}^i_{\mu\nu}(x) \quad ,$$

$$I_i = \int d^4x \; T\{J_\mu(x)J_\nu(0) - \mathcal{C}^i_{\mu\nu}(x)u_3(0)\} \; D^{\mu\nu}(x) \quad . \qquad (4.136)$$

Of course, the total electromagnetic energy density $\mathcal{H}_{em.}$ is observable, so it does not depend on i :

$$\mathcal{H}_{em} = -e^2 I_i + f_i u_3(0). \qquad (4.137)$$

For example, we may decide to define the subtraction condition i=7 by imposing the constraint

$$\langle p|I_7|p\rangle = \langle n|I_7|n\rangle \; , \qquad (4.138)$$

in which case f_7 is proportional to the proton-neutron mass difference :

$$f_7\{\langle p|u_3|p\rangle - \langle n|u_3|n\rangle\} = m_p - m_n \; . \qquad (4.139)$$

Note :
(i) As this example shows, the principles of renormalization are not confined to perturbation theory.
(ii) It is not possible to substitute the leading x^{-2} singularity in eq.(4.34) for $\mathcal{C}_{\mu\nu}(x)$, because that would introduce an infrared singularity in $\int d^4x$ at $x = \infty$.
(iii) As a rule (e.g. in perturbation theory), $\mathcal{C}^i_{\mu\nu}(x)$ should include the complete leading power of $T\{J_\mu(x)J_\nu(0)\}$ as $x \to 0$. In particular, if the theory is asymptotically free, terms in $T\{J_\mu(x)J_\nu(0)\}$ which are only logarithmically less singular than $x^{-2\mu}$ can also produce logarithmic infinities in $\int d^4x$, so they must be included in the subtraction.

(iv) For a given prescription i, $\mathcal{C}^{i}_{\mu\nu}(x)$ remains unspecified up to function $\Delta_{\mu\nu}$ which satisfy

$$\int d^4x\ D^{\mu\nu}(x)\Delta_{\mu\nu}(x) = 0 \quad ; \qquad \mathcal{C}^{i}_{\mu\nu} \to \mathcal{C}^{i}_{\mu\nu} + \Delta_{\mu\nu} \quad . \tag{4.140}$$

Now consider the amplitudes for $\eta \to 3\pi$ decay :

$$A = - \langle 3\pi|\mathcal{H}'_{em}|\eta\rangle \tag{4.141}$$

(The symbol A is used instead of T (eq.(4.121)) to avoid confusing the results implied by the interactions (4.117) and (4.135)). The u_3 subtraction does not affect the soft-$\pi^{\pm}$ limit because it commutes with the corresponding axial charges. Therefore we recover the desirable result (SU(2)xSU(2) approximation)

$$A(\eta \to 3\pi^{\circ}) \simeq A_0$$

$$A(\eta \to \pi^{+}\pi^{-}\pi^{\circ}) \simeq (1 - 2E_0/m\,)A_0 \ , \tag{4.142}$$

where the constant A_0 is given by

$$A_0 = A(\eta \to \pi^{\circ}(\text{soft})2\pi)$$

$$= e^2 \langle 2\pi,\pi^{\circ}(\text{soft})|I_i|\eta\rangle + (f_i/F_\pi)\langle 2\pi|v|\eta\rangle \ , \tag{4.143}$$

and is assumed not to be small.

It is evident that the question of whether Sutherland's argument is applicable or not has nothing to do with the convergence of the integral in (4.135). All of the integrals I_i are finite. If we consider a pair I_1,I_2 corresponding to inequivalent subtraction conditions ($f_1 \neq f_2$), eq.(4.137) implies

$$e^2(I_1-I_2) = (f_1-f_2)u_3 \tag{4.144}$$

$$e^2\langle 2\pi,\pi^{\circ}(\text{soft})|(I_1-I_2)|\eta\rangle = -(f_1-f_2)\langle 2\pi|v|\eta\rangle\ /F_\pi, \tag{4.145}$$

so Sutherland's argument cannot be valid for both of the convergent integrals I_1 and I_2. In fact, we can define a subtraction prescription i=S by requiring the validity of the "Sutherland condition"

$$\langle 2\pi,\pi^{\circ}\ (\text{soft})|I_S|\eta\rangle = 0 \tag{4.146}$$

or equivalently,

$$A_0 = (f_S/F_\pi)\ \langle 2\pi|v|\eta\rangle \quad . \tag{4.147}$$

The same reasoning can be applied to Dashen's rule, eq. (4.115). Obviously, its derivation assumes the convergence of $\int d^4x$ in eq.(4.114). For the energy density (4.135), the problem is that the u_3 term in eq.(4.144) does not satisfy the rule; according to eq.(4.119), it contributes to the total mass difference m defined in eq.(4.118). So there is no analogue of Dashen's rule for the subtracted integrals I_i except for a special subtraction prescription i=D in which the "Dashen condition"

$$\langle K^+|I_D|K^+\rangle - \langle K^\circ|I_D|K^\circ\rangle = \langle \pi^+|I_D|\pi^+\rangle - \langle \pi^\circ|I_D|\pi^\circ\rangle \tag{4.148}$$

is true by definition. The formula for the corresponding constant f_D is

$$\delta m^2 = f_D[\ \langle K^+|u_3|K^+\rangle - \langle K^\circ|u_3|K^\circ\rangle - \langle \pi^+|u_3|\pi^+\rangle + \langle \pi^\circ|u_3|\pi^\circ\rangle]\ . \tag{4.149}$$

The conclusion is that the interaction (4.135) produces the same results for $\eta \to 3\pi$ decay as the tadpole Hamiltonian (4.117) only if the subtraction prescriptions S and D happen to coincide (either exactly, or within the SU(3)xSU(3) approximation). There is no reason to suppose that this is the case : in general, there will be a renormalization mismatch :

$$f_S \neq f_D \tag{4.150}$$

The analogue of the previous SU(3) x SU(3) analysis can be carried out by observing that the SU(3) approximation to eq.(4.149) looks like eq.(4.120)

$$\delta m^2 \simeq 2f_D \langle K^+|u_3|K^+\rangle \simeq f_D \sqrt{3}\langle \pi^\circ|u_3|\eta\rangle \tag{4.151}$$

Combining eqs.(4.129), (4.147) and (4.151), we obtain the result

$$A_o \simeq (f_S/f_D)\ \delta m^2/\sqrt{3}\ F_\pi^2 \tag{4.152}$$

Apart from the extra factor f_S/f_D, the answer is the same as eq.(4.130). The obvious conclusion is that this factor is responsible for the discrepancy between the conventional answer (4.130) and experiment, with

$$|f_S/f_D|_{expt.} \simeq 1.8 \tag{4.153}$$

At present, an independent theoretical estimate of f_S/f_D is not available, but it is easy to see that a connection with operator-product expansions exists. A combination of eqs.(4.143) and (4.147) yields the result

$$F_\pi \langle 2\pi, \pi^\circ(\mathrm{soft})|I_D|\eta\rangle = e^{-2}(f_S - f_D)\langle 2\pi|v|\eta\rangle$$

$$= \lim_{q\to 0} i\int d^4y\, e^{iq\cdot y}\, T\, \langle 2\pi|\partial^\alpha \mathcal{F}^3_{5\alpha}(y)\, I_D|\eta\rangle \,, \qquad (4.154)$$

or equivalently,

$$\partial^\alpha_y\, T\{\mathcal{F}^3_{5\alpha}(y) I_D\} = T\{\partial^\alpha \mathcal{F}^3_{5\alpha}(y) I_D\} + ie^{-2}(f_S - f_D)\delta^4(y)v(0)$$

$$+ Q_\alpha \partial^\alpha \delta^4(y) + Q_{\alpha\beta}\partial^\alpha\partial^\beta\delta^4(y) + \ldots \,, \qquad (4.155)$$

where the series of derivative terms $Q_\alpha\partial^\alpha\delta^4(y)$, $Q_{\alpha\beta}\partial^\alpha\partial^\beta\delta^4(y)$, ... is finite, and does not contribute to the $q\to 0$ limit. The term proportional to $v(0)\delta^4(y)$ is not an equal-time commutator because the operator I_D (defined by eq.(4.136)) is bilocal, not local. Indeed, Sutherland's method, which works for the bilocal $\mathcal{H}_{em}$ (pure) in (4.114) when $\int d^4x$ converges, cannot be applied to I_D. If we suppose that the derivative ∂^α_y can be commuted with $\int d^4x$ in (4.136), the left-hande side of (4.155) becomes

$$\int d^4x\, D^{\mu\nu}(x)\, (\partial/\partial y)^\alpha\, T[\mathcal{F}^3_{5\alpha}(y)\, \{J_\mu(x)J_\nu(0) - \mathcal{C}^D_{\mu\nu}(x)u_3(0)]$$

$$= T\{\partial^\alpha \mathcal{F}_{5\alpha}(y) I_D\} + i\delta^4(y)v(0)\int d^4x\, D^{\mu\nu}(x)\,\mathcal{C}^D_{\mu\nu}(x) \,, \qquad (4.156)$$

because the equal-time commutators with J_μ, J_ν vanish, and the commutator with u_3 is given by eq.(4.123). The factor $\int d^4x D^{\mu\nu}\, \mathcal{C}^D_{\mu\nu}$ is infinite, so the method fails.

The simplest way to understand why the $\delta^4(y)\, v(0)$ term appears is to note that it must correspond to a singularity $O(y^{-3})$ in the trilocal $T\{\mathcal{F}_{5\alpha} I_D\}$. Eq.(4.155) and the identity

$$\partial^\alpha[y_\alpha/(y^2 - i\varepsilon)^2] = -2\pi^2\, i\delta^4(y) \qquad (4.157)$$

imply

$$\int d^4x\, D^{\mu\nu}(x) T[\mathcal{F}^3_{5\alpha}(y)\, \{J_\mu(x)J_\nu(0) - \mathcal{C}^D_{\mu\nu}(x)u_3(0)\}]$$

$$\sim -(2\pi^2 e^2)^{-1}\, (f_S - f_D)\{y_\alpha/(y^2 - i\varepsilon)^2\}\, v(0) \,, \quad (y_\alpha \to 0) \qquad (4.158)$$

for the pseudoscalar isoscalar sector of the expansion of the trilocal. The left-hand side of (4.158) can be analyzed in the same manner as the example discussed in eqs.(4.17-27) and fig. 12. The following regions must be considered :

(i) The limit $y\to 0$ for the product of two operators associated with the subtraction term :

$$T\{\mathcal{F}^3_{5\alpha}(y)\ u_3(0)\} \sim \{y_\alpha/2\pi^2(y^2-i\varepsilon)^2\}\ v(0). \qquad (4.159)$$

The normalization is fixed by eqs.(4.123) and (4.157).

(ii) The limit $x,y \to 0$ for the product of three operators

$$T\{\mathcal{F}^3_{5\alpha} J_\mu(x) J_\nu(0)\} \sim \mathcal{C}_{\alpha\mu\nu}(x,y)v(0) \quad . \qquad (4.160)$$

Eq. (4.160) is yet another example of the special case (4.71), so the equation

$$\mathcal{C}_{\alpha\mu\nu}(\rho x,\rho y) = O(\rho^{-5}) \ , \qquad (\rho \to 0) \qquad (4.161)$$

is valid for either broken scale invariance or asymptotic freedom. In general, equal-time commutators from the regions $x-y \ll x,y$ and $y \ll x$ also have to be considered, but for the present case, there is no contribution because F^3_5 commutes with J_μ and J_ν at equal times :

Substituting eqs.(4.159) and (4.160) into eq.(4.158), we arrive at the desired result :

$$\int d^4x\ D^{\mu\nu}(x)[\mathcal{C}_{\alpha\mu\nu}(x,y) - \mathcal{C}^D_{\mu\nu}(x) y_\alpha/2\pi^2(y^2-i\varepsilon)^2]$$

$$\sim\ -\ (2\pi^2 e^2)^{-1}(f_S-f_D)\ y_\alpha/(y^2-i\varepsilon)^2 \ , \qquad (y \to 0) \qquad (4.162)$$

In contrast with the situation in eq.(4.156), the integral $\int d^4x$ converges at $x=0$, because the x^{-2} singularity of $\mathcal{C}_{\alpha\mu\nu}(x,y)$ in the limit $x \ll y$ is precisely cancelled off by the subtraction term. Observe that if we count powers in the integral (-5 for $\mathcal{C}_{\alpha\mu\nu}$, -2 for $D^{\mu\nu}$, and +4 for $\int d^4x$), the result $-5-2+4 = -3$ agrees with the power $O(y^{-3})$ which appears on the right-hand side of eq.(4.162).

It must be emphasized that the only similarity between this analysis and the short-distance analysis[4,89] of $\pi^\circ \to 2\gamma$ decay is that both involve coefficient functions for products of three operators[130]. The anomalous constant[131] S for $\pi^\circ \to 2\gamma$ decay can be produced by the function $C_{\alpha\beta\gamma}(x,y)$ in

$$T\{J_\alpha(x)J_\beta(0)\ \mathcal{F}^3_{5\gamma}(y)\} \sim C_{\alpha\beta\gamma}(x,y)I \ , \quad (x,y, \to 0,\ I\text{=unit operator}) \qquad (4.163)$$

if it is sufficiently singular[4) :

$$\lim_{\rho\to 0}\ \rho^9(C_{\alpha\beta\gamma}(\rho x,\rho y) \neq 0 \qquad (x,y \neq 0) \ . \qquad (4.164)$$

Thus, when the operator $(\partial/\partial y)_\gamma$ is applied to the left-hand side of (4.163), the leading power $C_{\alpha\beta\gamma}$ can generate a contact term which scales as ρ^{-10} :*

$$\partial^\gamma C_{\alpha\beta\gamma}(x,y) = -(S/2\pi^2)\varepsilon_{\alpha\beta\mu\nu}\,\partial_x^\mu\partial_y^\nu\,\delta^4(x)\delta^4(y)\;; \qquad (4.165)$$

(compare with eq.(4.157)). On the other hand, the analysis of $\eta \to 3\pi$ decay does not involve $\delta^4(x)\delta^4(y)$ or its derivatives explicitly and should not be understood purely in terms of the failure of a naive manipulation (∂_y not commuting with $\int d^4x D^{\mu\nu}(x)$). The important point for $\eta \to 3\pi$ decay (irrelevant for $\pi^\circ \to 2\gamma$ decay) is that one must keep track of the initial subtraction prescription.

Unfortunately, there is no analogue of the fact that the answer for $\pi^\circ \to 2\gamma$ decay can be computed directly from the leading-power coefficient function $C_{\alpha\beta\gamma}(x,y)$. Whatever function is chosen for the leading power** . $\mathscr{C}_{\alpha\mu\nu}(x,y)$ in the integrand of (4.162), there is no way of sidestepping the difficulty that we do not know the correct expression for $\mathscr{C}^D_{\mu\nu}$. Since this function refers explicitly to the subtraction prescription D, there is no hope of calculating it in a resonably model-independent fashion : it depends on complicated non-asymptotic details of the strong interactions. Instead,one should look for measurable amplitudes involving integrals which can be related to the integral in eq. (4.162).

So far, the U(1) problem[132] of quark models has been ignored. For a long time, it has been known that free-quark models (quark-parton model and its abstractions) are not consistent with approximate chiral symmetry. Such models contain an observable axial baryonic current

$$\mathcal{J}^{\,\circ}_{5\alpha} = \bar\psi\, i\, \gamma_\alpha\gamma_5\psi \qquad (4.166)$$

(e.g. generated by U(6)xU(6) algebra[133]) which is partially conserved. The consequences of this are disastrous*** :
(i) Glashow[134] obtained the result

$$m_\eta^2 \simeq m_\pi^2\,, \qquad (4.167)$$

---

* Conserved δ-function ambiguities in $C_{\alpha\beta\gamma}$(eg.$\varepsilon_{\alpha\beta\mu\nu}\,(\partial_x)_\gamma\partial_x^\mu\partial_y^\nu\delta^4(x)\delta^4(y)$) are irrelevant : they cannot contribute to the $\partial_x\partial_y\delta^4(x)\delta^4(y)$ term. In the literature, beware of incorrect statements that the anomaly is controlled by short-distance singularities of T $\{J_\alpha J_\beta \partial^\gamma \mathcal{J}^3_{5\gamma}$

** For asymptotically free theories, the leading singularity of $\mathscr{C}_{\alpha\mu\nu}$(but not the entire leading power) coincides with the result for free-field theory.

*** This is the motivation for not assuming U(6)xU(6) algebra in ref.89.

which corresponds to the fact that a linear combination $K_{5\alpha}$ of $\mathcal{J}^{0}_{5\alpha}$ and $\mathcal{J}^{8}_{5\alpha}$ becomes conserved in the SU(2) x SU(2) limit ($c \to -\sqrt{2}$):

$$\partial^{\alpha} K_{5\alpha} = -(\sqrt{2} + c) v / \sqrt{3} \tag{4.168}$$

More generally, there must be an observable pseudoscalar isoscalar particle L with mass m_L given by[119,130]

$$m_L^2 \lesssim 3m_\pi^2 \tag{4.169}$$

(ii) Brandt and Preparata[135] pointed out that eq.(4.168) implies

$$\langle 2\pi | v | \eta \rangle = 0 \quad , \tag{4.170}$$

because the matrix element is evaluated at zero-momentum transfer. Their solution is to abandon approximate chiral symmetry, but I think that too many good chiral-symmetric predictions (e.g. for $K_{\ell 3}$ decay[136]) have to be dismissed as accidents for this approach to be believable.

For gauge theories, the divergence of $\mathcal{J}^{0}_{5\alpha}$ is anomalous, but there is a gauge non-invariant symmetry current

$$\mathcal{J}^{S}_{5\alpha} = \mathcal{J}^{0}_{5\alpha} - (g^2 S/8\pi^2) \varepsilon_{\alpha\beta\mu\nu} A^{\beta} F^{\mu\nu} \tag{4.171}$$

which is partially conserved, so v is still equal to a total divergence :

$$v \propto \partial^{\alpha} K^{S}_{5\alpha} \tag{4.172}$$

In order to obtain a non-vanishing result for $\langle 2\pi | v | \eta \rangle$, Kogut and Susskind[129] proposed the formula

$$\langle 2\pi | \mathcal{J}^{S}_{5\alpha}(0) | \eta \rangle \sim (\text{constant}) q_\alpha / q^2 \; , \qquad (q \to 0) \tag{4.173}$$

where q is the momentum carried by $\mathcal{J}^{S}_{5\alpha}$. To ensure that the zero-mass pole is unobservable experimentally, they suppose that it is a linear combination of positive and negative metric propagators, i/q^2 and $-i/q^2$ which cancel for observable operators but not for unobservable operators such as * $\mathcal{J}^{S}_{5\alpha}$. Weinberg[119] has shown that the Kogut-Susskind mechanism also invalidates the bad result (4.169).

---

* The gauge invariant charge $\int d^3x \, \mathcal{J}^{S}_{50}(x)$ may be observable

Remarks at the end of Chapters II and III about the infrared behaviour of current amplitudes can be extended to amplitudes involving $\mathcal{J}^s_{5\mu}$. Since $\mathcal{J}^s_{5\mu}$ is partially conserved, it is not multiplicatively renormalized, as the form factor $\pi_1(q^2)$ for the amplitude

$$i \int d^4x\, e^{iq.x}\, T\langle 0|\mathcal{J}^s_{5\alpha}(x)\, \mathcal{J}^s_{5\beta}(0)|0\rangle = (q_\alpha q_\beta - g_{\alpha\beta}q^2)\pi_1(q^2) + g_{\alpha\beta}\pi_2(q^2) \quad (4.174)$$

satisfies the same renormalization-group equations as $\pi(q^2)$ in eq.(2.46), (but with a different subtraction function $K(x) \to K_1(x)$). To obtain a pole in $\pi_1(q^2)$ at $q^2 = 0$,

$$\lim_{\eta\to 0} \eta^2\pi_1(q^2;\bar{g},\bar{m},\mu) = \text{finite} \neq 0 \ , \quad (4.175)$$

it is necessary to assume $\bar{g}\to\infty$, or $\bar{m}\to\infty$, or both; otherwise the basic assumptions of the renormalization-group method are violated. If we assume the infrared-singularity theorems[63,65] mentioned at the end of Chapter II to be applicable here, the effective coupling constant $\bar{g}^1$ obtained by omitting massive fields should also diverge as η tends to zero.

REFERENCES

1. See the lectures given by T. Appelquist at this School.
2. E.C.G. Stueckelberg and A. Petermann, Helv.Phys.Acta 26, 499 (1953).
3. M. Gell-Mann and F.E. Low, Phys.Rev. 95, 1300 (1954).
4. K.G. Wilson, Phys.Rev. 179, 1499 (1969).
5. C.G. Callan, Jr., Phys.Rev. D2, 1541 (1970); K. Symanzik, Comm.Math.Phys. 18, 227 (1970).
6. G. 't Hooft, Nucl.Phys. B61, 455 (1973).
7. S. Weinberg, Phys.Rev. D8, 3497 (1973).
8. H.D. Politzer, Phys.Rev.Letters 30, 1346 (1973); D.J. Gross and F. Wilczek, Phys.Rev.Letters 30, 1343 (1973); G. 't Hooft, unpublished remarks at the Marseille meeting on Yang Mills theories (June, 1972).
9. The standard textbooks are : N.N. Bogoliubov and D.V. Shirkov, "Introduction to the Theory of Quantized Fields", Interscience, New York (1959); S.S. Schweber, "Relativistic Quantum Field Theory", Harper and Row, New York (1961); J.D. Bjorken and S.D. Drell, "Relativistic Quantum Fields", McGraw-Hill, New York (1965). See also lectures by S. Coleman, "Renormalization and Symmetry", Proc. 1971 Int.Summer School of Physics "Ettore Majorana", Editrice Compositori, Bologna (1973).
10. F.J. Dyson, Phys.Rev. 75, 1736 (1949); A. Salam, Phys.Rev. 82, 217 (1951); ibid. 84, 426 (1951).
11. S. Weinberg, Phys.Rev. 118, 838 (1960).
12. N.N. Bogoliubov and O.S. Parasiuk, Doklady Akad.Nauk (USSR) 100, 25, 429 (1955); O.S. Parasiuk, ibid. 100, 643 (1955); N.N. Bogoliubov and O.S. Parasiuk, Acta Math. 97, 227 (1957); O.S. Parasiuk, Ukr. Math.J. 12, 287 (1960); K. Hepp, Commun. Math.Phys. 2, 301 (1966).
13. The procedure can be formulated without using regulators : W. Zimmermann, Comm.Math.Phys. 6, 161 (1967); and 15, 208 (1969); "Lectures on Elementary Particles and Quantum Field Theory", 1970 Brandeis Summer Institute in Theoretical Physics, ed. S. Deser, M. Grisaru and H. Pendleton, MIT Press, Cambridge, Mass.(1971), Vol.I, p.397.
14. However, the situation is not completely hopeless. Interesting suggestions have been made recently by : G. Parisi, "Recent Progress in Lagrangian Field Theory and Applications", Marseille, June 24-28 (1974), ed. C.P. Korthals-Altes, E. de Rafael, and R. Stora, p. 21; Nucl.Phys. B (to be published); D.I. Blokhintsev, A.V. Efremov, and D.V. Shirkov, JINR preprint E2-8027 (1974); K. Symanzik, DESY preprints 75/12 and 75/24 (1975); F. Jegerlehner, Freie Universität Berlin preprint HEP 75-11 (1975).
15. W. Zimmermann, Ann.Phys. (N.Y.) 77, 536, 570 (1973).
16. Cf. Sect. 19.9 of the book by Bjorken and Drell, ref. 9.

17. M. Astaud and B. Jouvet, Comt.Rend. 264, 1433 (1967); Nuovo Cimento 53A, 841 (1968) and 63A, 5 (1969); M. Astaud, Nuovo Cimento 66A, 111 (1970).
18. S. Coleman, "Dilatations", Proc. 1971 Int.Summer School of Physics "Ettore Majorana", Editrice Compositori, Bologna (1973).
19. See the review by P. Carruthers, "Broken scale Invariance in Particle Physics", Phys.Reports 1C, No.1 (1971).
20. The use of B rules and a regulator Λ is not essential. For example, J.H. Lowenstein, Commun.Math. Phys. 24, 1 (1971) uses Zimmermann's normal-product algorithm (refs. 13, 15).
21. This method for deriving the CS equation was suggested by S. Coleman, ref. 18; S. Coleman and R. Jackiw, Ann.Phys.(N.Y.) 67, 552 (1971).
22. Similarly, the Ward identities of conformal transformations involve an additional vertex $-i\beta_R \int d^4x\, x_\mu \varphi^4(x)$ in $\int d^4x\, x_\mu \theta^\nu_\nu$ B. Schroer, Nuovo Cimento Letters 2, 627 (1971); G. Parisi, Phys.Letters 39B, 643 (1972); C.G. Callan Jr., and D.J. Gross, "Broken Conformal Invariance", Princeton report (1972); N.K. Nielsen, Nucl.Phys. B65, 413 (1973); S. Sarkar, Phys. Letters 50B, 499 (1974) and Nucl.Phys. B83, 108 (1974); N.K. Nielsen, Nucl.Phys. B97, 527 (1975).
23. S.L. Adler and W.A. Bardeen, Phys.Rev. D4, 3045 (1971); D6, 734(E) (1972); S.L. Adler, Phys.Rev. D5, 3021 (1972); D7, 1948(E) (1973).
24. K. Pohlmeyer, DESY report 74/36 (1974).
25. This terminology was invented by K. Symanzik, Springer Tracts in Modern Physics 57, 222 (1971). There is some question as to whether eq. (2.7) is adequate when some of the momenta ℓ_i are timelike : K. Symanzik, Comm.Math.Phys. 34, 7 (1973) and Springer Lecture Notes in Physics, Vol. 32. The problem is formulated differently in ref. 24.
26. For the basic formulas, see J.D. Bjorken and S.D. Drell, "Relativistic Quantum Mechanics", McGraw-Hill, New York (1964), pp. 170-171, and the textbooks mentioned in ref. 9. The properties of Feynman-parametric integrals are discussed in : R.J. Eden, P.V. Landshoff, D.I. Olive, and J.C. Polinghorne, "The Analytic S-Matrix", Cambridge University Press (1966).
27. For a precise mathematical analysis of the logarithmic power $\beta(g)$, consult J.P. Fink, J.Math.Phys. 9, 1389 (1968).
28. Compare Chapter VIII of the book by Bogoliubov and Shirkov, ref. 9.
29. Two-loop calculation of Z_3 : R. Jost and J.M. Luttinger, Helv.Phys.Acta 23, 201 (1950). Three-loop calculation : E. de Rafael and J. Rosner, Ann.Phys.(N.Y.) 82, 369 (1974).
30. L.D. Landau, A.A. Abrikosov and J.M. Khalatnikov, Dokl.Acad. Nauk 95, 497, 773, 1177 (1954) and 96, 261 (1954); L.D. Landau and I. Pomeranchuk, Dokl.Acad.Nauk 102, 489 (1955); L.D.Landau in "Niels Bohr and the Development of Physics", Pergamon Press, London (1955), p.52.

31. F.J. Dyson, Phys.Rev. 85, 631 (1952); B. Simon, "Fundamental Interactions in Physics and Astrophysics," 1972 Coral Tables Conference, Plenum Press, New York (1973).
32. S.L. Adler, Phys.Rev. D5, 3021 (1972).
33. K.G. Wilson, Phys.Rev. D3, 1818 (1971).
34. R.J. Crewther, S.-S. Shei and T.-M. Yan, Phys.Rev. D8, 3396 (1973), Appendix B.
35. R.J. Crewther, Proc. 2nd Int.Conf.on Elementary Particles (Aix-en-Provence), Journal de Physique, Tome 34, Colloque C-1 (1973), p.111.
36. O. Nachtmann, Phys.Lett. 51B, 469 (1974) discusses asymptotic corrections to this result for the special case $g = g_\infty$.
37. For an extensive review, see H.D. Politzer, "Asymptotic Freedom: an Approach to Strong Interactions", Phys.Reports 14C, 129 (1974). Ref. 35 summarizes some early contributions to the subject.
38. S. Coleman and D.J. Gross, Phys.Rev.Letters 31, 851 (1973).
39. D.J. Gross and J. Wilczek, Phys.Rev. D8, 3633(1973).
40. M. Gell-Mann, Acta Phys. Austriaca Suppl. IX, 733 (1972).
41. E.S. Abers and B.W. Lee, Phys.Reports 9C, 1(1973); M. Veltman, Proc. 1973 Bonn Symposium on Electron and Photon Interactions at High Energies (North-Holland), H. Rollnick, W. Pfeil (ed.); S. Coleman, "Secret Symmetry", 1973 Int.Summer School of Physics "Ettore Majorana", Erice (unpublished).
42. G.'t Hooft and M. Veltman, "Diagrammar", CERN Report 73/9(1973).
43. J. Zinn-Justin, Lecture Notes in Physics, Vol.37 (Springer Verlag, Berlin 1973); C. Becchi, Lectures at this School.
44. R.P. Feynman, Acta Phys.Polon. 26, 697(1963); B.de Witt, Phys. Rev.Letters 12, 742 (1964) and Phys.Rev. 162, 1195, 1239 (1967); L.D. Faddéev and V.N. Popov, Phys.Lett. 25B, 29(1967); L.D. Faddéev, Kiev ITP Report 67-36 (trans. NAL-THY-57(1973)); S. Mandelstam, Phys.Rev. 175, 1580 (1968); E.S. Fradkin and I.V. Tyutin, Phys.Rev. D2, 2841 (1970).
45. G.'t Hooft, Nucl.Phys. B33, 173 (1971).
46. Dimensional regularization : G.'t Hooft and M. Veltman, Nucl.Phys. B44, 189 (1972) and ref. 42; C.G. Bollini and J.J. Giambiagi, Phys.Letters 40B, 566 (1972).
47. Covariant regularization : A.A. Slavnov, Kiev preprint ITP 71-83 E(1971); B.W. Lee and J. Zinn-Justin, Phys.Rev. D5, 3121 (1972).
48. A.A. Slavnov, ref. 47 and Teor.Mat.Fiz. 10, 153 (1972) (translation : Theor.Math.Phys. 10, 99 (1972); J. Taylor, Nucl.Phys. B33, 436 (1971).
49. G.'t Hooft and M. Veltman, Nucl.Phys. B50, 318 (1972); B.W. Lee and J. Zinn-Justin, Phys.Rev. D5, 3121, 3137, 3155 (1972) and D7, 1049 (1973).
50. C. Becchi, A. Rouet and R. Stora, Comm.Math.Phys. 42, 127(1975), and "Recent Progress in Lagrangian Field Theory and Application", Marseille, June 24-28 (1974), ed. C.P. Korthals-Altes, E. de Rafael, and R. Stora, p.6.

51. T. Appelquist and H. Georgi, Phys.Rev. D8, 4000 (1973); A. Zee, Phys.Rev. D8, 4038 (1973).
52. P.W. Higgs, Phys.Letters 12, 132 (1964); Phys.Rev.Letters 13, 508 (1964); Phys.Rev. 145, 1156 (1966); F. Englert and R.Brout, Phys.Rev.Letters 13, 321 (1964); G.S. Guralnik, C.R. Hagen and T.W.B. Kibble, Phys.Rev.Letters 13, 585 (1964); T.W.B. Kibble, Phys.Rev. 155, 1554 (1967). Review article : J. Bernstein, Rev.Modern Phys. 46, 7(1974); 46, 855(E) (1974).
53. T.-P. Cheng, E. Eichten and L.-F. Li, Phys.Rev.9, 2259 (1974).
54. S. Ferrara and B. Zumino, Nucl.Phys. B79, 413 (1974).
55. N.-P. Chang, Phys.Rev. D10, 2706 (1974); E. Ma, Phys.Rev. D11, 322 (1975).
56. H.A. Bethe and E.E. Salpeter, Phys.Rev. 82, 309 (1951); M. Gell-Mann and F.E. Low, Phys.Rev. 84, 350 (1951); E.E. Salpeter and H.A. Bethe, Phys.Rev. 84, 1232 (1951); D. Lurié, "Particles and Fields", Interscience, New York (1968), ch.9.
57. R. Dashen, B. Hanslacher and A. Neven, Phys.Rev.D10, 4114,4130, 4138(1974); D11, 3424(1975); R. Jackiw and J. Goldstone, Phys.Rev. D11, 1486 (1975); R. Jackiw, "Collective Phenomena in Quantum Field theory", Acta Phys.Polon. B (to be published); L.D. Faddéev, Institute for Advanced Study (Princeton) preprint (1975).
58. F. Englert and R. Brout, Phys.Rev.Letters 13, 321 (1964); H. Pagels, Phys.Rev. D7, 3689 (1973); R. Jackiw and K. Johnson, Phys.Rev. D8, 2386 (1973); J. Cornwall and R. Norton, Phys.Rev. D8, 3338 (1973); S. Sarkar, Nucl.Phys. B56, 493 (1973); J. Cornwall, Phys.Rev. D10, 500 (1974); E. Eichten and F. Feinberg, Phys.Rev. D10, 3255 (1974); E.C. Poggio, E. Iomboulis and S.-H. H. Jye, Phys.Rev. D11, 2839 (1975).
59. S. Coleman and E. Weinberg, Phys.Rev. D7, 1888 (1973); D.J. Gross and A. Neven, Phys.Rev. D10, 3235 (1974).
60. S. Weinberg, Phys.Rev.Letters 31, 494 (1973); Phys.Rev. D8, 4482 (1973); H. Fritzch, M. Gell-Mann, and H. Leutwyler, Phys.Letters 47B, 365(1973); H. Fritzsch and M. Gell-Mann, Proc.XVI International Conference on High-Energy Physics(1972), Vol. 2, p.135.
61. J. Bloch and A. Nordzieck, Phys.Rev. 52, 54 (1937); D.R. Yennie, S.C. Frautschi and H. Suura, Ann.Phys.(N.Y.) 13, 379 (1961); K.E. Eriksson, Nuovo Cimento 19, 1010 (1961); D. Zwanziger, Phys.Rev. D11, 3481, 3504 (1975).
62. S. Weinberg, Phys.Rev. 140, B516 (1965).
63. T. Appelquist and J. Carazzone, Phys.Rev. D11, 2856 (1975).
64. K.G. Wilson, Phys.Rev. D10, 2445 (1974), and "Recent Progress in Lagrangian Field Theory and Applications", Marseille, June 24-28 (1974) ed. C.P. Korthals-Altes, E. De Rafael and R. Stora, p.125; J. Kogut and L. Susskind, Phys.Rev. D11, 395 (1975); V. Baluni and J.F. Willemsen, MIT. preprint CTP-468 (1975).

65. K. Symanzik, Commun.Math.Phys. 34, 7(1973).
66. N.N. Bogoliubov and D.V. Shirkov, JETP (USSR) 30, 77(1956) (transl. Soviet Physics JETP 3, 57 (1956)).
67. J. Kinoshita, J.Math.Phys. 3, 650 (1962).
68. A. Sirlin, Phys.Rev. D5, 436 (1972).
69. M. Baker and K. Johnson, Phys.Rev. 183, 1292 (1969).
70. R.J. Crewther, S.-S. Shei, and J.-M. Yan, Phys.Rev.D8,1730(1973).
71. K. Symanzik, Commun.Math.Phys. 23, 49 (1971)
72. K.E.Eriksson, Nuovo Cimento 30, 1423 (1963).
73. C.G. Callan Jr., unpublished. Princeton report (1973), modified the prescription to make it applicable to theories containing scalar particles.
74. G. 't Hooft, Nucl.Phys. B61, 455 (1973).
75. J.F. Ashmore, Lett.Nuovo Cimento 4, 289 (1972); Commun.Math.Phys. 29, 177 (1973); G.M. Cicuta and E. Montaldi, Lett.Nuovo Cimento 4, 329 (1972); E.R. Speer, J.Math.Phys. 15, 1(1974); G. Leibbrandt, Rev.Mod.Phys. 47, 849 (1975).
76. K.G. Wilson, Phys.Rev. D7, 2911 (1973), appendix.
77. J.C. Collins, Nucl.Phys. B92, 477(1975); P. Breitenlohner and D. Maison, Max-Planck Institute reports, Munich, MPI-PAE/PTh 25/74 and 15/75.
78. M.J. Holwerda, W.L. van Neerven, and R.P. van Royen, Nucl.Phys. B75, 302(1974); J.C. Collins and A.J. Macfarlane, Phys.Rev. D10, 1201(1974); J.C. Collins, Phys.Rev. D10, 1213(1974); S.-Y.Lee, Phys.Rev. D10, 1103 (1974); L.-P. Yu, Nucl.Phys. B81, 458(1974).
79. G.'t Hooft and M. Veltman, ref.46; J.C. Collins, Nucl.Phys. B80, 341 (1974); unpublished proofs of G.'t Hooft and K. Symanzik (quoted by Collins).
80. K.G. Wilson, "On Products of Quantum Field Operators at Short Distances", unpublished Cornell report (1964).
81. W. Zimmermann, "Lectures on Elementary Particles and Quantum Field Theory," 1970 Brandeis Summer Institute in Theoretical Physics, ed. S. Deser, M. Grisaru and H. Pendleton, MIT Press, Cambridge, Mass.(1971), Vol.I, p. 397.
82. K.G. Wilson and W. Zimmermann, Commun.Math.Phys.24, 87(1972), P. Otterson and W. Zimmermann, Commun.Math.Phys. 24,107(1972).
83. R. Brandt, Ann.Phys. 44, 221(1967); 52, 122 (1969); Fortschritte der Physik 18, 249(1970).
84. Y. Frishman, Phys.Rev.Letters 25, 966(1970); G. Altarelli, R.A. Brandt, and G. Preparata, Phys.Rev.Letters 26, 42(1971); R.A. Brandt and G. Preparata, Phys.Rev.Letters 25, 1530(1970); Nucl.Phys. B27, 541 (1971).
85. H. Fritzsch, M.Gell-Mann, Proceedings of the Coral Tables Conference on Fundamental Interactions at High Energies, January 1971, in "Scale Invariance and the Light Cone", Gordon and Breach (1971); J.M. Cornwall and R. Jackiw, Phys.Rev.D4, 367 (1971).
86. C.G. Callan Jr., Phys.Rev. D5, 3202(1972) and 1973 Cargèse lectures (unpublished).

87. N. Christ, B. Haeslacher and A.H. Mueller, Phys.Rev.D6, 3543 (1972).
88. K. Wilson, Proceedings of the Coral Gables Conference on Fundamental Interactions at High Energies, January 1971, in Vol.II, "Scale Invariance and the Light Cone", Gordon and Breach (1971); Proceedings of the Symposium on Electron and Photon Interactions at High Energies, 1971, p. 116.
89. R.J. Crewther, Phys.Rev.Letters 28, 1421 (1972).
90. J.E. Clark, Nucl.Phys. 81, 263 (1974).
91. A.H. Mueller, Phys.Rev. D9, 963 (1974).
92. C.G. Callan, Jr., and M.L. Goldberger, Phys.Rev. D11, 1542, 1553 (1975); N. Coote, Phys.Rev. D11, 1611 (1975).
93. D.G. Gross and J. Wilczek, Phys.Rev. D9, 980 (1974); H. Georgi and H.D. Politzer, Phys.Rev. D9, 416 (1974).
94. B.L. Joffe, Zh.Eksp.Teor.Fiz.Pis'ma Red. 9, 163 (1969) (JETP Lett. 9, (1969)); Phys.Lett. 30B, 123(1969); L.S. Brown, in "High Energy Collisions of Elementary Particles", Lectures in Theoretical Physics, Vol. XII-B (1969), ed. K.T. Mahanthappa and W.E. Brittin. (Gordon and Breach, New York 1971), p.201; R.A. Brandt, Phys.Rev.Lett. 23, 1260(1969); Phys.Rev. D1, 2808 (1970).
95. R. Jackiw, R. van Royen and G.B. West, Phys.Rev. D2, 2473(1970); H. Leutwyler and J. Stern, Nucl.Phys. B20, 77(1970).
96. Originally, this connection was obtained in the light-cone-Bjorken-scaling framework (refs. 84,85), but it was immediately recognized (e.g. G. Mack, Nucl.Phys. B35, 592(1971)) that the Bjorken-scaling assumption is irrelevant.
97. O. Nachtmann, Nucl.Phys. B63, 237 (1973).
98. O. Nachtmann, Nucl.Phys. B78, 455 (1974).
99. G. Parisi, Nucl.Phys. B59, 641 (1973); C.G. Callan, Jr., and D.J. Gross, Phys.Rev. D8, 4383 (1973).
100. G. Parisi, Phys.Letters 43B, 207(1973).
101. G. Parisi, Phys.Letters 50B, 367(1974); D.J. Gross, Phys.Rev. Lett. 32, 1071 (1974); A. de Rujula, Phys.Rev.Lett. 32, 1143 (1974); D.J. Gross and S.B. Freiman, Phys.Rev.Lett. 32, 1145 (1974); A. de Rujula, S.L. Glashow, H.D. Politzer, S.B.Freiman, F. Wilczek and A. Zee, Phys.Rev. D10, 1649 (1974); A.de Rujula, H. Georgi, and H.D. Politzer, Phys.Rev. D10, 2141(1974).
102. For reviews of deep inelastic scattering, see the talks by D. Gross, A. de Rujula and F.J. Gilman, pages III-65, IV-90 and IV-149 of Proc. XVII International Conference on High Energy Physics, London 1974 (Science Research Council, Rutherford Laboratory, 1974).
103. C. Chang et al., Phys.Rev.Lett. 35, 901 (1975).
104. D.J. Gross and S.B. Treiman, Phys.Rev. D4, 1059 (1971).
105. R.J. Crewther, CERN preprint TH.
106. Thus eq.(4.94) is an example of a supersymmetric transformation: J. Wess and B. Zumino, Nucl.Phys. B70, 39 (1974); talks by B. Zumino and J. Iliopoulos, pages I-254 and III-89 of Proc.

XVII International Conference on High Energy Physics, London 1974 (Science Research Council, Rutherford Laboratory, 1974).
107. H. Kluberg-Stern and J.B. Zuber, Saclay preprint DPh-T/75/28 (1975).
108. W.S. Deans and J.A. Dixon, Oxford University preprint 20-75 (1975).
109. S.D. Joglekar and B.W. Lee, FERMILAB-Pub-75/50 THY (1975).
110. R. Arnowitt and S.F. Fickler, Phys.Rev. 127, 1821 (1962); S. Coleman, in "Laws of Hadronic Matter", Proc. 1973 International School of Subnuclear Physics, Erice, Sicily (Academic Press, New York and London, 1975), p. 139.
111. J.A. Dixon and J.C. Taylor, Nucl.Phys. B78, 552 (1974).
112. H. Kluberg-Stern and J.B. Zuber, Phys.Rev. D12, 467 (1975).
113. Some of these peculiarities have been discussed by J.M.Cornwall, Phys.Rev. D10, 500 (1974), appendix.
114. W. Kummer, Acta Phys. Austriaca 41, 315 (1975). Appendix A.
115. J.A. Dixon and J.C. Taylor, Oxford University preprint 74-74 (1974).
116. W. Kainz, W. Kummer and M. Schweda, Nucl.Phys. B79, 484(1974); R. Delbourgo, A. Salam and J. Strathdee, Nuovo Cimento 23A, 237 (1974).
117. S.L. Adler and R.F. Dashen, "Current Algebras and Applications to Particle Physics", W.A. Benjamin Inc. (New York and Amsterdam, 1968); B. Renner, "Current Algebras and their Applications", Pergamon Press Ltd. (London 1968).
118. As far as I am aware, the existence of this term has not been noted previously. The only hint in the literature seems to be a remark of D.G. Sutherland, Nucl.Phys. B2, 433 (1967), concerning Harari's analysis of electromagnetic mass differences : (H. Harari, Phys.Rev.Letters 17, 1303 (1966).
119. S. Weinberg, Phys.Rev. D11, 3583 (1975).
120. D.G. Sutherland, ref. 118 ; N. Cabibbo and L. Maiani, Phys. Rev. D1, 707 (1970), and in "Evolution of Particle Physics", ed. M. Conversi, Academic Press (New York 1970), p.50; A.J. Cantor, Harvard University Ph.D. Thesis, october 1969 (unpublished) and Phys.Rev. D3, 3195, 3205 (1971).
121. S.K. Bose and A.M. Zimerman, Nuovo Cimento 43A, 1165 (1966); R. Ramachandran, Nuovo Cimento 47A, 669 (1967); R.H. Graham, L. O'Raifeartaigh and S. Pakvasa, Nuovo Cimento 48A, 830(1967); Y.J. Chin, J. Schechter and Y. Ueda, Phys.Rev. 161, 1612(1967). These early references do not include the term $\mathcal{H}_{em}$ (pure) in eq.(4.117), so the results have to be modified (ref. 120) to take account of Dashen's rule, eq. (4.115).
122. M. Gell-Mann, Phys.Rev. 125, 1067 (1962); S.L. Glashow and S. Weinberg, Phys.Rev.Lett. 20, 224 (1968); M.Gell-Mann, R.J. Oakes and B. Renner, Phys.Rev. 175, 2195 (1968). See the review by H. Pagels, Phys.Reports 16C, 219 (1975).
123. D.G. Sutherland, Phys.Lett. 23, 384 (1966).

124. R. Dashen, Phys.Rev. 183, 1245 (1969).
125. S. Coleman and S.L. Glashow, Phys.Rev. 134, B671 (1964).
126. The amplitudes are taken to be linear in the pion energies. For a discussion of this approximation and of theoretical alternatives to the tadpole interaction (4.117), see J.S.Bell and D.G. Sutherland, Nucl.Phys. B4, 315 (1968).
127. C. Baglin et al., Phys.Lett. 29B, 445 (1969); D.W. Carpenter et al., Phys.Rev. D1, 1303 (1970).
128. See Particle Data Group, Phys.Lett. 50B, 1(1974) for the branching ratio. The total width is obtained from the recent Cornell result for $\eta \to 2\gamma$: A. Browman et al., Phys.Rev.Lett. 32, 1067 (1974).
129. Wilson's argument is reviewed by J. Kogut and L .Susskind, Phys.Rev. D11, 3594 (1975).
130. M. Yamada and N. Nakazawa, Miyazaki preprints MM-1 (1974) and MM-2 (1975), claim to derive a three-current effect for the Sutherland interaction (4.114). However, they make assumptions which are not consistent with the assumed convergence of (4.114).
131. J.S. Bell and R. Jackiw, Nuovo Cimento 60A, 47(1969); S.L. Adler, Phys.Rev. 177, 2426 (1969). For reviews, see : S.L. Adler, Lectures on Elementary Particles and Quantum Field Theory, Brandeis University summer Institute, MIT Press (Cambridge, Mass. 1970), Vol. I; R. Jackiw, in S.B. Treiman, R. Jackiw and D.J. Gross, Lectures on Current Algebra and its Applications, Princeton University Press (Princeton, N.J.,1972).
132. See the review by S. Weinberg, Proc. XVII International Conference on High Energy Physics, London 1974 (Science Research Council, Rutherford Laboratory, 1974), p. III-59.
133. R.P. Feynman, M. Gell-Mann and G. Zweig, Phys.Rev.Lett.13, 678 (1964).
134. S.L. Glashow, in "Hadrons and their Interactions", Academic Press Inc. (New York, 1968),p. 83; (I thank R. Jackiw for this reference). See also S.L. Glashow, R. Jackiw and S.-S. Shei, Phys.Rev. 187, 1916 (1969); M. Gell-Mann, in Proc.Third Topical Conference in Particle Physics, Honolulu (1969), ed. W.A. Simmons and S.F. Tuan, Western Periodicals (Los Angles 1970), p.1.
135. R.A. Brandt and G. Preparata, Ann.Phys.(N.Y.) 61, 119 (1970), Section IVB. Also, see K.G. Wilson, Phys.Rev. D2, 1478(1970), footnote 14.
136. G. Donaldson et al., Phys.Rev.Letters 31, 337 (1973).

QUARKS

F. BUCCELLA

Istituto di Fisica, Università di Roma

Roma, Italy

This concluding talk is about the concept of quark which has characterized the last fifteen years research on hadrons and came out also in the different topics discussed in this school.

In nuclear physics the two conserved quantum numbers A and Z are directly connected to the existence of the neutron and proton particles.

The presence of a third conserved quantum number, strangeness, has been connected by Sakata [1] to the existence of the Λ particle; in this framework a successful classification scheme arises for the pseudoscalar mesons thought as bound $(P\ n\ \Lambda)$ $(\bar{P}\ n\ \bar{\Lambda})$ states : the 3 pions and 4 caons already known have just quantum numbers which can be built in this way. The desease of this scheme was the existence of the Σ's and Ξ's particles with the same spin parity $1/2^+$ than P, n and Λ . This difficulty has been brilliantly overcome by Gell-Mann[2] by classifying the eight baryons as well as the 0^- mesons in the eightfold representation of SU(3). This algebra is effective also in classifying the $3/2^+$ baryonic resonances in the 10 representation and the 1^- meson resonances in the $1 \oplus 8$. More in general all the baryonic states presently known fit in the representations 1, 8 and 10 of SU(3) and the meson states in 1 and 8.
So one can say that the internal quantum numbers of the hadrons can be built from the fundamental representations 3 and $\bar{3}$ as the tensor products :

$$3 \otimes 3 \otimes 3 = 1 \oplus 8 \oplus 8 \oplus 10 \qquad \text{baryons}$$

$$3 \otimes \bar{3} = 1 \oplus 8 \qquad \text{mesons}$$

It is rather remarkable that, opposed to what happens in nuclear physics, there is no evidence in nature for the 3 representation (with fractional charge!). The situation gets still more intriguing, when realizing that the association to the SU(3) triplet of an additional degree of freedom of spin 1/2 brings to SU(6) [3], surprisingly good as a classification group; in fact, developing the analogous tensor products of the previous case, one gets

$$6 \otimes 6 \otimes 6 = 56 \oplus 70 \oplus 70 \oplus 20$$

$$6 \otimes \bar{6} = 1 \oplus 35$$

The 56 representation is just what is needed to put together the $1/2^+$ octet and the $3/2^+$ decuplet, while in the 35 happen to fall the pseudoscalar octet and the vector nonet. (It is to stress that also in this occasion an unlucky Japanese[4] scientist has been pushed by reasonable physical considerations to the wrong choice for the baryons : in fact spin 1/2 particles are expected to behave as fermions and therefore rather to combine their internal quantum numbers in the totally antisymmetric 20 representation rather than in the totally symmetric 56).
The negative parity resonant states can be classified respectively in the 35 L = 1 (mesons)[5] and 70 L = 1 (baryons)[6] of the enlarged classification group SU(6) $\otimes$ O (3) obtained by introducing an "angular orbital momentum" $\vec{L}$, which commutes with the "SU(6) spin" $\vec{S}$.

There is evidence also for a 56 L = 2 and a 35 L = 2. It is rather natural to understand the angular momentum $\vec{L}$ in terms of the relative motion of this fundamental objects which build up the hadrons.
The fact that the states of the baryon octet seem to be built of three quarks in S-wave makes reasonable to compute their axial charges (the vector are just given by C.V.C.) in terms of the axial charges of the quarks : if one assumes that, as leptons, only the left-handed quarks build up the weak current, the analogy with nuclear physics would imply that the Gamov-Teller operators concerning the baryon octet are just the corresponding SU(6) generators $A\left(\frac{\lambda_i \sigma_z}{2}\right)$. This assumption gives rise to the famous predictions D/F = 3/2 and $|G_A/G_V| = 5/3$ [7]; while the first one is in excellent agreement with experiment, the value 5/3 is still more far from the measured value 1.25 than the number 1 predicted by a naive analogy between leptons and nucleons. From the other side the analogy between quarks and leptons brings to a good prediction for G_A/G_V ; in fact, assuming that the hadron vector and axial charges have the same commutation rules than the lepton ones, one is lead to the SU(3) $\otimes$ SU(3) algebra[8]

$$[Q^i, Q^j] = i f^{ijk} Q^k$$

$$[Q^i, Q_5^j] = i f^{ijk} Q_s^k$$

$$[Q_5^i, Q_5^j] = i f^{ijk} Q_5^k$$

The first two sets of equations are the content of the Cabibbo theory[9] of weak interactions, while the third one has been successfully tested by Adler and Weissberger[10], which obtained $(G_A/G_V)^2$ in terms of the pion-nucleon cross-sections. This rather remarkable result has lead people to investigate the possibility of saturating the SU(3) ⊗ SU(3) algebra (in the $P_Z = \infty$ frame of reference as in the Adler calculation) within a finite set of states; the simplest idea of choosing single representations of SU(6) ⊗ O (3), in particular the 56 L = 0 for baryons, brings back to the SU(6) result[11]. Therefore people tried to saturate the algebra within wider sets of states, consisting of a finite [12] or even an infinite[13] number of representations of SU(6) ⊗ O (3). The satisfactory predicting power for the axial couplings of the mesons with L=0 and 1 [14] (in particular the right polarization properties found for the decays of the A_1 and B mesons) has inspired the proposal of the following form for the axial charge[15]:

$$Q_5^i = U\, A\, (\frac{\sigma_Z \lambda_i}{2}) U^+$$

with the unitary operator U given by

$$U = 1 + iZ - \frac{Z^2}{2} + \ldots$$

and

$$Z = (\vec{W} \wedge \vec{M})_Z$$

where $\vec{W}$ is the spin 1 unitary singlet of a 35 representation under SU(6) and $\vec{M}$ is a vector under O (3). The agreement with experiment is excellent. A definite break-through on the origin of the unitary operator has been achieved with the work of Melosh[16], who showed that even for free quarks such a transformation is expected to be there due to the relativistic motion of the quarks inside the hadrons.

The origin of the transformation can be easily understood in terms of group-theoretical arguments concerning the Poincaré group [17]. One one has defined the direction of quantization of the spin of a state belonging to an irreducible representation of the Poincaré group in the rest frame, there is a rather large arbitrariness in the choice of the operator to be diagonalized in an arbitrary frame; it depends from the particular Lorentz transformation chosen to reach the moving state starting from the one at rest. Therefore, when one builds up an irreducible representation of the Poincaré group combining with the proper Clebsch-Gordan coefficients two irreducible representations of the same group, these coefficients of course depend on the particular Lorentz transformation chosen to define the quantization axis in a moving frame. With a more intuitive language one can say that the direction of the spin, which matters, is no more related to the momenta of the two individual particles but to the centre of mass momentum. The part of the Clebsch-Gordan arising in this way is commonly called "Wigner rotation".

Since here one is interested to establish the transformation properties of the compound states under chiral algebra (at $P_Z=\infty$!) the more appropriate choice for the spin operator for a single quark is just the null-plane charge

$$\hat{Q}_3 = \int d^4X\, S(X^0+X^3)\psi^+(X)(1+\alpha_3)\frac{\sigma_3}{2}\psi(X)$$

To get from it the spin operator used in the centre of mass system (which is useful to classify states!) one needs the Wigner rotation

$$R(p,P) = e^{-i\theta(p,P)\frac{\vec{\sigma}\Lambda\vec{P}}{P_\perp}}$$

The strong similarity of R with the operator U so successfully introduced gives the confidence to have correctly understood the distinction between current and constituent quarks in purely kinematical terms[16]. The interpretation of the chiral mixing of the hadrons in terms of the internal momenta of the quarks inside them is a further proof of the fundamental role of these objects; needless to say another compulsing reason to consider them is the success of the quark parton model for the deep inelastic electron and neutrino induced phenomena.

For a long time, however, many physicists considered them rather as an useful mathematical tool without worrying about their existence: a good reason for this attitude is the undisputable fact that no quark has been observed (despite many attempts) while the related explanation in terms of high masses and binding energies is difficult to recouncile with the additive properties they show up.

A brilliant understanding of the original behaviour of quarks has been achieved with the introduction of colour[19]. The assumption is that quarks have an additional SU(3) quantum number (they are classified in the fundamental 3 representation of this new group), which cannot be observed; this superselection rule gives reason also of the fact that mesons seem to be built of a pair quark-antiquark and baryons of triplets of quarks (the singlet representation of SU(3) is contained both in the products $3 \otimes \bar{3}$ and $3 \otimes 3$!). Moreover since the singlet representation is obtained from three triplets in a completely antisymmetric way the symmetry of baryons in the traditional indices is just the consequence of Fermi statistics. Also a factor nine for the π^0 decay in two photons, which -as needed, has been achieved with colour.

The introduction of such an unconventional concept to characterize quarks seems rather appropriate due to their funny behaviour, which shows up also in the way one has achieved information on them; while for the traditional particles one has introduced first the momentum, then the spin and finally the internal quantum numbers, the opposite pattern has been followed for quarks.

REFERENCES

1) S. Sakata, Progr.Theor.Phys. 16, 686 (1956)
2) M. Gell-Mann, Phys.Rev.125, 1067 (1962)
3) F. Gursey and L.A. Radicati, Phys.Rev.Lett.13, 173 (1964).
4) B. Sakita, Phys.Rev. 136B, 1756 (1964)
5) E. Borchi and R. Gatto, Phys.Letters 14, 352 (1965)
6) R.H. Dalitz, Proc.Oxford Conf. on elementary particles (1965)
7) F. Gursey, A. Pais and L.A. Radicati, Phys.Rev.Letters 13, 299 (1964)
8) M. Gell-Mann, Physics 1, 63 (1964)
9) N. Cabibbo, Phys.Rev.Lett. 10, 531 (1963)
10) S.L. Adler, Phys.Rev.Lett. 14, 1051 (1965)
W.I. Weissberger, Phys.Rev.Lett. 14, 1047 (1965)
11) S. Bergia and F. Lannoy, CERN preprint (1965)
I.S. Gerstein, Phys.Rev.Letters 16, 114 (1966)
12) R. Gatto, L. Maiani and G. Preparata, Phys.Rev.Letters 16, 377 (1966) and 18, 97 (1967) - Physics (N.Y.) 3, 1 (1967).
G. Altarelli, R. Gatto, L. Maiani and G. Preparata, Phys.Rev. Letters 16, 918 (1966)
H. Harari, Phys.Rev.Letters 16, 964 (1966)
I.S. Gerstein and B.W. Lee 16 , 1060 (1966) and Phys.Rev. 152, 1418 (1966).
H.J. Lipkin, H.R. Rubinstein and S. Meshkov, Phys.Rev. 148, 1405 (1966).
D. Horn, Phys.Rev.Letters 17, 778 (1966).
N. Cabibbo and H. Ruegg, Phys.Letters 22, 85 (1966)
F. Buccella, M. De Maria and B. Tirozzi, Nucl.Phys. B8, 521(1968)
13) F. Buccella, E. Celeghini and E. Sorace, Lettere al Nuovo Cimento 2, 57 (1969)
14) C. Boldrighini, F. Buccella, E. Celeghini, E. Sorace and L. Triolo, Nuclear Phys. 22B, 651 (1970)
15) F. Buccella, E. Celeghini, H. Kleinert, C.A. Savoy and E. Sorace, Nuovo Cimento 69A, 133 (1970)
16) H.J. Melosh, Phys.Rev. D9, 1095 (1974)
17) F. Buccella, C.A. Savoy and P. Sorba, Lettere al Nuovo Cimento 10, 455 (1974)
18) E. Celeghini and E. Sorace, Florence TH 74.2 (submitted to Physics Letters)
Lettere al Nuovo Cimento 11, 166 (1974)
E. Celeghini, L. Lusanna and E. Sorace, Nuovo Cimento 25A, 331 (1975)
19) M. Gell-Mann, Acta Physica Austriaca, Suppl.IX 733 (1972).

INDEX